T0379046

Software Defined Radio

Theory and Practice

For a complete listing of titles in the
Artech House Mobile Communications Series,
turn to the back of this book.

Software Designed Radio

Theory and Practice

John M. Reyland

ARTECH HOUSE
BOSTON | LONDON
artechhouse.com

Library of Congress Cataloging-in-Publication Data
A catalog record for this book is available from the U.S. Library of Congress.

British Library Cataloguing in Publication Data
A catalogue record for this book is available from the British Library.

Cover design by Joi Garron

ISBN 13: 978-1-68569-039-7

Accompanying appendices can be found at:
https://us.artechhouse.com/assets/downloads/reyland_039.zip.

© 2024 ARTECH HOUSE
685 Canton Street
Norwood, MA 02062

All rights reserved. Printed and bound in the United States of America. No part of this book may be reproduced or utilized in any form or by any means, electronic or mechanical, including photocopying, recording, or by any information storage and retrieval system, without permission in writing from the publisher.
    All terms mentioned in this book that are known to be trademarks or service marks have been appropriately capitalized. Artech House cannot attest to the accuracy of this information. Use of a term in this book should not be regarded as affecting the validity of any trademark or service mark.

10 9 8 7 6 5 4 3 2 1

*To my wife, Diane, who lovingly took care of many things
so I could finish this book on time*

Contents

Preface   xv

Chapter 1   Introduction   1
  1.1   Origins of SDR   2
    1.1.1   Speakeasy   2
    1.1.2   Joint Tactical Radio System   2
  1.2   Software Communications Architecture   3
    1.2.1   Other SCA Considerations   4
    1.2.2   SDR Currently   4
  1.3   Radio Hardware Choices for SDR   5
  1.4   DSP Software Development Choices for SDR   5
    1.4.1   GNU Radio Companion   7
    1.4.2   Simulink   8
    1.4.3   SDR Sharp   9
    1.4.4   A Word About Interfaces   9
  References   10

Chapter 2   Communication System Fundamentals   11
  2.1   Introduction   11
  2.2   Basic Measurements   11
    2.2.1   dB, dBm, dBc   11
    2.2.2   dBu   13
    2.2.3   dBi, dBd   13
    2.2.4   VU   14
    2.2.5   Equivalent Isotropic Radiated Power   14
    2.2.6   Energy per Bit to Noise Power Spectral Density   14
  2.3   Other Measurements   16
    2.3.1   Error Vector Magnitude   16

	2.3.2	Voltage Standing Wave Ratio	17
2.4		Some Basic Procedures	19
	2.4.1	Impedance Matching	19
	2.4.2	Signal Transmission	21
2.5		Satellite Communications Measurements	22
	2.5.1	Antenna Noise Temperature	22
	2.5.2	Satellite Receiver Sensitivity	25
2.6		Link Budgets	25
	2.6.1	Satellite Receiver Link Budget	27
	2.6.2	Terrestrial Receiver Link Budget	30
2.7		Important Communications Concepts	32
	2.7.1	Nyquist Bandwidth Criteria	32
	2.7.2	Nyquist Pulse Criteria	32
	2.7.3	Shannon-Hartley Capacity Limit	34
2.8		Questions for Discussion	41
References			42

Chapter 3 Modulation Types 43

3.1		Introduction	43
3.2		Linear Modulation	43
	3.2.1	Basic Parameters	44
	3.2.2	Binary Phase Shift Keying	45
	3.2.3	Quadriphase Shift Keying	48
	3.2.4	BPSK Bit Error Analysis	53
	3.2.5	QPSK Bit Error Analysis	56
	3.2.6	BPSK and QPSK Bandwidth Efficiency	57
	3.2.7	Differential BPSK	63
	3.2.8	Quadrature Amplitude Modulation	66
	3.2.9	$pi/4$ DPSK	66
	3.2.10	OFDM	68
	3.2.11	Single-Carrier Frequency Division Multiplexing	81
3.3		Nonlinear Modulation	84
	3.3.1	Basic Parameters	86
	3.3.2	Frequency Shift Keying	89
	3.3.3	Continuous Phase Modulation	93
	3.3.4	Minimum Shift Keying	101
	3.3.5	Gaussian Minimum Shift Keying	106
	3.3.6	Pulse Position Modulation	119

3.4	Direct Sequence Waveforms		119
	3.4.1 Spread Spectrum		120
3.5	Questions for Discussion		125
References			127

Chapter 4 RF Channels — 129

- 4.1 Introduction . . . 129
- 4.2 RF Wave Basics . . . 129
  - 4.2.1 Polarized Antennas . . . 130
  - 4.2.2 RF Spectrum Regions . . . 134
- 4.3 RF Propagation . . . 136
  - 4.3.1 Fixed Propagation Environment . . . 138
  - 4.3.2 Multipath in a Fixed Environment . . . 144
  - 4.3.3 Moving Propagation Environment . . . 146
  - 4.3.4 Multipath in a Constant Velocity Moving Environment . . . 147
- 4.4 Multipath Mitigation . . . 154
  - 4.4.1 Spatial Diversity . . . 155
  - 4.4.2 Spatial Diversity for CDMA . . . 155
  - 4.4.3 Time Diversity . . . 157
  - 4.4.4 Frequency Diversity . . . 163
  - 4.4.5 Polarization Diversity . . . 163
  - 4.4.6 Space-Time Coding . . . 164
  - 4.4.7 Multiple Input-Multiple Output . . . 165
- 4.5 Questions for Discussion . . . 170
- References . . . 173

Chapter 5 Channel Equalizers — 175

- 5.1 Introduction . . . 175
- 5.2 Equalizers using Linear Regression . . . 179
  - 5.2.1 LMS Linear Adaptive Equalizer . . . 181
- 5.3 LMS Equalizer Theory . . . 186
  - 5.3.1 The Orthogonality Principle . . . 187
  - 5.3.2 Equalizer LMS Adaptation Equations . . . 189
- 5.4 Further Advances in Equalizer Design . . . 194
- 5.5 Questions for Discussion . . . 194
- References . . . 195

Chapter 6 Coding — 197

- 6.1 Introduction . . . 197

6.2 Source Coding 198
    6.2.1 Weather Station Encoding 199
    6.2.2 Lempel-Ziv Coding 200
6.3 Channel Coding 204
    6.3.1 Block Coding 204
    6.3.2 Block Coding Example 209
    6.3.3 Convolutional Coding 215
    6.3.4 Concatenated Coding 226
    6.3.5 Trellis-Coded Modulation 228
6.4 Questions for Discussion 238
References 240

Chapter 7 Analog Signal Processing   241
7.1 Introduction 241
7.2 Components 241
    7.2.1 RF Amplifiers 241
    7.2.2 RF Mixers 249
    7.2.3 Local Oscillators 256
    7.2.4 RF Filters 257
7.3 Receiver Configurations 260
    7.3.1 Nonzero IF Receiver 261
    7.3.2 Zero IF Receiver 275
7.4 Transmitter Configurations 287
    7.4.1 SDR Transmitters 287
7.5 Questions for Discussion 293
References 294

Chapter 8 ADC and DAC Technology   295
8.1 ADC Sampling Theory 295
    8.1.1 Time-Domain 295
    8.1.2 Frequency-Domain 296
8.2 ADC Specifications 297
    8.2.1 Signal-to-Noise Ratio 299
    8.2.2 ADC Nonlinearity 300
    8.2.3 ADC Measurements 301
    8.2.4 ADC Designs 304
8.3 Digital to Analog Converters 320
    8.3.1 DAC Comparison with ADCs 320
    8.3.2 DAC Specifications 324

		8.3.3 Interpolating DAC 325

    8.3.3   Interpolating DAC 325
8.4 Questions for Discussion 328
References 329

Chapter 9   Digital Signal Processing   331

9.1 Fundamental DSP Concepts 331
      9.1.1   Unit Delay 331
      9.1.2   Z-Transform 332
      9.1.3   Unit Circle 334
      9.1.4   Poles and Zeros 336
      9.1.5   Digital Filter Frequency Response 336
9.2 Digital Filter Examples 338
      9.2.1   Example 1: Single Pole on Unit Circle 338
      9.2.2   Example 2: Single Pole Inside Unit Circle 340
      9.2.3   Example 3: Exponential Averaging Filter 342
      9.2.4   Example 4: Cascade Integrator Comb 343
      9.2.5   Example 5: Resonator 348
      9.2.6   Example 6: Halfband Filters 350
      9.2.7   Example 7: Upsampling Filters 356
      9.2.8   Example 8: Down Sampling Filters 358
      9.2.9   Example 9: Standard Filters 358
      9.2.10 Example 10: Arbitrary Digital Filters 363
      9.2.11 Example 11: Hilbert Transform 363
      9.2.12 Example 12: Allpass Filter 367
9.3 Digital Signal Processing Circuits 374
      9.3.1   Example 1: Spectral Inversion of Sampled Signal 374
      9.3.2   Example 2: NZIF to Complex Baseband 375
      9.3.3   Example 3: High Resolution Sinewave Generation 385
      9.3.4   Example 4: Arctangent Approximation 385
9.4 Quantizing Noise 396
      9.4.1   Quantization Noise Due to Sampling 397
      9.4.2   Multiplier Roundoff Noise 398
      9.4.3   Overflow 399
9.5 Questions for Discussion 400
References 402

Chapter 10   Symbol and Carrier Tracking   403

10.1 Symbol Tracking 403
      10.1.1 Symbol Boundary Adjustment 403

    10.1.2 Time-Domain Based Timing Error Detector 404
    10.1.3 Frequency-Domain TED 408
    10.1.4 Timing Closed-Loop Dynamics 421
    10.1.5 Timing Resilience to Carrier Frequency Offset 427
10.2 Carrier Tracking 430
    10.2.1 Coarse Frequency Offset 430
    10.2.2 Symbol Rate Decision Directed Carrier Tracking 431
    10.2.3 Costas Loop 431
10.3 Questions for Discussion 437
References 438

Chapter 11 ADSB Digital Signal Processing   439

11.1 Introduction 439
11.2 ADALM-Pluto SDR Hardware 440
    11.2.1 AD9363 Transceiver Interface 440
    11.2.2 AD9363 Receive Antenna 445
11.3 ADSB Digital Signal Processing 447
    11.3.1 ADSB Preamble Detect 447
    11.3.2 ADSB Packet Samples Tabulation 451
    11.3.3 bitParser Block 453
11.4 ADSB PHY Layer Output Details 453
    11.4.1 Aircraft ID, Type Code = 1,2,3,4 456
    11.4.2 Aircraft Velocity, Type Code = 19 456
    11.4.3 Aircraft Position, Type Code = 9:18 460
    11.4.4 ADSB Message Logging 466
References 466

Chapter 12 APSK Digital Signal Processing   467

12.1 APSK Overview 467
12.2 Modem Overview 467
    12.2.1 Modem Front End 468
    12.2.2 Modem Back End 468
12.3 Modem Transmitter Signal 470
12.4 Modem Receiver Front End Blocks 472
    12.4.1 IF to Complex Baseband 472
    12.4.2 Sample Interpolator 472
    12.4.3 Cubic Farrow Interpolator, Fixed Point (CFAfixed) 476
    12.4.4 Sample Stacker 476
    12.4.5 Root Raised Cosine Matched Filter 477

		12.4.6 Timing Error Detector 482

        12.4.6 Timing Error Detector 482
        12.4.7 Timing Loop Filter 483
  12.5 Modem Back-End Blocks 486
        12.5.1 Carrier Tracking 486
        12.5.2 Equalizer 492
  12.6 APSK System Testing 506
        12.6.1 APSK Hardware Testing 507
  References 518

Chapter 13 IEEE802.11a Digital Signal Processing   **519**
  13.1 IEEE802.11a Overview 519
        13.1.1 IEEE802.11a OFDM Basics 519
        13.1.2 IEEE802.11a Frame Structure 523
        13.1.3 IEEE802.11a Transmit Spectrum 524
  13.2 IEEE802.11a Transmitter Overview 524
  13.3 IEEE802.11a Channel Overview 524
  13.4 IEEE802.11a Receiver Overview 528
  13.5 IEEE802.11a Transmitter Design Details 528
        13.5.1 OFDM Generator 530
  13.6 IEEE802.11a Receiver Design Details 538
        13.6.1 Receiver Front-End Signal Path 538
        13.6.2 Coarse Frequency Estimation 542
        13.6.3 Preamble Detection and Alignment 545
        13.6.4 Coarse Frame Timing 546
        13.6.5 Fine Frame Timing 552
        13.6.6 OFDM Receiver Back End 553
  13.7 IEEE802.11a System Testing 571
        13.7.1 Simulation Testing 571
        13.7.2 SDR Hardware Testing 571
  References 577

Chapter 14 More Fundamentals   **579**
  14.1 Introduction 579
  14.2 Fixed Point Number Formats 579
  14.3 Complex Number Review 583
        14.3.1 Euler's Formula 584
        14.3.2 Phase Modulation 585
  14.4 Amplitude Companding 588
        14.4.1 μ-Law Analog Companding 588

   14.4.2 μ-Law Digital Companding 589
 14.5 Power Amplifiers and PAPR 590
   14.5.1 Peak to Average Power Ratio 590
   14.5.2 Power Amplifiers 596
 14.6 Samples per Symbol Question 598
   14.6.1 $M = 2$ or 4? 598
   14.6.2 Resampler Output Noise Floor 600
   14.6.3 Timing Error Detector Considerations 600
   14.6.4 Equalizer Considerations 603
 14.7 Decision Detectors 603
   14.7.1 Maximum Likelihood Detector 603
   14.7.2 Maximum A Posteriori Detector 604
 14.8 Preamble Notes 606
   14.8.1 Existing Preambles 606
   14.8.2 Frequency Domain Detection 609
 14.9 Doppler Details 612
   14.9.1 Doppler Time Dilation 612
   14.9.2 Adjacent Channel Interference 613
 14.10 Low Cost SDR 613

About the Author   619

Index   621

Preface

Every book makes a promise to the reader. Some books promise to inform, some promise to entertain, and some make other promises. The promise of this book is to sharpen the software defined radio (SDR) skills of readers who use it.

For the working electrical engineer, this book contains a wealth of definitions, design ideas, and clarification of difficult concepts. For the SDR hobbyist, this text aims to provide deeper, more insightful knowledge of how signals can be processed to alleviate impairments and extract information.

An effort is made to replace mathematical derivations with intuitive explanations. There are a few places where equations are appropriate and essential. However, for the hobbyist, who just wants to see something work, most of the math can probably be skipped. A reader with a basic understanding of trigonometry and elementary calculus should have no trouble with the math in this book. Digital signal processing as well as some MATLAB® and Simulink® training is also a good preparation. Mathworks.com has plenty of information about training as well as free tutorials. Just add determination and you will be an expert soon.

Many MATLAB and Simulink software examples and fully developed simulations are explained. Although they have all been tested, the author takes no responsibility for their operation.

I sincerely hope that readers of this book obtain the knowledge and skills that they were promised when they chose it.

Chapter 1

Introduction

On Christmas Day 2004, the Cassini spacecraft sent the Huygens space probe coasting down towards the surface of Saturn's largest moon, Titan. The high magnitude of the descent velocity resulted in significant Doppler frequency shift. The received signal at the Huygens space probe was shifted outside the receiver symbol timing bandwidth, causing a signal detection failure.

Would a software defined radio (SDR) have provided an expedient solution? SDR allows both receiver parameter and algorithm changes on the fly. Indeed NASA has an SDR solution, called Space Telecommunications Radio System (STRS). Their SDR rational is summed up in the following statement from [1]:

> The traditional approach is to procure radios that meet requirements exactly. SDRs are designed to meet more than existing launch requirements; they are sized with additional resources and reconfigurability so that new capabilities can be added in flight.

The purpose of this book is to provide practical design details and theoretical background to facilitate the implementation of software defined radio signal processing. In other words, design approaches for algorithms that are software defined. The book relies on MATLAB® code and working Simulink® models to illustrate many practical details. To get started, a brief history of SDR is followed by a discussion of current hardware and software for SDR development.

By the way, the communications failure between the Cassini spacecraft and the Huygens space probe was fixed by changing the descent trajectory to reduce Doppler shift. An SDR solution may have been easier.

1.1 ORIGINS OF SDR

Joseph Mitola, a Mitre Corporation scientist, is usually credited with originating the software defined radio concept. He presented his idea in a 1992 paper called simply "The Software Radio" [2].

Mitola envisioned a radio that could be reprogrammed and reconfigured through software. At about the same time, the U.S. military saw this concept as a solution to their dissimilar and noninteroperable radio problem between different branches of the military services. The vision of a radio that could adapt itself to many different signals (even signals that were not defined when the radio was built: the ultimate SDR) fueled billions of dollars of spending from the first SDR project, called Speakeasy, until 2012 when funding for the ultimate SDR, the Joint Tactical Radio System (JTRS), was stopped [3] [4].

1.1.1 Speakeasy

Speakeasy, version 1, was tested in 1994. For an SDR, it had a lot of hardware; it took up the back of a truck. The bullet points below, from the *1998 International Symposium on Advanced Radio Technologies* clearly show the U.S. military's early vision for SDR.

- The PC of the Communications World;
- Fully Programmable Waveform and COMSEC for Voice, Multimedia, and Networking;
- Multiband . . . continuous from 2 MHz to 400 MHz;
- Open Modular HW Architecture;
- Open SW Architecture;
- Commercially Successful HW and SW;
- Legacy Systems Compatibility.

1.1.2 Joint Tactical Radio System

Started around 1998, the Joint Tactical Radio System (JTRS) was an ambitious attempt to standardize several U.S. military voice and data radios. Among them were:

- Multifunctional Information Distribution System (MIDS);
- Handheld, Manpack, Small Form Fit (HMS);
- Airborne, Maritime, Fixed Station (AMF);

CHAPTER 1. INTRODUCTION

- Ground Mobile Radio (GMR).

JTRS was restructured in 2005 under the Joint Program Executive Office (JPEO). Finally, in 2012, JTRS was canceled, and its funding allowed to expire. A letter dated October 12, 2011, from the Committee on Armed Service, U.S. House of Representatives, to the Under Secretary of Defense contained the following [5]:

> Our assessment is that it is unlikely that products resulting from the JTRS GMR development program will affordably meet service requirements, and may not meet some requirements at all. Therefore, termination of the program is necessary. . . .

JTRS was restructured into the Joint Tactical Networking Center (JTNC). The JTNC maintains a repository of former JTRS waveforms, some of which are:

- Mobile User Objective System (MOUS);
- Wideband Networking Waveform (WNW);
- Soldier Radio Waveform (SRW).

1.2 SOFTWARE COMMUNICATIONS ARCHITECTURE

JTRS's major accomplishment, Software Communications Architecture (SCA) is a software structure that describes how transmit and receive signal processing algorithms are to be assembled and connected to run on radio hardware. The SCA is a major SDR innovation that was meant to be the "software backbone" of JTRS radios. You could call it a specialized computer operating system for radios. The SCA details are outlined in several documents that tell the radio manufactures how their SCA compliant implementation must work (https://www.jtnc.mil/Resources-Catalog/Category/16990/sca/).

An SCA SDR starts its configuration process by reading the domain profile (DP) a set of XML (Extensible Markup Language) configuration files that decide what components this radio needs and how they should be connected.[1] The DP transfers the setup information to the core framework (CF).

The CF is the primary waveform (radio receiver or transmitter) deployment mechanism. CF reads the DP and provides methods to connect components, resources, and devices to set up a functioning waveform. CF is the set of SCA compliant interfaces that abstract the underlying platform hardware in a standardized manner so that multiple waveforms can use them.

---

1 XML is a plain English text file meant to allow a nontechnical user to compose a radio setup script.

For each waveform application, the CF sets up an application factory (AF) that requests allocation of resources and instantiates, configures, and connects components. The AF also establishes application component connections during the instantiation process. The required connections are described in the Software Assembly Descriptor (SAD) file. The term "factory" in SCA refers to the capability to initiate new resources or devices.

What does *connects components* mean in SCA software? Enter CORBA (Common Object Request Broker Architecture), an object-oriented client-server communications system for passing data around an SCA SDR. CORBA was first released in 1991 to allow multiple diverse computer systems to work together. Each CORBA device has an Object Request Broker (ORB) that keeps track of data origination and destination points. SCA no longer requires CORBA, in part because of CORBA complexity and size. A simpler connection system called Modem Hardware Abstraction Layer (MHAL) is still in use.

1.2.1 Other SCA Considerations

Full SDR reconfigurability, like the SCA radio was intended to do, has several overhead costs:

1. Cost to develop various built-in connection software such as CORBA and MHAL. These are generally not part of a single-purpose radio.
2. Cost to test the multitude of configurations the radio can take on.
3. Cost to make sure that the SDR cannot put its transmitter into an illegal power and/or bandwidth mode.

1.2.2 SDR Currently

The military's original vision of SDR as a radio that provides for reconfiguration based on a repository of signal processing components and configuration directives is still important. However, in modern parlance, there seems to be little difference between an SDR and any radio implemented in reconfigurable DSP software. Figure 1.1 represents what many readers of this book might consider an SDR setup, that is, a small RF receiver/transmitter with a data interface (typically USB) to a PC where all the digital signal processing (DSP) for the signal of interest occurs. The SDR receiver hardware is typically nothing more than an RF to complex baseband tune and filter. The SDR transmitter hardware, often contained on the same circuit card as the receiver, is the reverse of that. The ADALM-Pluto is an ANALOG DEVICES® product and the USRP E-310 is from ETTUS RESEARCH™.

CHAPTER 1. INTRODUCTION

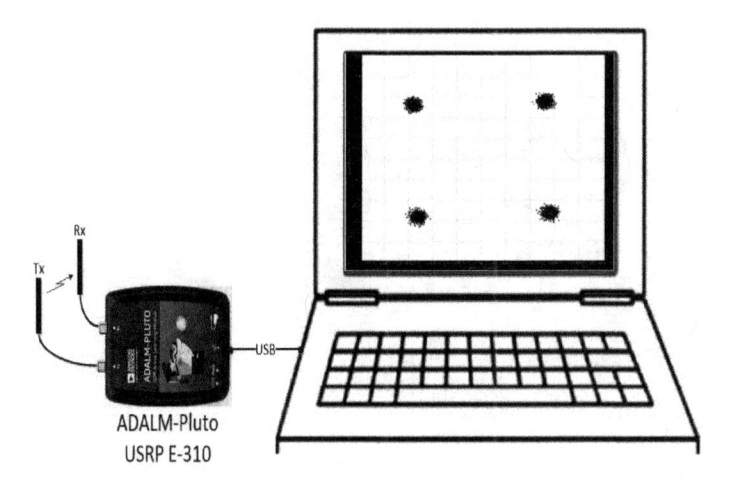

Figure 1.1 A typical modern SDR setup.

1.3 RADIO HARDWARE CHOICES FOR SDR

Table 1.1 shows some (but not all) currently popular SDR hardware choices from the most expensive to the cheapest. Before purchasing SDR hardware, it is important to understand the intended use. For example, if you intend to interface to GNU Radio Companion (GRC) or Simulink, find out if an interface support package is available. A good SDR choice for a very low-cost initial try at designing DSP receivers is the RTL-SDR. The ADALM-Pluto is another low-cost SDR with an ANALOG DEVICES AD9363 high performance radio transceiver integrated circuit. The other entries also have various strong points, for example, the SDRplayDuo and AirSpy have a set of RF preselect filters that should provide enhanced receive sensitivity.

The information in Table 1.1 was carefully checked, but it is not guaranteed to be accurate. I take no responsibility for errors or subsequent product changes. Readers are encouraged to double-check product specifications with the manufacturer.

1.4 DSP SOFTWARE DEVELOPMENT CHOICES FOR SDR

A DSP software development system for SDR is a program that facilitates DSP radio software implementation. The SDR provides the antenna interface and tune and filter hardware. However, the follow-on DSP is usually implemented in a PC. For

Table 1.1 Some Digital Signal Processing Hardware Choices for SDR, as of June 2023

Name	Tuning Limits	BW Max	Rx/Tx	Approx $	Source
USRP E310	70 MHz-6 GHz	56 MHz	2Rx,2Tx	$4,546	ETTUS RESEARCH™ Independent, no PC required
ADRV9361-Z7035	70 MHz-6 GHz	56 MHz	2Rx,2Tx	$1,581	ANALOG DEVICES Good for MIMO
Ettus B200	70 MHz-6 GHz	56 MHz	1Rx,1Tx	$1,109	ETTUS RESEARCH™
Sidekiq™	70 MHz-6 GHz	50 MHz	1Rx,1Tx	$649	Includes RF preselector
PicoSDR	70 MHz-6 GHz	56 MHz	4Rx,4Tx	Contact NUTAQ	NUTAQ™, Canada Great for MIMO
SDRplayDuo	1 KHz-2 GHz	10 MHz	2Rx	$370	SDRplay Includes RF preselector
HackRF One	1-6 GHz	20 MHz	1Rx,1Tx	$320	Great Scott Gadgets
LimeSDR	300 MHz-3.8 GHz	28 MHz	2Rx,2Tx	$315	Lime Microsystems, England Good for MIMO
ADALM Pluto	325 MHz-3.8 GHz	20 MHz	1Rx,1Tx	$229	ANALOG DEVICES Low cost educational tool
FUNcube	150 KHz-1.6 GHz	180 MHz	1Rx	$199	Radio Communications Foundation, Bedford, UK
AirSpy R2	125 KHz-8 MHz	10 MHz	1Rx	$169	AirSpy Includes RF preselector
Smart HF Bundle	100 Hz-1.75 GHz	2.56 MHz	1Rx	$114	NESDR, great for listening to amateur radio HF
RTL-SDR	24 MHz-1.70 GHz	2.56 MHz	1Rx	$39	RTL-SDR See Chapter 14

CHAPTER 1. INTRODUCTION 7

example, the laptop on the right side of Figure 1.1 contains DSP code to decipher ADSB messages (see Chapter 11). These include a block diagram environment (BDE); the user places component blocks on the PC screen and connects them with lines representing signal connections. The block diagram includes an interface block that controls the USB connection to the radio SDR hardware. When the "run" button is clicked, the BDE blocks are compiled into a program that runs on the host PC. There are two popular choices for SDR BDEs; these are briefly discussed below.

1.4.1 GNU Radio Companion

GNU Radio Companion (GRC) is a free and open-source[2] radio development block diagram environment (BDE). Users can download and install the GRC program, plug their SDR into a USB port and get started at no extra cost. Clicking on "run" will start an event-driven simulation. There are a few factors to consider:

1. GRC has a provision to generate USRP (Universal Software Radio Peripheral) interface blocks for USRP-compatible SDR hardware. The USRP is an invention of ETTUS RESEARCH and is used on many SDR hardware products.
2. GRC makes a great tool for research and education. In addition, GRC is featured on the fully deployable (i.e., no external PC required) USRP E310 SDR.
3. GRC is an event-driven simulation environment. This means that each block runs when the preceding block has data to process. Due to its event-driven structure, GRC may have trouble with recursive algorithms that are not contained within block code and require highly constrained feedback delay.
4. GRC blocks are typically written in C++. There are many preprogrammed blocks, but if you need to make a new block, you will have to be proficient in C++ programming language.
5. GRC questions are typically answered by emails to a support bulletin board. There are many volunteer experts who monitor this so you will likely get your questions answered, however there is no guarantee of an expedient answer.

---

[2] A comment was once made that open-source software is only free if your time has no value. A bit cynical, but sometimes true.

1.4.2 Simulink

Simulink is a block diagram environment (BDE) simulation tool that executes sample by sample in discrete time mode (the only mode used in this book).[3] In addition, there are Simulink toolboxes needed. These toolboxes are special purpose software add-ons. The only toolboxes needed for every example in this book are:

1. Communications System Toolbox;
2. DSP System Toolbox;
3. Fixed-Point Designer;
4. Signal Processing Toolbox.

Simulink is a time, not event, driven simulation environment. Block states and outputs are updated at every discrete time simulation step - starting at the inputs and moving to the outputs. There can be many sample rates in a Simulink simulation. These are generally integer multiples of each other. Signals at the highest sample rate can change every simulation step, signals at 1/2 the highest sample rate can change every other simulation step, and so on. Say, for example, a Simulink model fastest sample time is 1 second and some blocks only execute every 5 seconds, that is every 5 occurrences of the fastest sample time. The simulation update will occur every second and the slower blocks will only be updated every 5 seconds.

Like most USB-connected SDR hardware, the input to Simulink is a record of samples. While Simulink is processing the current sample record input in one buffer, radio hardware is collecting the next sample record in another buffer. The buffers are then swapped and Simulink can step through these recorded samples and calculate the step updates all the way to the model output very quickly. The numerical result is the same as if Simulink actually ran at the exact radio sample rate.

Using MathWorks® HDL coder tool, a Simulink simulation can be transformed into an HDL project compatible with popular FPGA synthesis tools. This ensures that data in a Simulink simulation is an exact match to the data that would flow through an FPGA[4] implementation of the circuit. The time alignment between the simulation samples at different sample rates in Simulink will match the time alignment between the samples at different sample rates in the FPGA. Changes to the simulation BDE are like making changes to the FPGA hardware implementation. This is sometimes

---

3   This book relies on MATLAB and Simulink to illustrate software defined radio designs.
4   Field programmable gate arrays are integrated circuits that implement large amounts of concurrently running digital hardware. FPGA are generally programmed by synthesis tools such as Xilinx Vivado. These tools convert VHDL coding (HDL means hardware descriptive language; the V has various meanings) into actual concurrently running digital logic.

CHAPTER 1. INTRODUCTION

called bit and cycle true. This feature makes Simulink useful for designing FPGA based products. Factors to consider for Simulink are:

1. Simulink offers ADALM-Pluto, RTL-SDR, and an Ettus USRP hardware interface blocks as hardware add-ons. There are some SDRs for which there is no Simulink built-in interface.
2. Unlike GRC, Simulink is not free. However, there are academic, student, and home licenses, which are low-cost ($500 range). These also have low-cost annual maintenance fees.
3. Simulink is a time-driven simulation environment. As explained above, time-driven simulations are very useful for FPGA DSP implementation.
4. Simulink top-level BDE blocks can usually be based on a hierarchy of simpler blocks, although some blocks are based on proprietary code that the user cannot change. Finally, there are several ways to write underlying code for custom blocks. The block hierarchy makes Simulink very self-documenting.
5. For users who have paid their annual license maintenance fees, Simulink and MATLAB questions are usually answered within one day by an email to support@MathWorks.com. In my experience, MathWorks will do whatever it takes to get questions answered quickly and to the user's satisfaction.

1.4.3 SDR Sharp

SDR Sharp is a receive signal analyzer, very useful for some of the receive-only SDRs in Table 1.1. SDR Sharp is a downloadable application that interfaces to an SDR through a PC's USB connector. The spectrum analysis features of SDR Sharp can be very helpful to understanding the frequency-domain big picture. There is a large of amount of helpful information about this product at: https://www.rtl-sdr.com/sdrsharp-users-guide/.

1.4.4 A Word About Interfaces

Most SDRs must set up and control a radio integrated circuit, usually the Analog Devices AD936x, through a PC's USB port. Writing a USB device driver to control a particular SDR is a formidable task. Fortunately, both GNU Radio and Simulink provide those for us. However, they will not give you access to every AD936x register. To see why users might not want this access, consider that the complete AD936x register reference is a 72-page document.

Setting parameters is much more easily done in a device driver GUI such as in Figure 11.3. MATLAB also has a system object-oriented interface discussed in Chapter 12.

The MATLAB system object interface has an easy way to change the AD936x receiver bandwidth. If this is important to your use of the SDR, then do not choose an interface that has no receiver bandwidth setting option. The point is, the SDR interface you choose determines what you can do with the radio, so study the choices carefully.

REFERENCES

[1] J. Mitola. "SCAN Testbed Software Development and Lessons Learned". *IEEE National Telesystems Conference*, 1992.

[2] J. Mitola. "The Software Radio". *IEEE National Telesystems Conference*, 1992.

[3] C. Adams. "Software Defined Radio". *Avionics Magazine*, 2006.

[4] C. Adams. "SDR Takes Flight". *IEEE National Telesystems Conference*, 2013.

[5] D. Ward. "Tactical Radios Military Procurement Gone Awry". *National Defense*, 2012.

Chapter 2

Communication System Fundamentals

2.1 INTRODUCTION

The purpose of this chapter is to review some fundamental communications systems principles and measurements.

2.2 BASIC MEASUREMENTS

2.2.1 dB, dBm, dBc

A dB (decibel) is simply the logarithm of a ratio. For example, given two measurements of electrical power:

$$N_{dB} = 10 log_{10}\left(\frac{P_a}{P_b}\right) = 10 log_{10}\left(V_a^2/V_b^2\right) \quad (2.1)$$

For V_a, V_b across the same resistance

Because a logarithm is really an exponent of 10 ($y = log_{10}(x) \Rightarrow x = 10^y$), the log of a number acts to compress the range of that number so we can express very large ratios in a compact form such that very large numbers can be represented by smaller numbers. For example, 100 dBm is 10 million watts. Figure 2.1 reveals this compressive property as x gets larger than 1. Another useful property of the log function is $log_{10}(1) = 0$. For example, a linear circuit with a gain of 0 dB has the output equal to the input. Finally, if the output is less than the input, the dB gain is negative.

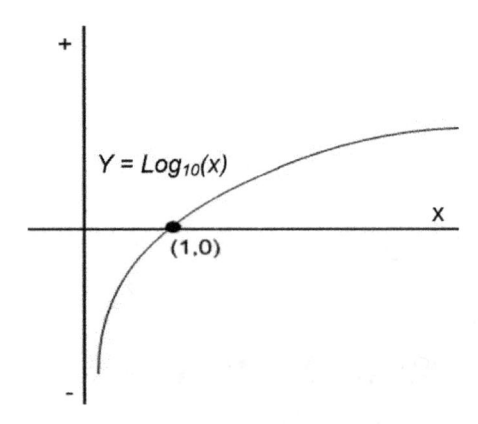

Figure 2.1 $y = log_{10}(x)$.

Equation (2.1) is useful for specifying ratios like gain or loss. However, for standardized single power measurements, we can make P_b a known constant, for example, 1 mW. Then we have (2.2). This is very useful for comparing power measurements.

$$N_{dBm} = 10log_{10}\left(\frac{P_a}{0.001}\right)$$
$$N_{dBW} = 10log_{10}\left(\frac{P_a}{1}\right) \quad (2.2)$$
$$N_{dBV} = 20log_{10}\left(\frac{V_a}{1}\right)$$

Equation (2.2) shows three commonly used logarithmic ratios. The first two are a compact indication of power. N_{dBV} uses a measurement of volts RMS. RMS stands for root mean square, calculated as shown in (2.3).

$$v_{RMS} = \sqrt{\frac{1}{N}\sum_{i=0}^{N-1} v_i^2} \quad (2.3)$$

These measurements are usually made on stationary signals. That means the signal statistics do not change. For example, the mean (or average) that is part of Equation (2.3) does not change (or practically speaking, the mean does not change very much) over the course of one measurement.

CHAPTER 2. COMMUNICATION SYSTEM FUNDAMENTALS

dB below carrier, dBc, is often used for testing analog-to-digital converters (ADC). These devices output a digitized version of their input as well as some extra low-level tones called spurs. The power difference (in dB) between the desired ADC output and the highest power spur is called the spur-free dynamic range (SFDR) in dBc. See Section 8.2.3.

2.2.2 dBu

Another measurement is dBu. dBu is used extensively by the U.S. Federal Communications Commission (FCC) to indicate electric field strength on radio station coverage maps. Electric field strength at a point in the receive field depends only on the transmitter power, distance from transmitter, and losses due to obstructions. dBu is calculated from E = electric field strength expressed in uV/meter: $dBu = 20log_{10}E$. In addition, dBu values on a coverage map are always at one specified carrier frequency for the entire map.

FCC coverage maps in dBu predict interference to other radio stations. However, a radio system designer also needs to know if the signal is strong enough for the intended receivers. Receiver sensitivity (i.e., the lowest received signal power that can be received) is often measured in dBm, not dBu. In [1], a formula is derived to convert dBu (field strength) to dBm (received power). This formula is repeated below (note that λ is received wavelength). Also, $G_{antenna}$ is the gain of the antenna used for the measurement for example, a half-wave dipole has a gain of 1.64.

$$dBm = dBu + 10log_{10}\left(\lambda^2 G_{antenna}\right) - 126.76 \qquad (2.4)$$

Say an FCC coverage map has a 60 dBu line of constant electric field strength around a frequency modulated (FM) broadcast transmitter (this is the same idea as lines of constant elevation on topographic maps). We want to know how well an FM receiver will pick up this signal at any point on this line. Assuming an antenna with a gain of 1.5, 89.5 MHz signal center, and speed of light = 300 million meters per second, the wavelength is 300/89.5 = 3.3520 meters. The formula tells us the receiver will see a -54.4930 dBm received power level, quite adequate for listening to FM radio.

2.2.3 dBi, dBd

An isotropic radiator is an antenna that radiates equally in all directions. This nondirectional antenna has a gain of 1. We call that $10log_{10}(G_{antenna}) = 0dBi$. In general, dBi is used to define the gain of an antenna relative to an isotropic radiator.

For example, a directional antenna may concentrate all transmit power in a 72° sector (think orange slice). The power in the sector is then five times that of an isotropic radiator. This antenna is said to have antenna gain $G_{antenna} = 5$, or 7 dBi gain, in the direction of the sector. Alternatively, dBd is sometimes used to define the gain of an antenna relative to a dipole radiator. Converting between dBd and dBi requires adding 2.15: dBi = dBd + 2.15.

2.2.4 VU

VU (volume units) is used for measuring audio frequency power on a VU meter. There is no direct conversion between dBm and VU. All that is required is that 0 VU on a meter is calibrated with a 1 KHz sine wave to +4 dBm or 1.23V in 600 ohms. This is +4 dBm power. People who monitor VU meters can tell by the ballistics of the needle (or modern digital display) how high the signal peaks are likely to be and how much distortion might be generated.

2.2.5 Equivalent Isotropic Radiated Power

For directional transmit antennas, equivalent isotropic radiated power (EIRP) is used to measure total power in the direction where the transmitted power is maximum.

Isotropic power is the same in every direction on a sphere with the radiator (transmit antenna) in the center.[1] However, many antennas concentrate their power in a fraction of this sphere.[2] For example, a directional transmit antenna may concentrate power in one-fourth of a sphere. The one quarter sphere gets the power from the other three nonradiating quarters. So, in that direction the antenna radiates four times the isotropic power and we say that the antenna has a gain of four. EIRP measures this directed power in the radiating quadrant only because that is where the transmit power is. In general, EIRP is the product of transmitter output power and transmit antenna gain.

2.2.6 Energy per Bit to Noise Power Spectral Density

$\dfrac{E_b}{N_0}$ is a widely used measurement in practical communications design.

---

1   Isotropic antennas are called omnidirectional if they do not transmit power up or down but only outward perpendicular to their vertical axis.
2   Transmit and receive antennas are reciprocal. That means a transmit antenna with directional power can also operate as a receive antenna with a directed sensitivity.

CHAPTER 2. COMMUNICATION SYSTEM FUNDAMENTALS

2.2.6.1 Energy per Bit

Energy is generally measured in (power)(time). For example, the electric company uses watt-hours to calculate your bill. A series of received bits at a constant rate (i.e., a cyclo-stationary signal sometimes called a bauded signal) has an energy per bit of E_b = received power multiplied by bit time duration = received power divided by bit rate (bits per second). For power measured in watts and time in seconds, E_b is in joules equal to the energy allocated to each bit.

Note that different communication systems can have the same E_b but different bit rates and different amounts of receive power. Some QAM systems receive bauded data symbols that carry multiple bits. For example, a 16QAM modem detects one of 16 symbols and then each symbol represents $M = 4$ bits; see Figure 12.23. In this case, we may work with E_s = received power multiplied by symbol time duration. We can also calculate $E_b = E_s/M$.

2.2.6.2 Noise Power Density

Noise power density, N_0, is the total noise power, P_{noise}, divided by the bandwidth of the noise, B_{noise}. N_0 can be measured with a low noise spectrum analyzer that has a band power measurement feature. As a noise density, N_0 is *normalized*, meaning it is the same for any bandwidth. This is the same concept as the density (weight/volume) of a metal is the same for any amount of the metal.

As noise power spectral density, N_0 has dimensions of P_{noise}/B_{noise} = (watts)/(cycles/second). We can also write N_0 = (watts)(seconds/cycle) = joules/cycle. Converting power to energy always implies a time interval (e.g., the longer you keep your air conditioner running, the higher is your energy use, and the larger is your energy bill). Thus, joules/cycle is the energy delivered by the noise source in the time of one cycle of bandwidth. In general, N_0 is not an absolute measure; it is always a ratio or rate.

2.2.6.3 $E_b N_0$

In Equation (2.5), T_{Symbol} is the symbol time and R_{Symbol} is the rate, which is the inverse of the symbol time.[3] E_b/N_0 is often expressed as a dB ratio, as shown. In some cases the symbol rate and the noise bandwidth are the same ($R_{Symbol} = B_{Noise}$).

---

3 R_{Symbol} can be considered the minimum bandwidth required to process a bauded signal.

$$\frac{E_b}{N_0} = \frac{P_{Signal} T_{Symbol}}{\frac{P_{Noise}}{B_{Noise}}} = \frac{\frac{P_{Signal}}{R_{Symbol}}}{\frac{P_{Noise}}{B_{Noise}}} = \left(\frac{P_{Signal}}{P_{Noise}}\right)\left(\frac{B_{Noise}}{R_{Symbol}}\right) \quad (2.5)$$

$$\frac{E_b}{N_0} dB = 10 \log_{10}\left(\frac{E_b}{N_0}\right)$$

Then the signal to noise power ratio (SNR = P_{Signal}/P_{Noise}) and the E_b/N_0 are the same number. In general, however, E_b/N_0 is not the same as SNR. As bandwidth B_{Noise} is increased, SNR will decrease because the total noise power will increase (SNR is a power ratio). However, E_b/N_0 will stay constant because N_0 is a noise density.

The reader may object: how can E_b/N_0 stay the same but the receiver's performance degrades, as it admits more noise? Because E_b/N_0 is a received signal parameter, that has nothing to do with the design quality of the receiver.

2.3 OTHER MEASUREMENTS

2.3.1 Error Vector Magnitude

Error vector magnitude (EVM) is used to quantify digital radio constellation accuracy. Figure 2.2 shows a QPSK point, $E(n)$, at time index n in quadrant zero. The ideal location for this point is C_0, a constant. For EVM, we are concerned with the power (length squared) of both $E(n)$ and C_0. Equation (2.6) summarizes this and also shows the EVM at time index n.

$$\begin{aligned} |C_0|^2 &= I_{C_0}^2 + Q_{C_0}^2 \\ |E_v(n)|^2 &= \left(I_{C_0} - I_{E(n)}\right)^2 + \left(Q_{C_0} - Q_{E(n)}\right)^2 \\ EVM(n) &= \sqrt{\frac{|E_v(n)|^2}{|C_0|^2}} \end{aligned} \quad (2.6)$$

$$EVM_{RMS} = \sqrt{\frac{1}{N} \sum_{n=1}^{N} \frac{|E_v(n)|^2}{|C_0|^2}} \quad (2.7)$$

CHAPTER 2. COMMUNICATION SYSTEM FUNDAMENTALS

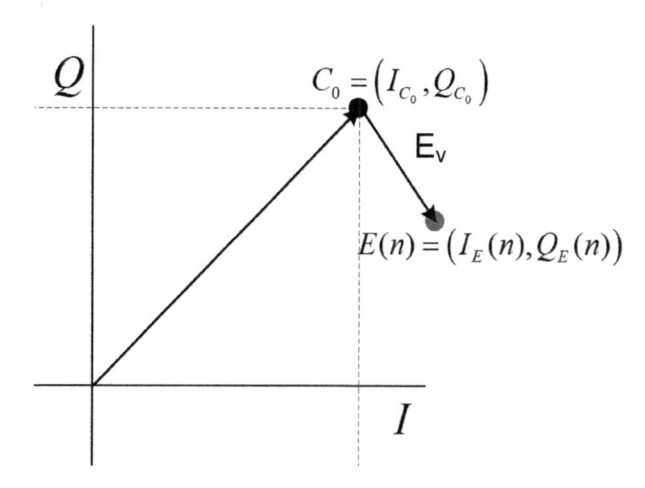

Figure 2.2 An illustration of EVM, symbol index n.

Equation (2.6) is correct; however, the EVM of a single point is not very useful. What we really need is shown in Equation (2.7). This root mean square (RMS) statistic is the average over N symbols. Laboratory instruments, such as the vector signal analyzer, often will show this number as a percentage (i.e., multiplied by 100).

Comparing EVM between receivers shows how one receiver might admit more noise than another. However, a transmitter can have high EVM if the constellation points do not have, for example, enough numerical precision.

2.3.2 Voltage Standing Wave Ratio

The voltage standing wave ratio (VSWR) is often used to indicate the amount of power reflected from an antenna back to a transmit power amplifier. To calculate VSWR, we start with the reflection coefficient, Gamma, shown in Figure 2.3. Z_s and Z_L are source and load impedances, respectively. These are complex quantities that vary with frequency, so the VSWR is generally specified at one frequency; the antenna resonant frequency. If Z_s and Z_L are matched (i.e., complex conjugates) then VSWR = 1 and reflected power is zero. This is the ideal situation. A VSWR greater than 1 indicates some reflected power. A VSWR of 1.5 means 4% of forward power is reflected back.[4] This wasted power is sometimes tolerable but can damage the transmitter in extreme cases. Some transmitters will provide a continuous reading of reflected power instead of VSWR.

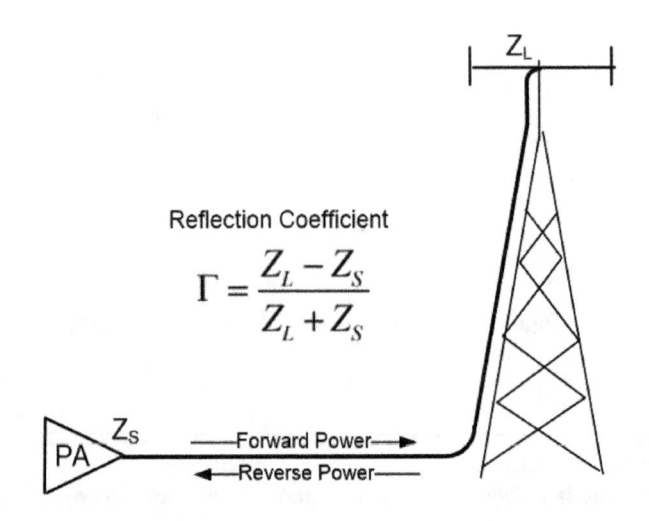

Figure 2.3 How to calculate the reflection coefficient.

Reflection Coefficient

$$\Gamma = \frac{Z_L - Z_S}{Z_L + Z_S}$$

Equation (2.8) shows how to convert Γ to VSWR. Note that using the absolute value results in a real number equal to 1 or higher. The forward and reverse power on the transmission line is constant. However, the forward and reverse voltages have drifting relative phase. This causes a composite wave, or standing wave, with a cyclically varying amplitude. The VSWR is the ratio between the highest and lowest amplitude of this standing wave. Also shown in Equation (2.8) is a calculation of VSWR directly from forward and reverse power in watts (not dBm or dBW).

---

4   In Equation (2.8), set $\dfrac{P_{reverse}}{P_{forward}} = 0.04$ and solve for VSWR.

CHAPTER 2. COMMUNICATION SYSTEM FUNDAMENTALS

$$VSWR = \frac{1 + |\Gamma|}{1 - |\Gamma|}$$

$$VSWR = \frac{1 + \sqrt{\frac{P_{reverse}}{P_{forward}}}}{1 - \sqrt{\frac{P_{reverse}}{P_{forward}}}} \qquad (2.8)$$

2.4 SOME BASIC PROCEDURES

2.4.1 Impedance Matching

There are two types of impedance matching: power matching and voltage matching.

2.4.1.1 Power Matching

Power matching maximizes the power transfer from source to load by matching the source to load impedance. These connections are often made with 50Ω or 75Ω coax cables carrying microwave transmit or receive signals. In many cases, the two impedances will be real.[5] Connections between radio frequency (RF) test equipment, from an RF power amplifier to an antenna or from antenna to an RF receiver, are generally matched this way. In those cases we are careful about impedance matching because we do not want to complicate the situation with reflections.

The maximum power transfer theorem states the power transferred from source to load is maximum when source and load impedances are equal. Figure 2.4 is an example matching circuit. Equation (2.9) finds the R_L, which maximizes P_{load}. Setting the derivative of P_{load} with respect to R_L to zero and solving for R_L works for this simple situation where there is only one maximum.

---

5 If matching complex impedances, they should be complex conjugates.

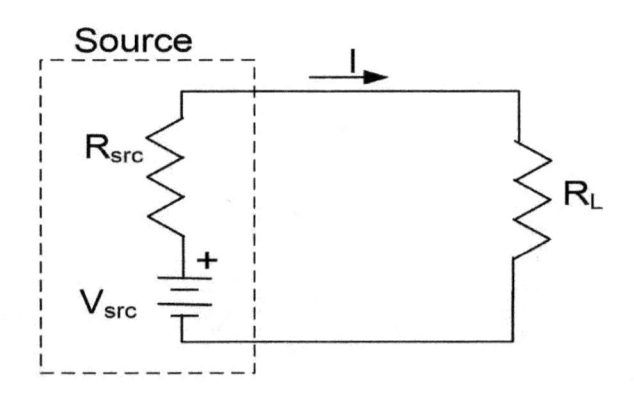

Figure 2.4 A circuit for demonstrating maximum power transfer.

$$P_{load} = I^2 R_L = \frac{V_{src}^2 R_L}{(R_L + R_{src})^2} = \frac{V}{U}$$

$$\frac{dP_{load}}{dR_L} = \frac{U'V - V'U}{V^2} = \frac{2(R_L + R_{src})V_{src}^2 R_L - V_{src}^2(R_L + R_{src})^2}{(V_{src}^2 R_L)^2}$$

To maximize P_{load}, we set the above derivative to 0 and solve for R_L

(2.9)

$$\frac{2(R_L + R_{src})V_{src}^2 R_L - V_{src}^2(R_L + R_{src})^2}{(V_{src}^2 R_L)^2} = 0$$

$$2(R_L + R_{src})V_{src}^2 R_L - V_{src}^2(R_L + R_{src})^2 = 0$$
$$2R_L^2 + 2R_L R_{src} - R_L^2 - 2R_L R_{src} - R_{src}^2 = 0$$

$$R_L = R_{src}$$

2.4.1.2 Voltage Matching

Instead of maximum power transfer, some audio frequency systems are maximum voltage matched. In this case the source, for example, a low-level microphone preamplifier output, is very low impedance, sometimes just a fraction of an ohm.

CHAPTER 2. COMMUNICATION SYSTEM FUNDAMENTALS

The load will be higher impedance, for example, a power amplifier input. This kind of impedance matching is also referred to as impedance bridging and maximizes the voltage transferred but not necessarily the power transferred. Often the load impedance is required to be 10 or more times the source impedance for maximum voltage transfer.

2.4.2 Signal Transmission

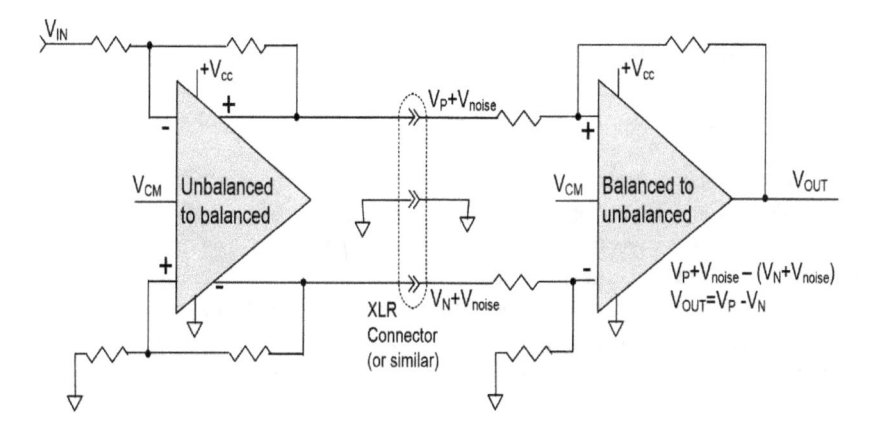

Figure 2.5 A circuit with unbalanced input to balanced cable driver to unbalanced output.

2.4.2.1 Balanced, Unbalanced

The left side of Figure 2.5 shows an unbalanced input V_{IN}. Unbalanced signals are referenced directly to circuit ground. So, noise from the signal source, as well as noise from ground currents, goes directly into the amplifier. The output of the first amplifier in Figure 2.5 is now balanced around V_{CM}, common mode voltage.[6] Noise picked up on both sides of the balanced connection is canceled in the second amplifier. At audio frequencies the cable between the two amplifiers will be a shielded twisted pair of wires and the connector will probably be an XLR or maybe

---

[6] Notice that both V_p and V_n are referenced to ground. However, they are balanced around a common mode voltage, V_{CM}, which could be zero. V_p and V_n are 180° apart and are centered on V_{CM}, hence the term common mode. Common mode voltage or noise pickup should be rejected by the subtraction in the balanced to unbalanced amplifier.

just screw terminals. Balanced connections like this are commonly used for long cable runs in high noise environments, such as automotive and factory applications.

Microphones and other transducers, such as thermometers and pressure sensors, have a very low-level output so their connecting cables are usually balanced to avoid noise pickup. Audio cable carrying stronger signals, such as a volt RMS or more, might be unbalanced. Microwave coax cables are a special case. Microwave BALUNs (balanced to unbalanced, or vice versa) circuits are often used to connect unbalanced coax cable to balanced dipole antennas or coaxial antenna cable to TV set balanced 300Ω twin-lead.

2.4.2.2 Ground Loops

The top plot of Figure 2.6 shows an audio frequency driver amplifier connected to another amplifier via a single wire shielded cable. These cables are commonly used in home audio systems and are called phono cables. Connecting the shield at both ends creates a ground loop between the two amplifiers; this is shown as a bold line. Magnetic fields across the ground loop can cause an induction current, as predicted by Faraday's law.[7] The induced current is typically 60 Hz, matching the power mains frequency. This can cause buzz or hum in the audio signal. A solution, is to connect the shield only at the driver amplifier side. This breaks the ground loop and prevents the induced current. Some home audio equipment is designed to operate without a safety ground, this also fixes the problem.

A better solution, shown in the lower plot of Figure 2.6, is to use a balanced shield cable. This separates the signal transmission from the ground loop. There may still be an advantage to disconnecting the shield ground from the receiving end of the cable.

2.5 SATELLITE COMMUNICATIONS MEASUREMENTS

2.5.1 Antenna Noise Temperature

To discuss antenna noise, we first need to define noise temperature. As shown in Figure 2.7, a sensitive power meter will measure a noise power density proportional to the physical temperature of the resistor. The value of the resistor (in ohms) is not critical [2]. The temperature must be measured using the Kelvin scale. At $T_{kelvin} = 0$, absolute 0, all thermal motion ceases and the noise power becomes

---

[7] A closed conductor placed in a varying magnetic field results in a current induced in the conductor.

CHAPTER 2. COMMUNICATION SYSTEM FUNDAMENTALS

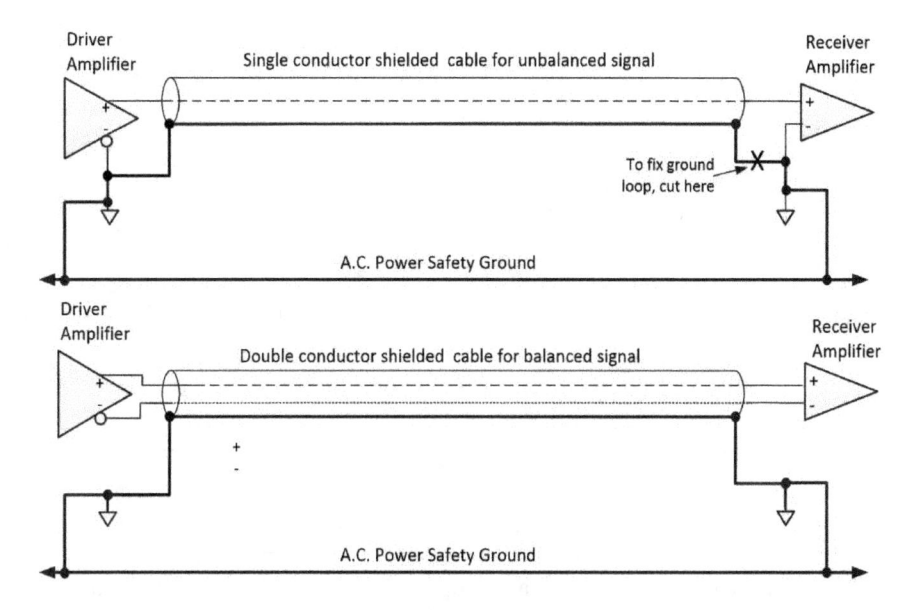

Figure 2.6 Ground loops in unbalanced and balanced cabling.

zero. Boltzmann's constant, k_B = 1.38e-23 joules/kelvin, is a physical constant describing the ratio between temperature and energy.[8] This constant was discovered in the late 1800s by Ludwig Boltzmann, an Austrian scientist.

Antenna noise temperature is not necessarily a temperature that can be directly measured but rather an effective temperature. Let's say we point a receive antenna and measure the noise power spectral density that it delivers to its receive terminals. To quantify the antenna noise temperature, we can heat up the resistor in Figure 2.7 until it delivers the same N_0. That temperature in kelvin will be the effective antenna noise temperature. The effective antenna noise temperature is a representation or model of noise generation. In the experiment just described, the resistor temperature could be measured, however, generally, antenna noise temperature cannot be measured with an instrument.

An actual antenna mounted outdoors is not as simple as the resistor in Figure 2.7. In addition to heat and radiation from the Sun, there are noise generators on Earth, such as power lines and machinery. There is also radiation from objects in

---

[8] Because Joules = (watts)(seconds), the units of Boltzmann's constant are also ((watts)(seconds))/kelvin, also known as (watts)/((hertz)(kelvin)).

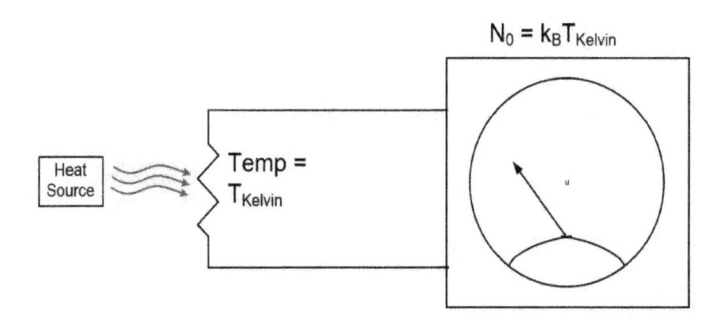

Figure 2.7 Definition of noise temperature.

space. These can come into the antenna field of view and affect the noise density. The effective antenna temperature of a dish will also vary with the pointing angle and between day and night. The antenna temperature is the average of effective antenna temperatures over all the noise sources which the antenna will be subject to.

The total receiving system noise temperature is the sum of antenna and receiver noise temperature. Unlike the receiver, the antenna is a one-port device so there can be no noise factor[9] ratio; the antenna has antenna temperature only. Thus, for receiving system temperature, the receiver noise factor must first be converted to a receiver noise temperature. Receiver, in this context, usually means simply the first low noise amplifier (LNA) and perhaps some cabling between the antenna and the LNA. As we will see, the LNA noise factor tends to dominate the receiver noise.

To understand the conversion of noise factor to noise temperature, consider Equation (2.10), where the receiver noise is used as an example. The reference temperature 290_{kelvin} represents a roughly room temperature noise source (about 17° Celsius or 62° Fahrenheit) at the receiver input. The noise factor then represents how much noisier the receiver is than the fixed reference temperature. For example, if $T_{receiver} = 0$, then $F_{receiver} = 1$, which means that the receiver does not generate any noise at all. If $T_{receiver} = 290$, then $F_{receiver} = 2$ or equivalently 3 dB. This means the receiver doubles the noise. For the history of this technique, see [3].

$$F_{receiver} = \frac{(SNR)_{out}}{(SNR)_{in}} = \frac{290_{Kelvin} + T_{receiver}}{290_{Kelvin}} \quad (2.10)$$

$$T_{receiver} = 290_{Kelvin}(F_{receiver} - 1)$$

---
9 See Section 7.2.1 for information about noise factor.

CHAPTER 2. COMMUNICATION SYSTEM FUNDAMENTALS

Now we can calculate $T_{system} = T_{antenna} + T_{receiver}$.

2.5.2 Satellite Receiver Sensitivity

Noise temperature analysis is often used to evaluate dish antenna installations. For example, the fixed satellite downlink dishes commonly seen at TV stations. Figure 2.8 shows the important parameters of a dish receiving antenna. For carrier wavelength λ, the antenna gain is shown in Equation (2.11). The antenna aperture, A_D, is shown in Figure 2.8. Due to antenna reciprocity, the gain applies to the same antenna used for either receiving or transmitting.

$$G_{antenna} = \frac{4\pi A_D}{\lambda^2} \qquad (2.11)$$

Receiver system refers to the combination of antenna, cable, and receiver input LNA. Satellite receiver system sensitivity is called G_{system}/T_{system} or just G over T. Of course, high G over T means high gain and low noise. This is a desirable situation for receiving signals with very little received power.

Figure 2.9 is a schematic of a receiving system front end for which we need the sensitivity G_{system}/T_{system}. The noise temperatures are calculated in Figure 2.9 so the only calculation left is shown in Equation (2.12). Notice that Equation (2.12) does not use dB. Also notice how raising the gain or lowering the noise temperature of any of the major three components improves the G_{system}/T_{system}. Thus, the system designer has a lot of options for meeting a G_{system}/T_{system} specification. For example, a higher quality, lower loss, waveguide or a higher gain LNA with the same noise factor.

$$\frac{G_{system}}{T_{system}} = \frac{(G_{antenna})(G_{wg})(G_{LNA})}{T_{antenna} + T_{wg} + T_{LNA}} = \frac{20 * 0.95 * 100}{300 + 15.3 + 145} = \frac{1900}{460.3} = 4.12 \qquad (2.12)$$

2.6 LINK BUDGETS

Consider a simple household financial budget. You start at the beginning of the month with $X. You deduct bills you have to pay before the end of that month. At the end of the month, you are either in the black (good) or in the red (bad). An RF link budget is similar. You start at the transmitter with P_{tx} watts of power. You deduct all the power reductions due to various physical phenomenon in the transmission path.

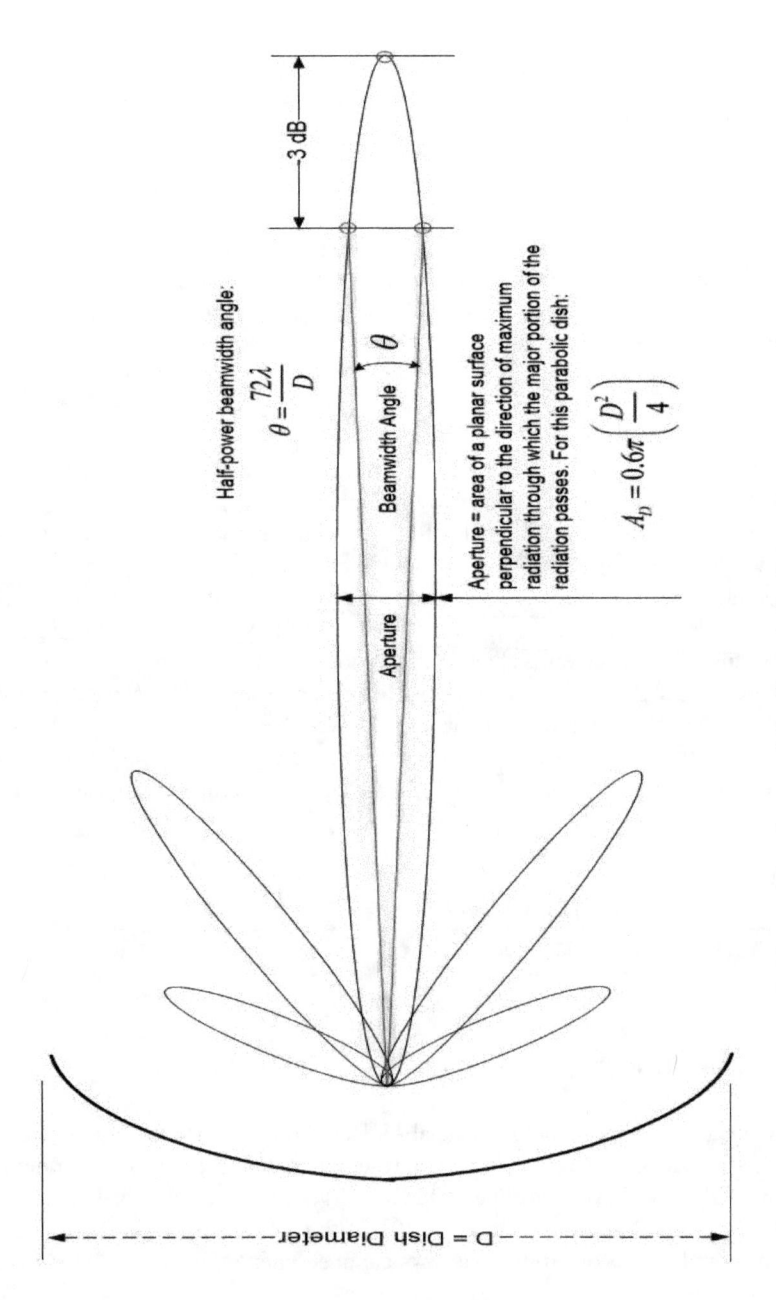

Figure 2.8 Commonly used parabolic dish antenna parameters.

Half-power beamwidth angle:
$$\theta = \frac{72\lambda}{D}$$

Aperture = area of a planar surface perpendicular to the direction of maximum radiation through which the major portion of the radiation passes. For this parabolic dish:

$$A_D = 0.6\pi \left(\frac{D^2}{4}\right)$$

D = Dish Diameter

CHAPTER 2. COMMUNICATION SYSTEM FUNDAMENTALS

$T_{antenna} = 300_{Kelvin}$

$G_{antenna} = 20$

$F_{LNA} = 1.5, \quad G_{LNA} = 100$

$T_{LNA} = 290(F_{LNA} - 1) = 145_{Kelvin}$

——————Waveguide——————LNA

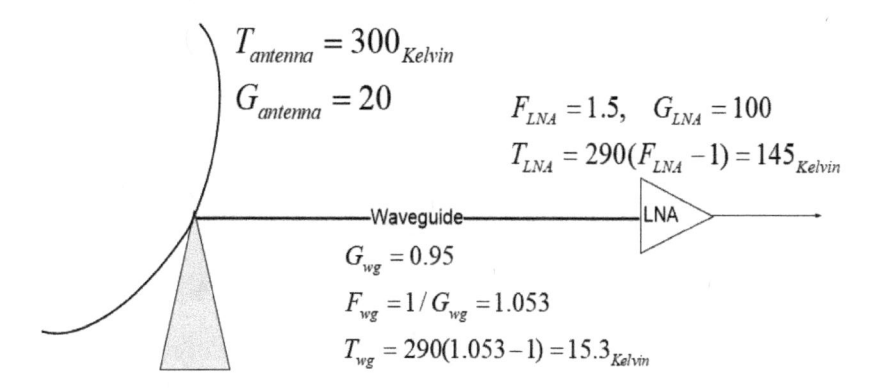

$G_{wg} = 0.95$

$F_{wg} = 1/G_{wg} = 1.053$

$T_{wg} = 290(1.053 - 1) = 15.3_{Kelvin}$

Figure 2.9 A satellite dish antenna system specification.

When the signal reaches the receiver, you either have enough power remaining for receiver processing (good) or you have an undetectable signal (bad).

You do not plan to spend your household budget down to $0 at the end of the month; you want to have a little money left in case something unexpected comes up. The same idea applies for RF link budgets, you want to have a little power (equivalently, $E_b N_0$ > minimum) left at the receiver in case some unexpected loss comes up in the propagation path. This extra power at the receiver is called a link margin. If the link margin power is greater than zero, the link functions properly and we say "The link is closed."

Figure 2.10 is a simple qualitative picture of a link budget. The dotted lines show losses in power and the solid lines show gains.

2.6.1 Satellite Receiver Link Budget

Table 2.1 is a link budget for a downlink (DL) satellite to parabolic dish receiving antenna. In this example, $\frac{G_{system}}{T_{system}}$ is not used directly to calculate link margin but is still an important design quality parameter.

2.6.1.1 Alternative Signal to Noise Indicators

Sometimes the received signal to noise ratio will be specified as in the last line of Equation (2.13). All four terms are in dB, including M, which is the link margin.

Table 2.1 An Example of Satellite Communications Link Budget Details

Link Budget Factor	Magnitude	Comment
Rx antenna gain	25 dBi	Determined by antenna design
Rx antenna gain	316.2	Ratio, not in dBi
Waveguide + LNA gain	10 dB	10 dB = gain of 10
Rx input noise factor	13	Waveguide and LNA noise factor, not dB
Rx noise temperature	3,480 Kelvin	290(13-1) = 3,480; see Equation (2.10)
Rx antenna temperature	300 Kelvin	Varies with installation
Rx system temperature	3,780 Kelvin	$T_{system} = T_{antenna} + T_{receiver}$
Rx isotropic power	-140 dBm	Receive power without receive antenna gain
Pointing loss	-1 dB	DL (downlink) antenna slightly misadjusted
Rx power (antenna gain)	-106 dBm	Receiver input power (-140+25+10-1 = -106)
System G/T	0.91	(Rx antenna gain)(LNA gain)/(Rx noise temp) = 3162/3480
Boltzmann constant	1.38e-23	k_B = (watts/Hz)/kelvin
N_0	-193.2 (dB/Hz)	$N_0(dB) = 10 log_{10} (k_B T_{system})$
Rx P_r/N_0	87.2 (dB)(Hz)	-106-(-193.2) = 87.2, subtract logs to divide
Data rate	18 Mbit/sec	Actual data rate
Data rate (log)	72.5 bit/sec	$10 log_{10} (DataRate)$
Rx $E_b N_0$	14.7 dB	87.2 - 72.5
Required $E_b N_0$	10 dB	Typical requirement for MPSK with FEC
Link Margin	4.7 dB	√ Additional loss we can incur and still close the link

CHAPTER 2. COMMUNICATION SYSTEM FUNDAMENTALS

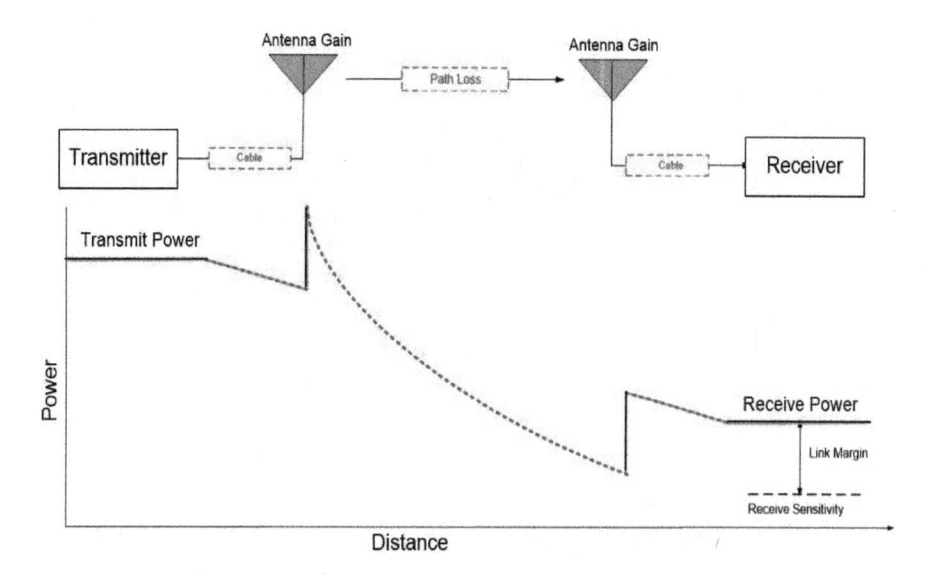

Figure 2.10 The basic idea behind link budgets.

Note that C is the received power, without regard to the bit rate. $kT = N_0$ so C/kT only requires measuring the receive power, using an RF power meter. Antenna temperature often defaults to 290 kelvin. C/kT is sometimes called the carrier power to noise density ratio and is measured in dB-Hz. The dB is for carrier power and the hertz signifies noise power in 1 Hz.

$$\frac{E_b}{N_0} = \frac{C}{kTR}$$
$$log_{10}\left(\frac{E_b}{N_0}\right) = log_{10}\left(\frac{C}{kTR}\right) = log_{10}\left(\frac{C}{kT}\right) - log_{10}(R) \quad (2.13)$$

From this we get :
$$\frac{C}{kT} = \frac{E_b}{N_0} + R + M$$

2.6.2 Terrestrial Receiver Link Budget

For receiving signals over the Earth, for example, receiving a signal from a broadcast station with a quarter-wavelength monopole whip receiving antenna mounted on a car or building, we might use a slightly different model for link budget.

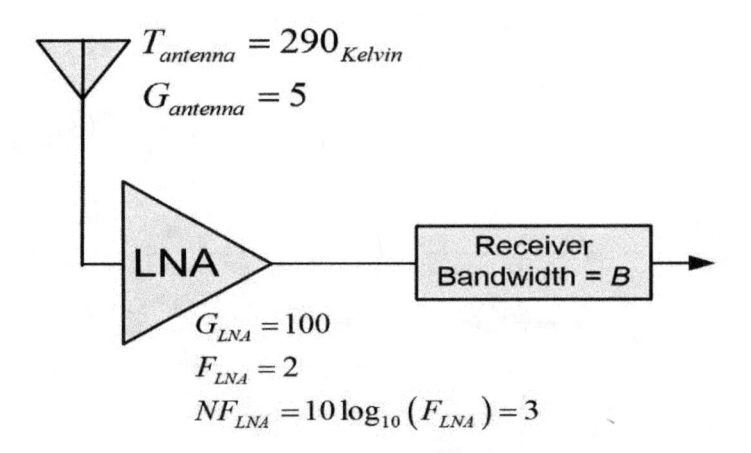

$$T_{antenna} = 290_{Kelvin}$$
$$G_{antenna} = 5$$
$$G_{LNA} = 100$$
$$F_{LNA} = 2$$
$$NF_{LNA} = 10\log_{10}(F_{LNA}) = 3$$

Receiver Bandwidth = B

Figure 2.11 Typical terrestrial antenna noise analysis.

As shown in Figure 2.11, we assume an antenna noise temperature fixed at 290_{kelvin}, noise power density into the receiver (in dB) is thus:

$$N_{0_{antenna}} = 10\log_{10}(k_B T_{antenna}) = \frac{-174dBm}{Hz} \quad (2.14)$$
$$N_{0_{receiver}} = N_{0_{antenna}} + NF_{LNA}$$

For data rate = R, we start with: $E_{b,received} = P_{received} - 10\log_{10}(R)$ and this leads to: $E_b N_0 = E_{b,received} - N_{0_{receiver}}$. Note that $P_{received}$ is the power delivered from the antenna to the LNA. Table 2.2 shows a detailed link budget for this situation. The bold items are intermediate sums.

CHAPTER 2. COMMUNICATION SYSTEM FUNDAMENTALS

Table 2.2 Example Terrestial Link Budget Details

Link Budget Factor	Magnitude	Comment
Tx power output	+60 dBm	Link budget starting point
Tx antenna gain	+5dB	Due to directivity, wavelength, aperture
EIRP	+65 dBm	+60-5 = 65, Total directed Tx power
Propagation loss	-160dB	Using free space loss, see Section 4.3
Link margin	-5dB	Maximum expected propagation loss uncertainty
Rx isotropic power	-100 dB	65-165 = power received at Rx isotropic antenna
Rx antenna gain	+5dB	Rx antenna gain, whip antenna
Rx signal power	-95 dBm	-100+5
Rx antenna noise temp	290 Kelvin	Approximate
RxLNA noise figure	2 dB	Low noise amplifier at receiver front end
Boltzmann constant	-198.6 dBm/(K*Hz)	Converts kelvin to noise power spectral density
$N_{0_{antenna}}$	-174 dBm/Hz	$10log_{10}(k_B T_{antenna})$
$N_{0_{receiver}}$	-172	Add RxLNA NF
Data rate	2 Mbit/sec	Assume a fixed data rate
Data rate (dB)	63 dB/sec	10*log10(Data rate)
Energy/bit (E_b)	-158 dBm/Hz	-95-63
Received $E_b N_0$	14 dB	-158-(-172)
Required $E_b N_0$	10 dB	Receive modem requirement
Link margin	4 dB	14-10 = +4 √ Additional loss we can incur and still close the link

2.7 IMPORTANT COMMUNICATIONS CONCEPTS

2.7.1 Nyquist Bandwidth Criteria

The Nyquist criteria simply states that to transmit R_s bits/sec, a channel bandwidth of $R_s/2$ is needed. This is known as the Nyquist bandwidth criteria. A typical pulse that meets the Nyquist criteria is shown in Figure 2.12.

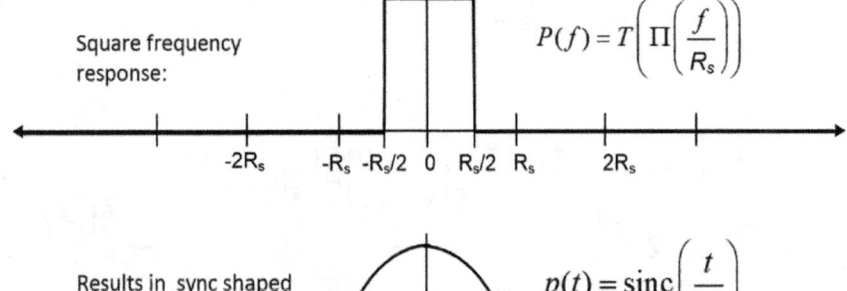

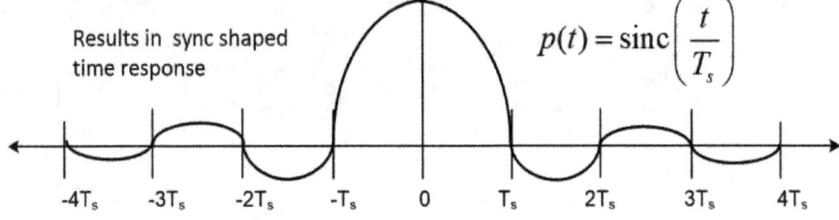

Square frequency response:
$$P(f) = T\left(\Pi\left(\frac{f}{R_s}\right)\right)$$

Results in sync shaped time response:
$$p(t) = \text{sinc}\left(\frac{t}{T_s}\right)$$

Figure 2.12 Square pulse bandwidth and corresponding time-domain pulse.

2.7.2 Nyquist Pulse Criteria

There is also a Nyquist pulse criterion that requires that each pulse in a string of pulses can be nonzero only at the pulse sampling time. The Nyquist signaling pulse shown in Figure 2.12 fulfills the Nyquist pulse criteria:

$$p(nT_s) = \begin{cases} 1 & n = 0 \\ 0 & n \neq 0 \end{cases} \tag{2.15}$$

CHAPTER 2. COMMUNICATION SYSTEM FUNDAMENTALS

A series of Figure 2.12 sync signaling pulses will not interfere with each other at the center. Figure 3.16 shows this clearly.[10] Section 3.2.6 describes a modified sync pulse used as a signaling pulse. This pulse meets the Nyquist criteria. This raised cosine pulse makes receiver symbol timing easier by using a little extra bandwidth.

2.7.2.1 Real Transmit Signals and Single Sideband

Let's take an example where the statement of the Nyquist bandwidth above refers to real signals. The sampled spectrum is said to need bandwidth $R_s/2$. However, as shown in Figure 2.13, the real baseband spectrum extends from $-R_s/2$ to $+R_s/2$ with the negative half identical to the positive half (a requirement for real valued signals). Because the two spectrum halves are identical, one of them is redundant and does not need to be transmitted. Deleting the redundant half results in a single sideband signal. Signal sideband transmission is an example of a real signal with the absolute minimum Nyquist required bandwidth of 0 to $R_s/2$ (or $-R_s/2$ to 0).

Because single sideband signal processing is complicated, a double sideband signal $-R_s/2$ to $+R_s/2$ is often transmitted to make the circuitry simpler (the AM broadcast band is an example, a receiver can just be a diode). Double sideband uses twice the Nyquist bandwidth criterion. A more complicated single sideband transmitter (and receiver) can generate a valid, useful, upper sideband AM (USB AM) signal from 0 to $+R_s/2$ (or lower sideband from $-R_s/2$ to 0). This baseband signal is frequency shifted to an RF carrier for a single sideband (SSB) transmission system. This signal does meet the Nyquist bandwidth criterion.

An example of the circuitry to generate this single sideband signal is shown in Figure 2.14. To expedite understanding, we show the signal spectra at each processing stage. The time dependance of the various signals is not shown. The circuit could be implemented in DSP or analog hardware. Starting on the right, the input $y(n)$ is modified in the 90° phase splitter to form complex $y_I + jy_Q$. The Hilbert transform[11] modifies y_Q to complex value y_{QH}. As shown, $y_{QH} + y_I = y_{SSB}$, the desired complex baseband single sideband signal. This can be shifted to the desired carrier frequency for transmission. Notice that the complex multiplication between y_{SSB} and the NCO is strictly a signal frequency shifter. Unlike traditional DSB AM, no carrier is generated. For SSB, the carrier (and its unmodulated power) is optional.

---

10 We should note that digital transmission often uses modified sync pulses with wider bandwidth because the receiver symbol timing recovery with sync pulses is difficult.

11 Note that we assume an ideal Hilbert transform filter and with an opposite rail delay to match the Hilbert transform delay. As described in Section 9.2.11, a practical Hilbert transform has a highpass frequency response. So, unlike the spectra shown in Figure 2.14, a practical Hilbert filter will need to work with signals having little or zero frequency content.

This signal could be AM voice or BPSK but not QPSK. SSB radios are popular for marine applications. SSB signals are often sideband data only so no transmitter power is allocated to a constant frequency carrier. All the power is applied to the modulated signal so it can travel a much further distance.

Real Baseband Signal
(spectrum must be identical on either side of 0 Hz)

Complex Baseband Signal
(spectrum can be different on either side of 0 Hz)

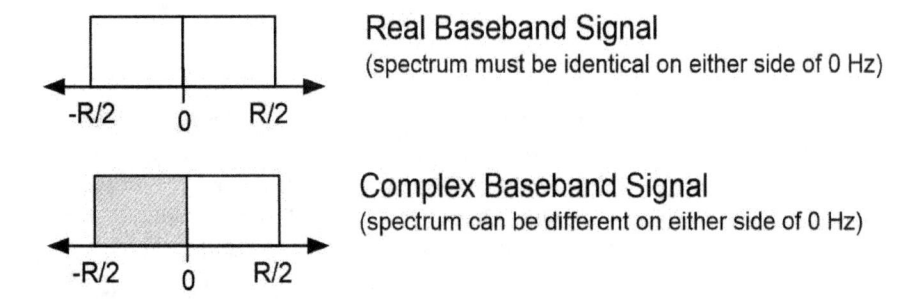

Figure 2.13 Spectrum of real versus complex baseband signals.

2.7.2.2 Complex Transmit Signals

The complex baseband spectrum extends from -R/2 to +R/2. As shown in Figure 2.13, the two spectral halves can be different. This requires separate real and imaginary signals, 90° apart.[12] The total spectrum width supports the two separately modulated, real and imaginary signals. This complex signal can support a two-dimensional signal confined to the complex plane, such as four-state QPSK. The real signal can only support two-state BPSK (see Chapter 3). Thus we have the statement: "A complex baseband signal doubles the available bandwidth." Translate the 0 frequency center to an RF carrier frequency and the same situation exists around the carrier frequency. Thus, QPSK meets the absolute minimum Nyquist required bandwidth. Single sideband is not needed.

2.7.3 Shannon-Hartley Capacity Limit

The Shannon-Hartley Capacity Limit helps us understand how many bits/second can be sent error-free through a bandwidth limited channel,[13] from [4]:

[12] We are not saying that the real part is to the right of the spectrum center and the imaginary part is to the left.
[13] A power-limited channel has practically unlimited bandwidth, for example, the link from Earth to the Voyager spacecraft.

CHAPTER 2. COMMUNICATION SYSTEM FUNDAMENTALS

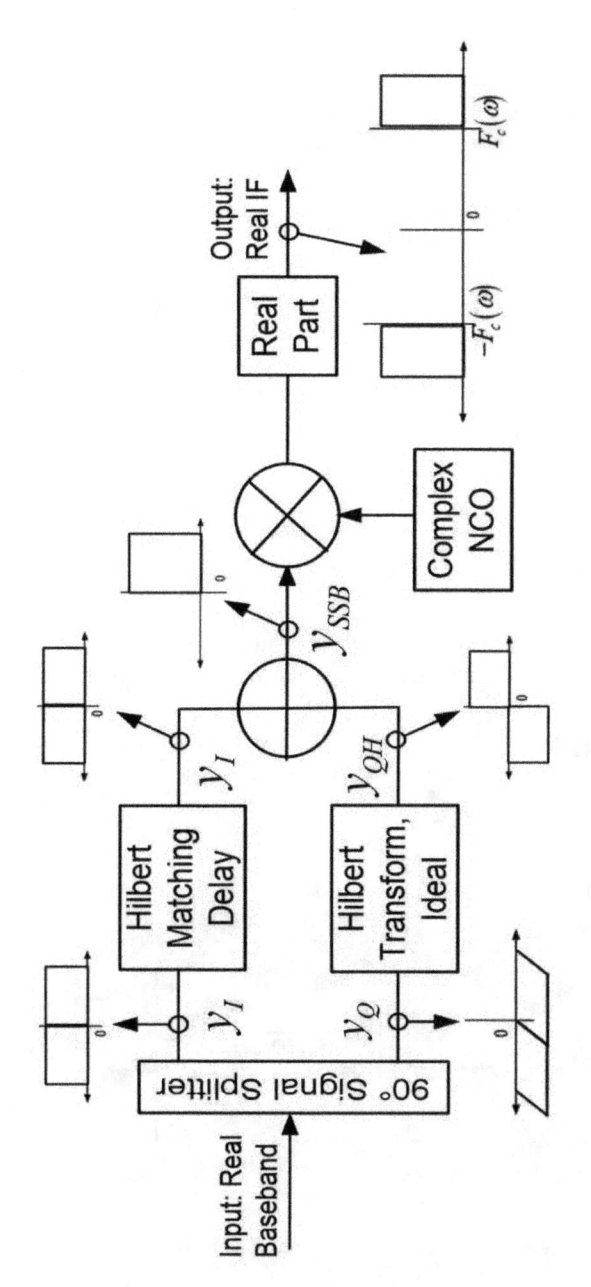

Figure 2.14 The phasing method of a single sideband generation.

Given a discrete memoryless channel [each independent signal symbol is perturbed by Gaussian noise independently of the noise effects on all other symbols] with capacity C bits per second, and an information source with rate R bits per second, where $R < C$, there exists a code such that the output of the source can be transmitted over the channel with an arbitrarily small probability of error.

$$C = Wlog_2\left(1 + \frac{P_s}{P_n}\right) = Wlog_2(1 + SNR)$$
$$\frac{C}{W} = log_2(1 + SNR) \tag{2.16}$$

W = Brick wall channel bandwidth limit (Hz). Signal and noise power are measured within W.
C = Max error-free input data rate (bits/second). If $R < C$, channel can be error-free.
R_{code} = Error correction coded data rate. Must fit in W. Shannon did not address this.
P_s = Total signal power (watts) within bandwidth W.
P_n = Total noise power (watts) within bandwidth W.
$SNR = \frac{P_s}{P_n}$ = signal to noise power ratio.

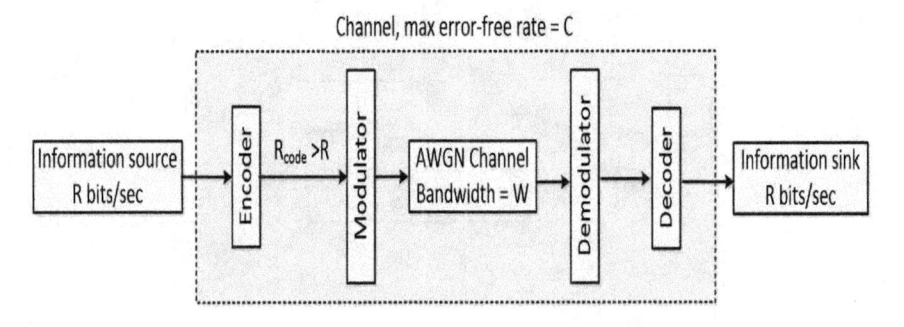

Figure 2.15 Shannon's channel model.

CHAPTER 2. COMMUNICATION SYSTEM FUNDAMENTALS

In Figure 2.15, the Shannon channel is in the shaded box. This channel has a brick wall band width of W and adds AWGN.[14]

Shannon's insight was that maximum bit rate is not limited by bandwidth alone but rather a combination of bandwidth and signal to noise power ratio. Shannon tells us that for $R < C$ an encoder exists that allows error-free transmission. He does not tell us how to design it. To understand the practicality of Shannon's equation let's plot bandwidth efficiency C/W verses $E_b N_0$.

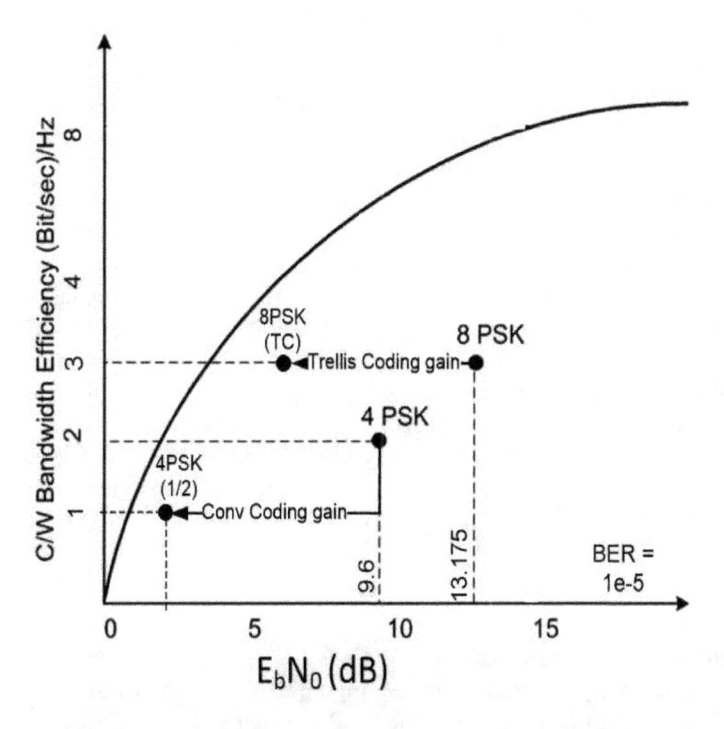

Figure 2.16 Shannon's channel capacity curve, BER = 1e-5.

Consider Figure 2.16, where Equation (2.16) is plotted. All the large dots in the figure represent digital modulations with performance of 1e-5 errors/bit (BER = 1 error for every 100,000 correct bits). Keeping BER constant allows for valid

---

14 Additive white Gaussian noise. (AWGN) is the noise model that we generally use for communications system simulation. Additive simply means the noise is added to the signal. White means the noise is uniformly distributed in frequency. Gaussian means the amplitudes of noise samples are Gaussian distributed. Thus, they are independent of each other and have a mean (usually zero) and a variance.

comparison of bandwidth efficiency verses $E_b N_0$. Let's consider two different cases of digital modulation.

As shown in Figure 2.16, for $E_b N_0$ = 13.175 dB, uncoded 8PSK will achieve a BER of 1e-5 and a bandwidth efficiency (C/W) of 3. Trellis coding of this modulation will allow for up to 6 dB coding gain without losing bandwidth efficiency. Movement along the horizontal line from 8PSK to 8PSK(TC)[15] means the same performance and same C/W at a lower $E_b N_0$, (i.e., coding gain). This might be a bandwidth limited channel where we must maintain high bandwidth efficiency without losing performance; the price for that is higher coding complexity.

Now consider 4PSK (QPSK). For $E_b N_0$ = 9.6 dB, uncoded 4PSK will achieve a BER of 1e-5 and a bandwidth efficiency (C/W) of two. As described in Chapter 3, QPSK symbols each carry two information bits, hence the uncoded C/W of 2. A convolutional code of rate 1/2 will use 1 bit of the 2 bit symbol as a coding bit and the other bit as an information bit. This lowers the bandwidth efficiency to 1, but provides coding gain. This might be a power-limited channel where we are trading bandwidth efficiency for BER performance.

The modulations shown in Figure 2.16 are designed for bandwidth-limited channels. The chart shows how the various coded and uncoded modulation types perform in relation to each other and to the limits discovered by Shannon. A similar curve is shown in [2] for power-limited, bandwidth unlimited channels. The modulation type in this case is noncoherent FSK. In 1948, when data signals were simple ASK and FSK, Shannon and Hartley were able to foresee the need for the complicated coded signals that we use today. Good references for the Shannon limit are [5], [6], and [7].

2.7.3.1 Adaptive Coding and Modulation

Adaptive Coding and Modulation (ACM) is a practical application of Shannon's Channel Capacity Theorem. Consider Figure 2.17. Here the BPSK required receive sensitivity is a BER of 1e-5. At time $t = 0$, we are receiving BPSK (see Chapter 3). The receive power is just enough to achieve an $E_b N_0$ of 9.6 dB which results in a BER slightly better than 1e-5. At time $t = 1$, the receive power increases as shown. Now we are well into the adequate region and are achieving much lower BER. The concept behind ACM is: what if we could tell the transmitter and receiver to increase the data rate and take advantage of the additional capacity predicted by Shannon? A video link, for example, may be much less likely to get behind its buffer filling. At

---

15 8PSK(TC) constellation looks like that of 16QAM (3/4). The C/W is still 3, but an extra bit is appended to each symbol to obtain the coding gain shown. See Chapter 6 for details.

CHAPTER 2. COMMUNICATION SYSTEM FUNDAMENTALS

time $t = 2$, we see that the link has adapted this way. Now we are achieving 13 dB EBNO and are able to send 8PSK, for a 50% increase in data rate. This is the basic idea behind ACM. Obviously, the flexibility of SDR can play an important role in ACM. For more detailed ACM information; see [8].

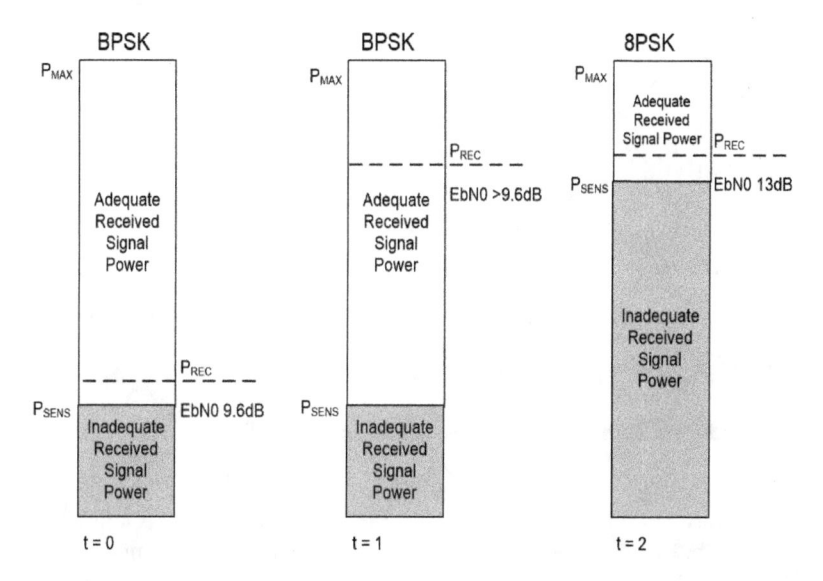

Figure 2.17 An application of Shannon's channel capacity curve, BER = 1e-5.

The obvious objection is: What happens if the link changes again and the receive power decreases or the noise increases? To control ACM, we need some kind of feedback or side channel from the receiver to transmitter. Another way is for both receiver and transmitter to know their GPS coordinates and monitor some kind of a service that predicts the link conditions. In either case, it seems obvious that this is not a system for combatting fast channel fading (see Chapter 4). Like any control system, the adaptation rate is inversely proportional to the loop delay.

DVB-S2 is a satellite communication system that depends on ACM for optimum performance; see [8] or [9]. On the left side of Figure 2.18 is the DVB-S2 Earth station that supplies time multiplexed packets intended for different Earth station receivers. Link B is subject to rain fading while link A is not. Each Earth station receiver reports its individual measurement of CNI (carrier to noise and interference) to the DVB-S2 ACM controller on the left. The controller has chosen to modulate the packets intended for receiver A using 16AOSK 2/3 (2/3 is the coding

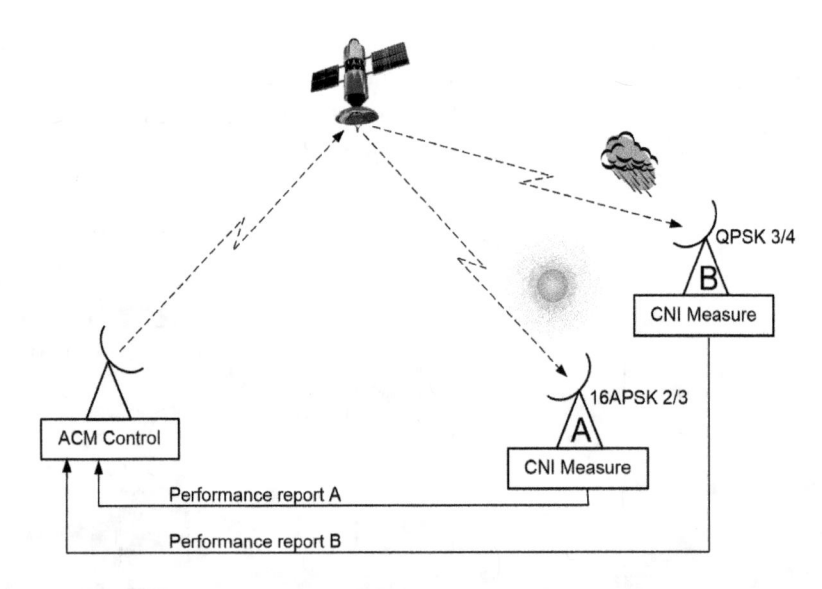

Figure 2.18 The basic idea behind adaptive coding and modulation.

rate, 2 information bits for every 3 total bits). Receiver B gets a lower bandwidth efficiency, but is more resilient to rain fading by employing a simpler modulation, QPSK 3/4.

2.8 QUESTIONS FOR DISCUSSION

1. Why must an isotropic transmit antenna always have a gain of 1 but an omnidirectional transmit antenna can have a gain greater than 1?

2. An analog telephone channel has a frequency response range $300 < f < 3,000 Hz$. Let's say the signal to noise power ratio is 35 dB. What is the maximum error-free channel capacity? This channel capacity was typical of the dial-up internet of the early 1990s.

3. What kind of signal power meter is only calibrated at one power level?

4. Say that we have a received $C/kT = 52$ dB-Hz. Let's also say we require a receive margin of 2 dB and a probability of bit error for BPSK of 10^{-5}. If this performance requires an $E_b N_0$ of 9.6 dB, what is the maximum bit rate that we can support?

5. The following link budget items below are randomly listed. Rearrange these into a correct link budget. Also what is the resulting link margin?
    1. -72 dBm received signal level
    2. +30 dBm transmit power
    3. -4 dB transmit cable loss
    4. -3 dB receive cable loss
    5. +10 dBi transmit antenna gain
    6. -82 dB received sensitivity
    7. +14 dBi receiver antenna gain
    8. -119 dB free space loss

REFERENCES

[1] J.K. Raines. "Simple Formula Relating dbm and dbu". *IEEE Broadcasting Technology Society Newsletter*, 2010.

[2] B. Sklar. *Digital Communications*. Prentice Hall, 1988.

[3] H.T. Friis. "Noise Figure of Radio Receivers". *Proceedings of the IRE*, 1994.

[4] C. Shannon. "A Mathematical Theory of Communication". *The Bell System Technical Journal*, 1994.

[5] T. J. Rouphael. *RF and Digital Signal Processing for Software-Defined Radio*. Newnes Press, 2009.

[6] E. McCune. *Practical Digital Wireless Signals*. Cambridge University Press, 2010.

[7] L. Hardesty. "Explained: The Shannon Limit". *MIT News Office*, 2010.

[8] E. Grayver. *Implementing Software Defined Radio*. Springer, 2013.

[9] ETSI. DVB-s2. *ETSI EN 302 307*, 2009.

Chapter 3

Modulation Types

3.1 INTRODUCTION

Consider the popular SDR setup of Figure 11.1 coupled via USB with one of the SDRs in Table 1.1. Many popular downloadable SDR software packages will tune, demodulate, and decode interesting signals, everything from aircraft tracking to tire pressure monitors.[1] In most cases, the underlying signal modulation is not discussed. This chapter does the opposite; instead of discussing interesting applications, we discuss the underlying signals.

Understanding signal modulation may make SDR more interesting to the hobbyist. However, for the engineer designing an SDR for critical applications (e.g., avionics) this understanding may be essential. This person will probably be using Simulink or GNU Radio block diagram environment to set up a new design for an existing signal or a completely new transmission system. Modulation choices and design will be critical. Chapters 11, 12, and 13 focus on this approach.

The chapter is separated these into two fundamentally different modulation categories, linear and nonlinear. Table 3.1 lists some differences. There is also an interesting catalog of signals and applications in [1].

3.2 LINEAR MODULATION

The defining characteristic of linear modulation is that the transmit signal amplitude is proportional to modulating signal amplitude. Here we study some details of the following linear modulations:

---

1 "The Hobbyist's Guide to the RTL-SDR," by Carl Laufer, self-published, has many examples.

Table 3.1 Major Differences Between Modulations

Characteristic	Linear (AM)	Nonlinear (FM)
Bandwidth	$\pm F_{symbol}/2$	Generally wider
Interference	Additive	Capture effect
Noise immunity	Poor	Can be very good
Complexity	Simple	Complex
Tx pwr amp	Class A, AB	Class C

1. Binary phase shift keying (BPSK),
2. Quadriphase shift keying (QPSK),
3. Differential phase shift keying (DPSK),
4. Quadrature amplitude modulation (QAM),
5. $\pi/4$ differential phase shift keying ($\pi/4$ DPSK),
6. Orthogonal frequency division multiplexing (OFDM),
7. Single-carrier frequency division multiplexing (SC-FDMA).

3.2.1 Basic Parameters

Bandwidth efficiency. Bandwidth efficiency is defined as the transmit bit rate divided by the required transmission bandwidth (BW). Required transmission bandwidth is based on the symbol rate and modulation type. Note that there can be more than one definition of required bandwidth [1].

Carrier frequency (CF). CF is the frequency in the RF spectrum where the modulated signal is shifted to for transmission.

Percent BW. Percent BW is the required transmission bandwidth divided by the carrier frequency. Units are (bits/second) per Hz. For example, a 20 MHz bandwidth signal transmitted at 2 GHz has a 1% bandwidth. Note the bandpass filters can also have a percent BW spec.

Power efficiency. One modulation type is more power-efficient than another if a lower $E_b N_0$ is required for the same bit error rate (BER). A power-efficient signal constellation can tolerate a higher noise level for the same BER. Sometimes power efficiency is expressed as: $\varepsilon_{pwr} = d_{\min}^2 / E_b$. Figure 3.12 shows an example of $d_{\min}$. Power efficiency is not the same as energy efficiency; the efficient use of battery power, for example.

Peak to average power (PAPR). Transmission signals with high PAPR require excess RF power amplifier capacity (sometimes called headroom) to avoid clipping

CHAPTER 3. MODULATION TYPES

peak power excursions. Increasing headroom can mean an expensive increase in SWAP (size, weight, and power). This is a big concern on a spacecraft.

Energy per bit divided by noise power spectral density ($E_b N_0$). The energy per bit is simply the signal power times the bit time; units will be Joules or watt-sec. The denominator is the noise power in 1 Hz of signal spectrum. Both of these are ideally measured at the receive detector input, although they can be derived in other ways such as using the received antenna power and the antenna noise temperature.

For modulations, such as QAM, where the power per symbol is easily measured we may use $E_s N_0$. Then $E_b N_0 = E_s N_0 / M$ where M is the number of bits carried by one symbol. Equation (3.1) shows the equivalence between $E_b N_0$ and SNR when the bit time equals the detector bandwidth.

$$\frac{E_B}{N_0} = \frac{S_P T_B}{N_P/B} = \frac{S_P}{N_P} = SNR \Leftrightarrow T_B = \frac{1}{B} \tag{3.1}$$

3.2.2 Binary Phase Shift Keying

Deep-space communications, such as the telemetry link from a spacecraft, often use binary phase shift keying (BPSK) because it is simple and resilient to errors when transmitted over additive white Gaussian noise (AWGN) channels. Binary phase shift keying is also constant envelope, near 0 PAPR, so class C RF power amplifiers (see Chapter 14) can be used and transmitter battery life can be longer.

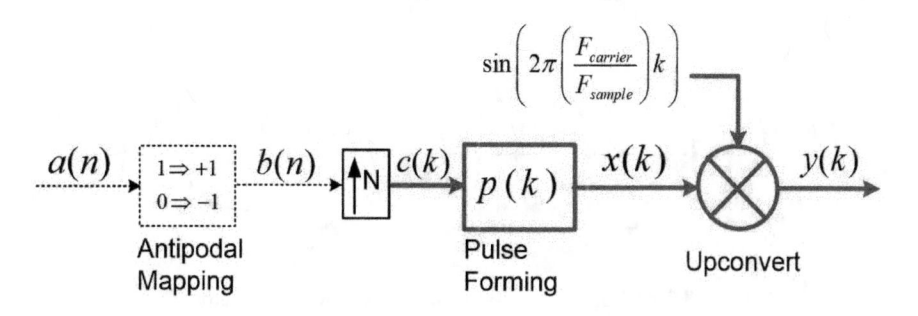

Figure 3.1 A basic BPSK modulator circuit.

Let's start with a very basic BPSK modulator circuit of Figure 3.1. The first block, antipodal mapping, simply changes 0 to -1 and passes through +1. The second block upsamples the $b(n)$ by N samples per symbol.[2] In the case of a rectangular pulse, for example, $p(k)$ in Equation (3.2), $b(n) = 1$ results in eight positive outputs for $N = 8$ and $b(n) = -1$ produces eight negative outputs on $x(k)$. More on pulse generators later.

$$x(k) = \sum_{n=-\infty}^{\infty} p(k-n)c(n)$$
$$p(k) = \left(\frac{1}{8}\right)[\,1\ \ 1\ \ 1\ \ 1\ \ 1\ \ 1\ \ 1\ \ 1\,] \quad (3.2)$$

There are only two sample rates in the circuit of Figure 3.1, the input bit rate (F_{bit}, dotted) and the output sample rate (F_{sample}, solid). Note that n changes at the bit rate and k changes at F_{sample}. There is also the output carrier frequency $F_{carrier} < \dfrac{F_{sample}}{2}$. The phase advance per cycle is shown in Equation (3.3).

$$\left(\frac{F_{carrier}}{1}\frac{\text{cycles}}{\text{second}}\right)\left(\frac{2\pi}{1}\frac{\text{radians}}{\text{cycle}}\right)\left(\frac{1}{F_{sample}}\frac{\text{seconds}}{\text{sample}}\right) = \frac{2\pi F_{carrier}}{F_{sample}}\frac{\text{radians}}{\text{sample}} \quad (3.3)$$

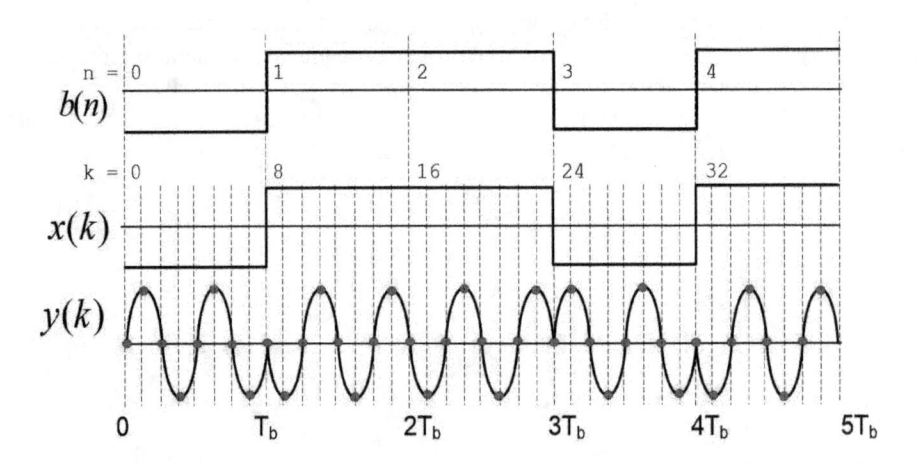

Figure 3.2 BPSK modulator details; dots show actual output samples.

Figure 3.2 shows signal details corresponding to Figure 3.1. Note the sample rate increase by N between $b(n)$ and $x(k)$. For $F_{bit} = 1$, $F_{sample} = N(F_{bit}) = 8$,

---

[2] In this context, upsampling means inserting $N - 1$ zeros between each input sample, resulting in a sample rate increase of N.

CHAPTER 3. MODULATION TYPES

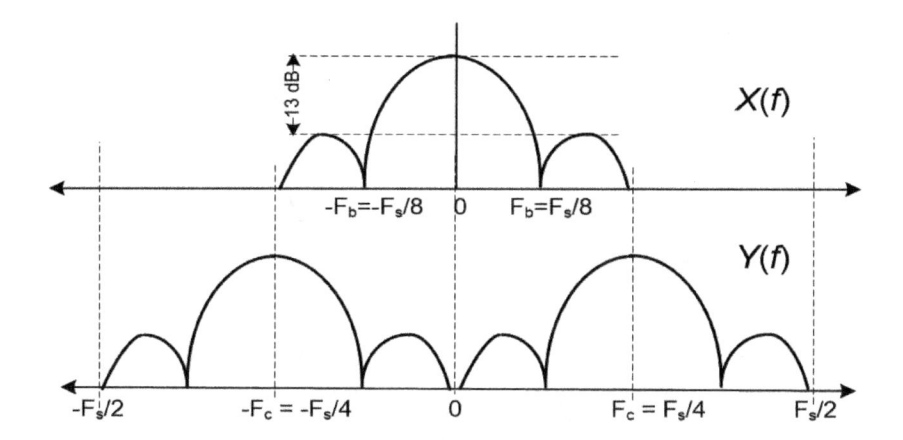

Figure 3.3 An example of a BPSK signal spectrum.

and $F_{carrier} = F_{sample}/4$, we obtain the output waveform, $y(k)$ corresponding to the five input bits of $b(n)$.

Figure 3.3 shows the spectrum of the BPSK signals before and after the upconverter. The real output spectrum, $Y(f)$, is symmetric for negative and positive frequencies. In Figure 3.2, note that although the carrier signal directly multiplies the data signal $x(k)$, it does not show up in the spectrum of $y(k)$ (lower plot in Figure 3.3). This is known as AM, double sideband suppressed carrier modulation (AM-DSBSC). DSBSC generation is also shown in the first line of Equation (3.4). The second line is called double sideband full carrier (DSBFC). The variable m is called the AM modulation index and controls the proportion of transmit power due to the modulation. DSBFC is sometimes used for modulated voice signals because the carrier power can be used to set the gain at the receiver. The rest of this book will only be concerned with DSBSC for digital communications. Another good overview of AM modulation is in [2].

$$y(k) = x(k) \left(\sin \left(k \left(\frac{2\pi F_{carrier}}{F_{sample}} \right) \right) \right)$$

$$y(k) = (1 + mx(k)) \left(\sin \left(k \left(\frac{2\pi F_{carrier}}{F_{sample}} \right) \right) \right)$$

(3.4)

3.2.3 Quadriphase Shift Keying

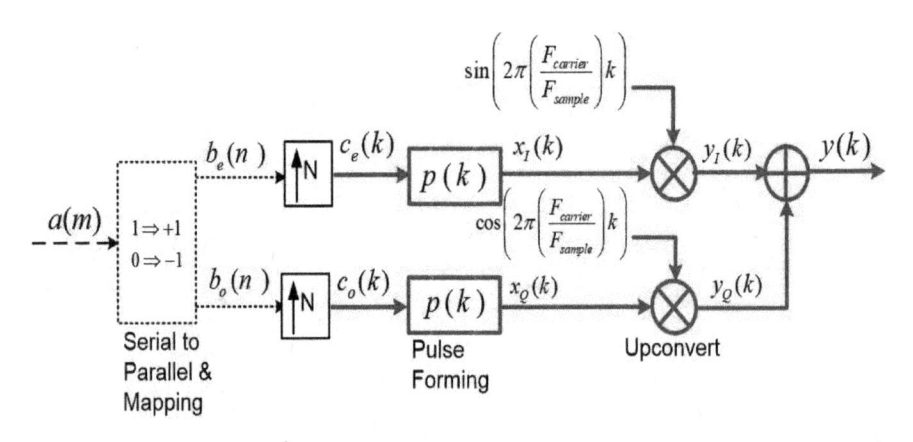

Figure 3.4 A basic quadriphase shift keying modulator circuit.

Let's expand the basic BPSK modulator circuit of Figure 3.1 to the quadriphase shift keying (QPSK) modulator circuit of Figure 3.4. The first block, serial to parallel and mapping, separates the input bit stream $a(m)$ into even and odd symbol streams $b_e(n)$ and $b_o(n)$, respectively. The same antipodal mapping as in BPSK is also applied. Even and odd separation implies a subsample by 2. This means the input $a(m)$ is changing at F_{bit}, where $F_{bit} = 2F_{symbol}$. Bit streams $b_e(n)$ and $b_o(n)$ are changing together at F_{symbol}.

The second block upsamples both $b_e(n)$ and $b_o(n)$ by N samples per symbol, similar to BPSK. In the case of a rectangular pulse, for example, $p(k)$ in Equation (3.2), $b_e(n) = 1$ produces eight positive outputs and $b_e(n) = -1$ produces eight negative outputs. Pulse forming is exactly the same for $b_o(n)$. Note the N can be any integer greater than 0 but it is often a power of two.

In the output Equation (3.5), $x_i(k)$ and $x_q(k)$ considered separately are independent BPSK signals changing polarity at the F_{symbol} rate. Two orthogonal sinusoidal carriers (sin and cosine, 90° phase difference) at the upconverter inputs can be used to generate four phases; see Equation (3.6). These four phases change every symbol time and each represents two input bits, see Figure 10.28. Interestingly, for the same input bit rate, BPSK and QPSK each require the same bandwidth and have the same bit error rate (BER). This is due to the lack of interaction between the two orthogonal carriers.

CHAPTER 3. MODULATION TYPES

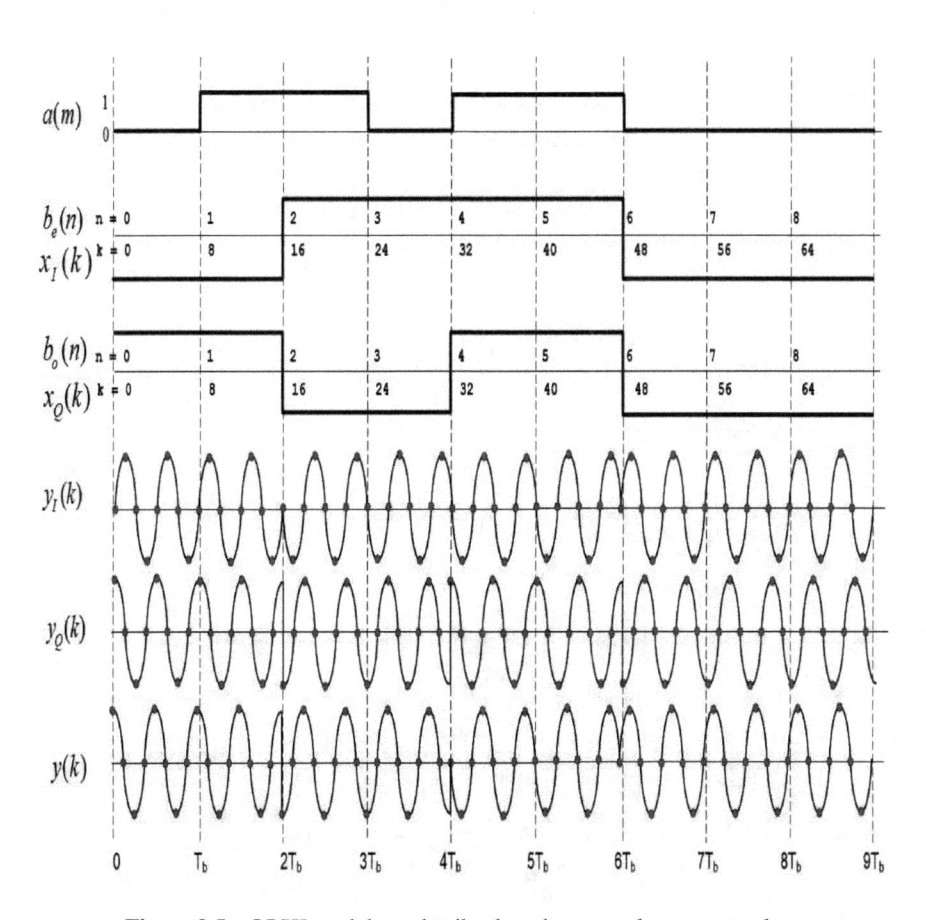

Figure 3.5 QPSK modulator details; dots show actual output samples.

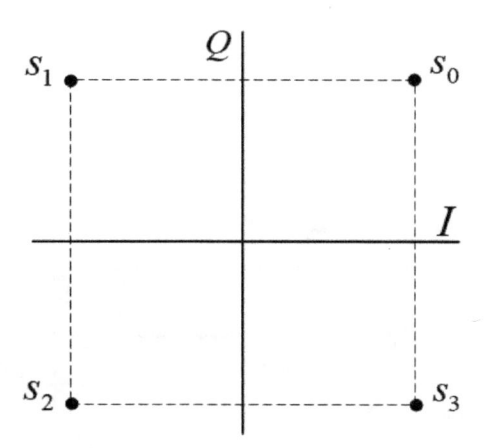

Figure 3.6 A complete set of QPSK symbol states.

$$y(k) = x_i(k) \sin\left(2\pi \left(\frac{F_{carrier}}{F_{bit}}\right) k\right) + x_q(k) \cos\left(2\pi \left(\frac{F_{carrier}}{F_{bit}}\right) k\right) \qquad (3.5)$$

$$\begin{aligned}
S_0 &: y(k) = \sin\left(2\pi \left(\tfrac{F_{carrier}}{F_{symbol}}\right) k\right) + \cos\left(2\pi \left(\tfrac{F_{carrier}}{F_{symbol}}\right) k\right) \\
&= \sqrt{2} \sin\left(2\pi \left(\tfrac{F_{carrier}}{F_{symbol}}\right) k + \tfrac{\pi}{4}\right) \\
S_1 &: y(k) = \sin\left(2\pi \left(\tfrac{F_{carrier}}{F_{symbol}}\right) k\right) - \cos\left(2\pi \left(\tfrac{F_{carrier}}{F_{symbol}}\right) k\right) \\
&= \sqrt{2} \sin\left(2\pi \left(\tfrac{F_{carrier}}{F_{symbol}}\right) k - \tfrac{\pi}{4}\right) \\
S_2 &: y(k) = -\sin\left(2\pi \left(\tfrac{F_{carrier}}{F_{symbol}}\right) k\right) + \cos\left(2\pi \left(\tfrac{F_{carrier}}{F_{symbol}}\right) k\right) \\
&= -\sqrt{2} \sin\left(2\pi \left(\tfrac{F_{carrier}}{F_{symbol}}\right) k + \tfrac{\pi}{4}\right) \\
S_3 &: y(k) = -\sin\left(2\pi \left(\tfrac{F_{carrier}}{F_{symbol}}\right) k\right) - \cos\left(2\pi \left(\tfrac{F_{carrier}}{F_{symbol}}\right) k\right) \\
&= -\sqrt{2} \sin\left(2\pi \left(\tfrac{F_{carrier}}{F_{symbol}}\right) k - \tfrac{\pi}{4}\right)
\end{aligned} \qquad (3.6)$$

Figure 3.5 shows signal details from Figure 3.4. For $N = 8$, $F_{bit} = 2$, and $F_{symbol} = 1$ we find that $F_{sample} = N(F_{symbol}) = 8$, and we set $F_{carrier} = F_{sample}/4$. Figure 3.5 shows the resulting output waveform, $y(k)$ corresponding to nine input

CHAPTER 3. MODULATION TYPES

bits of $a(m)$. As shown in Equation (3.3), the choice of $F_{carrier} = \dfrac{F_{sample}}{4}$ results in a 90° phase shift in each carrier cycle. This makes for a simple depiction in Figure 3.5, although other carrier frequencies could be used.

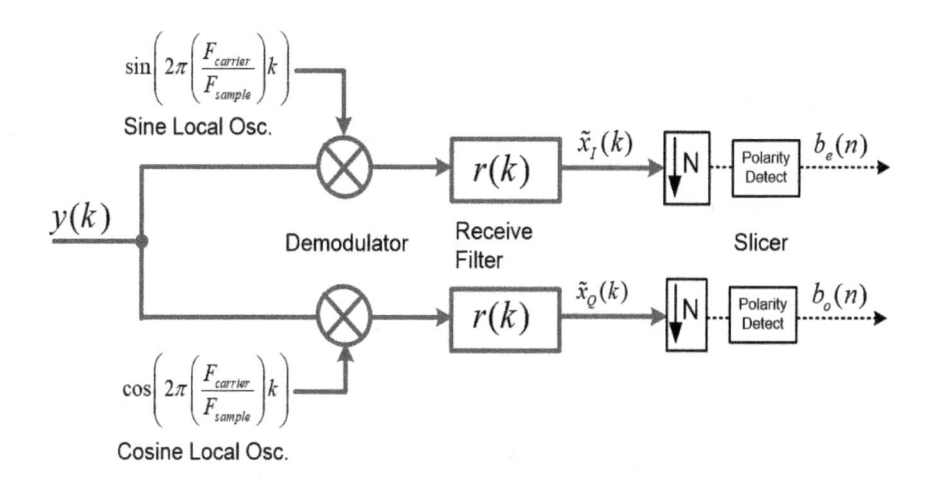

Figure 3.7 A basic QPSK demodulator circuit.

An important part of a QPSK communications system is the demodulator shown in Figure 3.7. We need the orthogonal local oscillators to separate the two orthogonal QPSK signal components. The receive filter $r(k)$ removes the mixer output double frequency term (e.g., $2cos(wt)cos(wt) = 1 + cos(2wt)$). The result is estimates for $x_I(k)$ and $x_Q(k)$. Finally, a subsample by N down to the symbol rate and a slicing around zero produces, in the low noise case, accurate estimates of $b_e(n)$ and $b_o(n)$. This is shown in Figure 3.8. For simplicity, symbol boundary tracking and carrier frequency tracking are not shown here. Chapter 12 describes a fully developed DSP receive modem that can be used for BPSK or QPSK.

Note that the receiver of Figure 3.7 should also be used for BPSK. Why, when BPSK only has two antipodal constellation points? Because more than likely the local oscillator carrier phase will not match the receive signal carrier phase and thus the two BPSK points will be at any angle across from each other in the constellation. Carrier tracking should be used to rotate the two points to the horizontal axis for good detection.

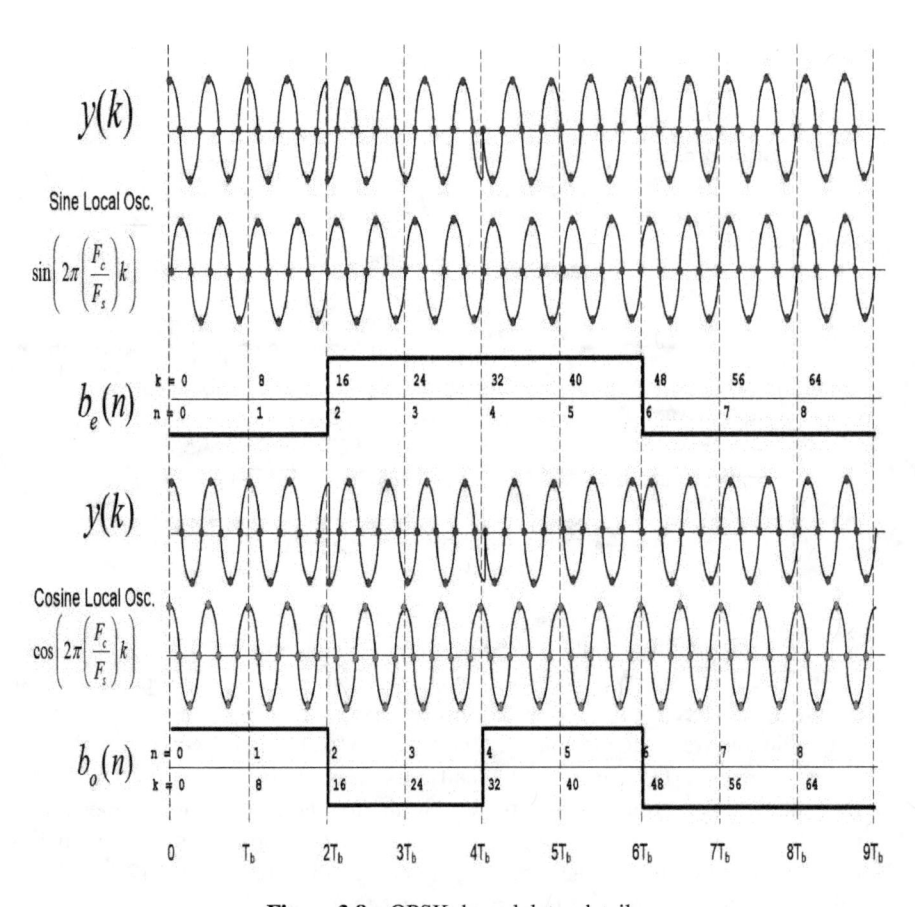

Figure 3.8 QPSK demodulator details.

CHAPTER 3. MODULATION TYPES

3.2.4 BPSK Bit Error Analysis

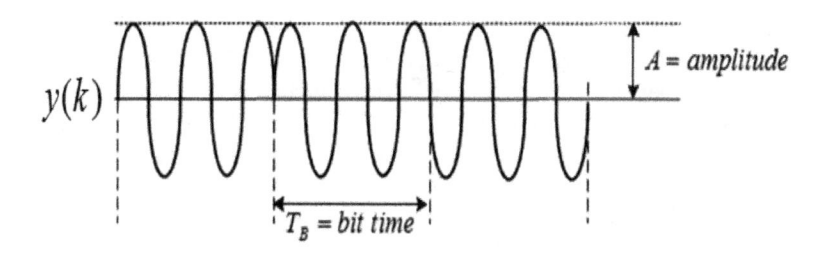

Figure 3.9 BPSK, QPSK waveform details.

This section, which gets a little more theoretical, could be skipped by those who just want to know the structure of these PSK modulations. BPSK and QPSK are constant envelope modulations. This means their transmitted signals have fixed amplitude and only their phase changes, at the bit rate (symbol rate for QPSK where each symbol carries 2 bits). This is shown in Figure 3.9. The average power of a sine wave (S_p) and the energy per bit (E_b) are shown in Equation (3.7).

$$S_p = \frac{A^2}{2}, \quad E_b = S_p T_b = \frac{A^2}{2} T_b \tag{3.7}$$

Energy E_b is generally measured in joules or watt-sec (same). Solving the E_b equation for amplitude and separating into energy and time factors, we obtain Equation (3.8).

$$A = \sqrt{\frac{2E_b}{T_b}} = \sqrt{E_b}\sqrt{\frac{2}{T_b}} \tag{3.8}$$

$$\phi_I(t) = \left[\sqrt{\frac{2}{T_b}} \cos(2\pi f_c t) \quad 0 \right], \quad \phi_Q(t) = \left[0 \quad \sqrt{\frac{2}{T_b}} \sin(2\pi f_c t) \right] \tag{3.9}$$

Unit energy basis vectors $\phi_I(t)$, $\phi_Q(t)$ are zero outside the bit time interval $[0 \quad T_b)$. Two-dimensional vectors in Equation (3.8) plotted in the I-Q plane are known as unit energy basis vectors. The unit energy is easy to see in Equation (3.10).

$$E_b = \frac{A^2}{2}T_b = \left(\sqrt{\frac{2}{T_b}}\right)^2 \frac{T_b}{2} = 1 \quad (3.10)$$

We define a constellation point as a scaling of $\phi_I(t), \phi_Q(t)$. If our required energy is E_b, then we must scale by $\sqrt{E_b}$; only an amplitude can scale a sine wave. Thus, our ideal BPSK constellation looks like Figure 3.10. We define the distance between these points as $d_{min} = 2\sqrt{E_b}$.

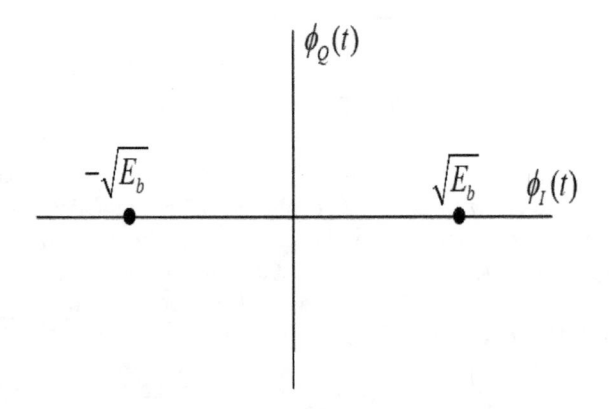

Figure 3.10 A BPSK constellation based on unit energy basis vectors.

3.2.4.1 The Q Function

To calculate the BPSK BER we need to understand the Gaussian distribution Q function. In Figure 3.11, consider antipodal transmit symbols ($S_0 = -1, \quad S_1 = +1$). If S_0 is transmitted, the probability of received symbol value r is shown in the left Gaussian probability density curve (sometimes called a standard normal distribution or bell curve). If S_1 is transmitted, the probability of received symbol value r is shown in the right Gaussian curve.

Consider $p(r|S_0)$ (the probability of r given S_0). The received symbol decision boundary is 0, so when S_0 is transmitted, if we receive $r < 0$, we estimate the transmitted symbol correctly. However, if $r > 0$, we made an error. The probability of error region is shaded. It turns out for σ = the Gaussian standard deviation (due to noise picked up on the channel) the probability of making an error is simply:

CHAPTER 3. MODULATION TYPES

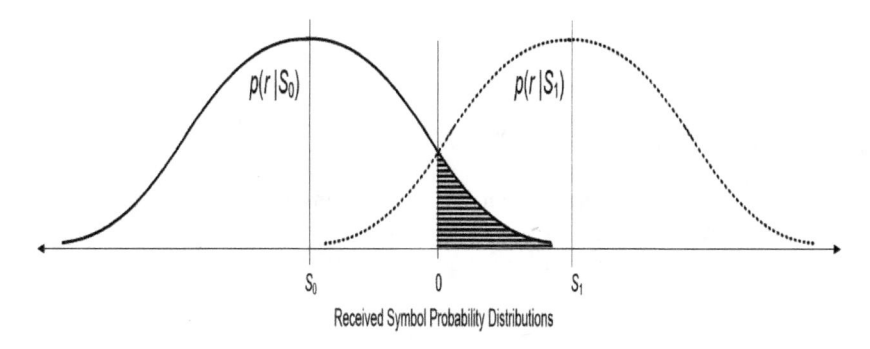

Figure 3.11 The Q function used to measure Gaussian distribution area.

Received Symbol Probability Distributions

$$p(error) = Q\left(\frac{S_1 - S_0}{2\sigma}\right) \quad (3.11)$$

Now if we make the substitutions $S_0 = -\sqrt{E_b}$, $S_1 = +\sqrt{E_b}$, and $\sigma^2 = \frac{N_0}{2}$. Assuming the bandwidth is 1, this is the standard relation between noise variance and noise spectral density for a double-sided signal with two sidebands, one on either side of the carrier frequency.[3] In general, $\sigma^2 = \frac{(B)(N_0)}{2}$ where B is the bandwidth. We end up with Equation (3.12) for the BPSK probability of bit error (BER).

$$P_{BPSK}(error) = Q\left(\frac{2\sqrt{E_b}}{2\sqrt{\frac{N_0}{2}}}\right) = Q\left(\sqrt{\frac{2E_b}{N_0}}\right) \quad (3.12)$$

```
%The Q function as a MATLAB function is shown below.
%Two different ways to calculate Q in MATLAB are:

y = qfunc(x);

function [y]= Qfunc(x)
```

[3] The carrier frequency can be zero, as in Figure 3.11.

```
y = 0.5*(erfc(x/sqrt(2)));
```

3.2.5 QPSK Bit Error Analysis

Figure 3.12 shows an example of an ideal QPSK constellation with four points. Each point has unit energy basis vector scaling $\sqrt{E_s}$. This allows us to easily calculate $d_{min} = \sqrt{2E_s}$; the minimum distance between points.

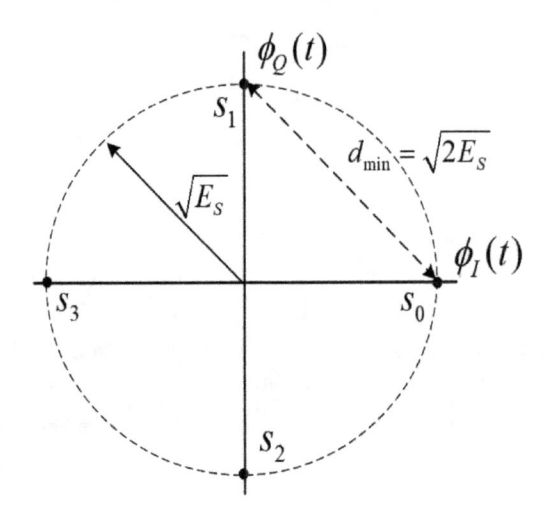

Figure 3.12 A QPSK constellation based on unit energy basis vectors.

For QPSK BER, we must consider that bit energy is half of symbol energy (because there are 2 bits per symbol for QPSK):

$$E_b = 0.5 E_s, \quad d_{min} = \sqrt{2E_s} = \sqrt{2(2E_b)} = 2\sqrt{E_b} \tag{3.13}$$

$$P_{QPSK}(err) = Q\left(\frac{d_{min}}{\sqrt{2N_0}}\right) = Q\left(\sqrt{\frac{2E_b}{N_0}}\right) \tag{3.14}$$

Equation (3.14) shows the final BER equation for QPSK. Note that the BER for Gray coded[4] BPSK and QPSK (with the same noise power) are the same. The

---

4 Gray coded means a QPSK symbol error, between adjacent symbols, will only cause one bit error.

CHAPTER 3. MODULATION TYPES

bonus is that QPSK, with its two orthogonal stream of bits, carries twice the number of bits as BPSK.

Figure 3.13 shows the decision regions for a rotated QPSK constellation, assuming Gray coding. The MSB is on the left and LSB on the right, for example $1x$ means MSB = 1, LSB = arbitrary. Notice that for each bit considered separately, the decision regions are half the signal space. Compare this with Figure 10.25 for BPSK to see why the BPSK and QPSK BERs are the same.

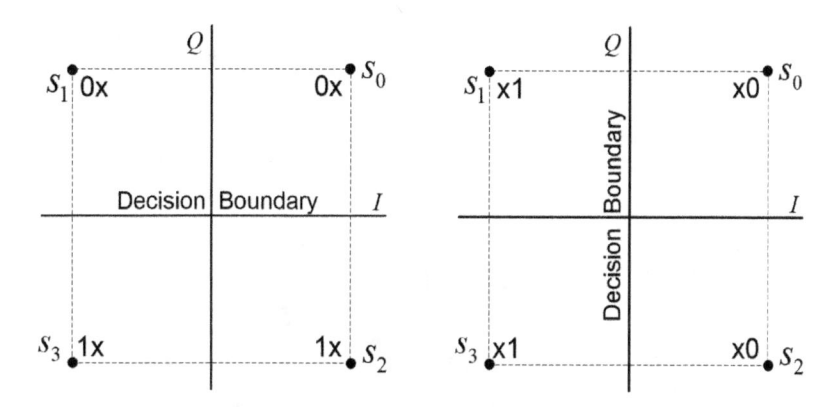

Figure 3.13 QPSK constellation MSB and LSB Gray-coded decision regions.

3.2.6 BPSK and QPSK Bandwidth Efficiency

Here we look at bandwidth efficiency: How much of the frequency spectrum do we need to reliably transmit our signal?

In Section 2.7.1, we saw that using sync pulses at R_s bits/sec, we can transmit a real only signal, such as BPSK, in a single sideband bandwidth at the Nyquist bandwidth limit of $R_s/2$ Hz. We also mentioned that BPSK generally is transmitted double sideband, not single sideband. This means that if the available bandwidth is F_{bit}, then the maximum signaling rate is F_{bit}, a bandwidth efficiency of 1.

At the other extreme, we can use square, unfiltered, pulses to transmit bits. Unfortunately, the square pulse spectrum can generate significant adjacent channel interference (ACI) because the first sidelobe is only 13 dB down from the main lobe. The spectrum that results from using square waves for signaling is shown in Figure 3.14.

Here we discuss a more commonly used practical pulse shape that also achieves good bandwidth efficiency, the raised cosine filter.

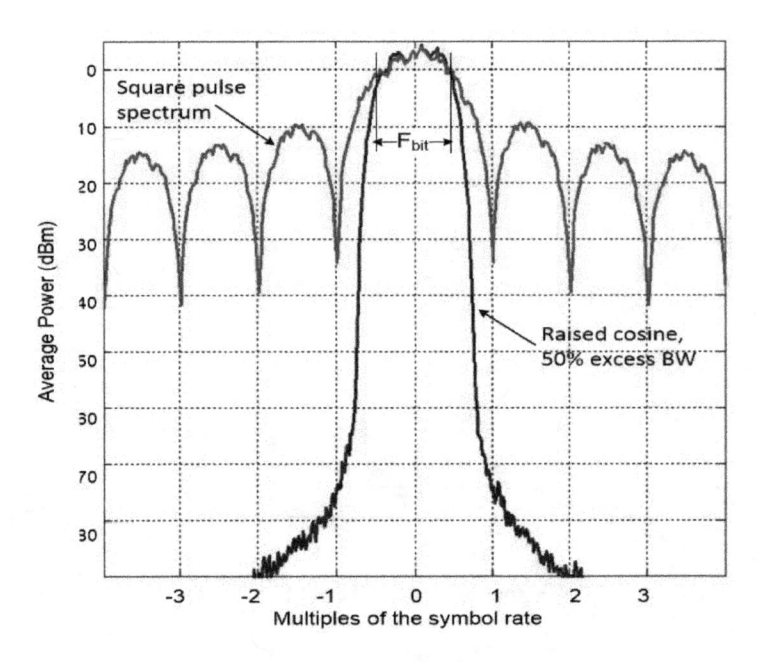

Figure 3.14 A comparison of square versus raised cosine signaling pulses.

Figure 3.14 shows the much tighter spectral control of the raised cosine (RC) versus square signaling pulses. Many signals used for commercial purposes are required to stay within a transmission mask (i.e., power limit at each allowed operating frequency) provided by the FCC or ETSI, the European standards organization. RC filtering is ideal for this purpose. Figure 3.15 is a detailed frequency domain view of RC filtering.

An interesting characteristic of RC pulses is that at the $\pm\dfrac{F_s}{2}$ frequency points the response is always 3 dB down from center. In fact, observing -3 dB markers on a laboratory spectrum analyzer displaying an RC signal will reveal the baud rate. Another characteristic is that transmit power goes to zero at $\pm F_s$. These points may be the center of the first adjacent channel so this characteristic reduces ACI. Both of these characteristics are demonstrated in Figure 3.15.

CHAPTER 3. MODULATION TYPES

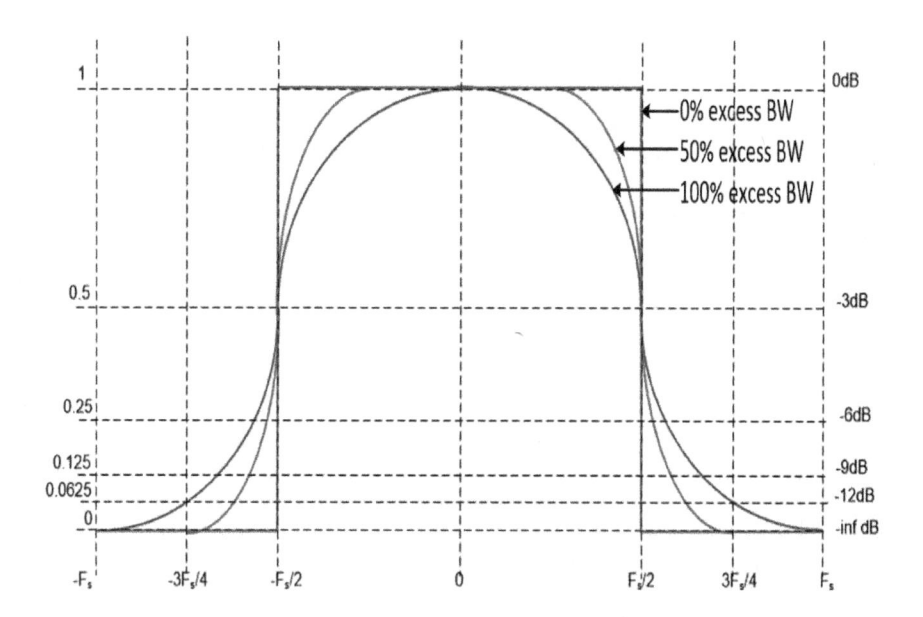

Figure 3.15 Frequency domain raised cosine pulse, F_S = symbol rate.

Excess bandwidth (ExBW) is an RC parameter that controls signal roll-off. We define F_{zero} as the point on the horizontal axis of Figure 3.16 where the RC response goes to zero. For example, 50% excess bandwidth results if $F_{zero} = 3F_s/4$. Equation (3.15) shows how to calculate excess bandwidth factor α. 100α = excess bandwidth percentage.

$$\alpha = \frac{2F_{zero}}{F_s} - 1 \qquad (3.15)$$

Excess bandwidth fraction α is generally between 0.25 and 0.5. ExBW = 1 is bandwidth wasteful and ExBW = 0 makes the time-domain waveform into the impractical sync function, previously discussed. The MATLAB code below generates and displays FIR RC filter coefficients.

In some cases we use a square root raised cosine at both transmitter and receiver to split the pulse shaping task between the two. The square root is taken in the frequency domain. The advantage is that the receive matched filter is optimally defined and the overall (transmit and receive RC filters combined) pulse shaping is RC [2].

```
% Should work with MATLAB version 2018 and forward
alph = 0.35; % RC Excess BW
M = 16; % Samples per symbol
N = 8*M; % Length of RC filter
h = ...
fdesign.pulseshaping(M,'Raised Cosine','N,Beta',N,alph);
Hd = design(h);
rcFilt = Hd.Numerator; % FIR filter, numerator only
stairs(rcFilt); grid
set(gca,'XLim',[1 N]);
set(gca,'XTick',1:M:N-1);
set(gca,'XTickLabel',0:M:N-1);
title('Raised Cosine Filter');
```

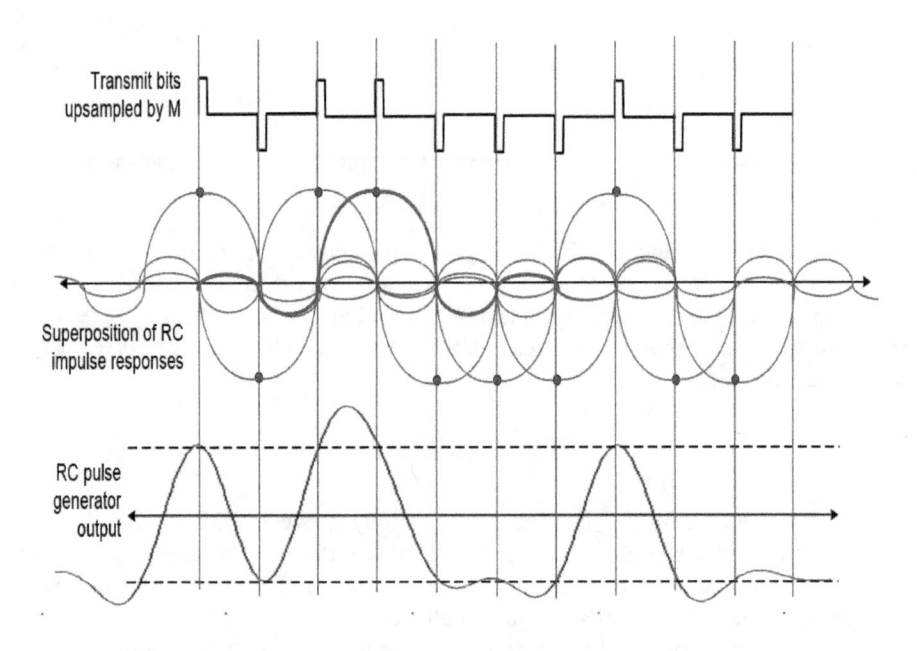

Figure 3.16 An example of raised cosine transmit signal generation.

The first plot of Figure 3.16 shows that the RC filter input is a series of impulses with $(N-1)$ zeros between each (N = number of samples per symbol).

CHAPTER 3. MODULATION TYPES

Each impulse that shifts through the RC filter results in an RC time response output. The superposition of these time responses becomes the complete pulse generator output, shown in the third trace. Notice how the pulse generator output is correct on either the plus or minus dotted lines at the symbol centers. Between the symbol centers there is a shaped excursion that is part of bandwidth determination. The second trace is not an actual signal, it is intended to show that each input bit does indeed result in an RC impulse response overlayed with that of several neighboring bits (depending on the design length of the RC filter). One of these is traced out in a heavy line to emphasize this.

3.2.6.1 Peak to Average Power

We mentioned earlier that BPSK and QPSK are constant envelope signals with 0 PAPR (peak to average power ratio). As shown in Figure 3.17, this is only true for square transmit pulses (heavy dotted line forming a square between the 4 QPSK constellation points). RC pulses will increase the PAPR and, in general, any transmit signal filtering has the potential to increase PAPR.

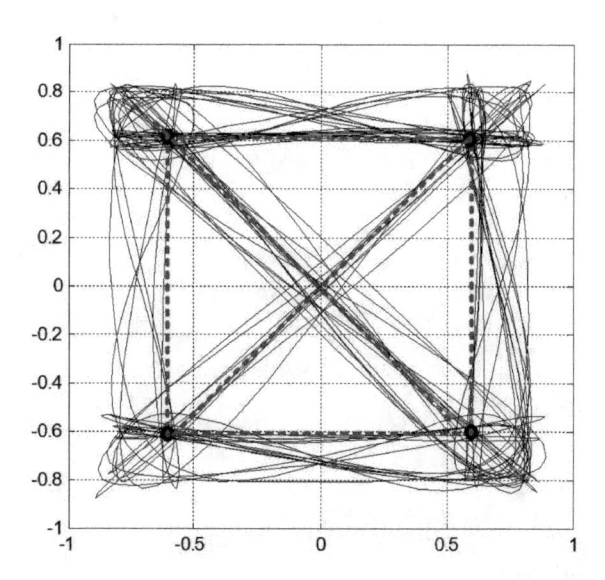

Figure 3.17 RC pulse shape for excess BW $\alpha = 0.5$ versus square pulses.

Sometimes the constellation shape itself can affect the PAPR. Consider Figure 3.18. Both constellations have the same bandwidth efficiency (ideally 4) and BER performance (i.e., same D_{min}). However, the more modern 16APSK on the right has lower PAPR. The MATLAB program below demonstrates this.

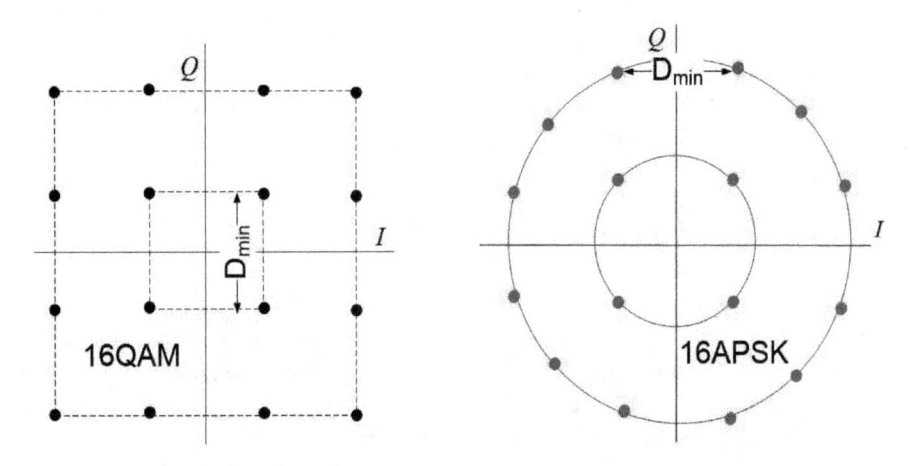

Figure 3.18 The effect of 16APSK constellation shape on PAPR.

```
% Comparison of dmin for 16APSK and 16PSK
% Find parameters for 16APSK
RingRatio = 0.414; % 16APSK r1/r0
rh = 0.8; % outer ring
rl = rh*RingRatio;
rr= rh*(1-RingRatio);
dl = 2*rl*sin(pi/4);
dh = 2*rh*sin(pi/12);
dmin = min(dl,dh);
peakpwr = rh^2 ;
avgpwr = (12*rh^2+ 4*rl^2)/16;
APSKpapr = peakpwr/avgpwr;
% fprintf(1,'16APSK: dmin= %g, PeakPwr= %g, AvgPwr= %g, 
PAPR = %g',dmin, peakpwr,avgpwr,APSKpapr);
```

CHAPTER 3. MODULATION TYPES 63

```
% Parameters for 16QAM, use ds = 0.5*dmin from 16APSK
ds = 0.5*dmin; %size of small box
peakpwr = (sqrt(2)*3*ds)^2;
avgpwr = ( 8*(sqrt((3*ds)^2 + ds^2 ))+4*((sqrt(2)*3*ds)^2)
+ 4*((sqrt(2)*ds)^2) )/16;
QAMpapr = peakpwr/avgpwr;
% fprintf(1,'16QAM: dmin= %g, PeakPwr= %g, AvgPwr= %g,
PAPR = %g,dmin, peakpwr,avgpwr,QAMpapr);
```

3.2.7 Differential BPSK

As shown in Figure 3.7, PSK receivers use a complex LO (local oscillator) to recover the baseband transmit signal. We assume here the that transmit signal frequency and the LO frequency are the same. However, phase differences between the two can still lead to ambiguous estimation of the transmitted bits. Figure 3.19 shows a BPSK transmitter and two receivers. The two receivers happen to have LOs that are 180° out of phase. So their outputs are inverted. How do we know which receiver output is correct?

One common solution is to use a preamble with a fixed start of message (SOM) pattern to force local oscillator phase alignment. Another solution is to use differential BPSK. The transmit encoder in Figure 3.20 goes between the antipodal mapping and upsample by N in Figure 3.1. The receive decoder connects to the polarity detector output. This output is at the bit rate and represents the receiver's differentially decoded estimate of the current antipodal bit.

Table 3.2 demonstrates the advantage of DBPSK. Down the first shaded column on the left are the source bits. The column headings match the signals in Figure 3.20. The next shaded column contains the receiver 1 decoded source bits (the first bit is missing due to unknown $b^{rx}_{diff}(n-1)$). The last shaded column contains the receiver 2 decoded source bits. The purpose of this chart is to show that both receivers get the correct decoded source bits even though their receive data, $b^{rx}_{diff}(n)$, columns are inverted with respect to each other. In Figure 3.20, we are using antipodal signals and multipliers. In some cases we need binary signals [0, 1]. Then we will replace the multipliers with exclusive-or gates.

We have one more important point to make about differential PSK. Note from Figure 3.20 that the receive decoder is comparing two noisy received bits, $b^{rx}_{diff}(n)$ and $b^{rx}_{diff}(n-1)$. The noise sources are shifted in time and are thus uncorrelated. Unfortunately, this means that the received noise power is doubled and results in a

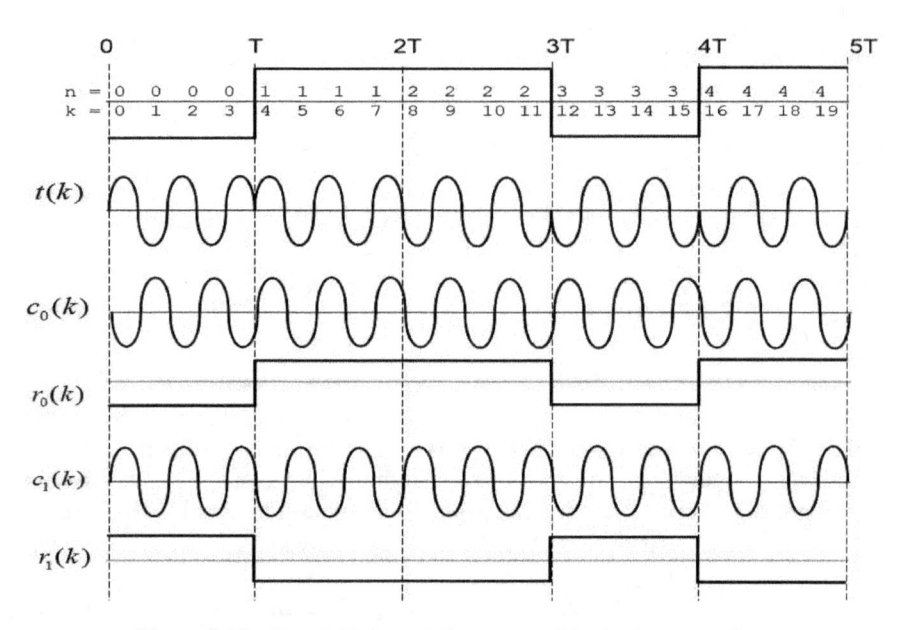

Figure 3.19 How LO phase shift causes ambiguity in receive data.

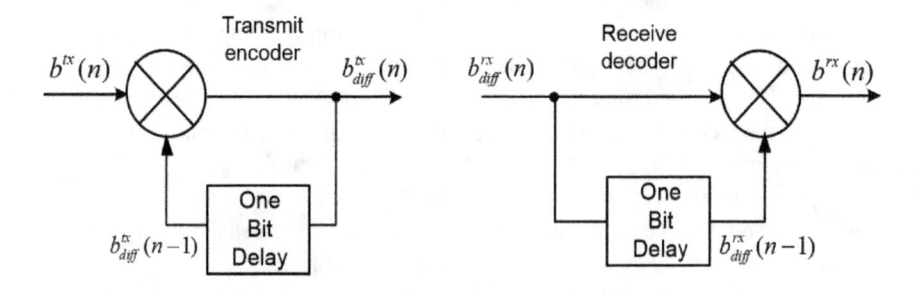

Figure 3.20 Differential coding to prevent ambiguity in receive data.

Table 3.2 A Differential BPSK Example

Transmitter			Receiver 1			Receiver 2		
$b^{tx}(n)$	$b^{tx}_{diff}(n)$	$b^{tx}_{diff}(n-1)$	$b^{rx}_{diff}(n)$	$b^{rx}_{diff}(n-1)$	$b^{rx}(n)$	$-b^{rx}_{diff}(n)$	$-b^{rx}_{diff}(n-1)$	$b^{rx}(n)$
+	+	+	+	+	−	−	−	−
−	−	+	−	+	+	+	−	+
+	−	−	−	−	+	+	+	+
+	−	−	−	−	−	+	+	−
−	+	−	+	−	−	−	+	−
−	−	+	−	+	−	+	−	−
−	+	−	+	−	+	−	+	+
+	+	+	+	+	−	−	−	−
−	−	+	−	+	+	+	−	+
+	−	−	−	−	+	+	+	+
+	−	−	−	−	−	+	+	−
−	+	−	+	−	−	−	+	−
−	−	+	−	+	−	+	−	−

BER degradation of 3 dB.[5] The original BPSK compares the received signal with a clean LO signal so the BER is not degraded. See Equation (3.16) below.

$$P_{BPSK}(err) = Q\left(\sqrt{\frac{2E_b}{N_0}}\right), \quad P_{DBPSK}(err) = Q\left(\sqrt{\frac{E_b}{N_0}}\right) \quad (3.16)$$

3.2.8 Quadrature Amplitude Modulation

A generalization of QPSK is called quadrature amplitude modulation (QAM). PSK modulations only use phase changes to represent symbols, QAM modulations use both phase and amplitude for a wider range of bandwidth efficiencies. See, for example, the QAM constellations in Figure 3.18.

Many QAM signals are raised cosine filtered, the channel bandwidth is then -3 dB at plus and minus half-symbol rate. This double sideband Nyquist bandwidth is where the theoretical bandwidth efficiency in Table 3.3 comes from. The practical bandwidth efficiency is roughly 10% less. Unfiltered QAM is widely used for the channels of OFDM signals. Figure 3.21 is a simplified QAM receiver. This is expanded and described in detail using a working Simulink simulation in Chapter 12.

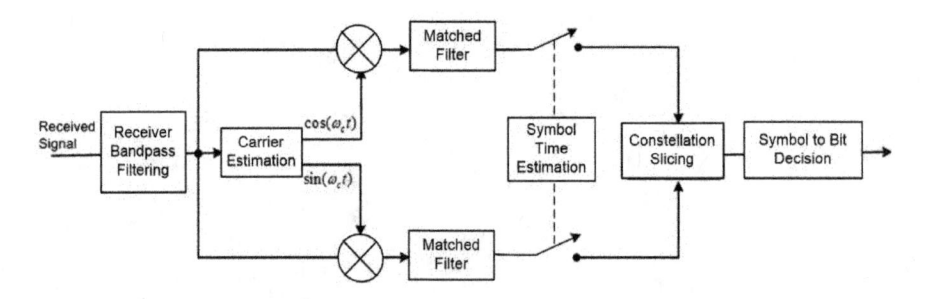

Figure 3.21 Basic construction of a quadrature amplitude modulation receiver.

3.2.9 pi/4 DPSK

The QPSK constellation in Figure 3.12 has the advantage of near ideal BER performance and the disadvantage that the signal on the channel can go to zero

---

[5] A BER degradation of 3 dB means that DBPSK needs 3 dB more $E_b N_0$ to match the performance of BPSK.

CHAPTER 3. MODULATION TYPES

Table 3.3 Bandwidth Efficiencies of Quadrature Modulations

ModType	Symbol Rate	Bit Rate	Theoretical Bandwidth Efficiency
BPSK	1	1	1
QPSK	1	2	2
8PSK	1	3	3
16QAM	1	4	4
16APSK	1	4	4
32QAM	1	5	5
32APSK	1	5	5

as the point-to-point baseband trajectories go through the origin (see also Figure 3.17). This increases the peak to average power ratio (PAPR) and can make the RF power amplifier design more difficult and less efficient. Practical applications, such as Bluetooth-EDR (enhanced data rate), typically use modified QAM constellations such as $\pi/4$DQPSK [3]. $pi/4$ DPSK is also used in TETRA (terrestrial trunked radio), a public safety radio standard.

$\pi/4$ DQPSK (differential quadrature phase-shift keying) is a differential format where the bits for a given symbol are determined by the phase change from the previous symbol. $\pi/4$ adds an $\pi/4$ offset to the phase changes compared with the phase changes in plain QPSK. As seen in Figure 3.22, there are a total of 8 ideal state positions (compared to the 4 for QPSK). The ideal state positions for symbols alternate between the four 45° states normally used by QPSK and four on-axis states. Due to this alternation, the ideal trajectory between symbols never crosses through zero. Thus the average transmit power is higher making the peak to average power lower; see Section 14.5.

$\pi/4$ DQPSK transmits 2 bits/symbol; there are only four phase changes between adjacent symbols to be decoded. These are shown in Table 3.4. Notice the switch between four point constellations on every symbol.

Table 3.4 List of $\pi/4$ DPSK Fundamental Phase Changes

Input Bit 0	Input Bit 1	Transmitted Phase Change
1	1	$\pi/4$
0	1	$3\pi/4$
0	1	$-3\pi/4$
1	1	$-\pi/4$

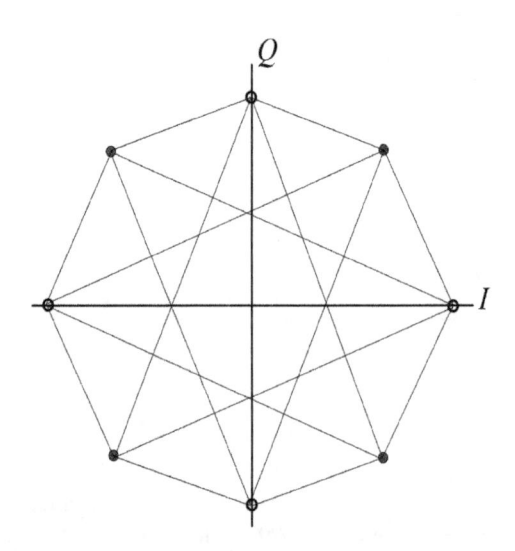

Figure 3.22 $\pi/4$ DPSK: open, closed circles separate a pair of four-point constellations.

$\pi/4$ DPSK is not difficult to detect; consider Figure 3.23. Let's say the two rails have signal $i(k)$ and $q(k)$. Now form the modified rail signals $\hat{i}(k)$ and $\hat{q}(k)$ shown in Equation (3.17). The $\pi/4$ DPSK decoded bit on the in-phase rail is 1 if $\hat{i}(k) > 0$ and 0 otherwise. Likewise, the decoded bit on the quadrature rail is 1 if $\hat{q}(k) > 0$ and 0 otherwise.

$$\begin{aligned}\hat{i}(k) &= i(k)i(k-1) + q(k)q(k-1)\\ \hat{q}(k) &= q(k)i(k-1) - i(k)q(k-1)\end{aligned} \quad (3.17)$$

3.2.10 OFDM

This section covers OFDM (orthogonal frequency division modulation) mostly from the transmitter point of view. A detailed example of an OFDM receiver is covered in Chapter 13.

The principle of orthogonal signals is fundamental to OFDM. Let's illustrate this with a simple example. Say our fundamental signaling rate is 1 Hz with a period of 1 second. Sine waves with frequencies that are an integer multiple of a fundamental frequency are orthogonal over an integer number of fundamental periods. Consider

CHAPTER 3. MODULATION TYPES

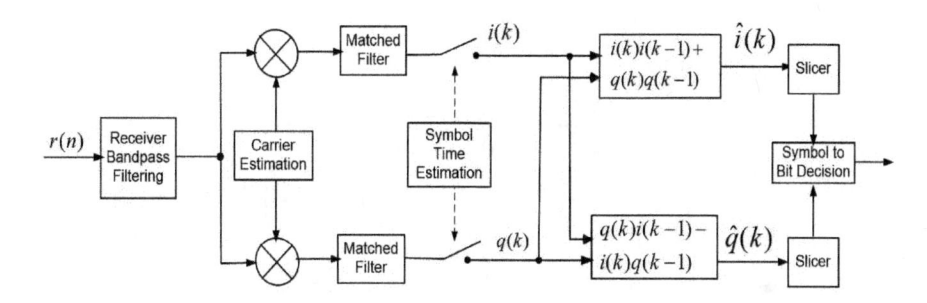

Figure 3.23 Example $\pi/4$ DPSK demodulator circuit.

two sine waves, $x_1(t)$ and $x_2(t)$ with $f_1 = 3$ Hz and $f_2 = 5$ Hz, respectively. For $T =$ any integer, the integral of the product of the two sine waves is zero:

$$\int_0^T \cos(2\pi 3t) \cos(2\pi 5t)\, dt = \frac{1}{2} \int_0^T \cos(2\pi 2t)\, dt + \frac{1}{2} \int_0^T \cos(2\pi 8t)\, dt$$

$$= \frac{1}{8\pi}(\sin(4\pi T) - \sin(0)) + \frac{1}{32\pi}(\sin(16\pi T) - \sin(0)) = 0$$

(3.18)

The OFDM modulator starts with a serial to parallel converter. The next four input symbols indexed by n at symbol rate F_{symbol} are shifted into a parallel frame. The IDFT (inverse discrete Fourier transform) is computed and the transformed symbols are shifted out. In this example M, the frame size, is 4. The IDFT converts each frame of 4 time domain inputs to 4 frequency domain outputs. The OFDM modulator ends with a parallel to serial converter as the 4 IDFT outputs are shifted out at the symbol rate. See Figure 3.24 and Table 3.5.

Table 3.5 Index Progression in an OFDM Serial to Parallel Converter

n	0	1	2	3	4	5	6	7	8	9	10	11	12	...
k	0	0	0	0	1	1	1	1	2	2	2	2	3	...

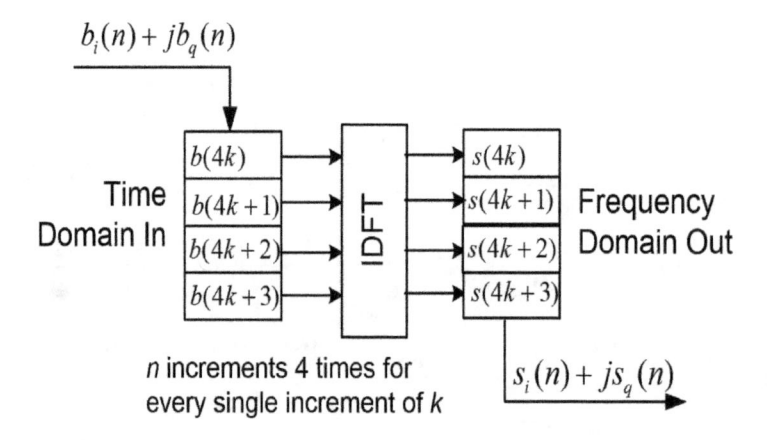

Figure 3.24 An OFDM generator using the discrete Fourier transform.

$$\begin{aligned}
s(4k) &= b(4k)e^{j0} + b(4k+1)e^{j0} + b(4k+2)e^{j0} + b(4k+3)e^{j0} \\
s(4k+1) &= b(4k)e^{j0} + b(4k+1)e^{j\pi/2} + b(4k+2)e^{j\pi} + b(4k+3)e^{-j\pi/2} \\
s(4k+2) &= b(4k)e^{j0} + b(4k+1)e^{j\pi} + b(4k+2)e^{j2\pi} + b(4k+3)e^{-j\pi} \\
s(4k+3) &= b(4k)e^{j0} + b(4k+1)e^{-j\pi/2} + b(4k+2)e^{-j\pi} + b(4k+3)e^{j\pi/2}
\end{aligned}$$
(3.19)

Equation (3.19) shows how the 4 outputs on the left are each calculated from the same 4 inputs to the right of the equal sign. The only difference is the progression of exponential factors. Let's introduce $m = 0, 1, 2, 3$ to keep track of the position of each symbol in the input and output blocks. For each k and subsequent $m = 0, 1, 2, 3$ we can convert Equation (3.19) to a more general equation:

$$s(4k+m) = b(4k)e^{j0} + b(4k+1)e^{j2\pi\left(\frac{m}{M}\right)} + b(4k+2)e^{j2\pi\left(\frac{2m}{M}\right)} + b(4k+3)e^{j2\pi\left(\frac{3m}{M}\right)}$$
(3.20)

With addition of $\frac{1}{M}$ scaling, we can change Equation (3.20) to Equation (3.21). This is the definition of the IDFT. So Figure 3.24 and Equation (3.19) are exactly what is needed for an OFDM transmitter.

CHAPTER 3. MODULATION TYPES

$$s(Mk+m) = \frac{1}{M}\sum_{n=0}^{M-1} b(Mk+n)\, e^{j2\pi\left(\frac{mn}{M}\right)} \quad (3.21)$$
$$m = 0, 1, \ldots, M-1$$

We mentioned the progression of exponential factors in Equation (3.19). We also showed that this equation implements an IDFT. Table 3.6 shows this equation in more detail. The first column shows the four output samples. Column 2 shows four copies of the first input sample modulated by four samples of a tone at 0 Hz. Column 3 shows four copies of the second input sample modulated by four samples of a tone at $\frac{F_{symbol}}{4}$ Hz. Column 4 shows four copies of the third input sample modulated by four samples of a tone at $\frac{F_{symbol}}{2}$ Hz. Finally, column 5 shows four copies of the fourth input sample modulated by four samples of a tone at $\frac{-F_{symbol}}{4}$ Hz.

The tone modulation just noted down four columns is also present across four rows. Summing the four products across each row results in a correlation between the four input samples and each orthogonal (noninterfering) tone frequency.

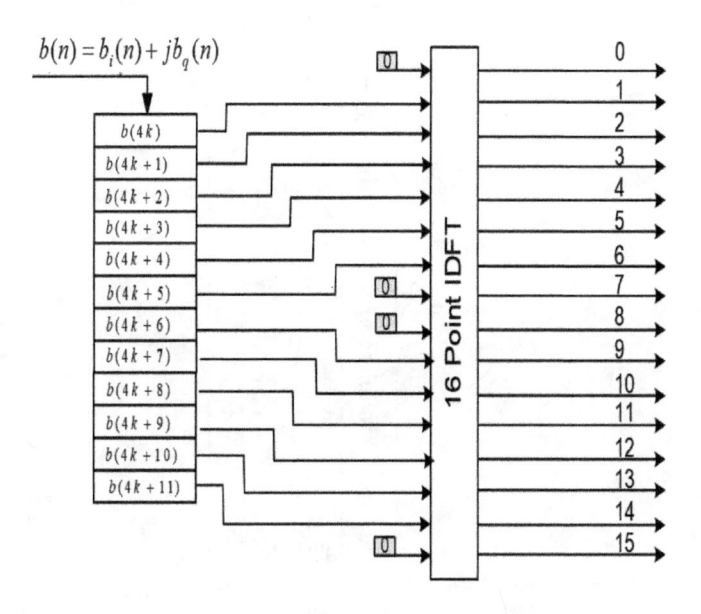

Figure 3.25 A practical OFDM transmit generator design.

Table 3.6 Example of the Modulation Effect of the IDFT

Four outputs	Modulate by 0	Modulate by $\dfrac{F_{symbol}}{4}$	Modulate by $\dfrac{F_{symbol}}{2}$	Modulate by $\dfrac{-F_{symbol}}{4}$
$s(4k) =$	$b(4k)e^{j0}$	$+b(4k+1)e^{j0}$	$+b(4k+2)e^{j0}$	$+b(4k+3)e^{j0}$
$s(4k+1) =$	$b(4k)e^{j0}$	$+b(4k+1)e^{j\pi/2}$	$+b(4k+2)e^{j\pi}$	$+b(4k+3)e^{-j\pi/2}$
$s(4k+2) =$	$b(4k)e^{j0}$	$+b(4k+1)e^{j\pi}$	$+b(4k+2)e^{j2\pi}$	$+b(4k+3)e^{-j\pi}$
$s(4k+3) =$	$b(4k)e^{j0}$	$+b(4k+1)e^{-j\pi/2}$	$+b(4k+2)e^{-j\pi}$	$+b(4k+3)e^{j\pi/2}$

CHAPTER 3. MODULATION TYPES 73

Figure 3.25 is a more practical OFDM modulator example. The input serial to parallel converter has $M = 12$ symbols. Thus, the input frame rate is $F_{block} = F_{symbol}/12$. The IDFT is 16 so that four of the output channels, at the ends and in the middle, can be set to 0. This is a common practice to make filtering the composite signal easier and avoid DC offsets in the middle. Thus, the output sample rate is: $F_{out} = 16 \left(\frac{F_{symbol}}{12} \right) = \frac{4 F_{symbol}}{3}$. This makes the output frequency bin spacing $F_{out}/16 = F_{symbol}/12$. The spectrum of this OFDM signal is shown in Figure 3.26. Like any sampled signal, spectral images repeat at integer multiples of sampling rate F_{out}, these are shown as dashed lines.

In Figure 3.26, the center of each orthogonal carrier is on a thin vertical line. The spectrum of each orthogonal carrier has a $sin(x)/x$ shape, with one main lobe at the center frequency and nulls at other integer multiples of $F_{symbol}/12$. Notice how, due to orthogonality, the $sin(x)/x$ nulls fall in the center of all adjacent carriers.[6]

3.2.10.1 OFDM Cyclic Prefix

Recall from Figure 3.26 that OFDM is a combination of orthogonal carriers. Two of these are shown in the time domain in the top half of Figure 3.27. Each carrier has an integral number of cycles to contribute to the aggregate. Because of this, even with the modulation applied, the aggregate phase will start and end the same for a given OFDM output block.[7] The lower half of Figure 3.27 shows how the last 15% or so of the OFDM frame is copied and prepended to the beginning of the same frame. This is called a cyclic prefix (CP) and is used to make frequency-domain equalization possible. Let's see how.

Say we have the discrete Fourier transforms (DFT) of the Figure 3.28 signals: $X(k)$, $Y(k)$, $H(k)$. Is multiplication in the frequency-domain, $Y(k) = X(k)H(k)$, equivalent to convolution in the time-domain $y(m) = \sum_{n=0}^{\infty} x(n)h(m-n)$? The answer is no.[8] Multiplication in the frequency domain is only equivalent to time-domain convolution if the channel and the transmit signal are circularly convolved:

$$Y(k) = X(k)H(k) \Rightarrow y(n) = \sum_{m=0}^{N-1} x(n)h(n-m)_{\text{mod } N} \qquad (3.22)$$

---

6   Recall from Figure 2.12 that the time-domain equivalent of a sync pulse is a square pulse. That is the reason the symbols input to the IDFT are square and unfiltered.
7   Even so, there can be a phase discontinuity between the ending phase of one block and the starting phase of the next. This can cause bandwidth expansion. Figure 13.14 shows an interesting way to reduce this discontinuity.
8   We mean the DFT of $y(m)$ will not equal $Y(K)$.

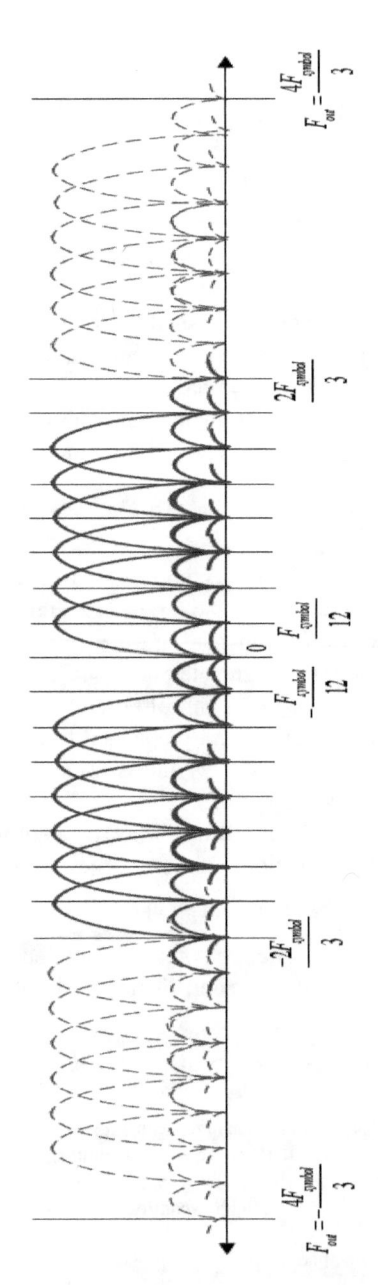

Figure 3.26 A practical OFDM transmit generator spectrum.

CHAPTER 3. MODULATION TYPES 75

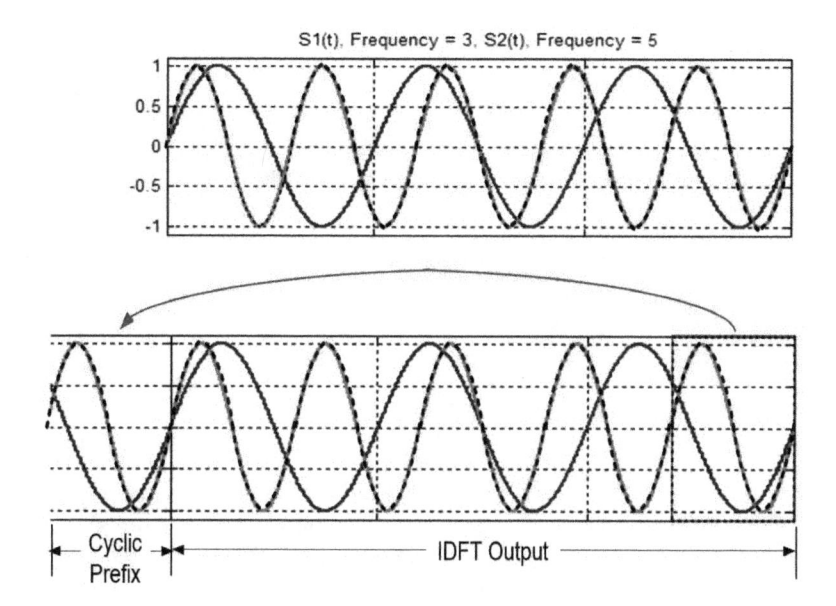

Figure 3.27 Use of the OFDM cyclic prefix.

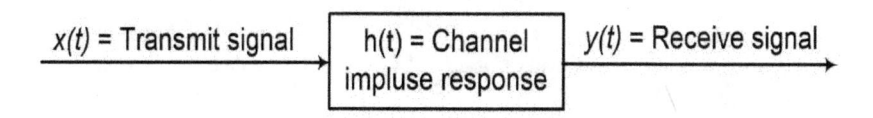

Figure 3.28 An OFDM channel for explaining the cyclic prefix.

Note that $h(n-m)_{\bmod N}$ means that when index $(n-m)$ reaches N, it wraps back to 0, that is the circular part.

Back to the cyclic prefix. The CP has no useful data; however, it does have an important function: With a CP the natural linear convolution between the transmit signal and the channel impulse response takes on the form of a circular convolution.

We should note that the transition between the end of one OFDM block and the beginning of the CP of the next OFDM block probably has a phase discontinuity. This impulse input to the channel will result in additional multipath interference at the receiver. This is perhaps why some references specify that the CP must be greater in length than the impulse response of the channel. This performance factor is in addition to the linear versus cyclic convolution discussed here.

Say that $X(k)$, $Y(k)$, and $H(k)$ are the DFTs of the transmit OFDM block, the receive OFDM block, and the transmission channel impulse response, respectively. Switching to the time-domain, say that the interaction between $x(k)$ and $h(k)$ is a circular convolution, as in Equation (3.22). If so, given $Y(k) = X(k)H(k)$, the DFT of $y(m)$ will equal $Y(K)$. Further, say that $x(k)$ in the time domain is a known pattern and we know that for a prefect channel $y(k) = x(k)$. Then we can calculate $H(k)$ and design a channel equalizer. For an equalizer, $x(k)$ will only have to be known for a short time, say, one IDFT block. Let's explore why the CP is so important for this situation.

Let's perform a simple MATLAB experiment to demonstrate linear and circular convolution. Say that the OFDM transmit block is $= [1, -2, 3, 2, 1]$ and the channel impulse response $= [3, 2, 1]$. The top box in Figure 3.29 shows a linear convolution between the IDFT block and the channel impulse response, and the second box shows a circular convolution. The third box shows the same circular convolution result accomplished by copying the two last symbols of the block (1, 2) and prepending them to the beginning of the block (if needed, review the shifting and adding operations of convolution; see [4] for a good description).

In summary, circular channel convolution is essential for OFDM equalization at the receiver; the channel's natural linear convolution is not helpful. The CP's job is to make the channel linear convolution act like a circular convolution. For more information, see [5]. The MATLAB code to generate the numbers in Figure 3.29 is shown also. In addition, a practical OFDM equalizer based on these ideas is demonstrated in Chapter 13.

```
h = [3 2 1 0 0]; %channel impulse response
x = [1 -2 3 1 2]; %transmit symbols, no CP
xcp = [1 2 1 -2 3 1 2]; %transmit symbols, CP
y = conv(x,h); %channel convolution, no CP
yct = conv(xcp,h); %channel convolution, CP

xf = fft(x);
hf = fft(h);
yf = hf .* xf;
% ifft result will always be circular convolution
yt = int16(ifft(yf)); %back to time domain
```

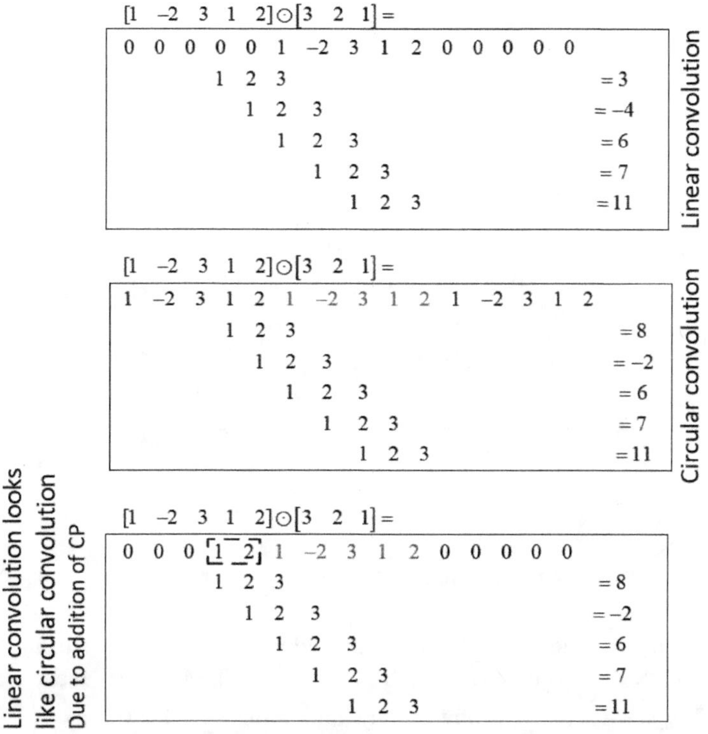

CHAPTER 3. MODULATION TYPES

$[1\ -2\ 3\ 1\ 2] \odot [3\ 2\ 1] =$

```
0  0  0  0  0  1 -2  3  1  2  0  0  0  0  0
            1  2  3                              = 3
               1  2  3                           = -4
                  1  2  3                        = 6
                     1  2  3                     = 7
                        1  2  3                  = 11
```
Linear convolution

$[1\ -2\ 3\ 1\ 2] \odot [3\ 2\ 1] =$

```
1 -2  3  1  2  1 -2  3  1  2  1 -2  3  1  2
            1  2  3                              = 8
               1  2  3                           = -2
                  1  2  3                        = 6
                     1  2  3                     = 7
                        1  2  3                  = 11
```
Circular convolution

$[1\ -2\ 3\ 1\ 2] \odot [3\ 2\ 1] =$

```
0  0  0 ⌈1 -2⌉ 1 -2  3  1  2  0  0  0  0  0
            1  2  3                              = 8
               1  2  3                           = -2
                  1  2  3                        = 6
                     1  2  3                     = 7
                        1  2  3                  = 11
```

Linear convolution looks like circular convolution Due to addition of CP

Figure 3.29 A demonstration of linear convolution versus cyclic convolution.

```
% Linear channel convolution without cyclic prefix:
disp(y);
% FFT multiplication results in circular convolution
disp(yt);
% Linear channel convolution with cyclic prefix:
disp(yct);
```

3.2.10.2 OFDM Bandwidth Efficiency

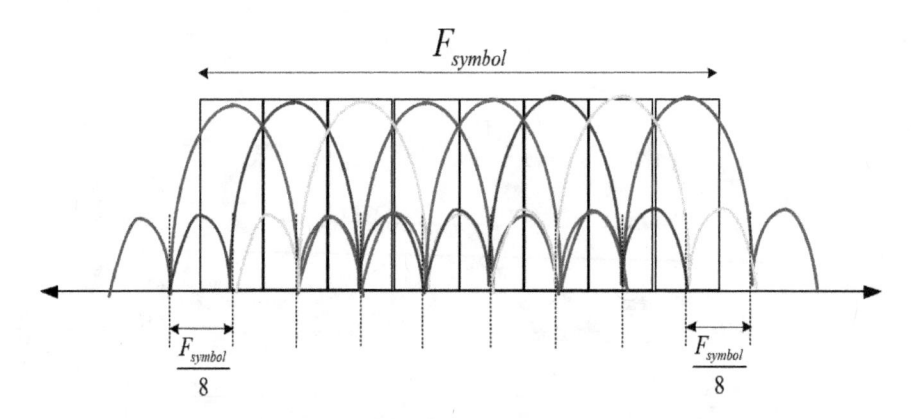

Figure 3.30 Details of OFDM bandwidth efficiency, an example.

Is multicarrier OFDM more bandwidth efficient than, for example, single-carrier QPSK? Consider Figure 3.30 with $M = 8$ OFDM carriers. If each channel has rate $F_{symbol}/8 = 1/8$. Required bandwidth for this signal is simply $8 \left(\frac{F_{symbol}}{8} \right) = 1$ for BPSK and 2 for QPSK. OFDM's primary advantage is not bandwidth efficiency[9] but rather resistance to multipath interference. As we have seen, thanks to the CP, channel frequency response estimates are easily and quickly made from known receive block patterns. Together with known pilot channels (see Chapter 13), OFDM has continuous protection against multipath interference. That is probably why

---

9 This assessment of bandwidth efficiency may be a little unfair because another source of bandwidth efficiency is the closeness with which the box like OFDM signals can be packed in the RF spectrum. See, for example, Figure 13.9. The LTE subchannels only change at 15 KHz, so their sidelobes fall rapidly compared with a single-carrier QAM signal of the same capability.

CHAPTER 3. MODULATION TYPES

OFDM is the preferred modulation for both 4G cell phone and some Wi-Fi channels. For an interesting look at this question, see [6].

3.2.10.3   OFDM Using MATLAB System Objects

Working with OFDM simulations in MATLAB is greatly facilitated by the MATLAB Communications toolbox system objects. In the code below, qpskMod is a MATLAB system object. It has internal data, parameters that can be set, and methods that can be used for working with the object data (to see the methods type methods(qpskMod)). This paradigm has its roots in object-oriented programming. The code below sets up and runs an OFDM generator with specified parameters and plots the spectrum versus time in Figure 3.31.

```
%OFDM Generation using MATLAB system objects
%Requires Communications Systems Toolbox
%First generate QPSK and OFDM system objects
qpskMod = comm.QPSKModulator;
ofdmMod = comm.OFDMModulator('FFTLength',128, ...
'PilotInputPort',true,...
'PilotCarrierIndices', [12 54 76 118], ...
'InsertDCNull',true, ...'NumTransmitAntennas',1);

%Frame Components making up 144 total IFFT out channels
% 112 payload data
% 4 Pilots
% 1 Center null (DC)
% 11 Guard bands
% 16 Cyclic Prefix

showResourceMapping(ofdmMod); %Make first plot
% Dimensions of the OFDM modulator using info method.
ofdmModDim = info(ofdmMod); %info is ofdmMod method
numData = ofdmModDim.DataInputSize(1); % Data
numSym = ofdmModDim.DataInputSize(2); % OFDM symbols
% Generate random 4 state (QPSK) data symbols
% to fill nframes OFDM frames.
nframes = 10;
```

```
data = randi([0 3],nframes*numData,numSym,1);
% Apply QPSK modulation to the random symbols
% and reshape to column vector
% Modulate frame data for nframes, not just one
modData = qpskMod(data(:));
modData = reshape(modData,nframes*numData,numSym,1);
%dataTOTAL collects nframes in a single record
dataTOTAL = [ ];
for k = 1:nframes
% Find 112 indices for kth OFDM frame random data input
indData = (k-1)*ofdmModDim.DataInputSize(1)+1:k*numData;
% Generate random OFDM pilot symbols for this frame only
pilotData = complex(rand(ofdmModDim.PilotInputSize), ...
rand(ofdmModDim.PilotInputSize));
% Modulate QPSK symbols using OFDM, also insert pilots
dataOFDM = ofdmMod(modData(indData,:,:),pilotData);
%Concatenate this frame to previous frames
dataTOTAL = [dataTOTAL; dataOFDM];
end

Fsss = 2; %Normalized sampling rate
NFFT = 1024; NFFT2 = NFFT/2;
HannWin = window(@hann,NFFT2);
HannWin = HannWin/sum(HannWin);
pSigCf=pwelch(dataTOTAL,HannWin,[],NFFT,[],'twosided');
fpSigCf = 10*log10(fftshift(pSigCf));
scale = ((-Fsss/2):((Fsss/16)):(Fsss/2));
stairs(fpSigCf,'b');
yh = ylabel('PSD');
set(gca,'XLim',[0 NFFT]);
set(gca,'XTick',0:((NFFT/16)):NFFT);
set(gca,'XTickLabel',scale,'FontSize',14);
fh = xlabel('Frequency (normalized)');
ff = title('OFDM Spectrum');
```

CHAPTER 3. MODULATION TYPES 81

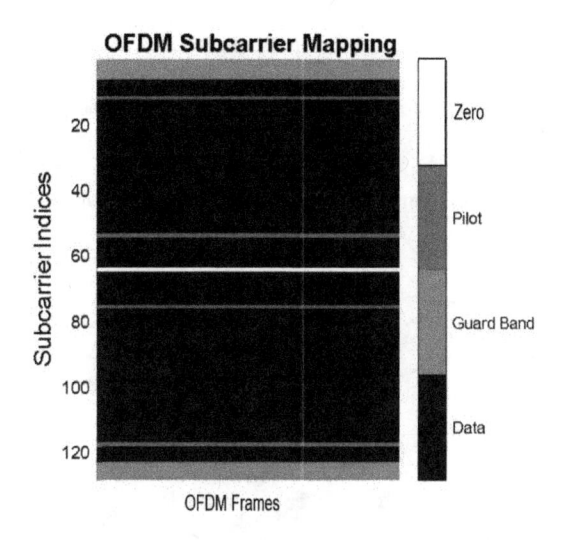

Figure 3.31 Example of OFDM subchannel allocation.

3.2.11 Single-Carrier Frequency Division Multiplexing

Here we describe an LTE uplink (UL) design enhancement, single-carrier frequency division multiplex (SC-FDMA). The UL requires a PA (power amplifier) in the handset to transmit OFDM up to a base station tower. Even a fraction of a dB reduction in peak to average transmit power ratio (PAPR) results in lower cost, longer battery life, and other benefits for handset design. See Section 14.5. A drawback is that SC-FDMA makes the pilot-based equalization described for IEEE802.11 more difficult; see Chapter 13. Therefore, the LTE UL does not use pilot-based equalization because it is more important to reduce PAPR at the handset transmitter.

The LTE downlink (DL) does not use SC-FDMA because the DL PA is not size and power-constrained in the same way as the handset. The downlink PAPR is close to the Gaussian curve in Figure 3.32. This is because the DL OFDM has many different users on separate 15 KHz OFDM channels; see Figure 3.26. These independent signals can be added because they are orthogonal. However, the central limit theorem tells us this sum will tend toward a Gaussian distribution.[10] Figure

---

10 The central limit theorem says that when independent random variables of any distribution are combined, their sum tends toward a normal distribution. Random variables here are the successive constellation points; they are independent because the probability of one does not affect the probability of any other.

3.32 shows the UL PAPR, dotted lines, as much lower than the Gaussian DL. This is due to SC-FDMA. A very readable description of SC-FDMA is [7]. Also the construction of the complementary cumulative distribution function (CCDF) is described in Section 14.5.

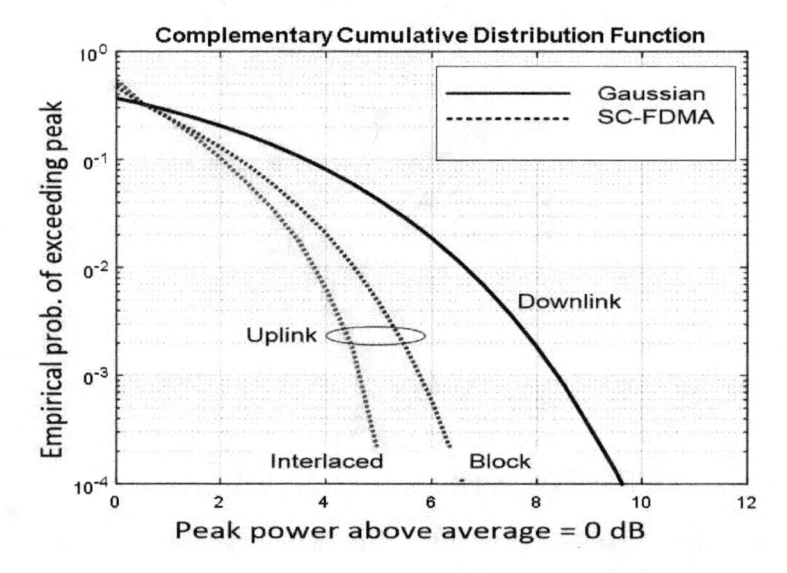

Figure 3.32 The complementary cumulative distribution function.

Figure 3.33 shows an example of an UL SC-FDMA circuit. This is meant to illustrate SC-FDMA while being simpler than the actual LTE UL.

On the left side of Figure 3.33, we have four users, $[\ a\ \ b\ \ c\ \ d\ ]$, whose symbols[11] are clocked in with clock m. Clock m runs at 64 KHz and every eight clocks latch m occurs, transferring the latest block of eight symbols to one of four FFT inputs.

The FFT output is a projection of the eight-symbol input frequency content into an eight-dimensional orthonormal space. The eight FFT outputs are transferred to eight points in the input frequency space of the IFFT. The time-domain IFFT output is then clocked out at clock n, which is four times clock m or 256 KHz. In SC-FDMA, the single-carrier part comes from the FFT outputs; that is, instead of eight independent QAM signals, we have one set of spectrum points (this is

---

11 By symbols, we mean constellation points such as one of the four points on a QPSK constellation. LTE can use up to 64QAM. These, as well as every signal in Figure 3.33, are complex valued.

CHAPTER 3. MODULATION TYPES

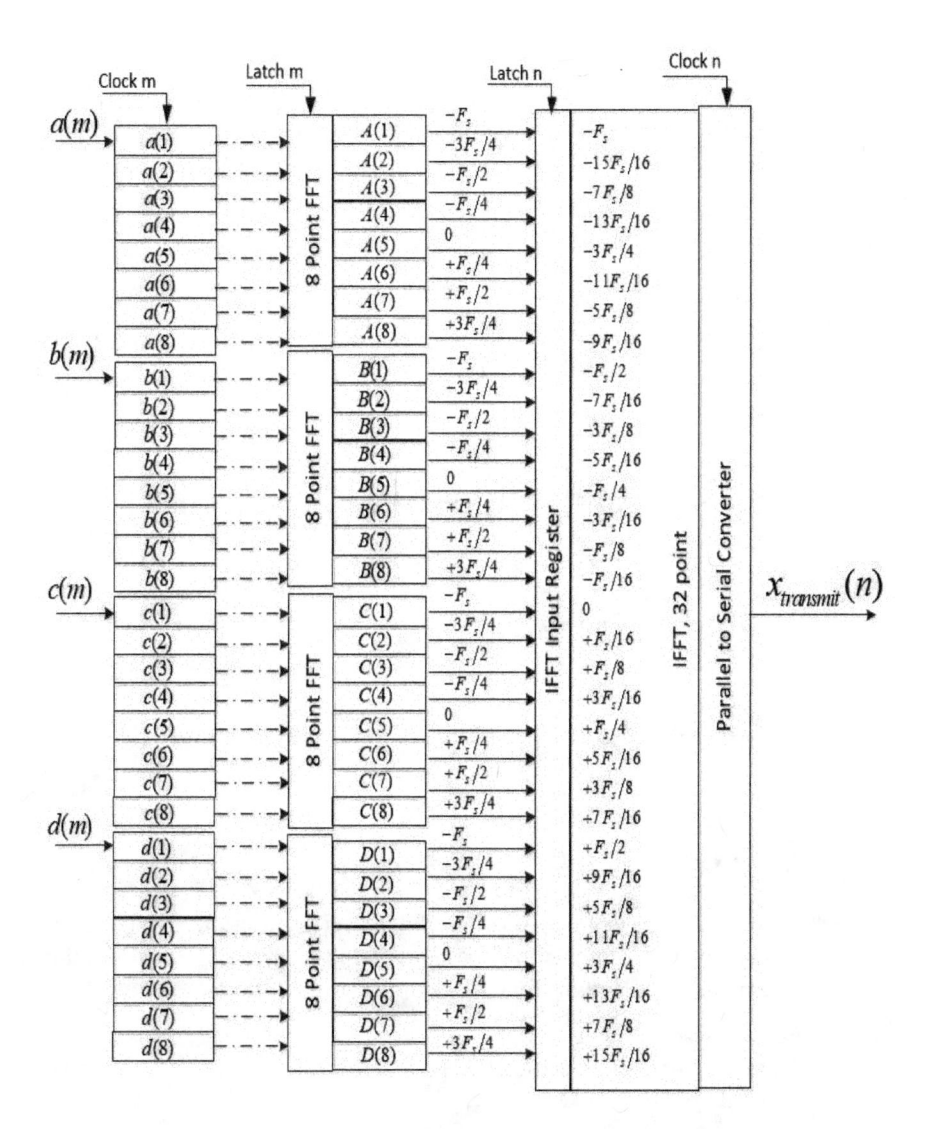

Figure 3.33 A simplified single carrier FDMA block circuit for uplink example.

sometimes called FFT precoding). The FDMA part comes from the frequency placement performed by the IFFT. Table 3.7 is a summary of the frequency points in Figure 3.33.

Figure 3.34 is the resulting transmit spectrum for our four-user example.[12] As an alternative, FFT outputs in Figure 3.33 can be interlaced (a(1), b(1), c(1), d(1), a(2), b(2), c(2), d(2), a(3), ...). As shown in Figure 3.32, this reduces PAPR even further.

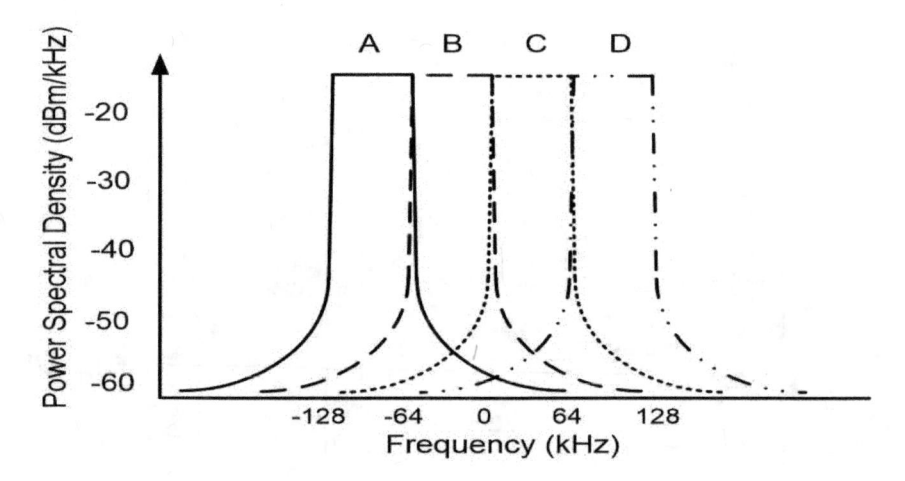

Figure 3.34 Single-carrier FDMA transmit example block spectrum.

3.3 NONLINEAR MODULATION

The defining characteristic of nonlinear modulation is that transmit signal amplitude is constant and phase or frequency is proportional to modulating signal amplitude. Here we study the details of the following nonlinear modulations:

1. Frequency shift keying (FSK),
2. Continuous phase modulation (CPM),
3. Minimum phase shift keying (MSK),
4. Gaussian minimum shift keying (GMSK),
5. Pulse position modulation (PPM),

[12] An actual LTE SC-FDMA UL spectrum would have more than four users.

CHAPTER 3. MODULATION TYPES   85

Table 3.7   SC-FDMA Frequencies, for the Figure 3.33 UL Example

Input Sym	FFT Output (kHz)	IFFT Output Sample	IFFT Output (kHz)
a(1)	-64	x(1)	-128
a(2)	-48	x(2)	-240
a(3)	-32	x(3)	-224
a(4)	-16	x(4)	-208
a(5)	0	x(5)	-192
a(6)	16	x(6)	-176
a(7)	32	x(7)	-160
a(8)	48	x(8)	-144
b(1)	-64	x(9)	-128
b(2)	-48	x(10)	-112
b(3)	-32	x(11)	-96
b(4)	-16	x(12)	-80
b(5)	0	x(13)	-64
b(6)	16	x(14)	-48
b(7)	32	x(15)	-32
b(8)	48	x(16)	-16
c(1)	-64	x(17)	0
c(2)	-48	x(18)	16
c(3)	-32	x(19)	32
c(4)	-16	x(20)	48
c(5)	0	x(21)	64
c(6)	16	x(22)	80
c(7)	32	x(23)	96
c(8)	48	x(24)	112
d(1)	-64	x(25)	128
d(2)	-48	x(26)	144
d(3)	-32	x(27)	160
d(4)	-16	x(28)	176
d(5)	0	x(29)	192
d(6)	16	x(30)	208
d(7)	32	x(31)	224
d(8)	48	x(32)	240

6. Spread spectrum,
7. Code division multiple access (CDMA).

3.3.1 Basic Parameters

The frequency modulation (FM) index is shown in Equation (3.23). For data signals such as FSK, the frequency deviation is simply the maximum difference between tones. The modulation bandwidth for FSK is the symbol rate.

$$MI = h_{fm} = \frac{f_{Delta}}{f_{mod}} = \frac{\text{Max frequency deviation}}{\text{Modulation bandwidth}} \quad (3.23)$$

The phase modulation (PM) index = $h_{pm} = 2\theta_\Delta/\pi$ where θ_Δ is the peak phase deviation of the carrier due to the data signal modulation. PM and FM are directly related in that carrier frequency modulation is given by the time derivative of the phase modulation.

The careful reader may point out that linear modulations BPSK and QPSK involve phase shifts, so can we consider those nonlinear? The answer is no. BPSK is a form of AM because the 180° phase shifts directly follow the antipodal swings of the input data bits. QPSK is similar because the I and Q orthogonal components are considered separately amplitude modulated (i.e., linear modulated).

3.3.1.1 FM Bandwidth

FM implies that the carrier frequency varies with the modulation amplitude. The rate of change of frequency varies with the modulation frequency. FM generates a series of sidebands. These are spaced around the carrier at multiples of the modulating frequency.

Carson's rule is sometimes used to estimate FM and PM bandwidth: $[\ 2f_\Delta\ \ 2(h_{fm}+1) f_{mod}\ ]$ where f_Δ is the maximum carrier frequency deviation, f_{mod} is the modulating frequency, and $h_{fm} = f_{Delta}/f_{mod}$ is the modulation index.

For digital communications using, for example, FSK, the bandwidth is usually restricted to $2f_\Delta$ = twice the symbol rate. The spectrum on a laboratory spectrum analyzer does not appear much wider than the linear modulations whose bandwidth is primarily due to pulse shaping (e.g., raised cosine) and symbol rate. This is referred to as narrowband FM (NBFM). All the digital signals discussed here are similar to NBFM.

FM broadcasting on channels that are 200 kHz apart (i.e., 88.1, 88.3, 88.5 ... MHz) is called wideband FM (WBFM). Maximum frequency deviation is set to

CHAPTER 3. MODULATION TYPES

f_Δ = 75 kHz and multiple sidebands are included to minimize distortion. Broadcast FM channels have much wider bandwidth than the audio signals they carry for two reasons. First, FM can trade off bandwidth for fidelity: wider modulation bandwidth equals better sound. Second, the FM channel is meant to carry other services such as HD channels and subscription services such as background music and reading to the blind.

The bottom plot of Figure 3.35 shows a simple WBFM spectrum example.[13] This is the spectrum of an FM broadcast station transmitting a 15 kHz steady tone. The carrier frequency can be set anywhere without change to the spectrum so here we set it to 0 Hz for convenience. There are three primary features of this spectrum as listed below. The MATLAB code that generated the figures is also shown.

1. The 15 kHz tone is causing the frequency of all spectral components to vary at 15 kHz. The modulation index is 75/15 = 5, so the frequency variation is 1/5 of the maximum possible. Frequency variation is difficult to see on Figure 3.35; the only clue is that the sideband spectrums may be a little wider.
2. The sidebands are all spaced at 15 kHz. Their amplitude is found from the Bessel functions in the top plot of Figure 3.35. Draw a vertical line through the modulation index of 5 and note where five of the Bessel functions intersect. Change negative intersection values to positive (noting that the phase of the associated sideband flips by 180°). More Bessel function could have been drawn; the limit to the number of sidebands is not equal to the modulation index.
3. For the maximum $f_{mod} = 15kHz$, Carson's rule gives us a bandwidth of 2(75 kHz + 15 kHz) = 160 kHz. Within that bandwidth modulating a nonstationary signal, such as voice, results in a constantly changing modulation index and sideband distribution. Thus, the sidebands fill in and the result is a very complicated signal spectrum.

```
% FM Modulate a 15MHz tone
Fs = 1500000; % Sample rate (Hz)
Ts = 1/Fs; % Sample period (s)
Fdmax = 75000; % Frequency deviation (Hz) for FM
Fmod = 15000;
ModIndex = Fdmax/Fmod;
%For a fixed FM, calculate sideband amplitudes.
```

---

13 A similar example can be found in [8].

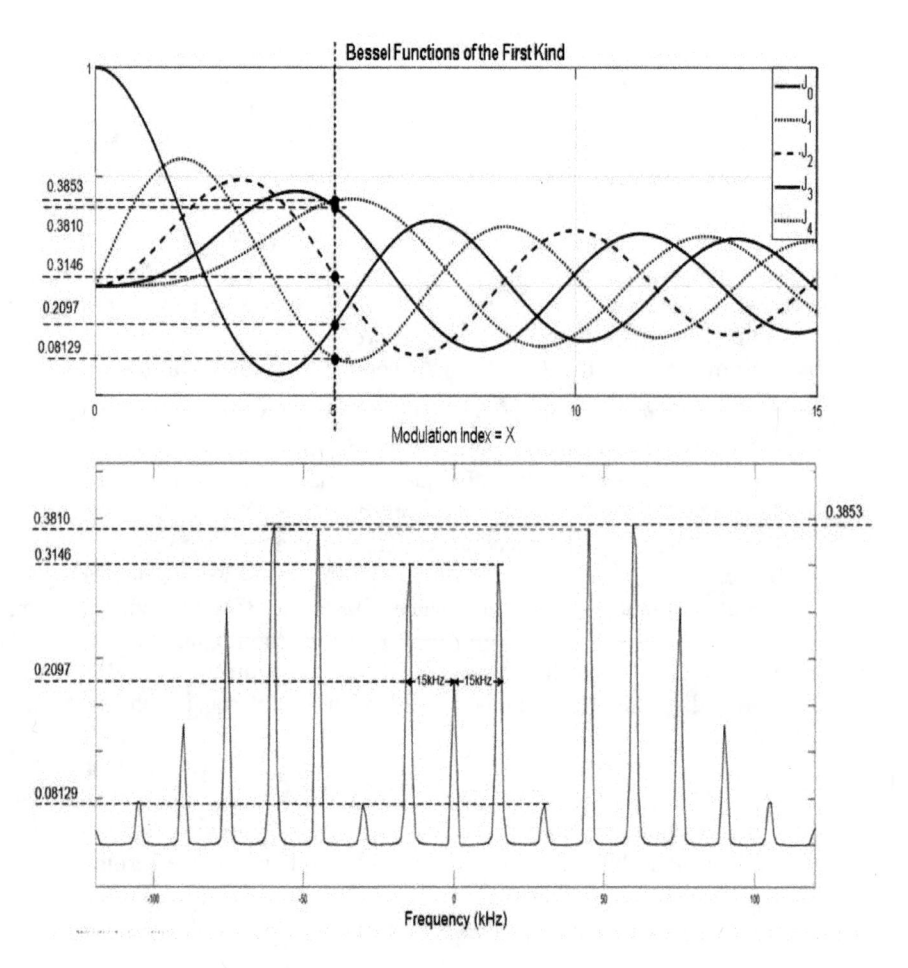

Figure 3.35 A wideband FM example, $f_{mod} = 15kHz$, Mod Idx = 5.

```
BessVals = GetBess(ModIndex);
disp(BessVals);
t = (0:Ts:5-Ts)'; % sample time progression, 5 seconds
x = sin(2*pi*Fmod*t); % FM Modulation tone

MOD = comm.FMModulator('SampleRate',Fs,...
```

```
'FrequencyDeviation',Fdmax);
%step is like a "run" command for the MOD object
y = step(MOD,x);
%plot(t,[x real(y)])
% Instantiate Spectrum Analyzer system object.
% Like a C++ object; has data and methods.
SA =...
dsp.SpectrumAnalyzer('SampleRate',Fs,'ShowLegend',true);
step(SA,[y]) %Analyze the spectrum of data record "Y"
%----------------------------------------------------
function bessrtn = GetBess(ModIndex)
% x = mod index horizontal scale
X = 0:0.1:20;
J = zeros(5,201); % Record of 5 Bessel Orders
for i = 0:4
J(i+1,:) = besselj(i,X);
end
% Return the 5 order magnitudes at the mod index
bessrtn = (J(:,round(10*(ModIndex))));
end
```

3.3.2 Frequency Shift Keying

Frequency shift keying (FSK) is a simple binary modulation where each source bit is mapped to one of two orthogonal sinusoids (see Section 3.2.10.1 for the definition of orthogonal). Equation (3.25) illustrates this.

$$\begin{aligned} b(n) = +1 &\Rightarrow s_1(t) = \sqrt{\frac{2E_b}{T_b}} \cos(2\pi f_1 t) & nT_b \le t \le (n+1)T_b \\ b(n) = -1 &\Rightarrow s_2(t) = \sqrt{\frac{2E_b}{T_b}} \cos(2\pi f_2 t) & nT_b \le t \le (n+1)T_b \end{aligned} \quad (3.24)$$

Equation (3.25) implies that the baseband antipodal data $b(n)$ effects a switching (or keying) between two orthogonal tones. Orthogonal here means $f_1 - f_2 = m\left(\frac{1}{2T_{bit}}\right)$, $m = 1, 2, 3, \ldots$. FSK modulation index is $h_{FSK} = |f_1 - f_2| T_b$.

Note also the amplitude factors, $\sqrt{\frac{2E_b}{T_b}}$ in Equation (3.25). For A = the amplitude scaling of the cosines, we can express these as:

$$A = \sqrt{\frac{2E_b}{T_b}} = \sqrt{E_b}\sqrt{\frac{2}{T_b}} \qquad (3.25)$$

Now we can plot the FSK $\sqrt{E_b}$ on the orthogonal basis vector axis; see Equation (3.9). We get the constellation of Figure 3.36 where the minimum distance is $d_{\min} = \|s_1 - s_2\| = \sqrt{2E_b}$.

Note that the FSK modulation index is not part of the minimum distance. FSK modulation index mostly has to do with bandwidth efficiency, not BER. Consider noncoherent FSK that is often noncoherently detected by two bandpass filters, one for each of the two tones. The detection is between the magnitude of the two BPF outputs. Deciding between the magnitude of two tones embedded in noise is not going to be much different if $h_{FSK} = 1$ or $h_{FSK} = 100$. However, the bandwidth efficiency will certainly be different.

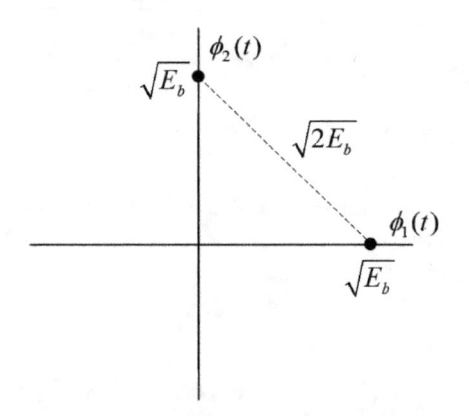

Figure 3.36 An FSK constellation showing minimum distance.

Compare the FSK constellation of Figure 3.36 with the BPSK constellation of Figure 10.25. For the same $\sqrt{E_b}$, it is easy to see that FSK orthogonal signals are closer together than BPSK antipodal signals. In Equation (3.12) we derived the BER for coherently detected BPSK. In [9] we find the second equation below

CHAPTER 3. MODULATION TYPES   91

for coherently detected FSK.[14] Equation (3.26) implies that, for the same BER performance as BPSK, we must double the FSK energy per bit.

$$P_{BPSK}(error) = Q\left(\sqrt{\frac{2E_b}{N_0}}\right), \quad P_{FSK}(error) = Q\left(\sqrt{\frac{E_b}{N_0}}\right) \quad (3.26)$$

3.3.2.1 FSK Noncoherent Detection

As shown in Figure 3.37, FSK can easily be demodulated noncoherently; carrier recovery is not needed. The demod output on the right side will toggle up and down at the symbol rate, but how do we correctly sample the symbols? For asynchronous character transmission, we only have to correctly decode individual 8-bit ASCII characters. This is done by detecting, at the demod output in Figure 3.37, a high to low transition known as a start bit. A low to high transition at the end of an 8-bit character burst is called a stop bit. Find the start bit, count eight more bits, and wait for the next start bit. That is all there is to it [10].

The design of Figure 3.37 is historically significant because it was used in the Bell 101 modem. This was introduced in 1958 by AT&T and was one of the first commercially available modems in the United States. The data rate was only 110 bits per second. The simplicity of FSK probably outweighed the BER performance advantage of BPSK. There is no need for complicated signal processing because the phase is not important and asynchronous characters were transmitted at a fixed baud rate.

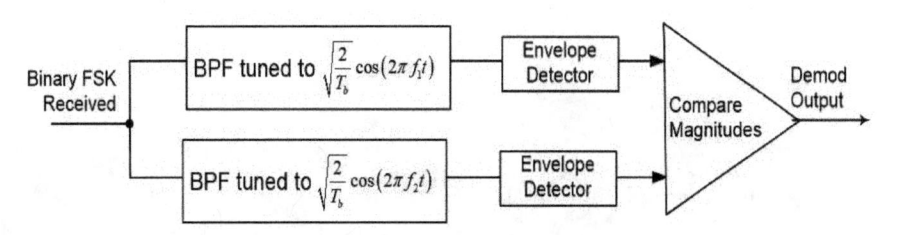

Figure 3.37 An FSK noncoherent demodulator example.

---

14 Note that points plotted on the Figure 3.36 complex plane have an amplitude and phase. BPSK coherent detection measures both. For a fair comparison in Equation (3.26), coherent detection is needed for both FSK and BPSK. However, in general, FSK does not need coherent detection. Measuring the amplitude at each tone frequency is generally adequate.

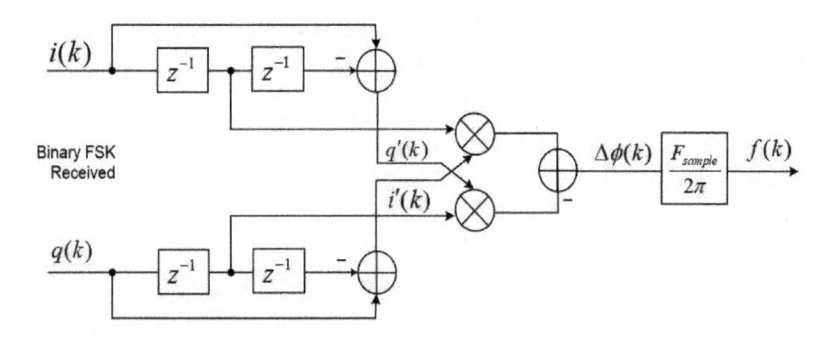

Figure 3.38 An FSK noncoherent demodulator, alternative design example.

An alternative, more modern, design is shown in Figure 3.38. This design may be easier to implement in the FPGAs that we have now. The input, $i(k) + jq(k)$ is complex baseband, 0 Hz, centered between the two tones. The index k is used to show that for this circuit $F_{sample} = M(F_{symbol})$ where M is an integer > 1. To get started analyzing this circuit, let's define several quantities:

$$\phi(t) = \tan^{-1}\left(\frac{q(t)}{i(t)}\right), \quad \tan(\phi(t)) = \frac{q(t)}{i(t)}, \quad a(t) = \frac{q(t)}{i(t)} \quad \phi(t) = \tan^{-1}(a(t))$$
(3.27)

To detect negative or positive baseband frequency, we differentiate the phase in Equation (3.28). The derivative of phase is frequency. A variable, $f(k)$, proportional to tone frequency is the desired result. Since the FSK is downconverted in frequency to complex baseband, the frequency changes are antipodal. Note that the expression for the derivative of $\tan^{-1}$ is listed in [11]. The end result of Equation (3.28) is implemented in Figure 3.38.

CHAPTER 3. MODULATION TYPES

$$\Delta\phi(t) = \frac{d\left(\tan^{-1}(a(t))\right)}{dt} = \frac{1}{1+a^2(t)}\left(\frac{i(t)\left(\frac{dq(t)}{dt}\right) - q(t)\left(\frac{di(t)}{dt}\right)}{i^2(t)}\right)$$

$$= \frac{1}{1+\frac{q^2(t)}{i^2(t)}}\left(\frac{i(t)\left(\frac{dq(t)}{dt}\right) - q(t)\left(\frac{di(t)}{dt}\right)}{i^2(t)}\right) = \frac{i(t)\left(\frac{dq(t)}{dt}\right) - q(t)\left(\frac{di(t)}{dt}\right)}{i^2(t) + q^2(t)}$$

(3.28)

The frequency discriminator circuit of Figure 3.38 has a wider bandwidth than the dual BPF in Figure 3.37. To reduce noise prior to the detector, a practical implementation will require a post detect filter tuned to the symbol rate.

3.3.3 Continuous Phase Modulation

The word keying in FSK is an old-fashioned concept that implies the potential for an abrupt phase discontinuity (like a telegraph key would produce). Continuous phase modulation (CPM) is an advancement that provides smooth frequency switching for FSK as well as many other modulation possibilities.

All the bit-by-bit phase progression possibilities of a continuous phase binary[15] FSK CPM signal are shown in the tree diagram in Figure 3.39 where h is the modulation index, generally equal to 1. From Figure 3.39, for $h = 1$, each new T spaced bit contributes a phase of $\pm\pi$.

3.3.3.1 CPM Transmitter

We will limit our CPM discussion to simpler designs. Figure 3.40 shows a modern continuous phase FSK (CPFSK) transmitter design. The phase ramp recursion block outputs phase transitions similar to Figure 3.39, at N samples per symbol. In Figure 3.40, customer bits are input at rate F_{bit}, the source bit rate indexed by n (dotted lines). The index n in Figure 3.40 is similar to T in Figure 3.39.

$F_{sample} = MF_{bit}$ = pulse sampling rate indexed by k, also called output sampling frequency. $F_{carrier}$ = carrier (or intermediate) frequency. By comparing Figures 3.4 and 3.40, we see that the latter introduces the phase ramp recursion and

---

[15] FSK can use more than two tones (e.g., quarternary FSK uses 4 tones).

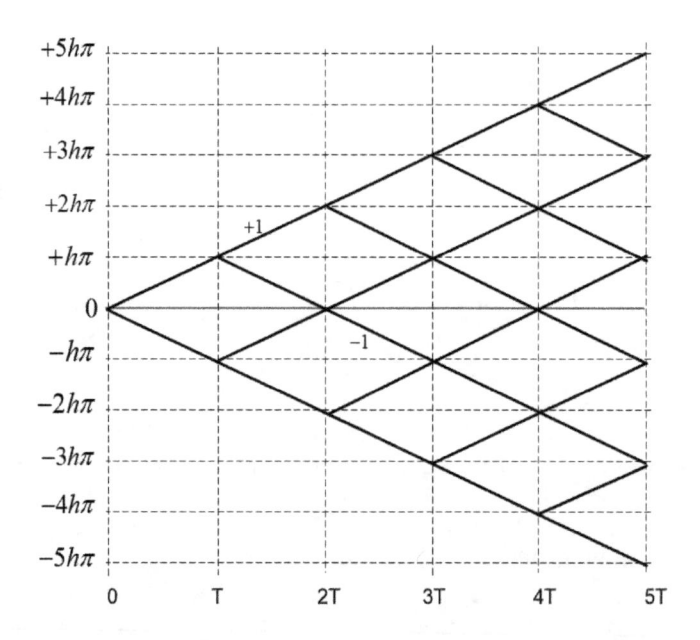

Figure 3.39 An example of an $h = 1$ CPM phase tree, T = bit times.

phase to frequency blocks. These are the amplitude to frequency blocks that make the CPFSK transmitter nonlinear.

Figure 3.41 is a closer look at the phase ramp recursion in Figure 3.40. Here $N = 8$. Again, note the similarity with Figure 3.39. We also show the complex baseband signal $x_i(k) + jx_q(k)$ as well as the output $y(k)$ upconverted in frequency to $F_{carrier}$. The resulting narrowband FM (NBFM) spectrum is shown in Figure 3.42.

3.3.3.2 CPM Receiver

The CPM transmit signal has a phase memory between symbols due to the continuous phase constraint. To take advantage of that, we study a receiver that makes a decision about the middle symbol of three consecutive continuous phase symbols. Many of these ideas can be found in [12].

We focus on an CPM receiver example for modulation index $h = 1/4$. The received signal is a full response signal, called REC1. REC1 means that the symbol

CHAPTER 3. MODULATION TYPES

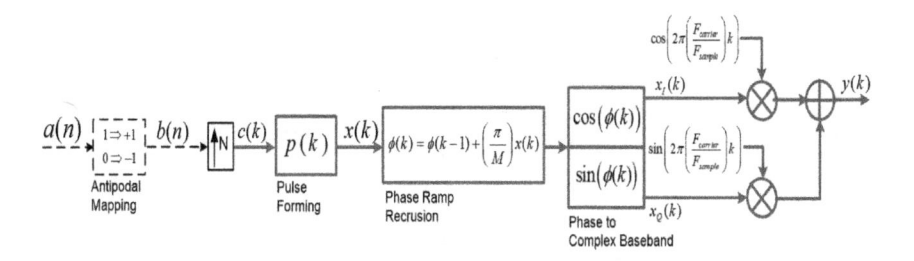

Figure 3.40 A CPFSK transmitter example.

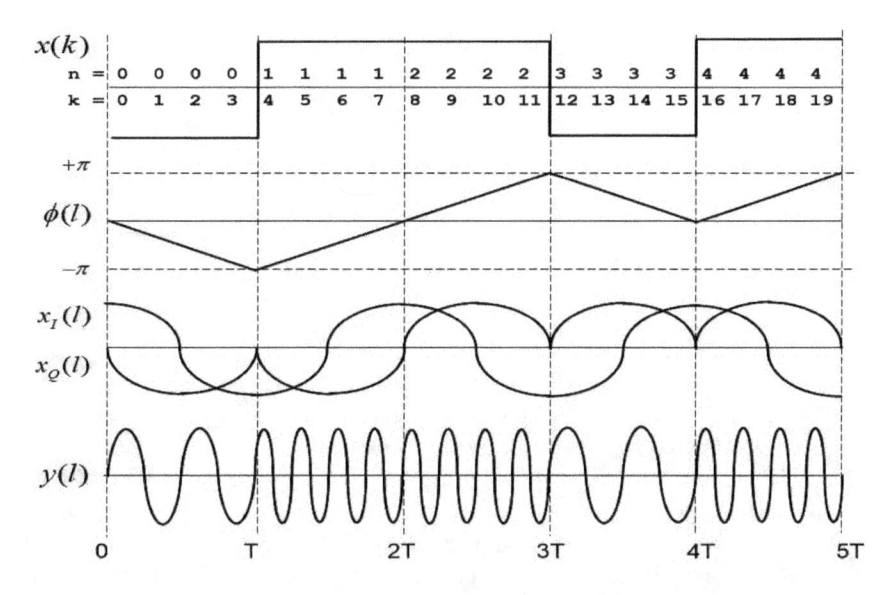

Figure 3.41 Details of CPFSK pulse forming and phase ramp recursion, $h = 1$.

shape is rectangular (i.e., unfiltered) and each symbol is independent of all other symbols; see $x(k)$ in Figure 3.41. However, the starting and ending phase of each symbol does depend on the previous symbols. From the tree diagram of Figure 3.39, we deduce that there are eight different starting phases for $h = 1/4$. These are shown on the left side of Figure 3.43. The trellis diagram on the right side is similar to a tree diagram except the phases are modulo 2π. The nomenclature $P(1a, n)$, for example, means phase transition counterclockwise from phase 1 ($\pi/4$), symbol

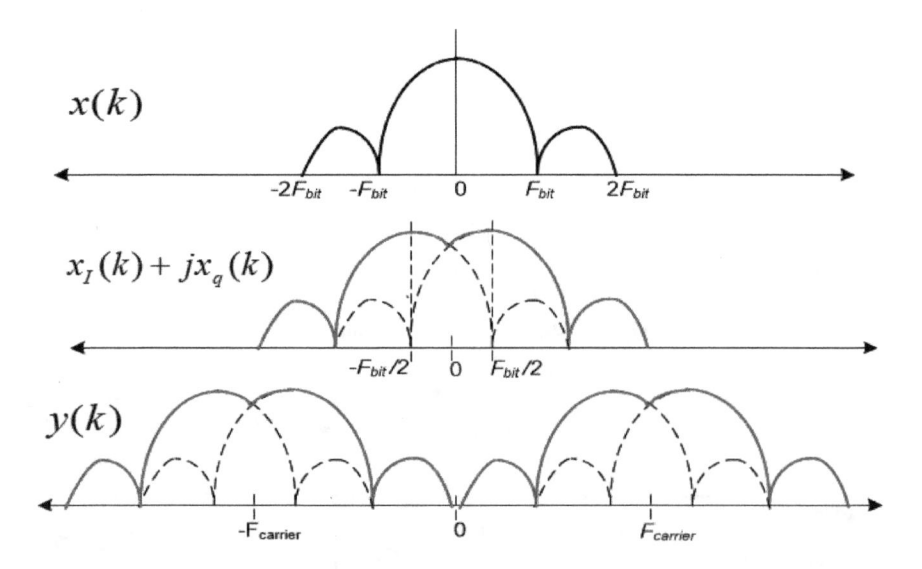

Figure 3.42 An example CPFSK pulse forming and phase ramp recursion spectrum.

number n. $P(1b, n)$ means the same except clockwise. This gives us 16 possible phase transitions for each symbol. Each phase transition has an ideal reference path that is a set of N complex points on the unit circle between starting and ending phases. N is the number of samples per symbol.

Table 3.8 shows three different symbol times. Figure 3.44 shows the three consecutive trellis structures corresponding to Table 3.8. Some careful examination reveals that there are only 64 valid paths over the three symbol span. The numbers in each column of Table 3.8 represent correlations between the N reference samples for each path and the symbols samples received at n, $n-1$ and $n-2$. For each new symbol, the procedure shown below is followed.

1. Column 2 shifted to and overwrites column 3.
2. Column 1 shifted to and overwrites column 2.
3. Column 1 is updated with correlations between N new symbol samples and 16 reference paths.

Let's assume that each new symbol is represented by $N = 4$ reference samples and that the symbol timing alignment is correct. These 4 reference samples will represent one of the 45° transitions on the left side of Figure 3.43. The equations

Table 3.8 CPM Three Symbol Receiver Example, $h = 1/4$

Current Symbol	Previous Symbol	Previous to Previous Symbol
P(0a,n)	P(0a,n-1)	P(0a,n-2)
P(0b,n)	P(0b,n-1)	P(0b,n-2)
P(1a,n)	P(1a,n-1)	P(1a,n-2)
P(1b,n)	P(1b,n-1)	P(1b,n-2)
P(2a,n)	P(2a,n-1)	P(2a,n-2)
P(2b,n)	P(2b,n-1)	P(2b,n-2)
P(3a,n)	P(3a,n-1)	P(3a,n-2)
P(3b,n)	P(3b,n-1)	P(3b,n-2)
P(4a,n)	P(4a,n-1)	P(4a,n-2)
P(4b,n)	P(4b,n-1)	P(4b,n-2)
P(5a,n)	P(5a,n-1)	P(5a,n-2)
P(5b,n)	P(5b,n-1)	P(5b,n-2)
P(6a,n)	P(6a,n-1)	P(6a,n-2)
P(6b,n)	P(6b,n-1)	P(6b,n-2)
P(7a,n)	P(7a,n-1)	P(7a,n-2)
P(7b,n)	P(7b,n-1)	P(7b,n-2)

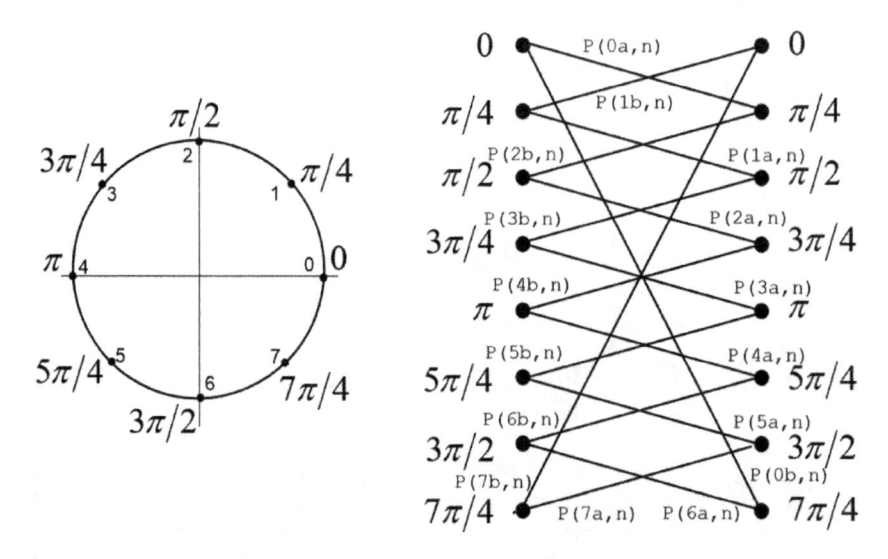

Figure 3.43 An example CPM trellis structure, $h = 1/4$.

that follow can be used to calculate all 16 sets of reference symbol samples. Note that the e^{jp} is simply one of the eight starting phases on the unit circle of Figure 3.43.

$$\begin{matrix} e^{jp} \\ e^{jp} \end{matrix} \begin{bmatrix} e^0 & e^{j\pi/16} & e^{j2\pi/16} & e^{j3\pi/16} \\ e^0 & e^{-j\pi/16} & e^{-j2\pi/16} & e^{-j3\pi/16} \end{bmatrix} \begin{matrix} CCW(a) \\ CW(b) \end{matrix} \qquad (3.29)$$

For that particular correlation, for example, P(0a,n), the four reference samples are complex multiplied by each of the four received complex symbol samples, the four products are summed, and, after columns 2 and 3 have been shifted forward, the complex correlation result is stored in the first column of Table 3.8. The next 15 correlations are calculated in the same way, only the reference samples are different.

To detect the middle symbol in column 2, combinations of three columns corresponding to 64 valid paths are added together to make 64 three-symbol correlations. The highest magnitude correlation is used to select the middle symbol. Notice that the detection filter is effectively 12 samples long. A symbol-by-symbol version would only be 4 samples long and would admit more noise. Thus, by taking advantage of known memory dependencies between symbols, we have improved the symbol error rate. Table 3.9 shows the 64 valid path correlation sums.

CHAPTER 3. MODULATION TYPES

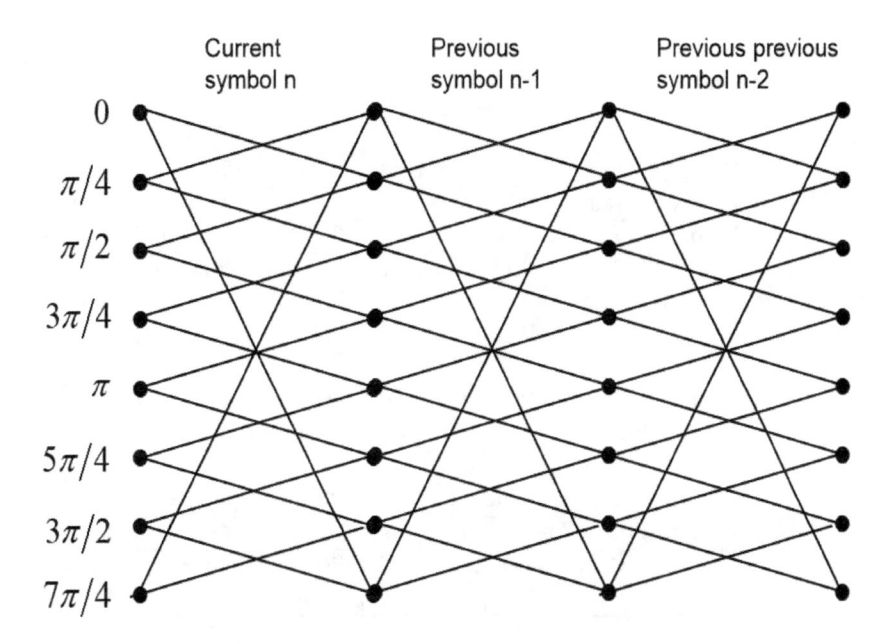

Figure 3.44 An example CPM three trellis structure, $h = 1/4$.

The reader might be thinking; why not just implement the Viterbi algorithm? The Viterbi algorithm produces coding gain because the convolutional coder makes some of the paths impossible. In this example, all paths are possible so no benefit would be obtained from a complete Viterbi implementation. Here, the trellis structure is simply a way to organize the phase trajectories for the demodulator.

3.3.3.3 CPM Generalization

From [12], CPM can be generalized to:

$$y(t) = \sqrt{\frac{2E_s}{T}} \cos\left(\omega_0 t + 2\pi \sum_{i=0}^{n} \alpha_i h q \left(t - iT\right)\right) \quad (3.30)$$

In Equation (3.30), α_i is a series of multilevel data symbols, h is the modulation index, $q(t)$ is the phase function that shapes the phase changes between symbols, and T is the time of a single symbol; see Figure 3.45. Note that the $cos()$ function

Table 3.9 Sixty-Four Valid CPM Path Combinations, Three Stage Trellis, $h = 1/4$

0	P(0a,n)+P(1b,n-1)+P(0a,n-2)	32	P(4a,n)+P(5b,n-1)+P(2a,n-2)
1	P(0a,n)+P(1b,n-1)+P(0b,n-2)	33	P(4a,n)+P(5b,n-1)+P(2b,n-2)
2	P(0a,n)+P(1a,n-1)+P(2b,n-2)	34	P(4a,n)+P(5a,n-1)+P(4b,n-2)
3	P(0a,n)+P(1a,n-1)+P(2a,n-2)	35	P(4a,n)+P(5a,n-1)+P(4a,n-2)
4	P(0b,n)+P(7b,n-1)+P(0a,n-2)	36	P(4b,n)+P(3b,n-1)+P(2a,n-2)
5	P(0b,n)+P(7b,n-1)+P(0b,n-2)	37	P(4b,n)+P(3b,n-1)+P(2b,n-2)
6	P(0b,n)+P(7a,n-1)+P(6b,n-2)	38	P(4b,n)+P(3a,n-1)+P(4b,n-2)
7	P(0b,n)+P(7a,n-1)+P(6a,n-2)	39	P(4b,n)+P(3a,n-1)+P(4a,n-2)
8	P(1a,n)+P(2b,n-1)+P(1a,n-2)	40	P(5a,n)+P(6b,n-1)+P(5a,n-2)
9	P(1a,n)+P(2b,n-1)+P(1b,n-2)	41	P(5a,n)+P(6b,n-1)+P(5b,n-2)
10	P(1a,n)+P(2a,n-1)+P(3b,n-2)	42	P(5a,n)+P(6a,n-1)+P(7b,n-2)
11	P(1a,n)+P(2a,n-1)+P(3a,n-2)	43	P(5a,n)+P(6a,n-1)+P(7a,n-2)
12	P(1b,n)+P(0b,n-1)+P(7a,n-2)	44	P(5b,n)+P(4b,n-1)+P(3a,n-2)
13	P(1b,n)+P(0b,n-1)+P(7b,n-2)	45	P(5b,n)+P(4b,n-1)+P(3b,n-2)
14	P(1b,n)+P(0a,n-1)+P(1b,n-2)	46	P(5b,n)+P(4a,n-1)+P(5b,n-2)
15	P(1b,n)+P(0a,n-1)+P(1a,n-2)	47	P(5b,n)+P(4a,n-1)+P(5a,n-2)
16	P(2a,n)+P(3b,n-1)+P(2a,n-2)	48	P(6a,n)+P(7b,n-1)+P(0a,n-2)
17	P(2a,n)+P(3b,n-1)+P(2b,n-2)	49	P(6a,n)+P(7b,n-1)+P(0b,n-2)
18	P(2a,n)+P(3a,n-1)+P(2b,n-2)	50	P(6a,n)+P(7a,n-1)+P(6b,n-2)
19	P(2a,n)+P(3a,n-1)+P(2a,n-2)	51	P(6a,n)+P(7a,n-1)+P(6a,n-2)
20	P(2b,n)+P(1b,n-1)+P(0a,n-2)	52	P(6b,n)+P(5b,n-1)+P(4a,n-2)
21	P(2b,n)+P(1b,n-1)+P(0b,n-2)	53	P(6b,n)+P(5b,n-1)+P(4b,n-2)
22	P(2b,n)+P(1a,n-1)+P(2b,n-2)	54	P(6b,n)+P(5a,n-1)+P(6b,n-2)
23	P(2b,n)+P(1a,n-1)+P(2a,n-2)	55	P(6b,n)+P(5a,n-1)+P(6a,n-2)
24	P(3a,n)+P(4b,n-1)+P(3a,n-2)	56	P(7a,n)+P(6b,n-1)+P(5a,n-2)
25	P(3a,n)+P(4b,n-1)+P(3b,n-2)	57	P(7a,n)+P(6b,n-1)+P(5b,n-2)
26	P(3a,n)+P(4a,n-1)+P(5b,n-2)	58	P(7a,n)+P(6a,n-1)+P(7b,n-2)
27	P(3a,n)+P(4a,n-1)+P(5a,n-2)	59	P(7a,n)+P(6a,n-1)+P(7a,n-2)
28	P(3b,n)+P(2b,n-1)+P(1a,n-2)	60	P(7b,n)+P(0b,n-1)+P(7a,n-2)
29	P(3b,n)+P(2b,n-1)+P(1b,n-2)	61	P(7b,n)+P(0b,n-1)+P(7b,n-2)
30	P(3b,n)+P(2a,n-1)+P(3b,n-2)	62	P(7b,n)+P(0a,n-1)+P(1b,n-2)
31	P(3b,n)+P(2a,n-1)+P(3a,n-2)	63	P(7b,n)+P(0a,n-1)+P(1a,n-2)

CHAPTER 3. MODULATION TYPES

has two terms in its argument: the carrier phase, $\omega_0 t$, and the excess phase due to the transmit symbols. Let's focus on the second term. Given any value of time t, symbol counter i counts up until $iT < t < (i+1)T$. At that point phase shaping function $q(t - iT)$ becomes nonzero because its argument is in the range between 0 and T, as shown in Figure 3.45. Thus, for every t, the term $\alpha_i h q(t - iT)$ contributes a constant slope ramp phase change to $y(t)$. The constant slope results in a linear phase increase and a constant frequency during each symbol time.[16]

Note from Figure 3.45 that after symbol time T the phase stays at 0.5. Recall that in Figure 3.39, for $h = 1$, each new T spaced bit contributes a phase of $\pm\pi$. If in Equation (3.30), $h = 1$ and $\alpha_i = 1$, then when $(t - iT = T)$, an additional phase of π will be accumulated. Finally, note that an FSK tone is produced only when the phase is ramping up or down. Phase going horizontal after time T in Figure 3.45 turns off the current symbol tone and provides a starting point for the next symbol phase to ramp up or down and start a new tone. The interested reader is referred to [12] where additional examples of CPM waveforms are discussed.

As an aside, for many years there has been a quest to use multilevel symbols to achieve bandwidth efficiency. For example, cable television signals are often encoded with 256 QAM to achieve a bandwidth efficiency of about 8. This results in a high peak to average power linear modulation. In contrast, constant envelope CPM signals can turn those multilevel symbols into inner phase complexity to achieve similar bandwidth efficiency. Judging by the standards that currently govern TV signals and also Gen 4 LTE cellular systems, CPM seems to have lost the bandwidth efficiency contest. Interestingly, however, although the BER of antipodal PSK is theoretically better than orthogonal CPFSK, there are applications where CPFSK is preferred due to survivability in, for example, nuclear radiation environments [9].

3.3.4 Minimum Shift Keying

Minimum shift keying (MSK) is a form of CPM that uses differential phase modulation conforming to the trellis of Figure 3.39 with $h = 0.5$. MSK is well known for its ability to tolerate hard limiting in satellite transmitters. This is because, being confined to a phase circle, the signal is strictly constant envelope.

Figure 3.46 shows the MSK constellation as well as the bit by bit ± 90 phase changes. Consider, for example, the two possible phase changes shown in Figure 3.46. In effect, the decision region is either the upper or lower part of the signal

---

[16] There are other, more complicated choices for phase-shaping functions that may result in better bandwidth efficiency.

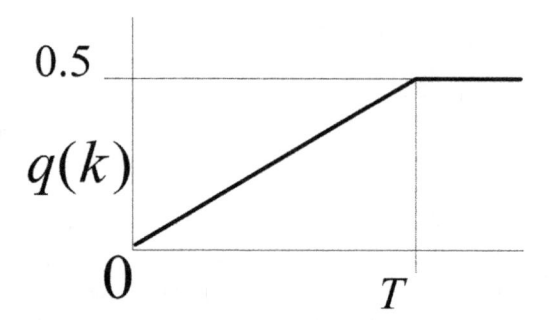

Figure 3.45 CPM phase shaping function for a rectangular pulse shape.

space. By comparing that with the analysis of one of the QPSK bits in Figure 3.13, it is easy to see that the MSK and QPSK BER performance is the same.

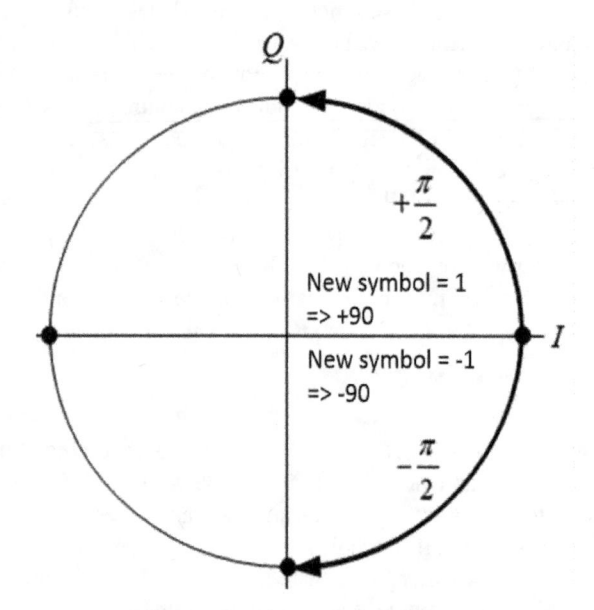

Figure 3.46 The MSK constellation.

Figure 3.47 is a simplified block diagram of an MSK transmitter. Figure 3.48 is the corresponding timing diagram. Bit input $a_{Tx}(m)$ is upsampled to $a_{ev}(k)$ and

CHAPTER 3. MODULATION TYPES

$a_{od}(k)$, which are ±1 followed by $(M-1)$ zeros for every symbol. Note that pulse $a_{ev}(k)$ matches the polarity of even bits of $a(m)$ and $a_{od}(k)$ matches the polarity of odd bits of $a(m)$. This is equivalent to offset QPSK (OQPSK), a variation on QPSK that avoids traversing through zero on the output.

The two stored half-cycle sine waves are both positive. Switches $s_{ev}(k)$ and $s_{od}(k)$ cause inversion of alternating half-cycles to bring about a complete sine wave. The ±90 rotations of $x_I(k)$ and $x_Q(k)$ are shown in the series of circular phase change diagrams near the bottom of Figure 3.48. Rotation is counterclockwise for a +1 input and clockwise for a -1 input of $a_{Tx}(m)$ at the top of the figure. The dotted traces overlaying $x_I(k)$ and $x_Q(k)$ in Figure 3.48 indicate the sine waves that bring about this rotation. To produce the plus and minus 90° phase changes of MSK modulation we simply invert half cycles of these sine waves.

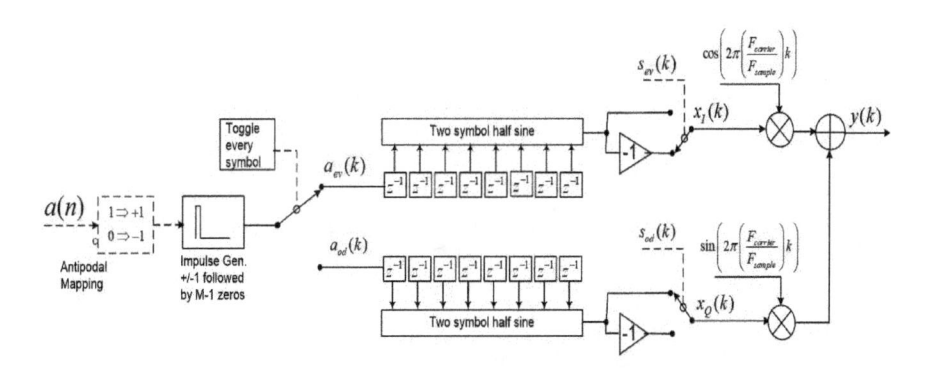

Figure 3.47 A typical MSK transmitter.

The MSK receiver in Figure 3.49 detects what phase change, plus or minus 90°, was transmitted. This phase change information reveals the current symbol value and allows updating of the next symbol starting phase. The symbol delay stores the previous symbol correlation so that the phase direction can be calculated and the symbol values decided.

Let's illustrate this by an example using the four symbols in Figure 3.50. A few assumptions are that the received symbols are all equal to -1, $M = 4$, the starting phase for symbol A is +90°, and finally that carrier and symbol timing have converged correctly. In Figure 3.49, the phase rotation correlator will correlate the next M symbol samples and then the phase rotation detect logic will sample $y_I(k)$ under the assumption that $y_Q(k) = 0$. Because clockwise rotation is detected, the symbol value is set to -1. The next symbol start phase is set to 0° and the next

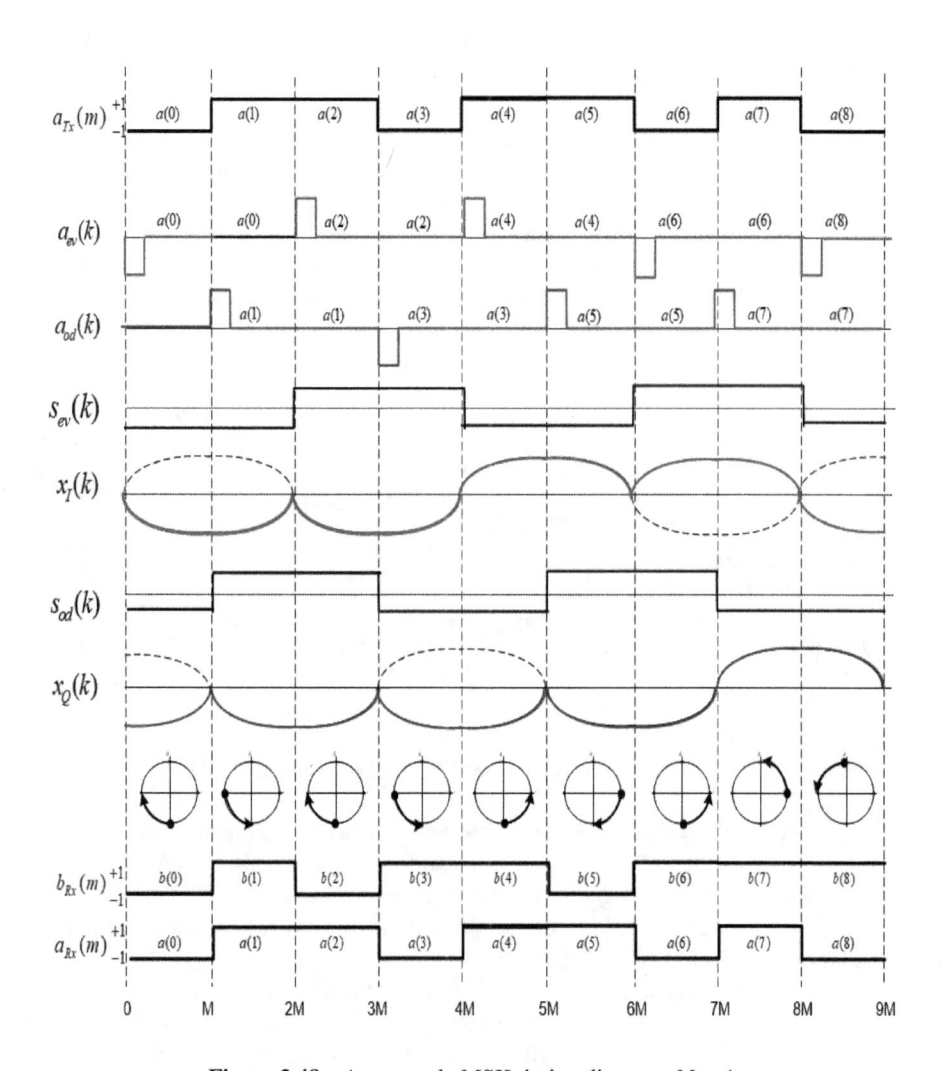

104 Software Defined Radio: Theory and Practice

Figure 3.48 An example MSK timing diagram, $M = 4$.

CHAPTER 3. MODULATION TYPES

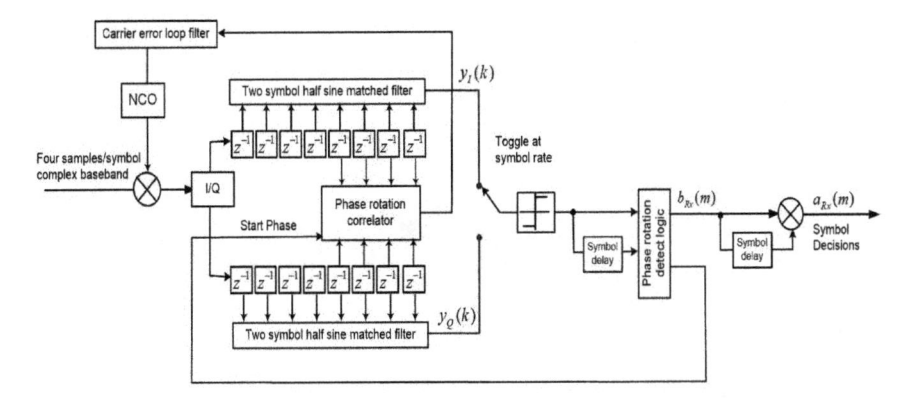

Figure 3.49 A typical MSK receiver.

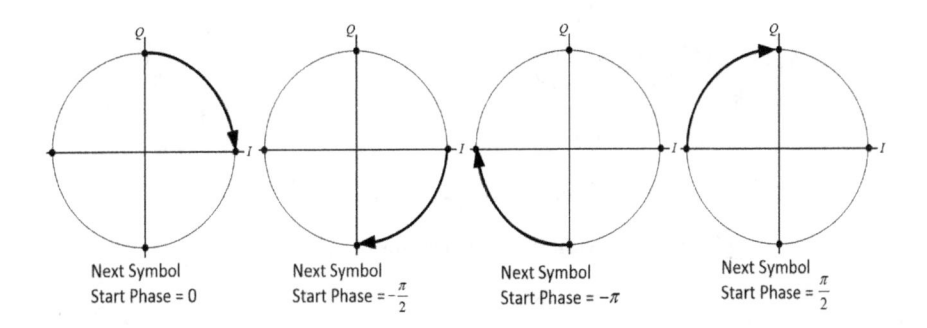

Figure 3.50 Typical MSK receiver phasing.

M symbol samples are correlated. Now the phase rotation detect logic will sample $y_Q(k) = -1$ under the assumption that $y_I(k) = 0$. Note from Figure 3.48 that the $b_{Rx}(m)$ detected symbols must be differentially decoded to produce an $a_{Rx}(m)$ bit stream that matches $a_{Tx}(m)$ at the top of the figure.

The absolute value of the maximum expected I or Q correlation can be used as an objective function to steer a symbol timing algorithm. Note from Figure 3.48 that when I is max, Q is zero and vice versa.

Figure 3.51 shows that raised cosine filtering has a significant bandwidth advantage over MSK. If the hard limiting, constant envelope performance of MSK is not needed, the raised cosine signal is probably a better choice. Some wireless phone handset systems, such as DECT, use MSK because it is very simple and bandwidth efficiency is not a major concern for such a short-range signal.

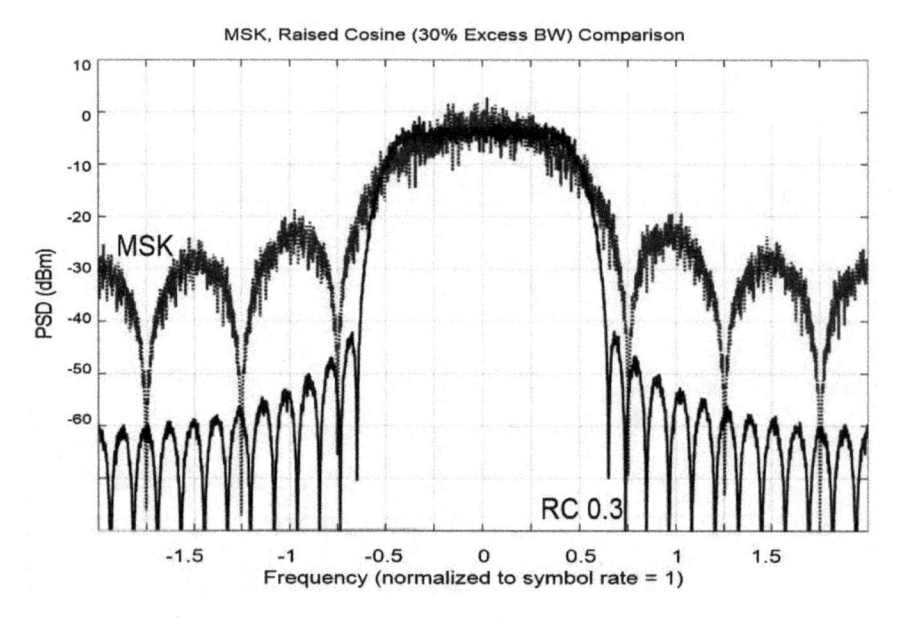

Figure 3.51 MSK bandwidth versus raised cosine bandwidth.

3.3.5 Gaussian Minimum Shift Keying

Gaussian minimum shift keying (GMSK) is well known for use in the second-generation GSM (Global System for Mobile Communications) cellular phone system, originally deployed in Europe. GMSK allowed GSM to transmit 270.8 Kbps in a 200 kHz radio channel for a bandwidth efficiency of about 1.35 bps/Hz. A further advantage of GMSK is that it is constant amplitude. That means that it can be amplified by a nonlinear amplifier (for example, a class C power amplifier in a satellite transponder) and remain undistorted; no elements of the signal depend on amplitude variations.

In the United States, GSM networks have been mostly replaced by LTE, fourth generation, systems. GMSK, however, still has important applications in both military and deep-space communications.

CHAPTER 3. MODULATION TYPES

3.3.5.1 GMSK Transmitter

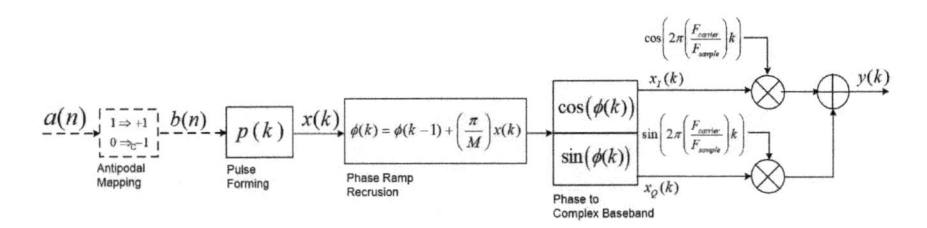

Figure 3.52 A typical GMSK transmitter.

Figure 3.52 is a standard GMSK transmitter block diagram. Let's discuss just the frequency to phase converter; we will come back to the Gaussian filter. We first upsample the input antipodal symbols by inserting $M - 1$ zeros times in the time domain (M = samples per symbol). As shown in the top trace of Figure 3.53, the Gaussian filter output $x(k)$ is ± 1; in this example, the Gaussian filter is assumed square to show the phase trajectory clearly. The frequency to phase converter integrates $x(k)$ to form the phase ramp $\phi(k)$. Notice the difference between n, for bits, and k, for multiple samples representing 1 bit, in Figure 3.53.

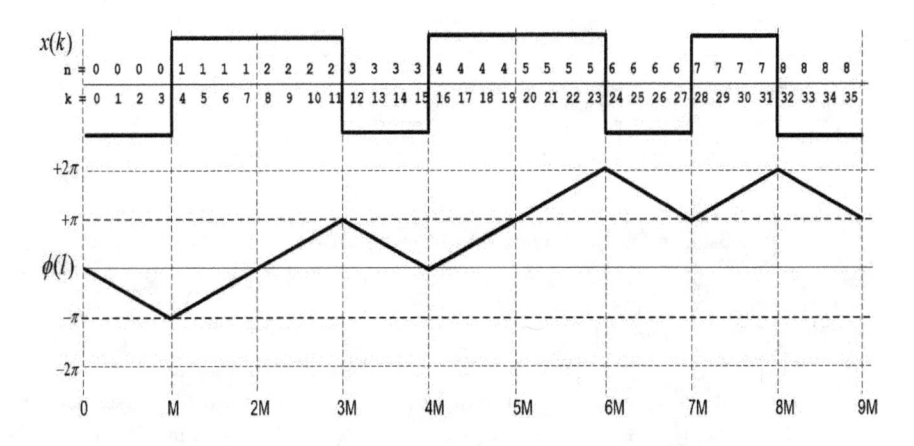

Figure 3.53 GMSK frequency to phase converter, $M = 4$.

Let's go back and look at actual Gaussian filters. This unique filter trades off transmit bandwidth and intersymbol interference (ISI). For B_{3dB} = one-sided 3 dB Gaussian filter bandwidth in Hertz, the fundamental defining parameter of the Gaussian filter is:

$$BT = \frac{B_{3dB}}{F_{symbol}} \qquad (3.31)$$

Figure 3.55 shows the GMSK transmit spectrum for three values of BT. The vertical lines show one side of the 99% power bandwidth.[17] Figure 3.56 compares GMSK, BT = 0.3, with BPSK, raised cosine excess bandwidth = 0.6. GMSK, BT = 0.3, has obvious advantage for channelization.

Table 3.10 is a comparison of four different values of BT. The first line is actually MSK. Notice the MSK double-sided bandwidth does not reach 99.9% total power until 2.4T, using a normalized scale as in Figure 3.55. Thus, the bandwidth efficiency of MSK is $0.41 = T/2.4T$.[18]

Many GMSK systems use $BT = 0.3$ because, as seen in Table 3.10, a very compact spectrum is achieved with minimal ISI. For example, the European GSM cellular phone system, Gen 2, achieves 270.8 Kbps in a 200 KHz channel for a practical GMSK bandwidth efficiency of about 1. The 2.35% ISI is not significant compared with the other distortions present on the typical cellular phone channel.

Table 3.10   GMSK Signal Characteristics for Different Values of BT

BT	99.9% BW	Adjacent Channel Pwr	ISI (Symbols)	ISI (%)
inf	2.4T	−25dB	1	0
1	2T	−28dB	3	0%
0.5	1.3T	−33dB	3	0.016%
0.3	1.1T	−45dB	3	2.35%

Figure 3.54 shows three Gaussian filter impulse responses and respective I,Q baseband waveforms $x_I(k), x_Q(k)$. MSK is the last row but notice how similar the I,Q baseband waveforms of the other two rows are. The first row, first graph, shows the ISI that is an acceptable price to pay for the narrow bandwidth of BT = 0.3.

At the beginning of this chapter we looked at independent symbol transmissions such as BPSK and QPSK. These are called full response because all symbol energy is confined to only one symbol time and the symbols

---

17   99% power bandwidth means the bandwidth that contains 99% of the total averaged double-sided signal power. This measurement is sometimes used as a signal bandwidth specification. There are other bandwidth measures [1].

18   Recall that bandwidth efficiency is the symbol rate divided by the required bandwidth.

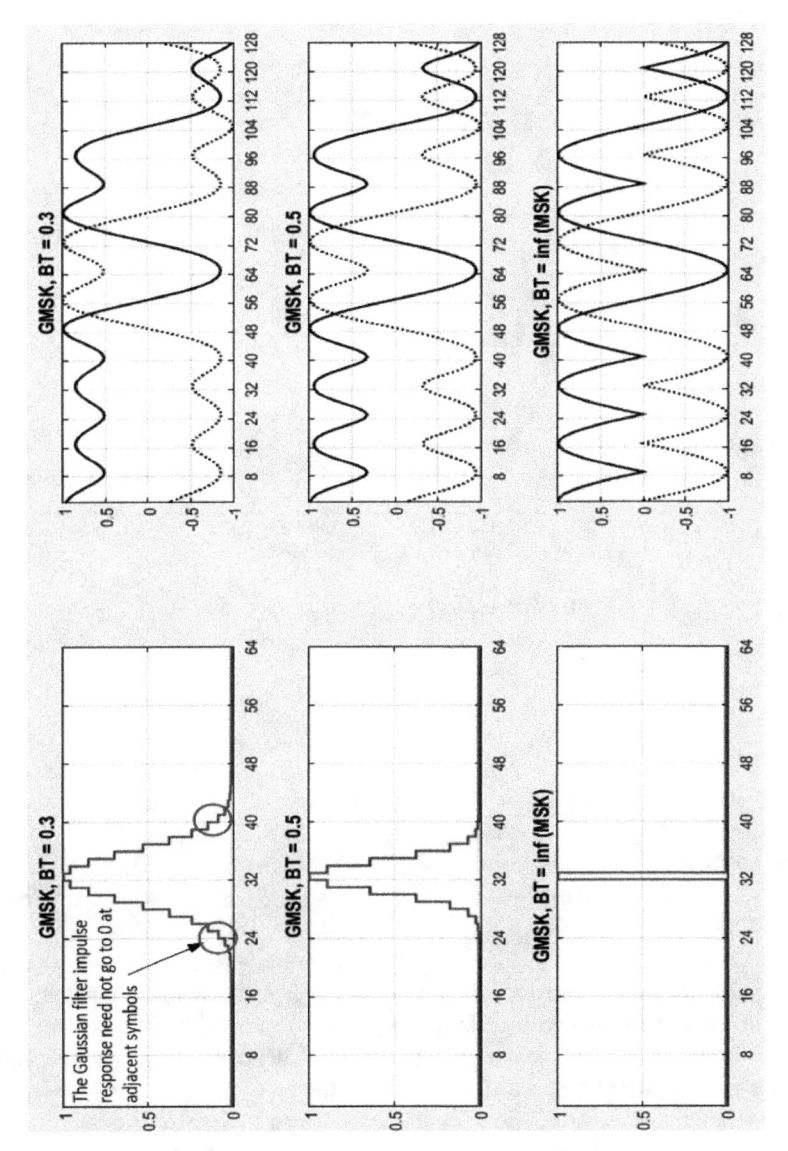

Figure 3.54 GMSK filter impulse response and IQ signals for three values of BT.

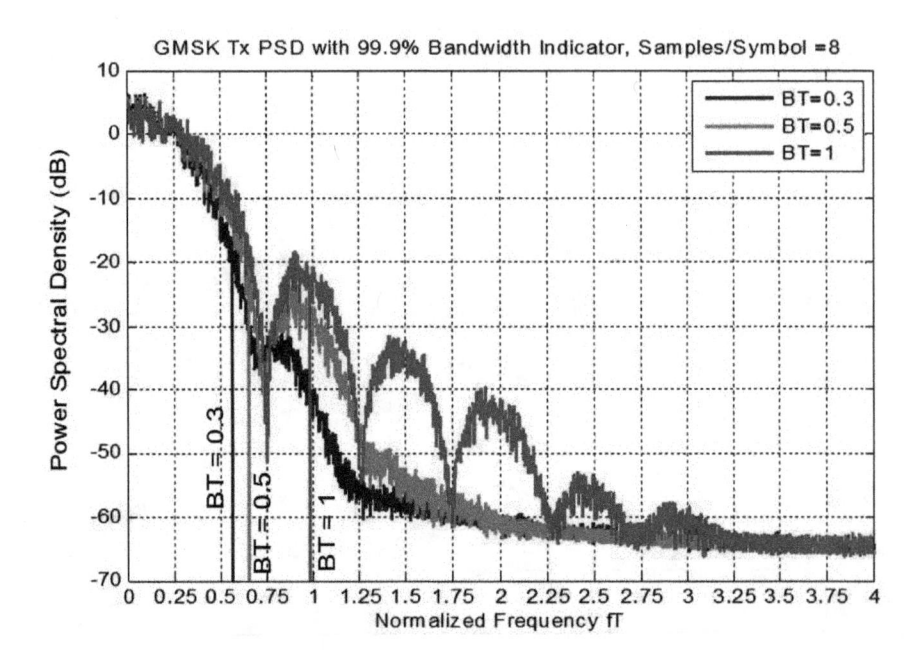

Figure 3.55 An example GMSK spectrum.

are independent of each other. There is also partial response where symbol $x(T) = \alpha y(T-1) + y(T) + \beta y(T+1)$ (i.e., transmit filter output $x(T)$ is a combination of past, present, and future transmit symbols). GMSK is an example of a partial response signal. This is especially obvious for $BT \leq 0.3$; see the top left of Figure 3.54.

GMSK as generated in Figure 3.52 has differentially continuous phase. As shown in Figure 3.53, a zero-input bit produces $-\pi/2$ phase change and a 1 input bit results in $+\pi/2$ phase change. This kind of signal also has modulo 2π phase memory in that the absolute phase in any symbol interval depends on all the preceding symbols and the starting phase. To detect bits, the receiver must compare current and previous bit phases; this doubles the noise since both estimations are noisy. GMSK solves this problem with transmit precoding. We cover precoding types 1 and 2.

CHAPTER 3. MODULATION TYPES

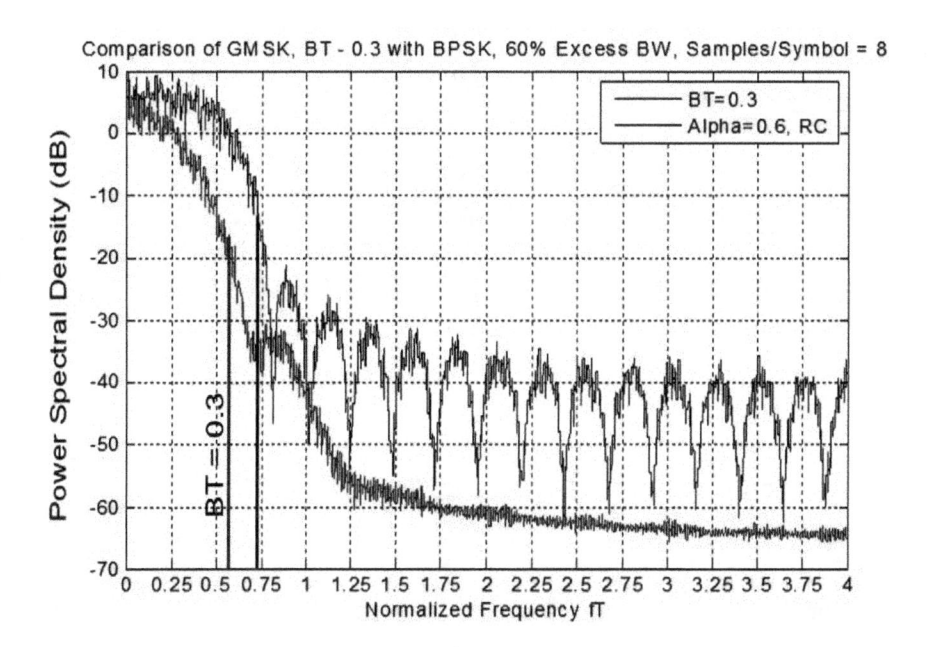

Figure 3.56 GMSK spectrum compared with a raised cosine spectrum.

3.3.5.2 GMSK Precoding Type 1

Precoding type 1 is shown in Figure 3.57. Instead of the transmitting phase corresponding directly to bits ($\theta(k)$ in Figure 3.53), precoding type 1 transmits a phase that indicates if a bit change has occurred. See $\theta(k)$ in Figure 3.58.

For a closer look at precoding type 1, compare Figures 3.57 and 3.58. In Figure 3.58, source bits and precoding bits are shown in the first two traces. The next two traces correspond to $x(k)$ and $\theta(k)$ in Figure 3.52. Finally, the complex baseband signal in the last trace of Figure 3.58 corresponds to the complex output on the right side of Figure 3.58.

Notice that in Figure 3.58, prior to the vertical dashed line A, the source bits are constant and the complex baseband signal represents a sample traveling clockwise around the unit circle. Rotation switches to counterclockwise when the source bits start toggling after vertical dashed line A. The center of Figure 3.58 shows rotation changes that follow the precoded random source bits. Finally, between lines B and C, the source bits start to toggle and the output goes back to a counterclockwise

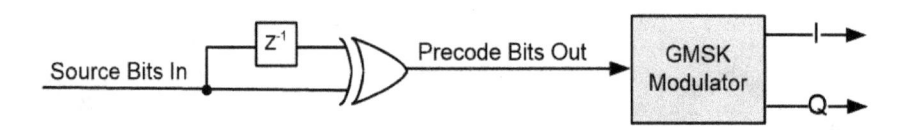

Source Bits all zero (or all ones) => Precode bits all zero => CW

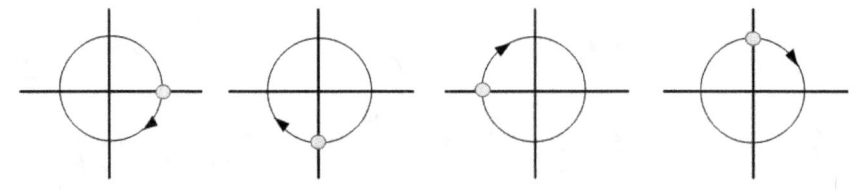

Source Bits toggling one to zero => Precode bits all one => CCW

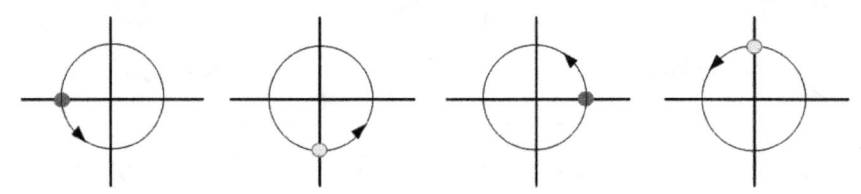

Figure 3.57 Definition of GMSK precoding type 1 circuit.

rotation followed by a clockwise rotation when the source bits stay fixed starting at line C.

3.3.5.3 GMSK Precode Type 2

For a closer look at precoding type 2, compare Figures 3.59 and 3.60. In Figure 3.60, source bits and precode bits are shown in the first two traces (same as Figure 3.58). The next two traces correspond to $x(k)$ and $\theta(k)$ in Figure 3.52. Finally, the complex baseband signal in the last trace of Figure 3.60 corresponds to the complex output on the right side of Figure 3.60.

Notice that in Figure 3.60, prior to the vertical dashed line A, the source bits are constant and the complex baseband signal represents a phase change alternating CW and CCW. Rotation alternates CCW to CW when the source bits start toggling after vertical dashed line A. The center of Figure 3.60 shows rotation changes that

CHAPTER 3. MODULATION TYPES 113

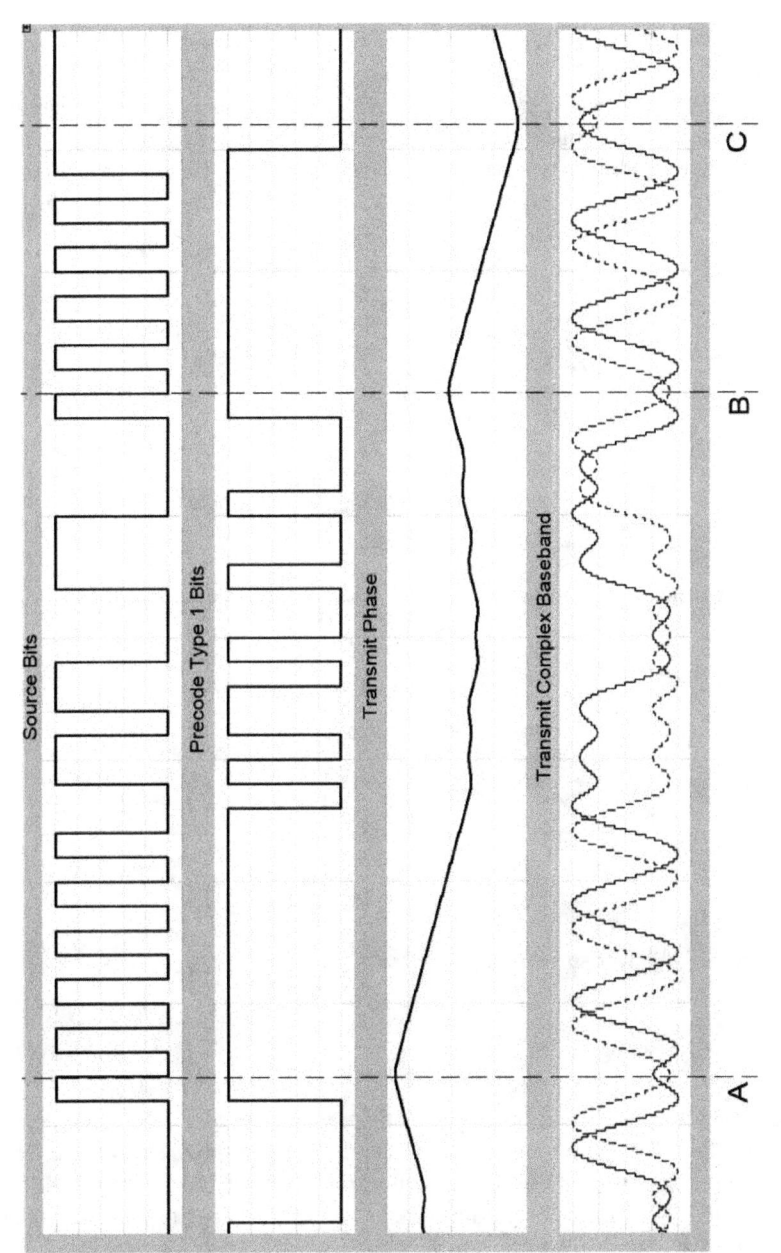

Figure 3.58 Example of GMSK precoding type 1 signals.

follow the precoded random source bits. Finally, between lines B and C, the source bits start to toggle and the output goes back to alternating CCW to CW.

Notice the dots in the first part of the last line of Figure 3.60. Compare these carefully with the transmit data bits in the first line. Precoding type 2 produces a complex output that directly indicates the transmit data. See [13], column 7, line 35. This fact will be used to design a simple GMSK receiver.

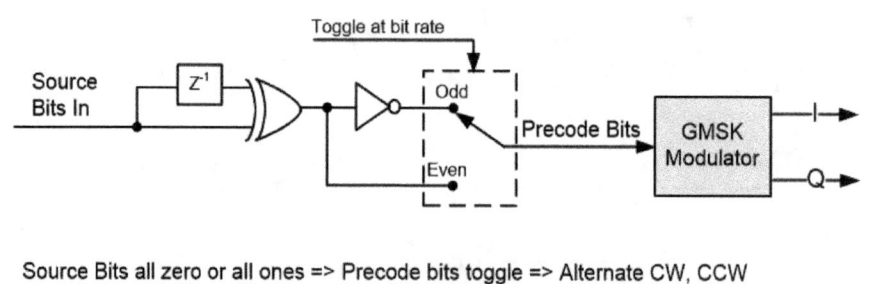

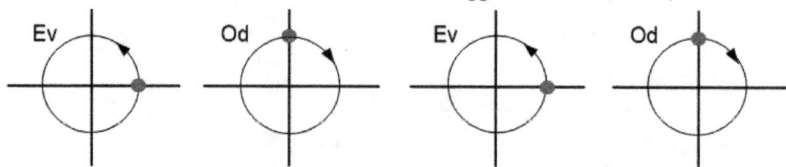

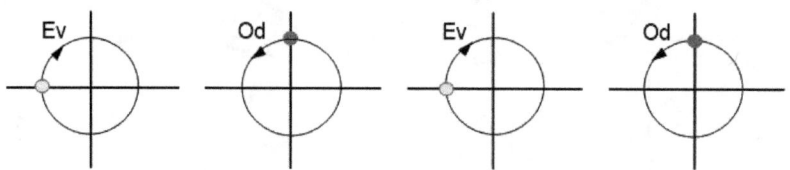

Figure 3.59 Definition of GMSK precoding type 2 circuit.

3.3.5.4 GMSK Receiver

If the transmit signal is precoded type 2, the circuit of Figure 3.61 can be used to detect the data. The input is complex baseband, the same signal as output in Figure 9.42. The real and imaginary inputs are first predetect filtered by the GMSK filter described in [14]. Fortunately, MathWorks has already calculated this equation for us; all we have to do is type the following in the MATLAB workspace:

CHAPTER 3. MODULATION TYPES 115

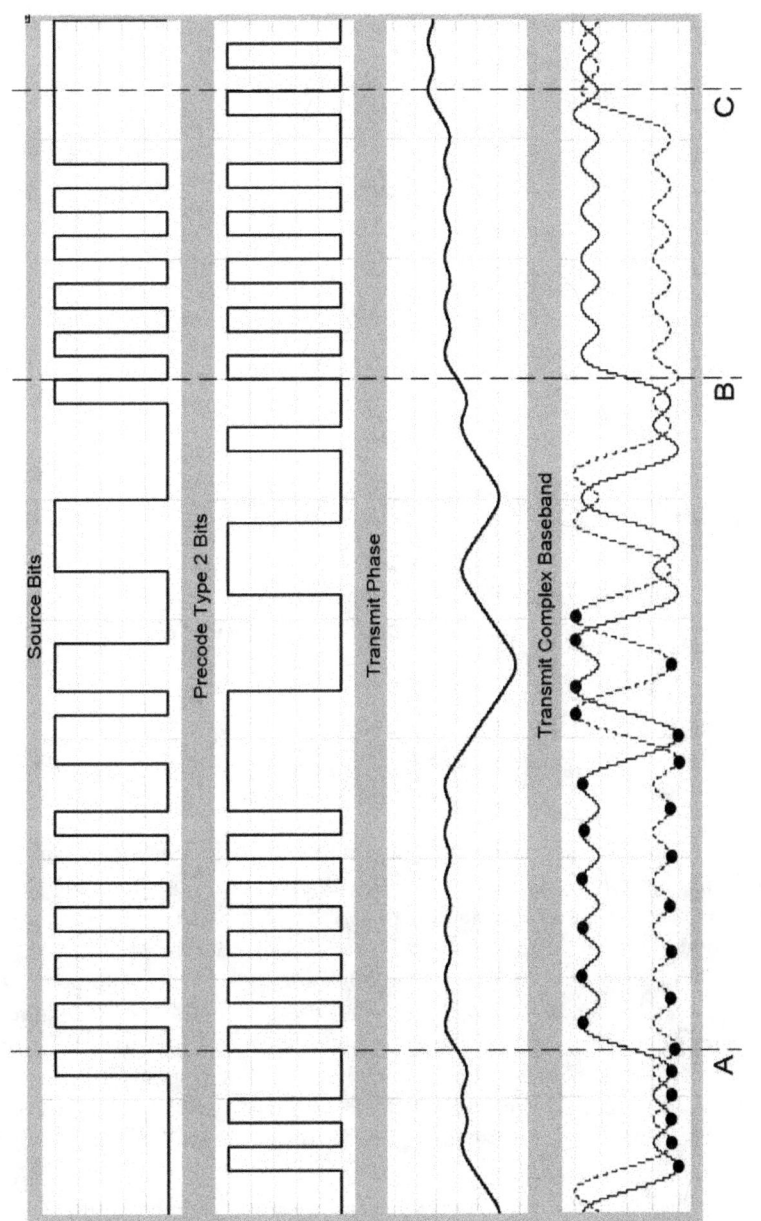

Figure 3.60 Example of GMSK precoding type 2 signals.

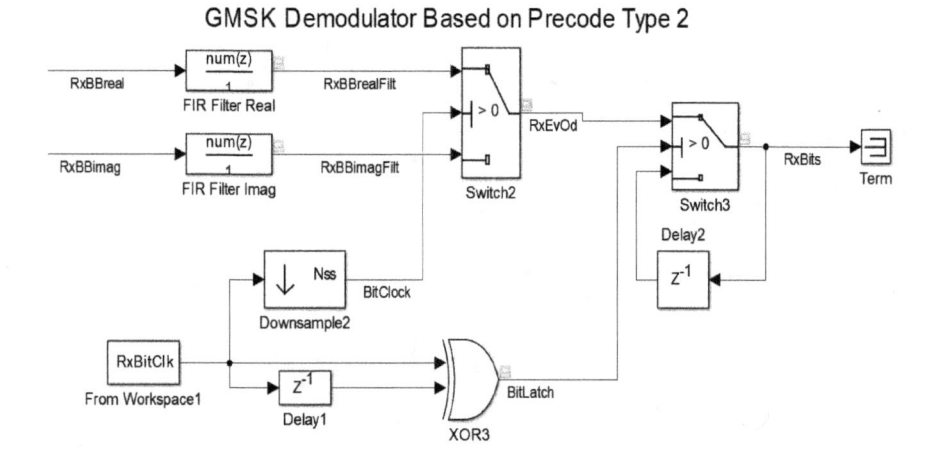

Figure 3.61 GMSK receiver Simulink diagram.

```
BT= 0.3; % GMSK BT product
L = 3; % GMSK partial response pulse length, in symbols
% MATLAB supplied function to generate a
% GMSK predetect filter
C0 = commgmsksoftdecision_genC0(BT, L);
(hb,wb) = freqz(C0,1,1024); % Calculate freq response
plot(10*log10(abs(hb))); % plot frequency response
```

The predetect filter response and the incoming GMSK signal are shown together in Figure 3.62. Figure 3.63 shows the received complex baseband and filtered received complex baseband in traces 2 and 3, respectively. Note in trace 4 the even-odd alternate switching at the bit rate.[19] Compare this with the last trace in Figure 3.60. Note there how the dots, indicating the correct bit value, alternate between real and imaginary signals. Also shown in the fourth trace of Figure 3.63 are the latch pulses that sample the even and odd segments of received waveform.

In a complete design, preamble detection would be responsible for carefully lining up the first latch pulse with the first sample of the first symbol. For long messages, there may need to be some closed-loop bit tracking following the preamble.

---

19 We could refer to bit rate or symbol rate. We use bit rate for GMSK to emphasize that there is only 1 bit per symbol.

CHAPTER 3. MODULATION TYPES

Because this signal has little energy at $F_{bit}/2$, the frequency-domain techniques of Chapter 10 will not work. Time-domain transition tracking may be a good choice, however. For short messages, open-loop bit tracking may work. Here the preamble lines up the first data bit and the bit tracking proceeds at the nominal rate with no closed-loop control.

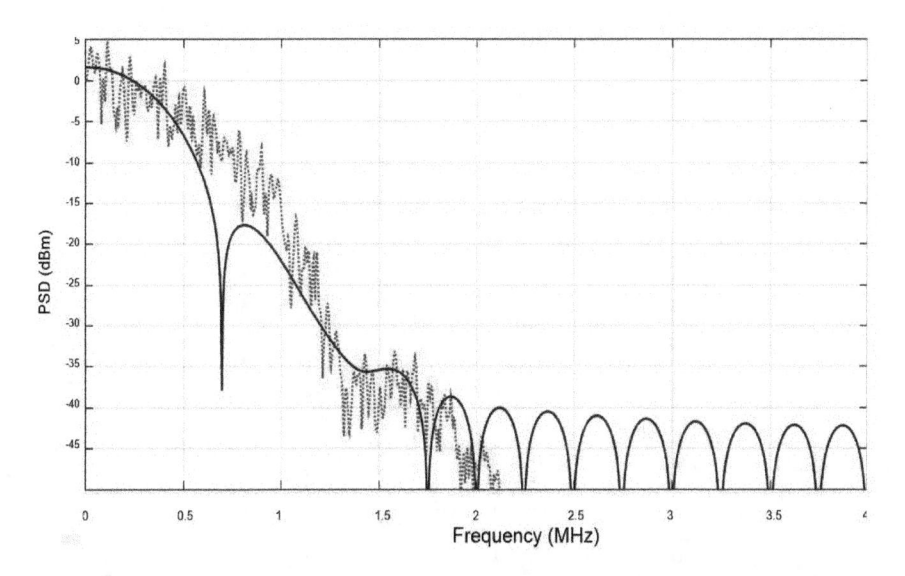

Figure 3.62 GMSK receive signal unfiltered (jagged) and filtered (smooth).

3.3.5.5 Note to the Careful Reader

The careful reader will notice that the CPM receiver in Section 3.3.3.2 attempts to use inherent CPM phase memory over three symbols and the GMSK receiver uses precoding to eliminate phase memory. If, as claimed in Section 3.3.3.2, using phase memory improves performance then can we receive GMSK using a trellis approach? The answer is yes. However, the GMSK phase transitions are much more complicated than the unit circle FSK phase transitions in Figure 3.44. The equations for calculating the GMSK reference phases are covered in [15] and [16].

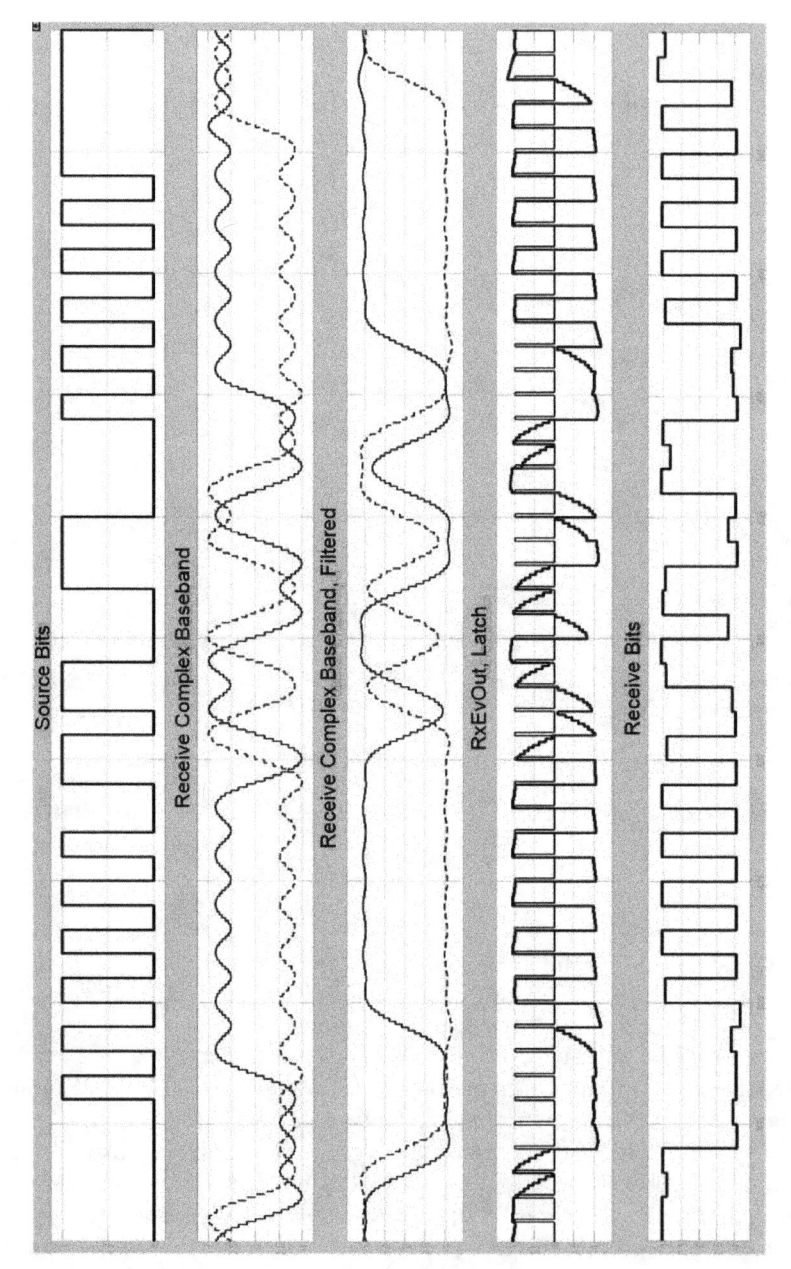

Figure 3.63 Examples of GMSK simple receiver signals.

CHAPTER 3. MODULATION TYPES

3.3.6 Pulse Position Modulation

Pulse position modulation (PPM) is a fairly simple modulation that has gained considerable importance because of its use in automatic dependent surveillance-broadcast (ADSB). ADSB is a system of short bursts of data that aircraft send to airport towers and each other on a regular repeating basis. An example of an ADSB signal is shown in Figure 3.64. When the receiver detects the preamble sync pattern, the data cells can be recorded at exactly 1 microsecond per cell. Demodulation is a simple matter of determining if the signal power is in the first half (for a one) or the second half (for a zero) of each cell. For these short messages and simple modulation, no symbol tracking is needed. Being an energy and position detection only, PPM is also naturally resistant to moderate carrier shift due to Doppler. This is an obvious advantage for aircraft in flight.

Chapter 11 of this book presents a complete PPM modem receiver design. ADSB is in a class of very simple and practical short message communication designs. For example, the FSK signal broadcast to the pocket pagers of the 1990s. A pager preamble provided a carrier frequency offset estimate and enabled the receiver to find the first sample of the first message symbol. Then the symbols were demodulated with no additional carrier and/or symbol tracking.

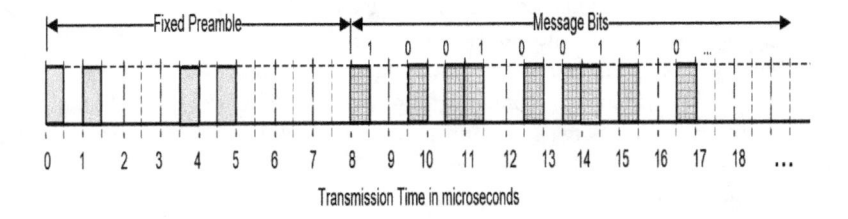

Figure 3.64 Aircraft communications ADSB pulse position modulation example.

3.4 DIRECT SEQUENCE WAVEFORMS

Direct sequence refers to the use of a pseudonoise feedback circuit to generate a clocked series of random bits called chips. An example is shown in Figure 3.65. We call this random series of chips pseudonoise because it looks like noise but it has a definite start, a known pattern, and repeats after a predetermined number of chips. In other words, a receiver with the correct knowledge can reproduce and synchronize to it. The uses of these chips will be described below. There is additional information in [17].

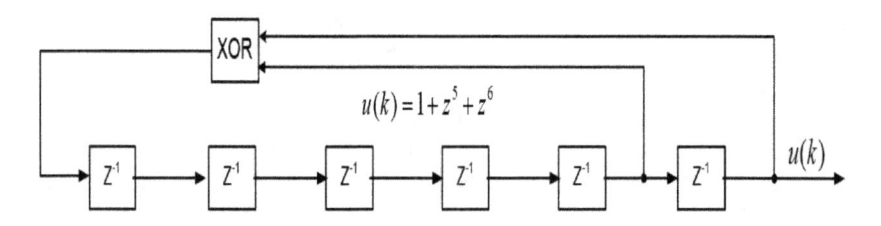

Figure 3.65 Direct sequence pseudonoise generator.

3.4.1 Spread Spectrum

Figure 3.66 is a simple example of spread spectrum (SS). The top trace is the input baseband data, at $F_{bit} = 1/T_{bit}$. The middle trace is a random series of chips at $F_{chip} = 1/T_{chip} = M(F_{bit})$, M is an integer greater than 1, in this case four. If these two streams are represented by ± 1, they can be multiplied. If they are represented by 0 and 1, they must be processed through exclusive or logic.

Multiplication in the time-domain is equivalent to convolution in the frequency-domain. The spreading signal bandwidth dominates the overall product bandwidth. The frequency domain graphs on the right of Figure 3.66 show the frequency expansion between the unspread bandwidth, main lobe between $\pm F_{bit}$, and the spread bandwidth, main lobe between $\pm F_{chip}$.

Figure 3.67 illustrates despreading when the intended receiver knows the spreading pattern and when an unintended receiver is ignorant of the spreading pattern. Thus, the transmitter can change the spreading pattern to select which receiver is able to pick up the message. This is known as code division multiple access (CDMA). The first digital cell phone base stations used this to separate users.

Figure 3.68 illustrates an early use of SS, jamming mitigation. The fundamental rule of SS is that whatever is in the receiver's spread bandwidth undergoes one of two transformations. If it was spread using the known SS random chip sequence, the receiver can recover the underlying data and the bandwidth is reduced back to $\pm F_{bit}$. If it was not spread using the known SS random chip sequence, or not spread at all, the receiver will expand the bandwidth to $\pm F_{chip}$.

Note how in Figure 3.69 the narrow solid black jammer signal in trace two is despread at the receiver; its bandwidth is increased by 4 between traces two and

CHAPTER 3. MODULATION TYPES

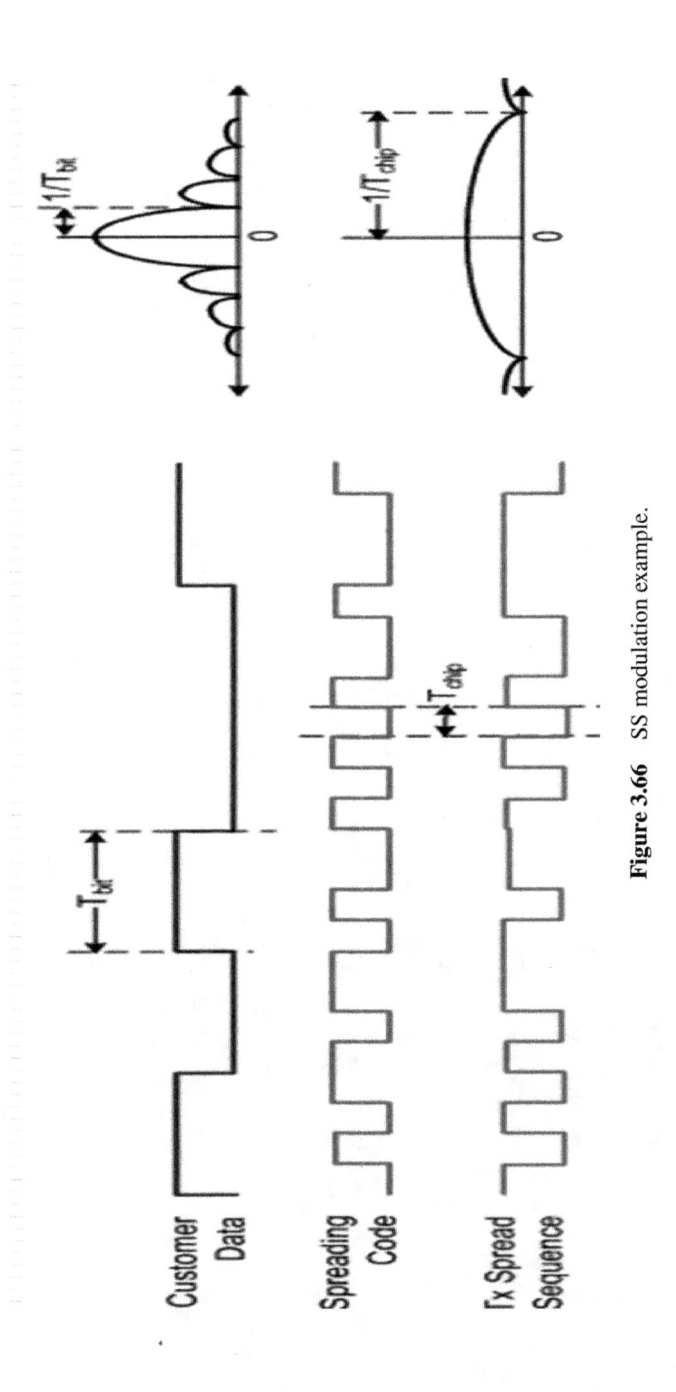

Figure 3.66 SS modulation example.

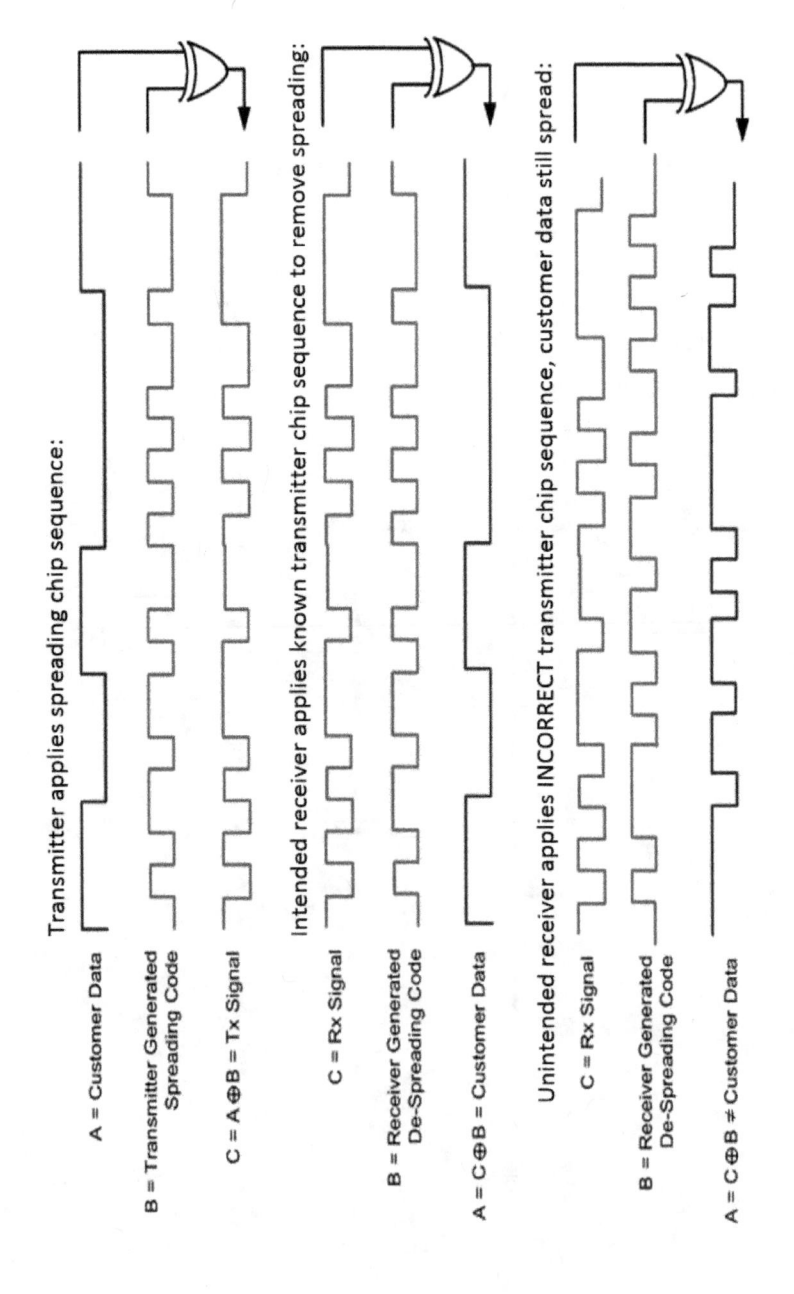

Figure 3.67 Basic idea behind code division mulitple access.

CHAPTER 3. MODULATION TYPES

three. However, when the intended signal is despread, its bandwidth is decreased by 4 between traces two and three. Because we only want the despread intended signal, the receiver applies a lowpass filter to eliminate the expanded part of the bandwidth. Figure 3.69, trace 4, shows how the jammer power after the filter is reduced considerably. For a jammer to be effective against SS signals, it must jam over a wider bandwidth than the source signal.

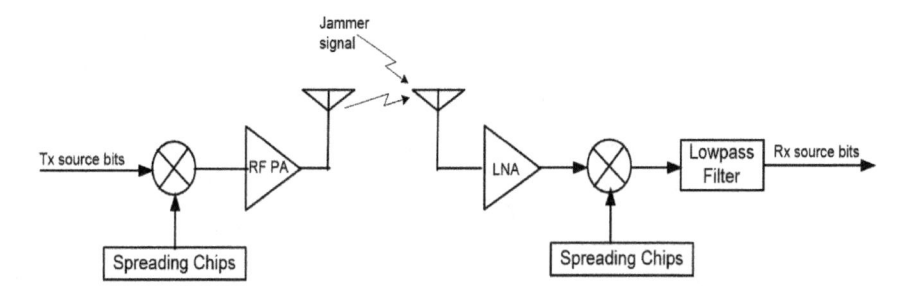

Figure 3.68 A circuit used to mitigate jamming.

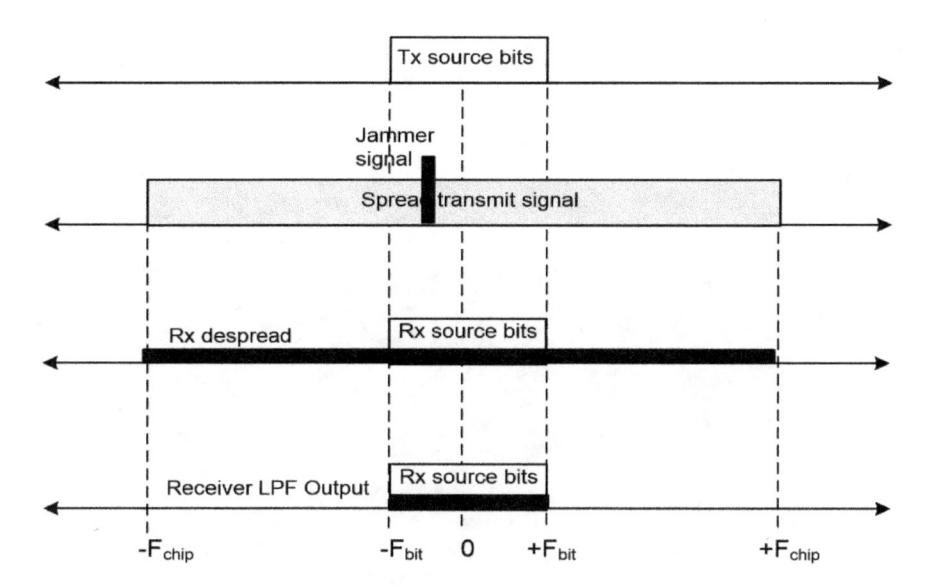

Figure 3.69 Basic idea behind despread processing gain.

Even without jamming, SS can result in a processing gain. Consider that instead of a jammer the received SS signal has broadband AWGN with total power between $\pm F_{chip}$ of P_{noise}. The noise density will be $N_0 = P_{noise}/(2F_{chip})$. However, after despreading, the total noise power is shown in Equation (3.32). This noise reduction is known as processing gain. This hints at another use of SS. If the desired signal is spread wide enough in frequency it can be "hidden" below the channel noise. At the intended receiver, most of the extra noise will be filtered out.

$$N_0 = \frac{P_{noise}}{2F_{chip}} = \text{Received Noise Power Density, Watts/Hz}$$

Noise after despread :
$$P'_{noise} = 2F_{bit} N_0 = \frac{2F_{bit} P_{noise}}{2F_{chip}} = \frac{P_{noise}}{4}$$

(3.32)

CHAPTER 3. MODULATION TYPES 125

3.5 QUESTIONS FOR DISCUSSION

1. There are many laboratory instruments that will display constellations like those in Figure 3.13. However, the constellation in Figure 3.36 is generally not displayed. Comment on the reasons for this.

2. For the same random data transmit symbols, list the following digital modulation types in order of lowest to highest peak to average transmit power: BPSK, MSK, $\pi/4$ DQPSK.

3. A linearly modulated, raised cosine filtered, digital communications signal with symbol rate T is received. The signal spectrum extends from 0 to $\pm 1/T$. The ADC sampling rate is $8/T$. Practically speaking, how many decibels of processing gain can be achieved?

4. Comment on a factor in Equation (3.28) that is not shown implemented in Figure 3.38. What is this factor for? What is a reason this factor might have been left out? What might be a simpler way to accomplish the same function?

5. Review the advantages of using raised cosine filtering for single channel data communications. Why is RCI filtering typically not used for the multiple PSK channels transmitted in OFDM schemes like IEEE802.11a? (Hint: consider the symbol rate of the individual OFDM channels verses the symbol rate required for an equivalent single channel PSK signal.)

6. An M-FSK constant envelope waveform generally has higher bandwidth efficiency than an M-PSK amplitude modulated signal (M = number of bits/symbol). True or false?

7. A raised cosine filtered digital communication signal must have -3 dB power at the carrier frequency:

a. +/- the symbol rate;
b. +/- the symbol rate divided by 2;
c. +/- two times the symbol rate;
d. None of the above.

8. The two constellations in Figure 3.70 each depict four state PSK signaling. The two systems are the same except for the symbol representation shown. Which constellation has greater average power? Which system is more appropriate for a limited dynamic range power amplifier? Why?

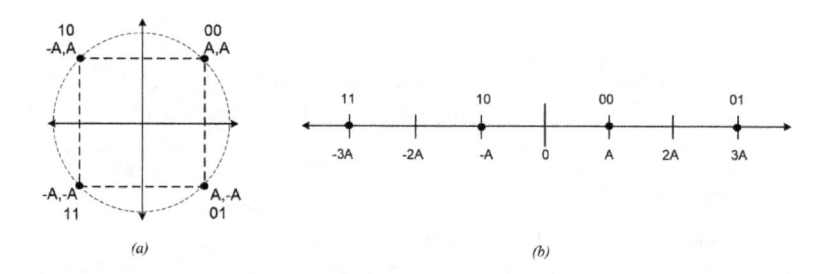

Figure 3.70 Comparison of QPSK and 4-PAM constellations.

REFERENCES

[1] E. McCune. *Practical Digital Wireless Signals*. Cambridge University Press, 2010.

[2] J.G. Proakis. *Communication System Engineering*. Prentice-Hall, 2002.

[3] J.W. Mark and W. Zhuang. *Wireless Communications and Networking*. Pearson Education, 2003.

[4] R.G. Lyons. *Understanding Digital Signal Processing*. Prentice Hall, 2011.

[5] F. Pancaldi. "Single-Carrier Frequency Domain Equalization". *IEEE Signal Processing Magazine*, 2008.

[6] E. McCune. "This Emperor has no Clothes?". *IEEE Microwave Magazine*, 14(4), 2013.

[7] M. Rumney. "3GPP LTE, Introducing Single-Carrier FDMA". *Agilent Measurement Journal*, 2013.

[8] J. C. Whitaker. *SBE Broadcast Engineering Handbook*. McGraw Hill, 2016.

[9] B. Sklar. *Digital Communications*. Prentice Hall, 1988.

[10] D.P. Nagpal. *Computer Fundamentals*. S. Chand Limited, 2008.

[11] D. Zwillinger. *CRC Standard Mathematical Tables and Formulae*. Chapman and Hall, 2003.

[12] J.B. Anderson and C.E. Sundberg. *Digital Phase Modulation*. Plenum Press, 1986.

[13] G.L. Lui and K. Tsai. "Gaussian Minimum Shift Keying (GMSK) Precoding Communication Method", U.S. Patent 7072414, 2006.

[14] P. Jung. "Laurent's Representation of Binary Digital Continuous Phase Modulated Signals with Modulation Index 1/2 Revisited". *IEEE Transactions on Communications*, 42(2-4), 1994.

[15] A. Linz and A. Hendrickson. "Efficient Implementation of an I-Q GMSK Modulator". *IEEE Transactions on Circuts and Systems-II: Analog and Digital Signal Processing*, 43(1), 1996.

[16] K. Murota and K. Hirade. "GMSK Modulation for Digital Mobile Radio Telephony". *IEEE Transactions on Communications*, 29(7), 1981.

[17] A.J. Viterbi. "Spread Spectrum Communications - Myths and Realities". *IEEE Communications Magazine*, 1979.

Chapter 4

RF Channels

4.1 INTRODUCTION

This chapter will hopefully be useful to the SDR hobbyist who wants a good review of RF propagation. There are a lot more visual examples than tedious math. There is also discussion of real-world statistical propagation studies used to plan, for example, broadcasting facilities. The working SDR engineer will probably already know most of this material or will have access to someone who does.

We start with the fixed propagation situation, describing fairly simple cases. Then we add moving elements, transmitter or receiver, or both. Finally, we look at techniques that have been invented to correct the hazards that the signal can encounter in either situation.

4.2 RF WAVE BASICS

In Figure 4.1, c = 300 million meters per second is the speed of light. λ is the wavelength in meters and, the frequency is:

$$f = \frac{c}{\lambda} = \frac{\text{cycles}}{\text{second}} \qquad (4.1)$$

In the figure, solid lines are the electric field and dashed lines are the magnetic field. This wave is said to be vertically polarized because the electric field is vertical to the Earth. This is also a transverse wave because the fields are at right angles to

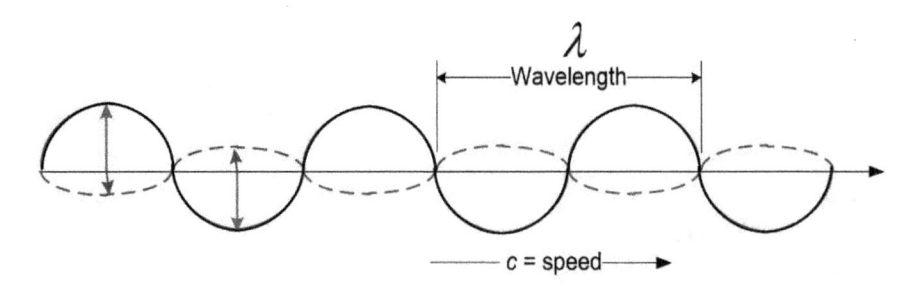

Figure 4.1 Structure of the basic transverse wave.

the direction of propagation.[1] Transverse wave propagation is sometimes referred to as transverse electric magnetic (TEM) mode.

If the electric and magnetic fields are moving perpendicular to the direction of travel and to each other, how does the wave move forward? The answer is the Poynting vector $S = E \times H$ (Figure 4.2). This is a vector cross-product that results in directional power S moving forward perpendicular to both E and H vectors. The direction of S can be verified using the right-hand rule for vector cross-products.[2]

As mentioned, the wave of Figure 4.1 is vertically polarized. A horizontally polarized wave will have an electric field parallel to the Earth. There are also two types of circular polarization. These are right-hand circular (RHC), the electric field rotates clockwise once per wavelength, and left-hand circular (LHC), the electric field rotates counterclockwise once per wavelength.

4.2.1 Polarized Antennas

For maximum power transfer through the receive antenna, the antenna polarization must match the wave polarization. For example, a vertically polarized wave, such as Figure 4.1, works best with a vertical antenna, such as Figure 4.3.

4.2.1.1 Halfwave Dipole

Figure 4.3 is simply a half-wavelength long antenna made from a pair of straight conductors such as a wire or tube. The conductors are colinear and the antenna is fed

---

[1] Another common type of wave is called longitudinal; an example is sound waves where the medium (air) moves parallel (instead of transverse) to the direction of propagation.
[2] Right-hand rule: Index finger in the electric field direction, middle finger in the magnetic field direction, and then the thumb shows resultant vector direction.

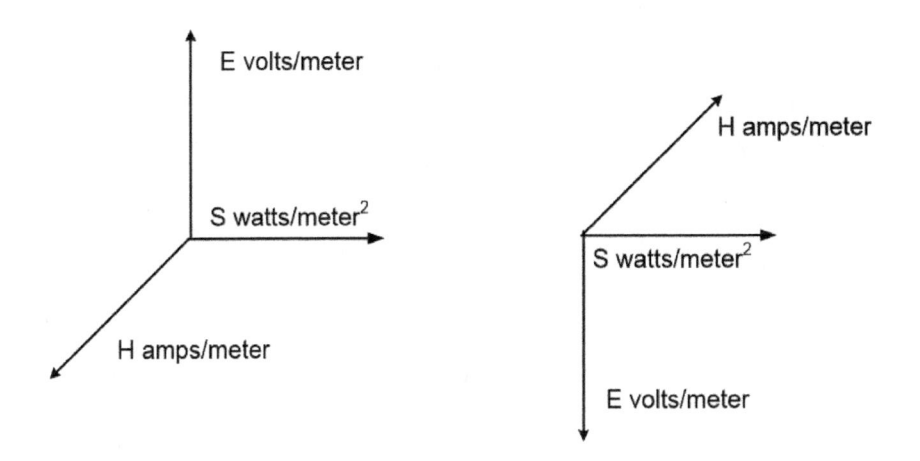

Figure 4.2 The Poynting vector construction for transverse waves.

in the middle by a balanced signal. The half-wave dipole antenna input impedance is about 73 ohms, with very little reactive component. If the antenna were shortened or lengthened, then this impedance would become reactive at the original intended frequency.

For a transmit half-wave dipole, the radiation pattern in the view on the right side of Figure 4.3, is toroidal shaped with maximum power perpendicular to the center. Due to the reciprocal nature of antennas, the receive version of the same antenna has a sensitivity pattern with the same shape.

4.2.1.2 Monopole

Figure 4.4 is a quarter-wavelength long antenna. This is similar to the (apparently) old-fashioned car radio antenna, where the antenna is a stiff wire of about 80 cm (quarter-wave in the FM band) and the ground plane is formed by the car body. For a well-constructed ground plane, there is effectively another quarter wave image below the ground plane. Thus, the antenna acts like a half-wave dipole with propagation on only one side.

When the antenna is shorter than the intended quarter-wavelength, these antennas can be formed into a helix. Now the termination impedance is inductive due to the helix and capacitive due to the shortening. Thus, the entire antenna can be made to resonate at the intended frequency. When the helix is covered with thick

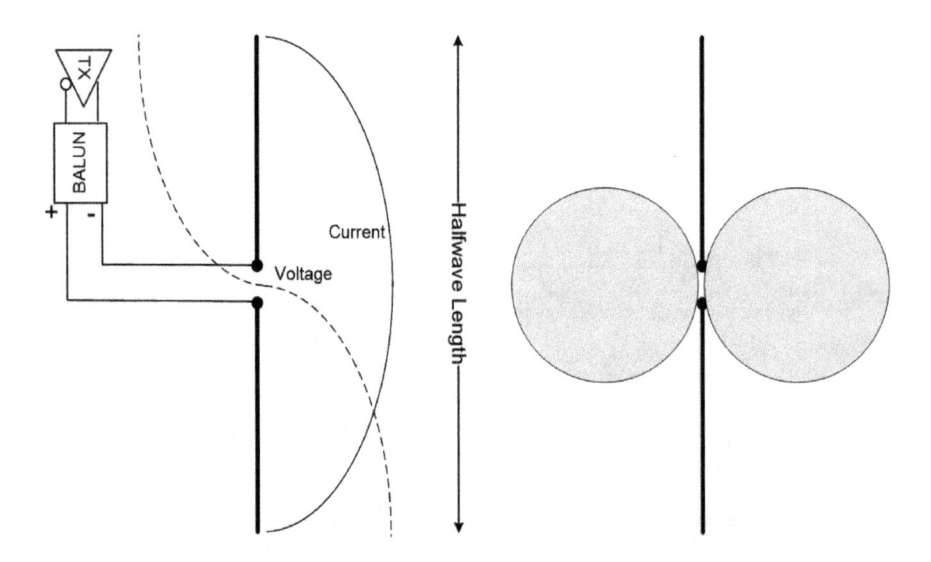

132   *Software Defined Radio: Theory and Practice*

Figure 4.3 A half-wave dipole vertical antenna.

rubber, it becomes springy and able to withstand bending. These antennas offer convenience and durability at the cost of slightly reduced performance.

4.2.1.3 Circular Polarization

A signal transmitted from a vertically polarized transmit antenna (i.e., a vertical antenna, Figure 4.3) cannot be picked up very well by a horizontally polarized receive antenna (i.e., parallel to the Earth instead of perpendicular to the Earth) and vice versa.[3] Circular polarization solves this problem and is commonly used. Signals transmitted with left hand circular polarized (LHCP) polarizations or right hand (RHCP) polarizations can be received by a horizontal or vertical polarization antenna with no concern for the actual receive antenna position. This is true even if, as sometimes happens, reflections change the LHCP signal into an RHCP signal, or vice versa.

The left side of Figure 4.5 is a circular polarized transmit antenna constructed from two half-wave dipoles. The transmitter and 90° power splitter outputs are

---

3 Practically speaking, reflection and diffraction can change polarization so a signal transmitted with, for example, vertical polarization, may have a random polarization when received.

CHAPTER 4. RF CHANNELS

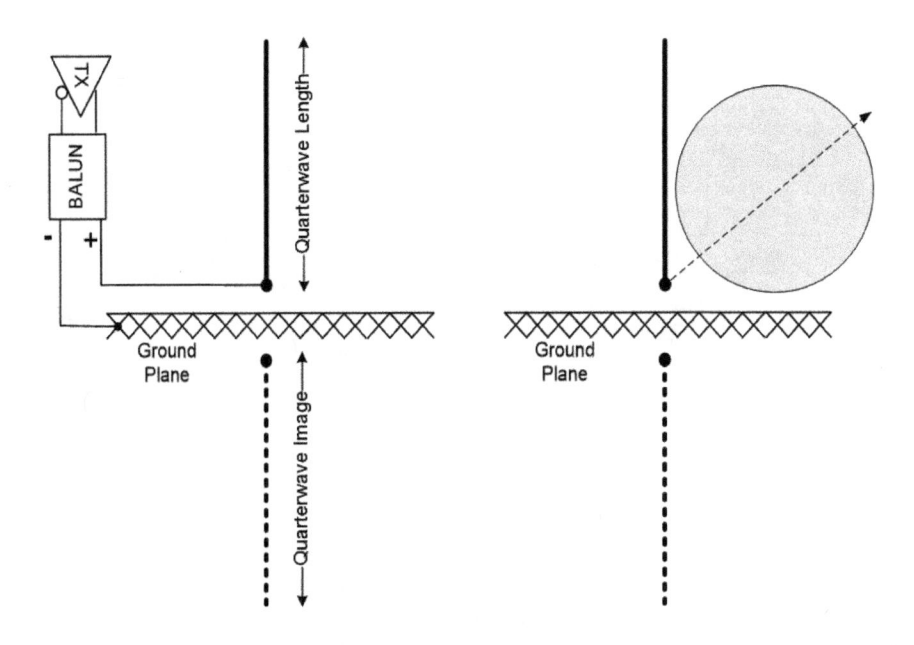

Figure 4.4 A quarter-wave dipole antenna.

unbalanced coax cable. The BALUN (balanced to unbalanced, although here we have unbalanced to balanced) converts both splitter outputs to a pair of 90° apart balanced signals. One of these is phase-shifted 90° making it orthogonal to the other. Thus, a horizontally polarized receive antenna will line up with the X axis, with no contribution from the Y axis. Likewise, a vertically polarized receive antenna will line up with the Y transmit axis with no contribution from the X axis. A receive antenna with a 45° tilt (slant polarization) will receive the same signal power from both X and Y axis. The composite receive power will be the sum of the squares of both contributions.

The right side of Figure 4.5 is a circular polarized antenna constructed of copper patches on a microwave substrate material. The substrate acts like a dielectric (a dielectric is a nonconductor that can support an electric field). The reverse side conductor (e.g., copper) acts like a ground plane, forcing the radiation outward. The BALUN is implemented on the reverse side. This antenna is about 3 inches square.

When selecting or designing these antennas and components such as power splitter and BALUN, bandwidth and center frequency are important. For example, a

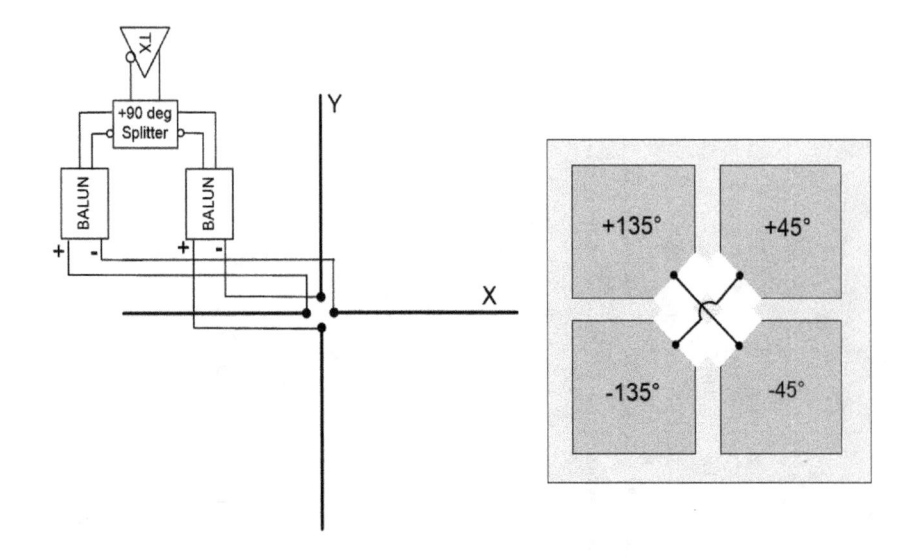

Figure 4.5 A circular polarization antenna.

90° splitter is only accurate over a specified frequency range, often only 10% of the carrier frequency. Another important factor is power handling capability.

4.2.1.4 Dual Slant Polarization

Figure 4.6 shows a type of circular polarization called dual slant polarization. Due to the slants, a strictly horizontal or vertical polarized received signal will incur a loss at either antenna. The diversity combining circuit will cophase and sum these two signals to make up this loss. Like circular polarization, this antenna system can work with any receive polarization. This dual slant system also provides some spatial diversity; see Section 4.4.1.

4.2.2 RF Spectrum Regions

Table 4.1 is a list of commonly used RF bands. Off-the-shelf SDRs are typically designed to work over VHF (very high frequency), UHF (ultrahigh frequency), L and/or S bands. A reason for this may be that the AD9361, AD9363, and AD9364 transceiver parts, upon which many SDRs are based, have a tuning range of 70 MHz

CHAPTER 4. RF CHANNELS

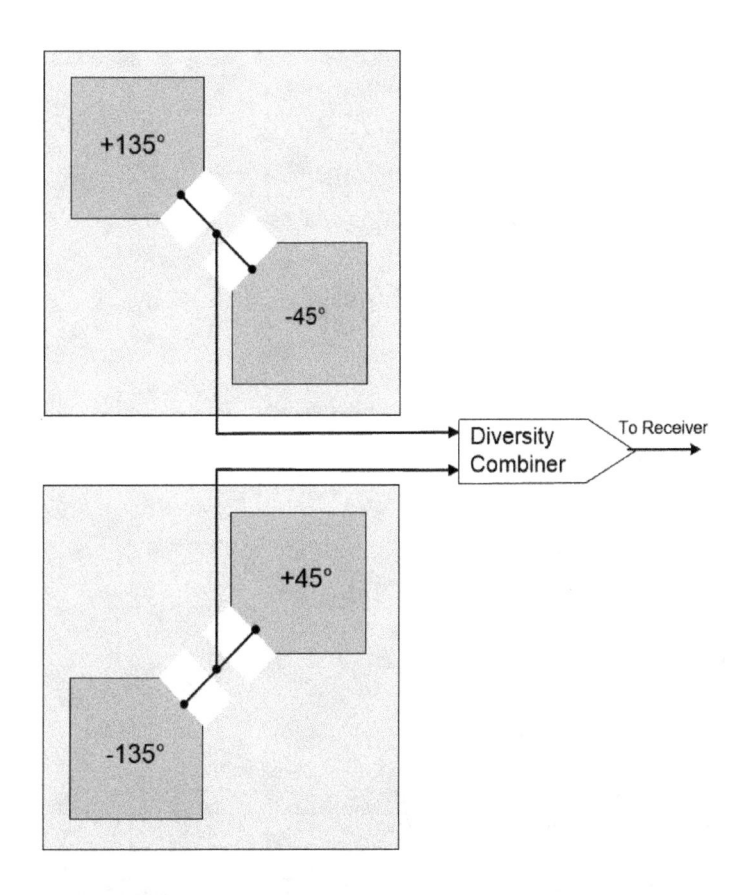

Figure 4.6 A dual slant polarization antenna system.

Table 4.1 Examples of the IEEE Electromagnetic Spectrum

Band	Frequency Range	Wavelength	Typical Use
MF	300 kHz to 3 MHz	1 km to 100m	AM, amateur radio
HF	3 to 30 MHz	100m to 10m	AM, long-range comm.
VHF	30 to 300 MHz	10m to 1m	TV and FM, weather radio
UHF	300 MHz to 1 GHz	1m to 0.3m	TV, cell phones, ADSB
L	1 to 2 GHz	30 cm to 15 cm	Phones, air traffic control
S	2 to 4 GHz	15 cm to 7.5 cm	Radar, microwave ovens
C	4 to 8 GHz	7.5 cm to 3.75 cm	Radar

to 6 GHz. The HF (high frequency), where there is a lot of amateur radio activity, is covered by fewer readily available SDRs.

4.3 RF PROPAGATION

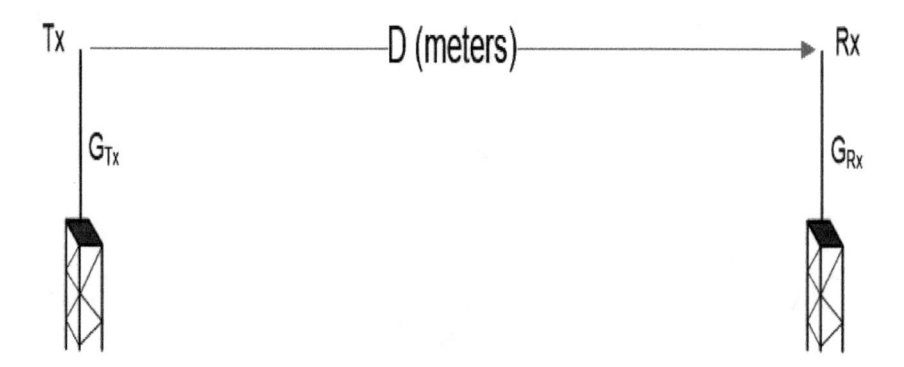

Figure 4.7 A simple RF propagation setup.

Table 4.2 RF Propagation Parameters

P_{Tx}	Transmit power (watts)
P_{Rx}	Received power (watts)
G_{Tx}	Transmit antenna gain with respect to isotropic gain
G_{Rx}	Receive antenna gain
n	2 for line-of-sight (LOS), no obstructions
λ	Signal wavelength (meters)
c	Speed of light (300 million meters/second)

Figure 4.7 is called large-scale line-of-sight propagation and has a transmit to receive signal loss (also called fading). The Friis equation, shown as Equation (4.2), gives us the receive power due to this loss. Note that this power is not in dB. Note also the exponent n of D^n. For line-of-sight, $n = 2$. For more complicated propagation paths, such as inside a building, $n > 2$. See [1] for a comprehensive list and discussion.

CHAPTER 4. RF CHANNELS

$$[!htbp] P_{Rx} = P_{Tx} G_{Tx} G_{Rx} \left(\frac{\lambda^2}{16\pi^2 D^n} \right) = P_{Tx} G_{Tx} G_{Rx} \left(\frac{c^2}{16\pi^2 f_{carrier}^2 D^n} \right)$$
(4.2)

In addition to the dependence on distance D of Equation (4.2), there is also a frequency scaling. Figure 4.8 shows where this comes from. The transmitter power spreads out like light from a light bulb. The amount of power in any fixed area decreases as D^2. This is what is meant by power loss although it is not actually a loss and it is not dependent on frequency. The receiver antenna aperture (effective area = A_{Rx}) is shown on the right side. This aperture gets smaller as frequency gets higher. As the aperture gets smaller, it collects a smaller amount of the spread transmit power. This creates the frequency dependence in the Friis equation.

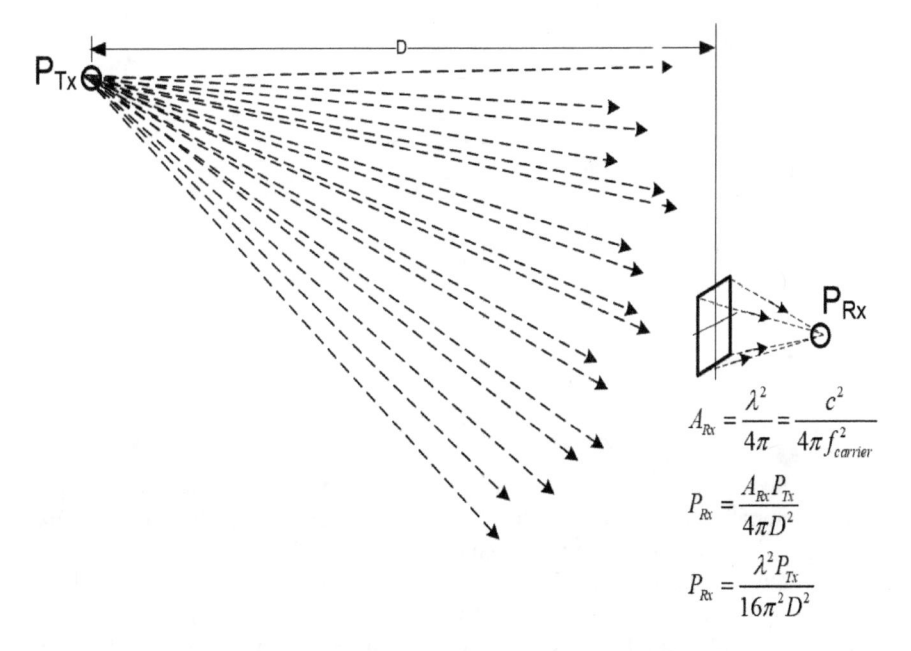

$$A_{Rx} = \frac{\lambda^2}{4\pi} = \frac{c^2}{4\pi f_{carrier}^2}$$

$$P_{Rx} = \frac{A_{Rx} P_{Tx}}{4\pi D^2}$$

$$P_{Rx} = \frac{\lambda^2 P_{Tx}}{16\pi^2 D^2}$$

Figure 4.8 Derivation of the Friss propagation loss equation.

4.3.1 Fixed Propagation Environment

Here we are concerned with a propagation environment where nothing is moving. The transmitter, environment, and receiver are all fixed. We will discuss the effect of these large-scale path loss mechanisms, starting with just one transmit wave and one reflection.

4.3.1.1 Specular Reflection

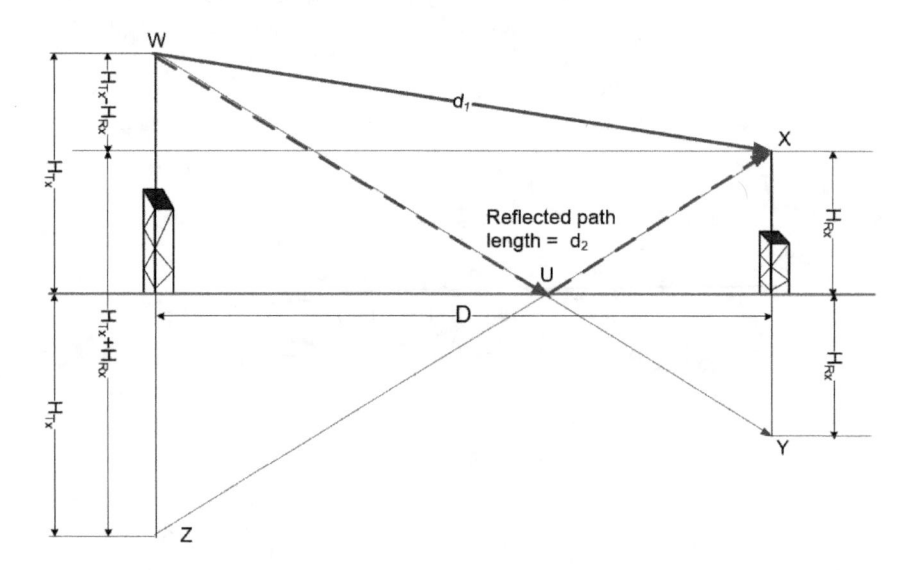

Figure 4.9 A two-ray propagation model.

Specular reflection occurs when waves hit an object whose dimensions are large compared to the wavelength. This results in a constant reflection angle, like a mirror reflects an image. Depending on the distances, the effect of reflection can be constructive or destructive. Figure 4.9 is an example where the receiver combines two versions of the transmit signal, one direct and one reflected. To calculate the received signal strength, we need to know the difference in the path lengths. Path 1 is the length of $\overline{WX}$ and path 2 is the length of $\overline{WUX}$.[4] The difference is calculated

---
4 Note that $\overline{WUX}$ has the same length as $\overline{ZUX}$.

CHAPTER 4. RF CHANNELS

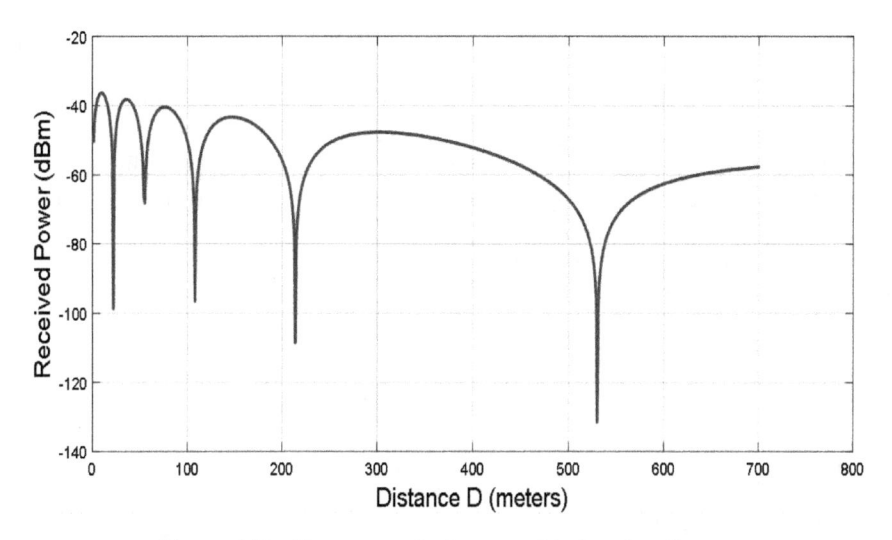

Figure 4.10 Two-ray received power with changing distance.

in Equation (4.3) and the received power due to the composite signal is calculated in Equation (4.4).

$$\Delta_d = d_2 - d_1 = \sqrt{(H_{Tx} + H_{Rx})^2 + D^2} - \sqrt{(H_{Tx} - H_{Rx})^2 + D^2} \quad (4.3)$$

$$P_{Rx} = 4P_{Tx}G_{Tx}G_{Rx} \left(\frac{\lambda^2}{16\pi^2 D^2}\right) \left(\sin\left(\frac{\pi(d_2 - d_1)}{\lambda}\right)\right)^2 \quad (4.4)$$

As distance D changes, the phase difference of the two signals received at X in Figure 4.9 changes as $\Delta\phi = 2\pi(d_1 - d_2)/\lambda$ (notice the dependence on wavelength). This combination results in a phase cancelation that depends on distance. The composite receive power variation with distance is shown in Figure 4.10. The nulls happen when the difference between the two signals is an integer multiple of the wavelength. Notice that, in addition to the nulls, there is an overall average power loss with distance.

4.3.1.2 Diffuse Reflection (Scattering)

When waves impinge on a rough surface, with roughness dimensions less than wavelength, reflected energy is diffused in random directions. Rough concrete walls, leafy trees, and even some street signs can produce this effect. The additional propagation effects due to scattering can be modeled by increasing exponent n in Table 4.2. For an unobstructed line-of-sight, $n = 2$, however, n can go as high as 6 for transmission inside a building with obstructions [1].

4.3.1.3 Refraction

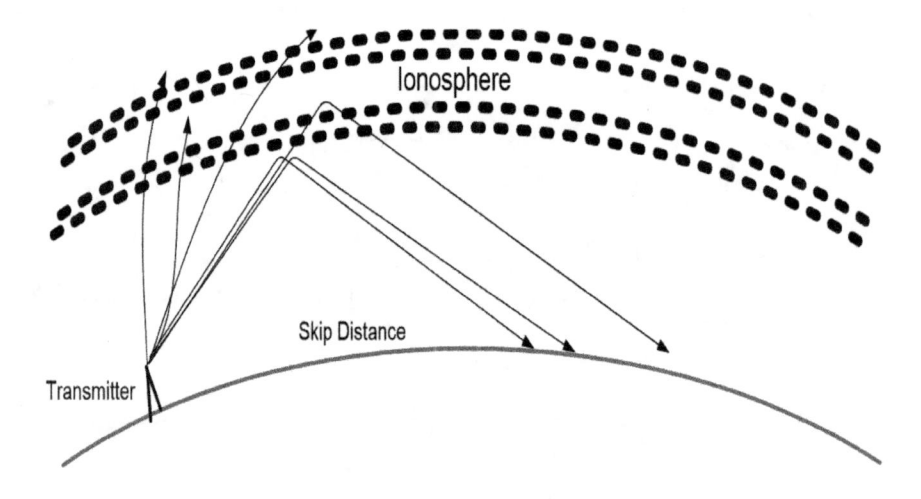

Figure 4.11 Radio waves refracted by ionosphere.

The ionosphere is a layer of the atmosphere with a high concentration of ions (atoms with a net charge). This layer can refract (bend) radio waves and send them back toward Earth. The ionosphere extends from about 50 to 600 miles above the surface of the Earth. Figure 4.11 shows how signals leaving the transmit antenna at angles close to vertical travel right through the ionosphere, whereas signals transmitted at more shallow angles bounce back to Earth after a skip distance. Receive polarization of bounce signals is unpredictable. Here are a couple of examples where refraction from the ionosphere was turned into an advantage.

A famous use of ionospheric skip is the communication with Admiral Richard Byrd in 1929, from the Antarctic base Little America. The transmission was in

CHAPTER 4. RF CHANNELS

the HF band (3 - 30 MHz) and covered about 9,000 miles. At various times, relay stations in Buenos Aires and Hawaii were used. The public could hear Admiral Byrd on CBS radio thanks to this important mode of communications, and the engineers who figured out how to use it.

During the Vietnam War, Near Vertical Incidence Skywave (NVIS) used ionospheric refraction to communicate several hundred miles across dense jungles where line-of-sight communications was impossible. Frequencies between 1.8 MHz and 8 MHz were transmitted almost (but not quite) straight up to the ionosphere, where they were refracted back into a circular region.

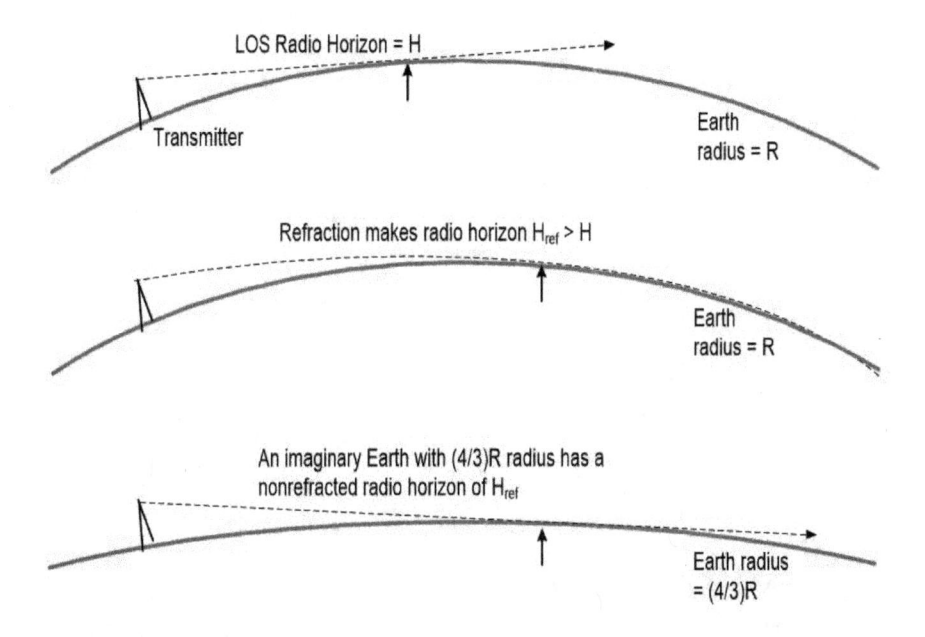

Figure 4.12 Radio horizon increase due to refraction.

Another interesting effect of refraction is radio horizon increase. As shown in the top part of Figure 4.12, the line-of-sight (LOS) radio horizon is the point on the Earth where a direct ray becomes tangent to the Earth (the visual horizon is similar but not identical). The middle of the figure shows how refraction bends the radio wave so that it tends to follow the Earth, effectively increasing the radio horizon. The bottom picture shows that a fictional Earth with a radius four-thirds times that of the real Earth will increase the radio horizon to make it equal to the refracted radio horizon. Let's try an example. For an observer at a height of 2m, and Earth radius of 6,371m, the optical and radio line-of-sights are calculated in Equation (4.5).

$$d_{optical} = \sqrt{2hR_{Earth}} = \sqrt{2(2)(6.371)} = 5000\,meters$$

$$d_{radio} = \sqrt{2h(4/3)R_{Earth}} = \sqrt{2(2)(4/3)(6.371)} = 5829\,meters$$

(4.5)

4.3.1.4 Diffraction

Diffraction means an incomplete blocking of the wavefront. In this discussion, the concept of knife-edge diffraction and diffraction are considered the same thing. Figure 4.13 shows how the power level changes gradually around the knife-edge on the left side of the solid object. Figure 4.14 shows how these large-scale effects can sometimes work together in unexpected ways. Here we have the transmit signal undergoing diffraction around building 1 and creating a large shadow zone on the right side. As the wave moves away from building 1, it spreads out and reaches the receiver by reflecting from the left side of building 2.

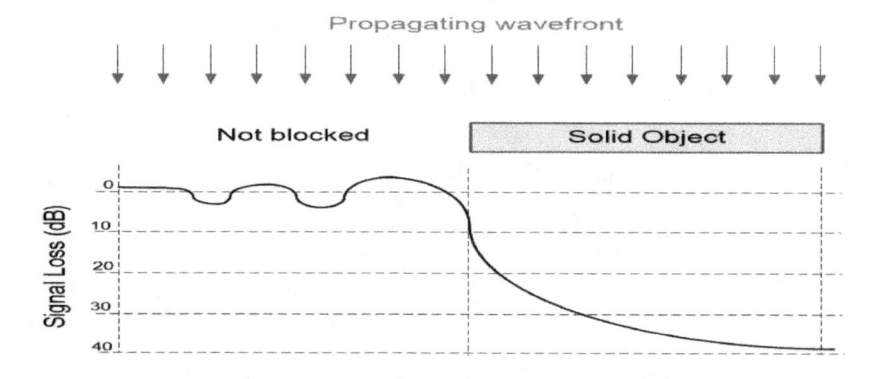

Figure 4.13 Example of knife-edge diffraction loss.

In Figure 4.7 we assume optical and radio line-of-sight, thus, we can use Equation (4.2) to calculate the amount of flat fading. In general, how do we know there is no diffraction between the transmit and receive antennas? Diffraction is predicted by Fresnel zone radius.[5]

---

5 Augustin-Jean Fresnel (1788-1827) was a French physicist who studied optics and invented a famous lighthouse lens.

CHAPTER 4. RF CHANNELS

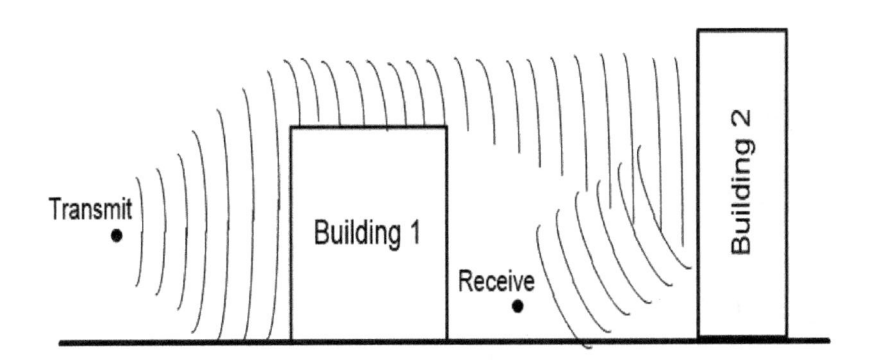

Figure 4.14 Example of knife-edge diffraction.

Figure 4.15 is a more real-world picture of Figure 4.7. There are two separate first Fresnel zones shown,[6] one for a higher frequency f_{high} and the second for a lower frequency f_{low}. Each Fresnel zone has the two antennas as the foci of an ellipsoid. Equation (4.6) is the radius of the first Fresnel zone for a distance A from the transmit antenna and distance D between antennas. The carrier signal wavelength is λ.

$$R_1 = \sqrt{\frac{\lambda A (D - A)}{D}} \qquad (4.6)$$

The f_{high} Fresnel zone has both optical and radio line-of-sight whereas the f_{low} Fresnel zone has optical but not radio line-of-sight due to knife-edge diffraction around the tall building in the center. In practice, up to half of the first Fresnel zone can be obstructed before the link is considered not line-of-sight for radio. To summarize our observations, the best way to design a point-to-point RF link is to use high frequencies and tall antenna towers.

Circular polarized signals tend to handle diffraction and reflection better than fixed (e.g., horizontal or vertical) polarized signals. This is because reflections of a fixed polarized signal can interfere with the direct signal and cause phase cancelation. The same reflections can occur with circular polarization; however, the constantly rotating polarization tends to reduce this phase cancelation.

---

6 There are actually a series of Fresnel zones, each Fresnel zone has a maximum reflected path phase change of $n\lambda/2$. The first Fresnel zone is of the most practical importance.

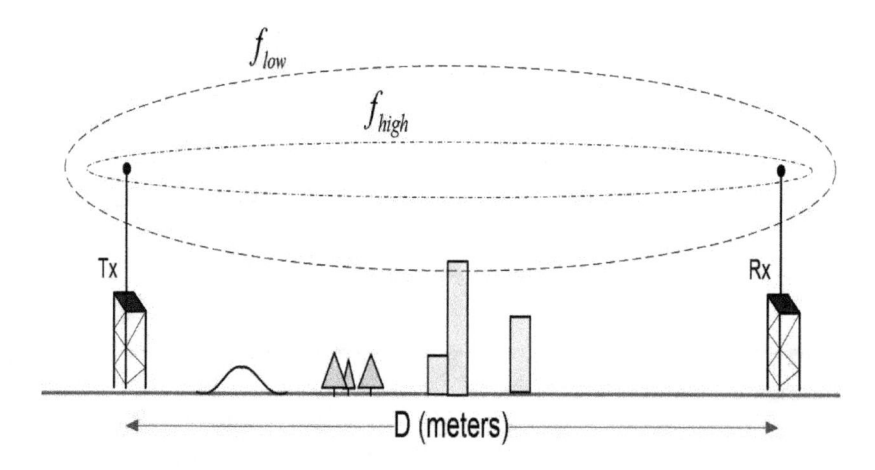

Figure 4.15 Fresnel zone one showing potential for knife-edge diffraction.

An example of a fixed polarization that does work well is horizontally polarized television transmit and receive antennas. The TV transmit antenna height often exceeds two hundred feet. This signal is transmitted down to horizontally polarized TV receive antennas on rooftops.

4.3.1.5 Absorption

Absorption is a function of the carrier frequency. Interaction between radio waves and oxygen molecules or water can make radio propagation difficult at certain absorption peaks (e.g., 60 GHz); see Figure 4.16. This absorption is the cause of rain fade, a loss that is sometimes built into link budgets. Interestingly, fading due to snow is less than fading due to rain [2].

4.3.2 Multipath in a Fixed Environment

The previous section considered large-scale effects on one receive signal in a fixed environment. In this section, we still have a fixed environment; however, we consider what happens when multiple reflected copies of the signal combine at the receiver. Imagine a cluttered office environment after all the workers have left and nothing is moving. The Wi-Fi signals form a fixed multipath pattern as they spread out and reflect from various objects on their way to each desk computer. Typically, one

CHAPTER 4. RF CHANNELS

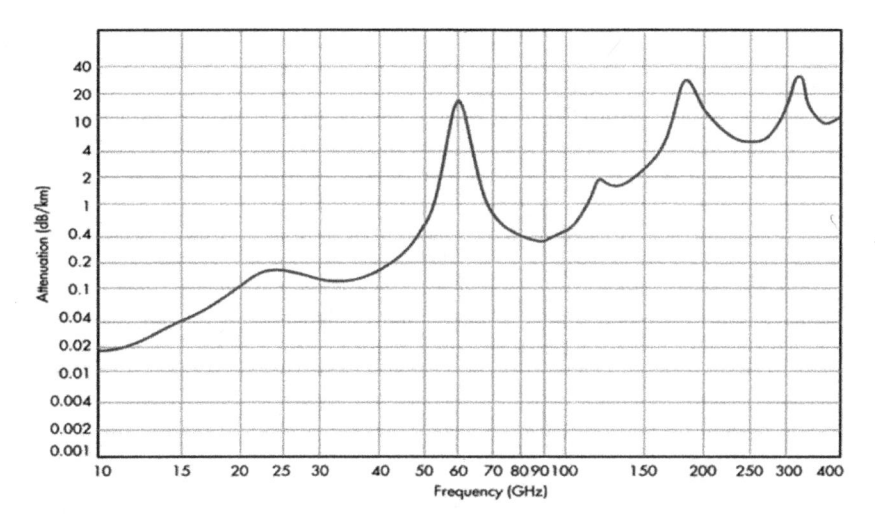

Figure 4.16 Radio absorption versus frequency.

impulse transmitted at t_0 arrives at the receiver over one direct path, delayed by t_1 and several reflected paths at various delays. As shown in Figure 4.17, the paths also have different amounts of attenuation. This is called a fixed delay spread profile. Equation (4.7) shows how the RMS delay spread, σ_t, is calculated.[7]

The most fundamental characteristic of our fixed multipath channel is called RMS delay spread. This describes the time-dispersive nature of the channel at a given stationary position. A related parameter is coherence bandwidth, B_{coh}. This describes the extent of channel flat frequency response. B_{coh} is range of frequencies over which two receive components are affected the same by the channel (i.e., coherently). We say a finite B_{coh} makes the channel frequency selective. Equation (4.8) shows an estimate of coherence bandwidth [1].

B_{coh} varies inversely with RMS delay spread in a similar way to how a low-pass filter bandwidth varies inversely with impulse response length (i.e., longer delay spread results in smaller bandwidth). In Figure 4.17, RMS delay spread is from a stationary set of time-invariant measurements; thus, B_{coh} is time-invariant in this example.

---

7   RMS is defined as the square root of the average of squared values.

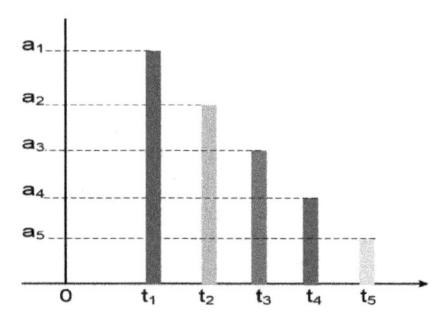

Figure 4.17 A fixed delay spread profile.

$$t_{avg} = \frac{\sum_k a_k^2 t_k}{\sum_k a_k^2} = \text{average excess delay}$$

$$[t^2]_{avg} = \frac{\sum_k a_k^2 t_k^2}{\sum_k a_k^2} = \text{second moment} \qquad (4.7)$$

$$\sigma_t = \sqrt{[t^2]_{avg} - (t_{avg})^2}$$

$$\frac{1}{50\sigma_t} \leq B_{coh} \leq \frac{1}{5\sigma_t} \qquad (4.8)$$

The top trace of Figure 4.18 is an example where the delay spread is smaller than the symbol period. The signal spectrum easily fits in the channel coherence bandwidth and the intersymbol interference (ISI) is manageable. This is called flat fading. The lower trace shows the same delay spread with a faster symbol rate. The signal bandwidth is starting to exceed the channel coherence bandwidth, and the ISI is becoming more of a problem. This is referred to as frequency-selective fading because the signal is subjected to bandwidth restrictions. Also note that this is still a fixed environment, so the terms fast or slow fading do not apply.

4.3.3 Moving Propagation Environment

Now we are going to let some combination of the receiver, transmitter, or the objects in the propagation path move.

CHAPTER 4. RF CHANNELS

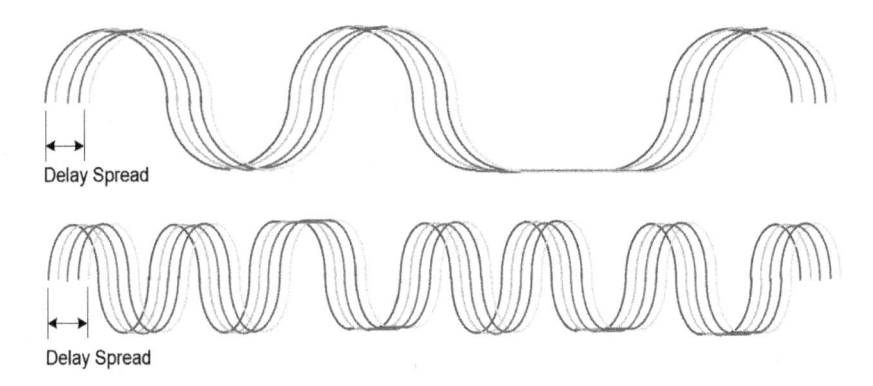

Figure 4.18 A fixed delay spread ISI effect on two different symbol rates.

4.3.4 Multipath in a Constant Velocity Moving Environment

As soon as anything in the propagation path starts to move, we have to account for Doppler shift. Also, since we have a nonfixed environment, delay spread profile is changing and we now have a fading rate. Let's start by defining Doppler frequency shift.

Figure 4.19 has RF carrier signal $s(t)$ received at two different points, A and B. Expression $(\Delta L)/\lambda$ is simply the number of carrier signal wavelengths (i.e., cycles) contained in ΔL. The factor 2π radians per cycle expresses Equation (4.9) in radians. Because frequency is phase advance per unit time, for a constant velocity a constant Doppler shift results in Equation (4.10).

$$\Delta\phi = \frac{2\pi (\Delta L)}{\lambda} = \frac{2\pi (v\Delta t) \cos(\theta)}{\lambda} \qquad (4.9)$$

$$f_d = \frac{1}{2\pi}\frac{\Delta\phi}{\Delta t} = \left(\frac{2\pi (v\Delta t) \cos(\theta)}{\lambda}\right)\left(\frac{1}{2\pi\Delta t}\right) = \frac{v \cos(\theta)}{\lambda} \qquad (4.10)$$

Figure 4.20 shows two reflectors in addition to the direct path. Each reflected signal has a different δL and θ and results in a different Doppler frequency shift at the car; see Equation (4.11).

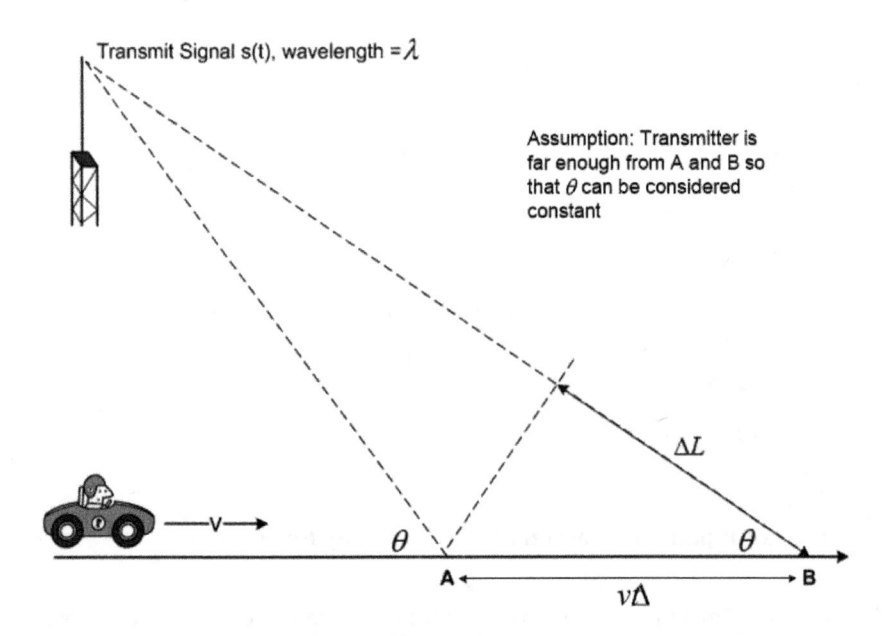

Figure 4.19 An RF environment for deriving Doppler frequency shift.

$$f_{direct} = \frac{1}{2\pi}\frac{\Delta\phi_{direct}}{\Delta t} = \frac{v\cos(\theta_{direct})}{\lambda}$$

$$f_{reflect0} = \frac{1}{2\pi}\frac{\Delta\phi_{reflect0}}{\Delta t} = \frac{v\cos(\theta_{reflect0})}{\lambda} \quad (4.11)$$

$$f_{reflect1} = \frac{1}{2\pi}\frac{\Delta\phi_{reflect1}}{\Delta t} = \frac{v\cos(\theta_{reflect1})}{\lambda}$$

4.3.4.1 Doppler Spectrum

The composite effect of multiple components arriving from different directions, each with their own Doppler shift, is called Doppler frequency spread, a measure of the spectral broadening caused by relative motion. The extent of this effect on the RF center frequency is: $\begin{bmatrix} f_{ctr} - f_{d\,max} & f_{ctr} + f_{d\,max} \end{bmatrix}$. A commonly used model for theoretical Doppler power spectrum is called Clarke's model. Assuming Rayleigh

CHAPTER 4. RF CHANNELS

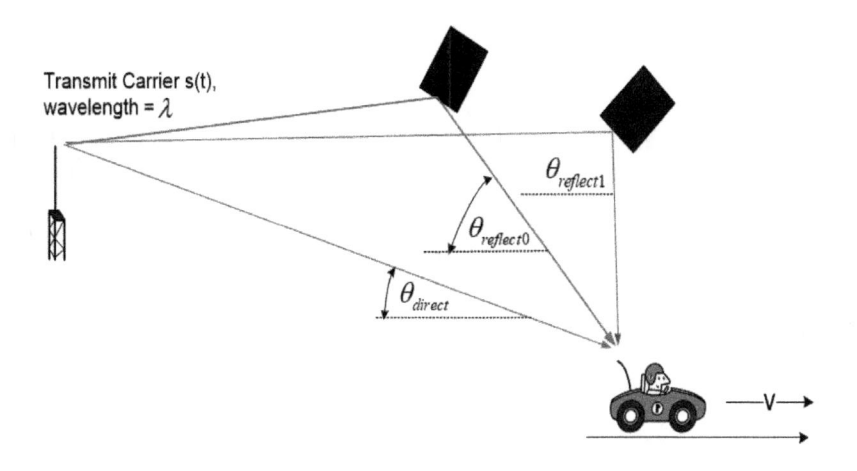

Figure 4.20 An RF environment for describing Doppler frequency spectrum.

fading, the theoretical Doppler spectrum at an omnidirectional receive antenna is shown in Equation (4.12). $f_{d\,\max}$ is the maximum expected Doppler shift [1].

$$S(f_d) = \frac{1.5}{\pi f_{d\,\max} \sqrt{1 - \left(\dfrac{f_d}{f_{d\,\max}}\right)^2}} \qquad (4.12)$$

Figure 4.21 is a plot of Equation (4.12). In a heavily populated urban area, an experimental measurement of many Doppler carrier shifts should match up closely with the bathtub-shaped curve of Figure 4.21 if the maximum Doppler shift of $f_{d\,\max}$ = 100 Hz is correctly estimated. Note that the Doppler spectrum model is undefined beyond the frequency limits of $\pm f_{d\,\max}$. MATLAB channel visualization toolbox provides a useful tool to display Doppler spectrum and many other RF propagation phenomena.

4.3.4.2 Coherence Time

Coherence time is the range of time over which two receive signal components are affected the same by the channel. Coherence time describes the time-varying nature

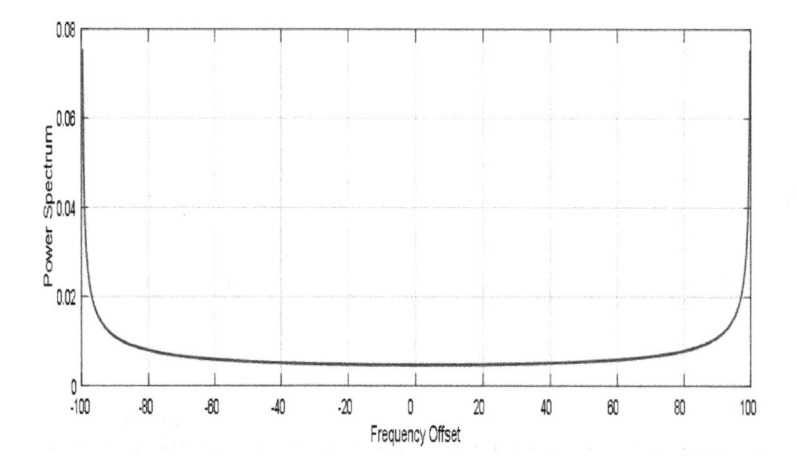

Figure 4.21 A Doppler frequency spectrum example.

of the channel as a vehicle moves through it. If there is no time variation (e.g., the car in Figure 4.20 has stopped) then coherence time is infinite.

A correct intuition from Doppler frequency spectrum at the receiver is that signal distortion will be minimal for a bandwidth (inverse of symbol time) much wider than the Doppler frequency spread. Signal spectrums more narrow than Doppler spectrum will be adversely affected. This is the same as saying that signal distortion will be minimal for a symbol time much less than the coherence time. Signals with symbol times larger than the coherence time will be adversely affected. Thus, coherence time is inversely related to Doppler frequency spread (i.e., large Doppler spectrum imples small coherence time). This inverse is usually scaled, for example Equation (4.13) is from [1]. Other sources may recommend slightly different scaling than 0.423.

$$\frac{0.423}{f_{d\,max}} \leq T_c \leq \frac{1}{f_{d\,max}} \qquad (4.13)$$

4.3.4.3 Fading Rate

The rate of signal power dips received by a moving vehicle is called the fading rate. Small coherence time implies fast fading: channel variations are faster than baseband signal variations (symbol time), receiver phase coherence can be lost over the symbol period. Large coherence time implies slow fading: channel variations are

CHAPTER 4. RF CHANNELS

slower than baseband signal variations. The receiver has an easier time maintaining phase coherence. Fading rate is illustrated in Figure 4.22.

For many moving receiver environments, received envelope variations can be described by the Rayleigh probability distribution in Figure 4.23.[8] Note that in Figure 4.23 the average receive power is $\sigma = E\left[r^2\right]$. Note also that sometimes this power can be much greater than average. That is when the multipaths happen to line up in phase.

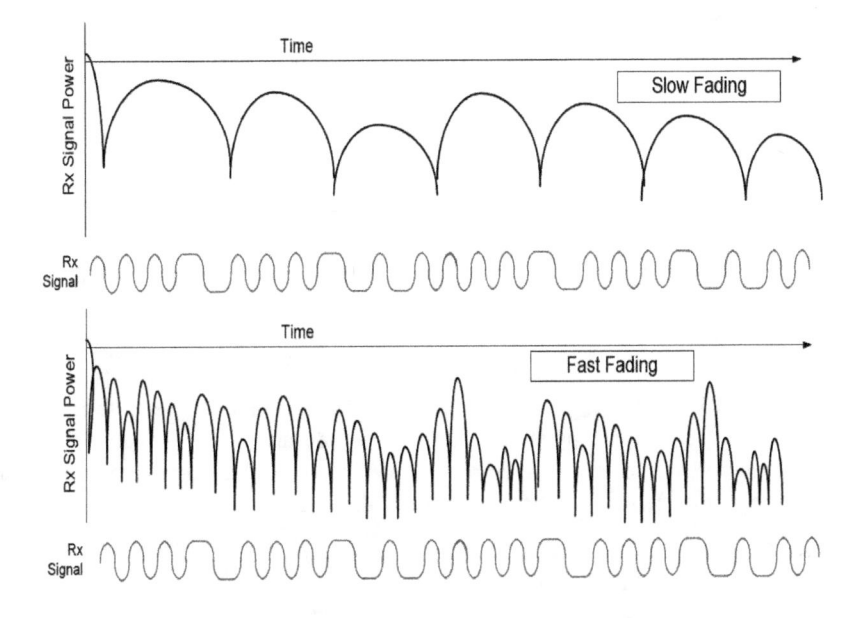

Figure 4.22 A comparison of slow and fast fading.

The Rayleigh faded received signal-to-noise ratio changes along with the power fluctuations. The familiar PSK BER expression, $P_{PSK}(err) = Q\left(\sqrt{\frac{2E_b}{N_0}}\right)$, no longer applies to a Rayleigh faded signal. If α is the average received envelope voltage of the received signal, the modified BER for coherently detected PSK is discussed in [3].

[8] Rayleigh fading is for the case where there is no line-of-sight path; the received signal is a composite of only reflected paths. If this composite includes a direct line-of-sight path, then the received signal power distribution is called Rician. A Rician received envelope power distribution applies to Figure 4.19.

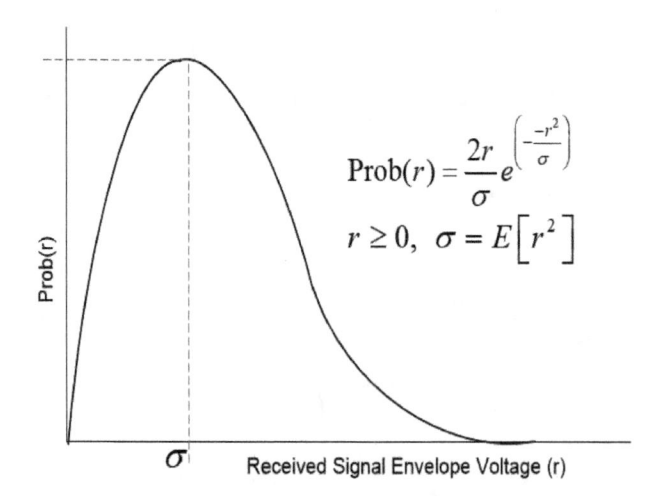

$$\text{Prob}(r) = \frac{2r}{\sigma} e^{\left(\frac{-r^2}{\sigma}\right)}$$

$$r \geq 0, \quad \sigma = E\left[r^2\right]$$

Figure 4.23 The Rayleigh probability distribution.

Table 4.3 summarizes the fixed propagation, no Doppler, situation from Section 4.3.1 and the moving propagation environment from Section 4.3.3. Note that the first two no Doppler rows do not include any variable fading. However, although the Doppler rows are dominated by fast or slow fading rate, there can still be a noninfinite coherence BW due to the multiple paths.

4.3.4.4 Advanced Propagation Models

The idealized models discussed so far can be difficult to successfully apply to real-world RF propagation problems. A big missing factor is the weather. Because rain, snow, ice, hail, and other impairments affecting RF propagation are unpredictable, a useful RF propagation model must be statistical. Another important factor is accuracy. AM, FM, TV broadcast services generally have to operate simultaneously and continuously. Prior to issuing a broadcast license, the FCC has to make sure that the new station does not interfere with and is not interfered with by existing stations. This coverage and interference analysis affects broadcast station authorized transmit power, market size, and thus profits. The Institute for Telecommunications Services (ITS) maintains an irregular terrain model, commonly called the Longley-Rice model. From [4]:

Table 4.3 Fading Summary for Nonmoving and Moving Environments

		Frequency Selective Fading due to Multipath Time Dispersion	
No Doppler	Flat fading	Frequency selective fading
	Signal BW < coherence BW	Signal BW > coherence BW
		Time Selective Fading due to Multipath Frequency Dispersion	
Doppler	Slow fading	Fast fading
	Symbol time < coherence time	Symbol time > coherence time

The ITS model of radio propagation for frequencies between 20 MHz and 20 GHz (the Longley-Rice model) is a general-purpose model that can be applied to a large variety of engineering problems. The model, which is based on electromagnetic theory and on statistical analyses of both terrain features and radio measurements, predicts the median attenuation of a radio signal as a function of distance and the variability of the signal in time and in space.

MATLAB antenna toolbox has a well-documented Longley-Rice model. This is an excellent place to start learning to use this important propagation model.

4.4 MULTIPATH MITIGATION

A common form of multipath mitigation is the technique of receiving the same signal from different sources, different points in time, or on different frequencies. This is called space, time, or frequency diversity. The received signals are then combined to produce a composite with more power and less distortion than the separate signals.

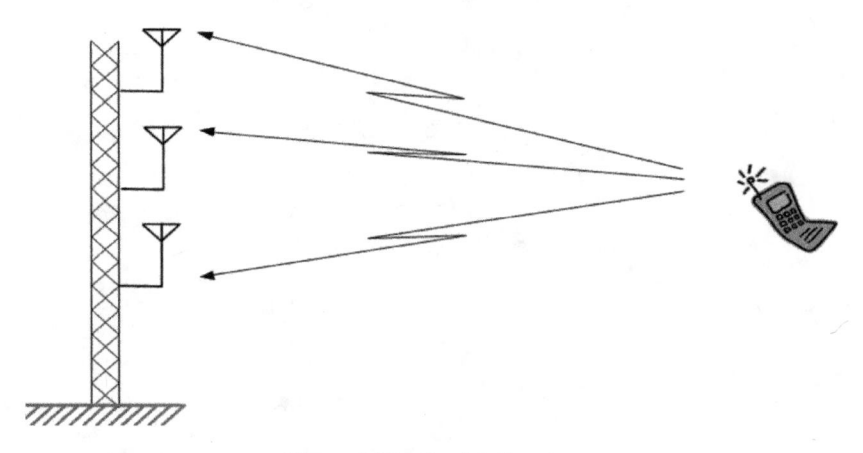

Figure 4.24   Spatial diversity.

We will look at several different approaches to multipath mitigation.

4.4.1 Spatial Diversity

Figure 4.24 illustrates a simple means of diversity using three signals undergoing different multipath delay profiles (multipath details not shown). The receive antennas on the tower are several wavelengths apart for the purpose of decorrelating the receive signals. Decorrelated signals have no fixed phase relative to each other so their covariance is zero.[9] This means that the total average power is the sum of the average power from each receive antenna. Some techniques for combining the received signals are:

1. **Maximum ratio combining (MRC)**: Measure (S+N)/N of the received signal at each antenna. Composite signal is the weighted sum; better signals get more weight, poor quality signals get less weight. Performance is best, but complexity is high because signals must be cophased (carrier signals lined up in phase to avoid cancelation) prior to sum.
2. **Equal gain combining (EGC)**: Composite signal is the unweighted sum, still necessary to cophase. EGC is less complex and almost as good as MRC.
3. **Scan and switch**: Cycle through (S+N)/N measurements and select the best. When the selected (S+N)/N goes below a threshold; start cycling again. No cophasing so less complex hardware is required.

We have seen previously how HF communications (i.e., 3 to 30 MHz) over long distances make use of ionospheric skip. Spatial diversity is used to mitigate some of the unpredictability of ionospheric skip and make reliable HF data communication possible. Figure 4.25 is a practical 1950s embodiment of this technique. Notice the three receivers. This system made transmission of critical data, such as defense and banking, much more reliable. The difference in power consumption and size between this unit and a modern design with similar functionality is truly astounding.

4.4.2 Spatial Diversity for CDMA

Code division multiple access (CDMA) was used by second and third generation cell phone systems to separate users by base station. Here we will consider CDMA to be the transmission of a random bit stream called a chipping sequence. The receiver correlates these chips with an a priori known reference pattern to extract data for this

9 $COV(x, y) = \frac{\sum_N (x_i - \bar{x})(y_i - \bar{y})}{N-1}$. With zero covariance, each signal changes independently of one another.

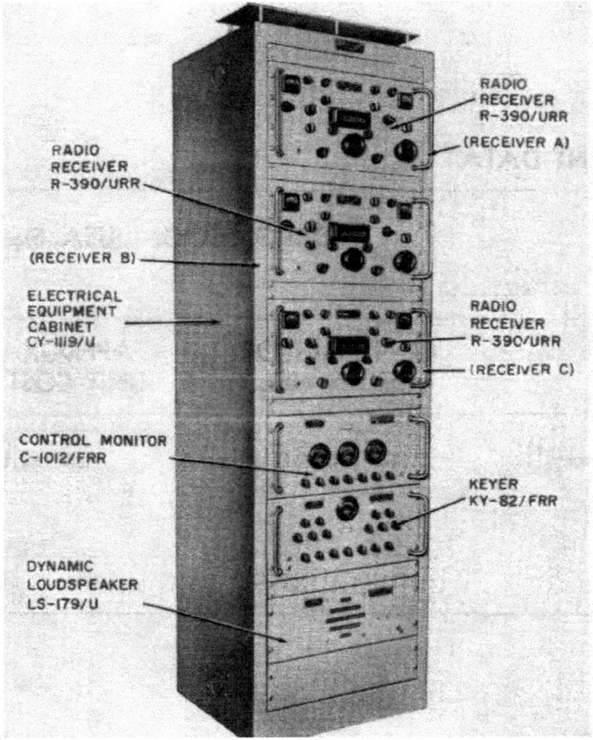

FUNCTIONAL DESCRIPTION:

Radio Receiving Set AN/FRR-34 is a triple-diversity equipment used for the reception of radiotelephone, radiotelegraph, and radioteletype signals.

This equipment consists essentially of three receivers, a control monitor, a keyer, and a loudspeaker, all mounted in an equipment cabinet.

By means of switching circuits in the keyer, the following modes of operation can be selected: each receiver separately; two in dual diversity or three in triple diversity, to minimize fading; or two in dual diversity while the third is operated as a search receiver on other frequencies.

TECHNICAL CHARACTERISTICS:

Frequency Range in Mc: 0.5 to 32
Type Modulation: AM
Type of Signal: Cw, mcw, voice
Power Output: 5 mw (phones); 10 mw (600-ohm balanced line); 500 mw (600-ohm unbalanced line)
Power Requirements: 980 w, 115/230-v 50/60-cycle 1-phase ac

Figure 4.25 Spatial diversity system using three paths (from MIL-HDBK-161).

CHAPTER 4. RF CHANNELS

user; see Section 3.4.1. This transmission can be reflected off various objects in the propagation field, creating multipath distortion. Thus, the actual received signal may be a composite of multiple delayed copies. These are paths 2, 3, 4, and 5 in Figure 4.26. Direct path 1 correlates perfectly. Paths 2, 3, 4, and 5 are delayed by more than one chip time relative to the reference pattern so they look like other users. Their correlation fails. However, the rake receiver can realign these path signals in time and phase and combine them in a form of spatial diversity.

Figure 4.27 shows the basic idea for diversity combining of CDMA chip sequencies in a rake receiver. First, the variable time delay blocks line up the received sequencies in time. Then the correlator averages the product of each sequence with the local known CDMA chip pattern. This is a complex correlation and all six outputs will need to be cophased in the multipliers right below the correlator. Finally, maximum ratio combining is used to generate a new composite signal with lower noise and distortion than any of the six individual path signals.

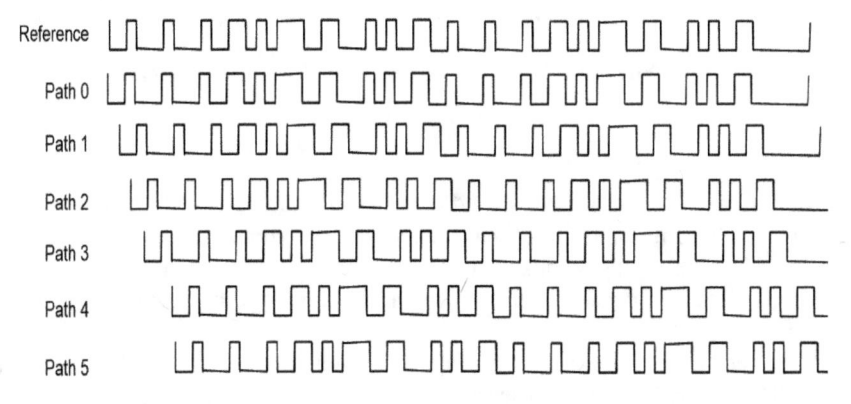

Figure 4.26 CDMA direct signal, path 0, and five reflected paths.

4.4.3 Time Diversity

Time diversity transmits a sequential bit stream at nonsequential points in time. The receiver knows the rearrangement sequence and can put the received bits back in order. Borrowing some terminology from [5], the interleaver block in Figure 4.28 has N rows and B columns. Symbols are written in row by row sequentially and read out by columns. At either the transmit side or receive, the delay to load and unload the interleaver or deinterleaver, respectively, is NB.

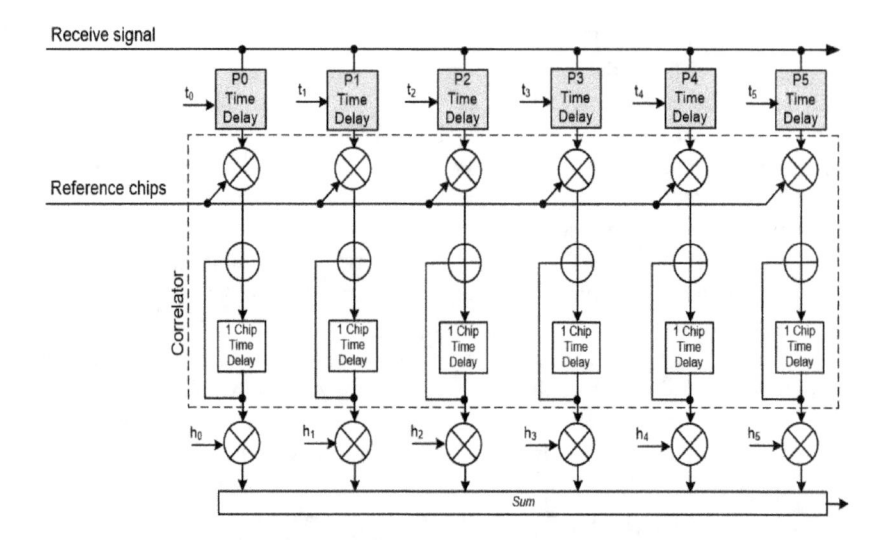

Figure 4.27 Block diagram of a CDMA rake receiver.

When used with block coded data, B is often one or more greater than the number of bits in a code word. Say a noise burst corrupts a sequence of less than B bits. When the corrupted bits are rearranged, they may be easier to correct because the effect of the noise will also be spread out among different uncorrupted code words.

Let's look at an example. Say a block coder converts a series of sequential bits into Hamming(7,4) blocks of seven total bits: four message bits and three parity bits. This code can detect two errors and correct one error; see Section 6.3.1. First, these code blocks are written into an $N = 7$, $B = 8$ interleaving matrix. In Figure 4.28, the message bits are shaded and the parity bits are not. Note that the total end-to-end delay is $2NB = 112$.

In the first row of Figure 4.29, the Hamming blocks are transmitted in time order, with no interleaving. Channel nulls corrupt a pair of message bits, 44 and 45. Our block code can only correct one error per block, so these message bits are not correctable. The center trace shows the time interleaved transmission read out by columns and subject to the same nulls. Instead of bits 44 and 45, now bits 22 and 30 bits are corrupted. These happen to be message bits from different Hamming blocks. So when the receiver rearranges the bit sequence back into correct time

CHAPTER 4. RF CHANNELS 159

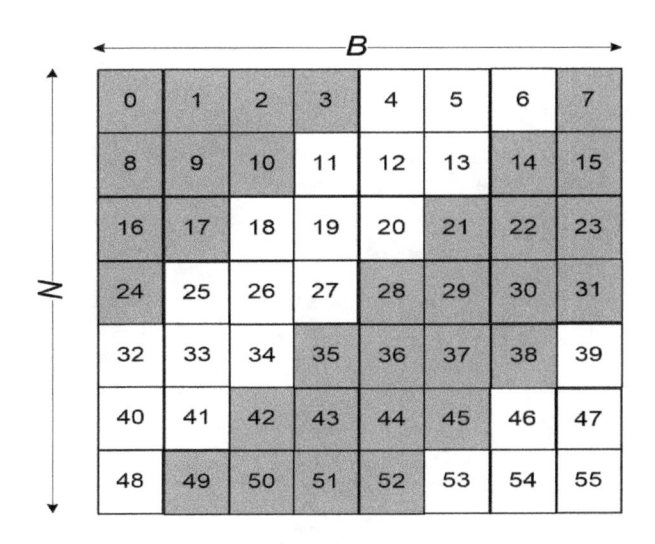

Figure 4.28 A time diversity block interleaving matrix.

order, last trace, only one corrupted bit per Hamming block remains and all errors can be corrected.

To reduce the block interleaving delay, convolutional interleaving was developed; see [6]. Convolutional interleaving is a continuous process; it does not have the fill and empty cycle of block interleaving. Figure 4.30 shows a simple example of convolutional interleave and deinterleave circuits at the transmitter and receiver, respectively. We define B as the number of switch positions, in this case 4. M is the number of additional delays at each switch position (in this case 1) and finally $N = MB = 4$. Figure 4.30 shows one cycle of B symbols, using the symbol numbers as inputs. On the right side, we have symbol numbers 12, 13, 14, and 15 input to the transmit side interleaver. The order of transmission becomes 12, 9, 6, and 3. On the right side, we have symbol numbers 0, 1, 2, and 3 exiting from the receive deinterleaver. The contents of various delay blocks at each switch are shown. Table 4.4 shows the interleave and deinterleave processing of 24 symbols, including the four-symbol example from Figure 4.30. Note that the end to end transmit to receive delay is $N(B - 1) = 12$. Compare this with the block interleaver delay, discussed above, of $2NB = 112$.

An important consideration for both types of time diversity approaches is synchronization. At the receiver, frame detection using, for example, a preamble is

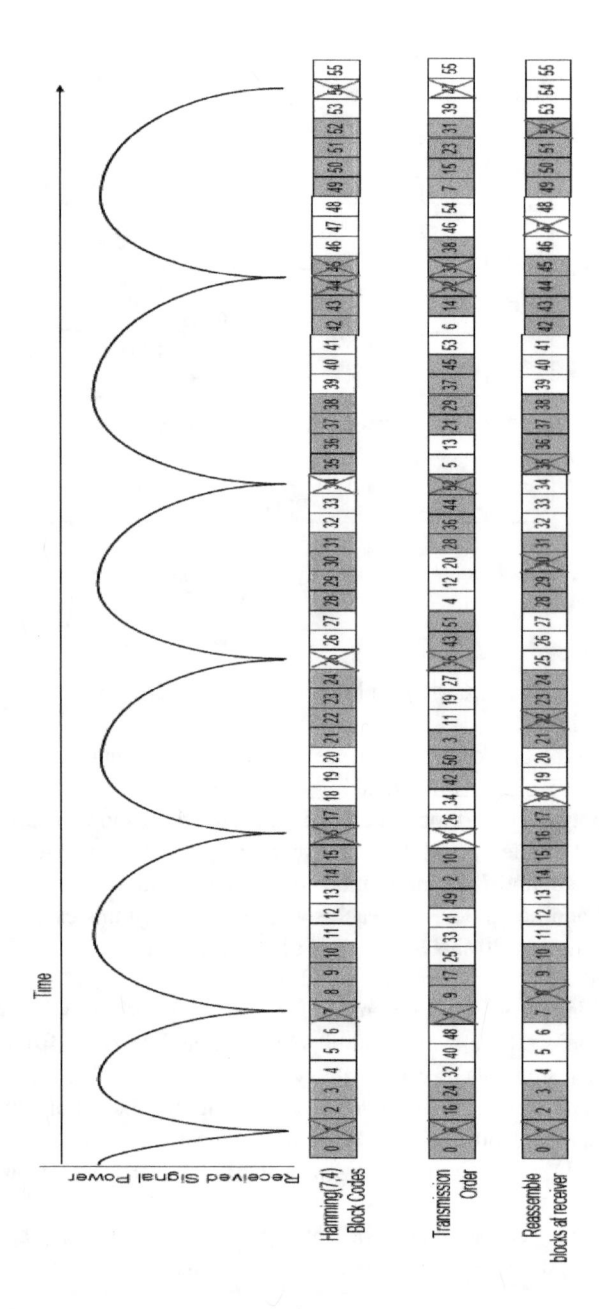

160   *Software Defined Radio*: *Theory and Practice*

Figure 4.29 A time diversity block interleaving example.

CHAPTER 4. RF CHANNELS 161

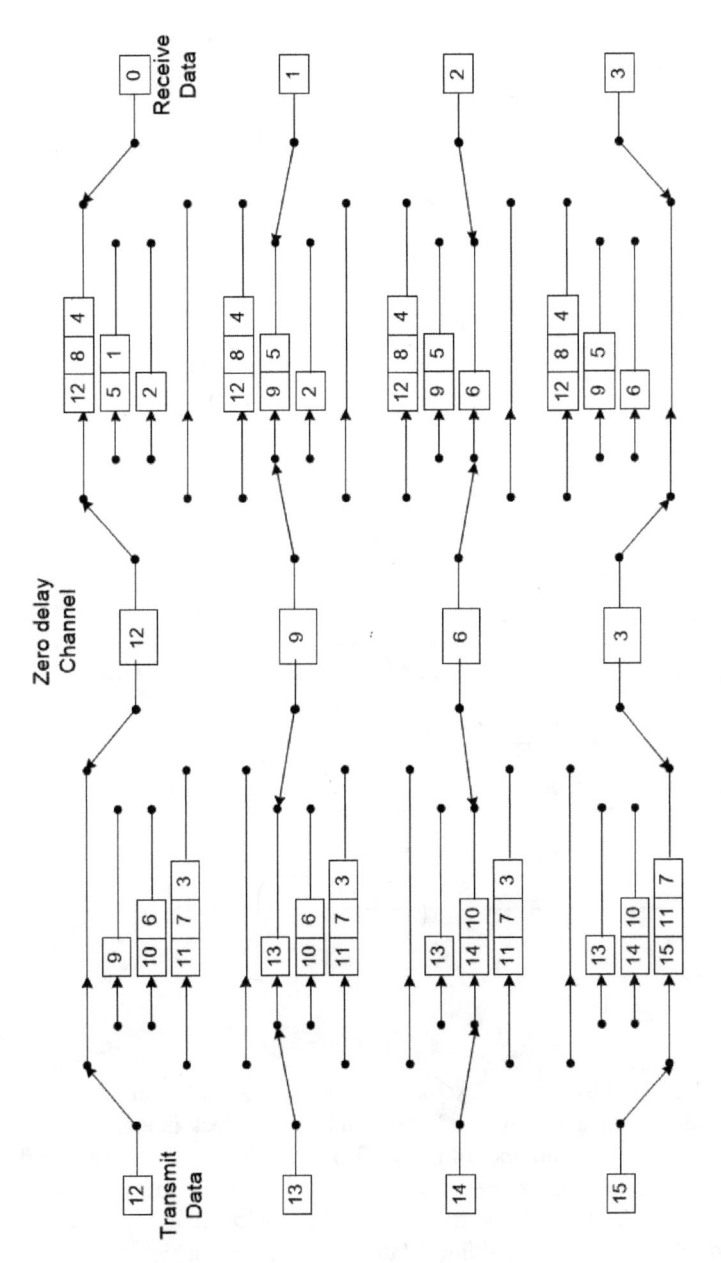

Figure 4.30 A time diversity convolutional interleaving example.

Table 4.4 Convolutional Interleaver Input Symbol, Output Symbol

Tx Symbol In	Tx Symbol Out	Rx Symbol In	Rx Symbol Out
0	0	0	–
1	–	–	–
2	–	–	–
3	–	–	–
4	4	4	–
5	1	1	–
6	–	–	–
7	–	–	–
8	8	8	–
9	5	5	–
10	2	2	–
11	–	–	–
12	12	12	0
13	9	9	1
14	6	6	2
15	3	3	3
16	16	16	4
17	13	13	5
18	10	10	6
19	7	7	7
20	20	20	8
21	17	17	9
22	14	14	10
23	11	11	11

often used. If the start of the packet is correctly detected, the number of symbols in each deinterleaver frame can simply be counted. For block interleaving, the frame may contain an integer number of blocks. Likewise, for convolutional interleaving, the frame may contain an integer number of switch cycles. Some tail symbols at the end may be needed to flush out the last N(B-1) symbols (these may also be useful for Viterbi decoding). Some additional details on this are in [5].

CHAPTER 4. RF CHANNELS

4.4.4 Frequency Diversity

Frequency hopping is a form of frequency diversity. The receiver tuning cycles through frequency channels and reassembles bits. As in time diversity, errors due to nulls on one channel only will tend to affect isolated bits. Figure 4.31 shows a very simple example where channels are simply cycled through. Frequency hopping where the channel selection is random may be more common.

There are two variations on this technique. Fast hopping means each new bit is assigned a new channel (shown here) and slow hopping assigns a new channel for groups of bits. Frequency hopping was invented during World War II as a means to increase signal security. In 1942, Hollywood actress Hedy Lamarr received U.S. Patent 2,292,387 for a Secret Communications System based on frequency hopping. It has been said that she got the idea while watching a moving piano roll [7].

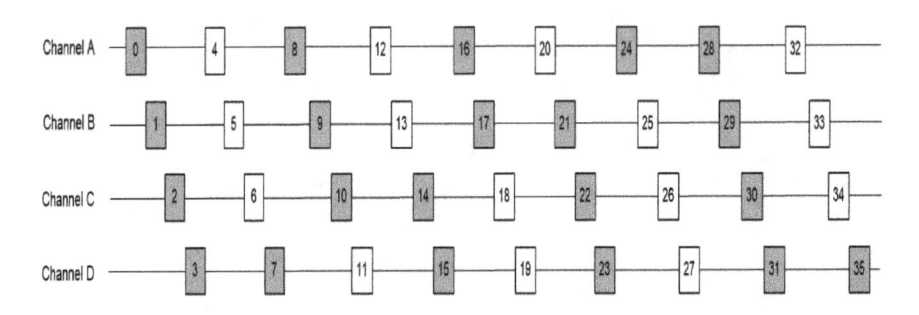

Figure 4.31 A frequency diversity example.

4.4.5 Polarization Diversity

The dual slant antenna described in Section 4.2.1.4 provides spatial diversity as described. It also provides polarization diversity. Consider a wireless microphone (transmitter built into the microphone) being used by a speaker who likes to move around a lot. The UHF band microphone signal transmitted has linear polarization that becomes random at the receiver. In addition, the microphone signal must often traverse a room, such as a lecture hall, full of objects that can generate multipath. A polarization diversity receiver is a must to avoid dropouts. In case all the paths are weak, maximum ratio combining will also improve this situation.[10]

4.4.6 Space-Time Coding

Spatial diversity combining is not difficult to implement on uplink (UL); see Figure 4.24. What about downlink (DL) from tower to handset? The handset cannot have multiple receive antennas spaced about one half wavelength apart (about 16.6 cm at 900 MHz). For the downlink we have space-time coding (STC), first proposed in [8]. STC requires two separated base station antennas transmitting the same symbols but with different coding. The handset receiver has only one antenna. STC performance is equivalent to two receive antenna diversity combining, using maximum ratio combining. STC is used in both LTE [9] and Wi-Fi [10].

Let's consider an example based on Figure 4.32. Two transmit antennas, mounted on a tower, communicate to one receiver through two complex channel responses, h_0 and h_1, assumed constant. Also assume the receiver has learned the complex channel responses.

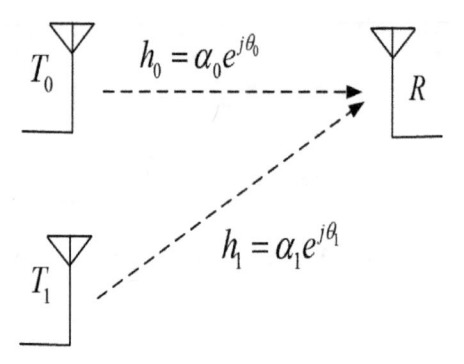

$$T_0 \quad h_0 = \alpha_0 e^{j\theta_0} \quad R$$

$$T_1 \quad h_1 = \alpha_1 e^{j\theta_1}$$

Figure 4.32 Space-time coding RF downlink environment.

Table 4.5 shows the ordering of symbols to the two transmit antennas. Note that no additional time is required, only an additional antenna. Equation (4.14) shows the signal at the receive antenna during the first two symbol times. Equation (4.15) solves for $s(0)$ and $s(1)$.

---

10 An interesting and well documented example can be found at: www.rfvenue.com/products/diversity-fin.

CHAPTER 4. RF CHANNELS

Table 4.5 Space-Time Coding Transmission Details

Symbol Time	Symbols to Transmit	Tx Antenna 0	Tx Antenna 1
0	$s(0), \ s(1)$	$s(0)$	$s(1)$
1		$-s(1)^*$	$s(0)^*$
2	$s(2), \ s(3)$	$s(2)$	$s(3)$
3		$-s(3)^*$	$s(2)^*$
...	...	...	...

$$\begin{aligned} r(0) &= h_0 s(0) + h_1 s(1) \\ r(1) &= -h_0 s(1)^* + h_1 s(0)^* \end{aligned} \quad (4.14)$$

$$\begin{aligned} \tilde{s}(0) &= h_0^* r(0) + h_1 r^*(1) \\ \tilde{s}(1) &= h_1^* r(0) - h_0 r^*(1) \end{aligned} \quad (4.15)$$

Having already estimated the channels, the receiver forms a final estimate for the first two symbols. This process continues for all remaining pairs of symbols.

$$\begin{aligned} \tilde{s}(0) &= h_0^* r(0) + h_1 r^*(1) = h_0^* \left(h_0 s(0) + h_1 s(1) \right) + h_1 (-h_0 s(1)^* + h_1 s(0)^*)^* \\ &= \alpha_0^2 s(0) + h_0^* h_1 s(1) - h_0^* h_1 s(1) + \alpha_1^2 s(0) = \left(\alpha_0^2 + \alpha_1^2 \right) s(0) \\ \tilde{s}(1) &= h_1^* \left(h_0 s(0) + h_1 s(1) \right) - h_0 (-h_0 s(1)^* + h_1 s(0)^*)^* = \left(\alpha_0^2 + \alpha_1^2 \right) s(1) \end{aligned}$$
$$(4.16)$$

4.4.7 Multiple Input-Multiple Output

A very flexible radio configuration with many applications is called multiple input-multiple output (MIMO); see [11]. There are two ways to make use of MIMO:

Diversity: Data rate is not increased because the same data is sent over multiple paths. Path combining at the receiver results in multipath mitigation and hence reliability.

Capacity enhancement: Transmit data is divided into N streams multiplying the overall data rate by N. Capacity is increased but multipath can still be a problem. For example, 2×2 MIMO allows the input data rate to be doubled. This is sometimes called evading Shannon (see Section 2.7.3).

Here we focus on using MIMO for diversity. Transmit symbols are sent over multiple streams. Each stream is transmitted from a different transmit antenna (separated by about one half-wavelength). Each receive antenna picks up a weighted sum of all the transmit signals. The MIMO receiver recombines the multiple propagation paths, achieving spatial diversity. Figure 4.33 is a 2×2 MIMO system. Consider a single channel system with transmit power P and data rate R. Each 2×2 MIMO Tx gets power $P/2$ and data rate R. In this case, we have improved the multipath mitigation, but not the input data rate.

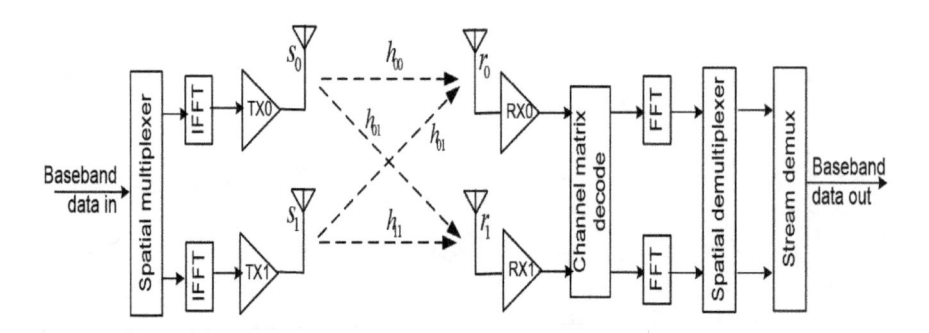

Figure 4.33 MIMO RF data transport system.

4.4.7.1 MIMO Channel Quality

MIMO effectiveness depends on the channel matrix, H, shown in Equation (4.17) and Equation (4.19). Equation (4.19) is often written as $r = Hs$. Let's assume for now that the receiver has identified the channel matrix. What makes a good one; that is, what are the characteristics of a channel matrix that provides reliable multipath mitigation or capacity increase?

A poor channel matrix might have rows or columns that are scaled versions of each other. A channel matrix this bad is easy to spot because it is rank-deficient; it has a matrix rank of 1 instead of 2 (see [12] if you need a matrix theory review). How do we make this channel better? Do whatever can be done to increase the difference in reflections between the two channels. MIMO channels with different multipath characteristics are known as micro-diversity channels in contrast to the macro-diversity looked at earlier in the chapter (e.g., Figure 4.25). The fascinating aspect of MIMO is that with multiple paths with different impairments, moving the

CHAPTER 4. RF CHANNELS

MIMO receiver, or putting objects in its way, does not degrade the signal the way a single channel system might be degraded.

Below is a channel matrix example. The matrix rank is 2, however, what else can we say about it? The answer is to understand the matrix condition. The channel matrix is a set of simultaneous equations. For a system where the number of receivers equals the number of transmitters, such as our 2×2 example, there is a unique solution to these equations if H is full rank. In this case, the receiver can go from $r = Hs$ to $s = H^{-1}r$ to arrive at the transmit signals.

$$H = \begin{bmatrix} 0.8 & 0.3 \\ 0.2 & -0.9 \end{bmatrix} \quad (4.17)$$

The usual MIMO quality metric is called the condition number. The MATLAB snippet below shows how to calculate it. Briefly, singular values of H are the square roots of the positive eigenvalues of $H^T H$. Right, but what does it mean? The SVD is described in detail in [12].

```
% Channel Matrix example
% Find Condition Number
H = [4 5; 2 7];
p = svd(H); % singular value decomposition
if(p(1)>p(2))
condnum = p(1)/p(2);
else
condnum = p(2)/p(1);
end
```

The condition number of the channel matrix is always greater than zero and a higher condition number indicates a higher-performance MIMO channel. As a modification of the equation in Section 2.7.3, we have the MIMO channel capacity theorem in Equation (4.19). The bandwidth efficiency C/W is plotted in Figure 4.34 for several SNR(dB) values and a range of channel matrix condition numbers. The MATLAB code to generate the figure data is below also. Compare with the single-channel case in Figure 2.16. Also note that bandwidth efficiency here is the Shannon limit, C, in bits/second divided by the Shannon channel bandwidth, W, see Figure 2.15. No particular modulation type is implied.

$$C = W \left[log_2 \left(1 + S\rho_1^2\right) + log_2 \left(1 + S\rho_2^2\right) \right] \quad (4.18)$$

C = channel capacity in bits/second;
S = SNR ratio (not in dB);
ρ_1, ρ_2 = channel matrix singular values.

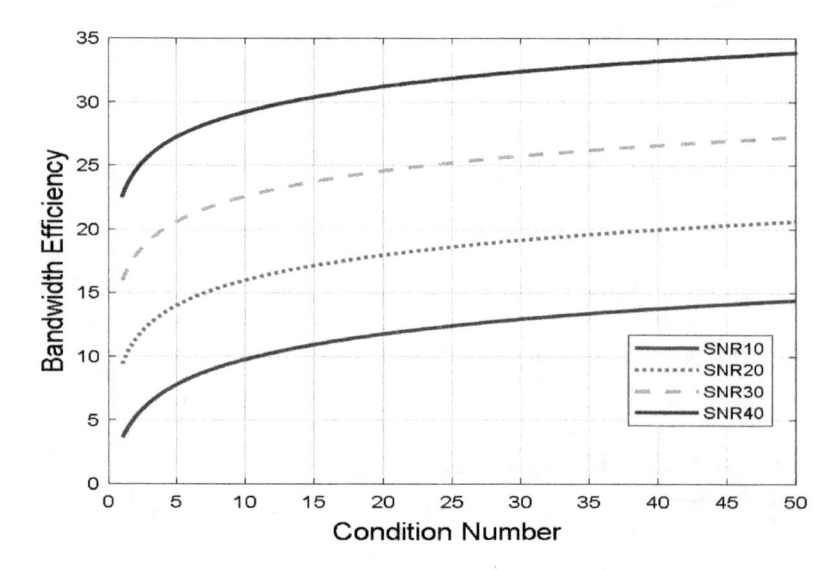

Figure 4.34 MIMO bandwidth efficiency versus condition number and SNR.

```
SnrLog = [10 20 30 40]; %dB
L = length(SnrLog);
SNR = 10.^(SnrLog/10);
p1= 0.5; % singular value
p2 = p1:0.1:25; % singular value
M = length(p2);
cn = zeros(1,M); % condition number
bell = zeros(L,M); % bandwidth efficiency
j = 1;
for k = 1:M
if(p1>p2(k))
cn(k) = pi/p2(k);
else
cn(k) = p2(k)/p1;
```

CHAPTER 4. RF CHANNELS

```
end
   for j=1:L
   bell(j,k) = (log2(1+SNR(j)*p1²) +log2(1+SNR(j)*p2(k)²));
   end
end
```

4.4.7.2 Learning the Channel Matrix

MIMO, as discussed here, depends on identification of the channel matrix, H, shown in Equation (4.19). The receiver generally learns H from a training pattern embedded in the packet preambles; see Figure 4.35. For this training process to work, there are several pre-conditions necessary. First, packet synchronization must be achieved so that the MIMO training sequences can be accurately detected. The receiver uses a priori knowledge of these MIMO training sequencies to estimate H. Second, H must be invertible.[11] This means that rows and columns must be decorrelated (i.e., zero covariance). This is the same as saying that the channel matrix has full rank. Full matrix rank is the same as saying vectors made up of the rows are not parallel to each other, likewise for columns. Both antenna spacing and independent multipath interference at each antenna help with this. Notice how the channel matrix is reestimated every packet. If the transmitter and/or receiver are moving, this can be very important. See [12] for details on the least mean square solution to simultaneous equations used to estimate the channel matrix.

$$\begin{bmatrix} r_0 \\ r_1 \end{bmatrix} = \begin{bmatrix} h_{00} & h_{01} \\ h_{10} & h_{11} \end{bmatrix} \begin{bmatrix} s_0 \\ s_1 \end{bmatrix} \quad (4.19)$$

| Preamble Short training sequence | Preamble Long training sequence | Signal description | MIMO training S₀ | MIMO training S₁ | Payload data |

Figure 4.35 Some Wi-Fi MIMO training patterns.

An example of a set of four MIMO training patterns is shown in Table 4.6. These patterns are orthogonal to each other in that the sum of any four symbol-time

---

11 As explained in [12], matrix A is invertible if a matrix B of the same dimensions as A exists such that $AB = I$, where I is the identity matrix of the same dimensions as A.

Table 4.6 The IEEE 802.11n, Long Training Field

Transmit Antenna	Symbol 0	Symbol 1	Symbol 2	Symbol 3
0	1	-1	1	1
1	1	1	-1	1
2	1	1	1	-1
3	-1	1	1	1

by symbol-time products will be zero. For example, try transmit antennas 0 and 1, the sum of $(1)(1) + (-1)(1) + (1)(-1) + (1)(1) = 0$. Each of the four receivers has a copy of Table 4.6. Orthogonality ensures that each of the four receivers can measure the complex gain of each individual path back to each of four transmitters. Thus, the channel matrix H is four by four. Since the received MIMO training pattern is a noisy observation and the ideal pattern is known, least mean squares estimation can be used to estimate H.

4.5 QUESTIONS FOR DISCUSSION

1. Comment on the differences between circular polarized signals undergoing reflection and fixed polarized signals undergoing reflection. How can this difference affect overall link loss?

2. Consider a multiuser system where a CDMA base station is communicating with scattered users. The near-far problem is when a close-in high power, adjacent channel or on-channel signal, blocks a further away lower power on-channel desired signal. From a system engineering point of view, what is the best way to prevent this problem?

3. Given $P_{Tx} = 20W$, $f_{carrier} = 500MHz$, $G_{Tx} = G_{Rx} = 0dB$, and $D = 2 miles$. Find P_{Rx}. Use Equation (4.2) with D exponent $n = 2$; be careful with units.

4. In Figure 4.36, $\lambda = 0.0556$m, D = 2000m, and A = 700m. We assume that the Earth between the antennas is a flat field. To avoid knife-edge diffraction, the tree

should be trimmed when it grows into the first Fresnel zone. At what height should the tree be trimmed?

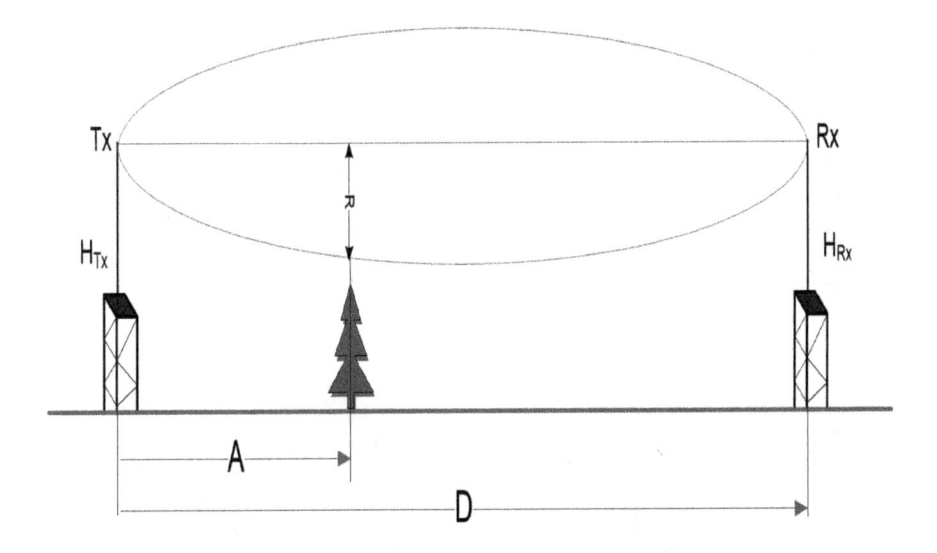

Figure 4.36 Potential knife-edge diffraction and the first Fresnel zone.

5. For line-of-sight transmission at a distance of 10 km, and $f_{carrier} = 1GHz$, what is the maximum radius of the first Fresnel zone?

6. Doppler shift always requires transmitter and receiver relative motion. True or false? Please explain your answer.

7. Consider Figure 4.37, the channel frequency response for two different channels. For an OFDM signal covering the range of subcarriers shown, which response has the largest coherence bandwidth?

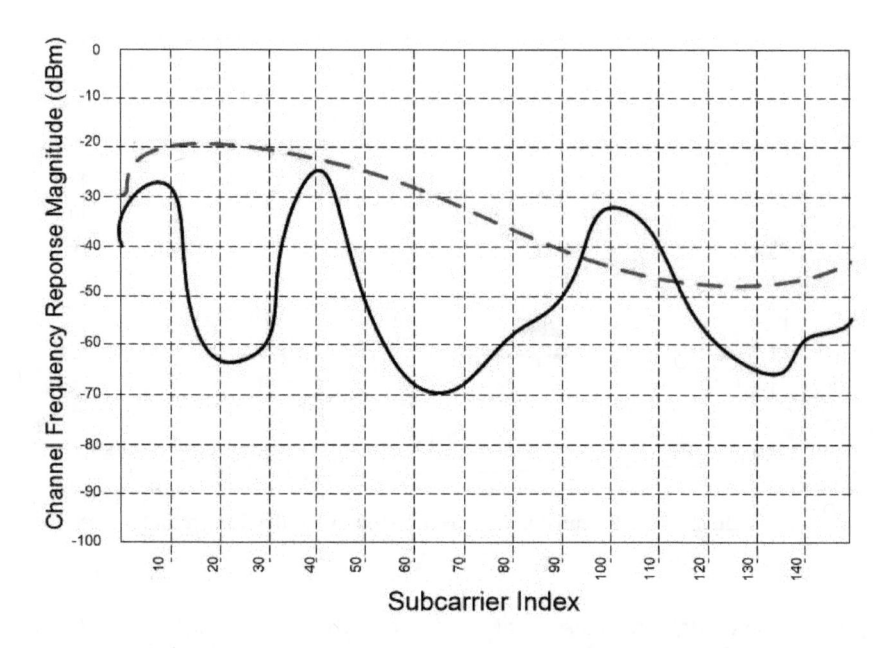

Figure 4.37 Frequency response of two different RF channels.

REFERENCES

[1] T. S. Rappaport. *Wireless Communications: Principles and Practice.* Prentice Hall, 2001.

[2] B. Sklar. *Digital Communications.* Prentice Hall, 1988.

[3] J. G. Proakis. *Communication System Engineering.* Prentice Hall, 2002.

[4] G. A. Hufford and A. G. Longley. *A Guide to the Use of the ITS, Irregular Terrain Model in the Area Prediction Mode.* BiblioGov, 2013.

[5] G. C. Clark and J. B. Cain. *Error-Correction Coding.* Plenum Press, 1981.

[6] G. D. Forney. "The Viterbi Algorithm". *Proceedings of the IEEE*, 61(3), 1973.

[7] M. Wenner. "Hedy Lamarr: Not Just a Pretty Face". *Scientific American*, June 2008.

[8] S. M. Alamouti. "A Simple Transmit Diversity Technique for Wireless Communications". *IEEE Journal on Select Areas in Communications*, 16(8), 1973.

[9] C. Gebner. "Long Term Evolution, A concise introduction to LTE and its measurement requirements". *Rohde & Schwarz GmbH Co., Germany*, 1973.

[10] T. Paul and T. Ogunfunmi. "Wireless LAN Comes of Age: Understanding the IEEE 802.11n Amendment". *IEEE Circuits and Systems Magazine*, 2008.

[11] R. S. Kshetrimayum. *Practical Digital Wireless Signals.* Cambridge University Press, 2017.

[12] G. Strang. *Linear Algebra and its Applications.* Saunders College Publishing, 1988.

Chapter 5

Channel Equalizers

5.1 INTRODUCTION

Consider a fixed frequency-selective channel with inadequate coherence bandwidth. Recall that smaller coherence bandwidth implies multiple reflected propagation paths spread out further in time and vice versa. See Section 4.3.2. These multiple paths cause intersymbol-interference (also called ISI). A simple example is a lecture hall or church with poor acoustics. The listeners hear the speaker over multiple echo paths added together. If the speaker talks too fast, and the reflected echoes are slow, his or her words are difficult to make out. For digital communication we have a transverse RF wave, not a sound wave. At the demodulator output, we have a sampled waveform, for example, Figure 5.1.

The top trace of Figure 5.1 is a repeated series of ideal baseband symbols. We call this an eye diagram. The symbol value decision points, 0, T, and 2T, are noise and distortion-free. Between these points, the raised cosine filtering causes variation; compare with Figure 3.17. The job of the equalizer is to remove the ISI in the lower trace so that the equalizer output looks like the top trace. Let's back up and start simple.

A simple channel's time response might look like Figure 5.2. The direct path response is in the center. The response at $n - 1$ is called a precursor or leading echo. The response at $n + 1$ is called a postcursor or trailing echo. Recall from the previous chapter that this is referred to as a delay spread profile.[1] This channel is simple enough that we can calculate the ISI directly. The example transmit data is in the third column of Table 5.1. The effect of the ISI is calculated in the last column, using

---

1   Actually, it is a delay spread profile sampled at the symbol rate.

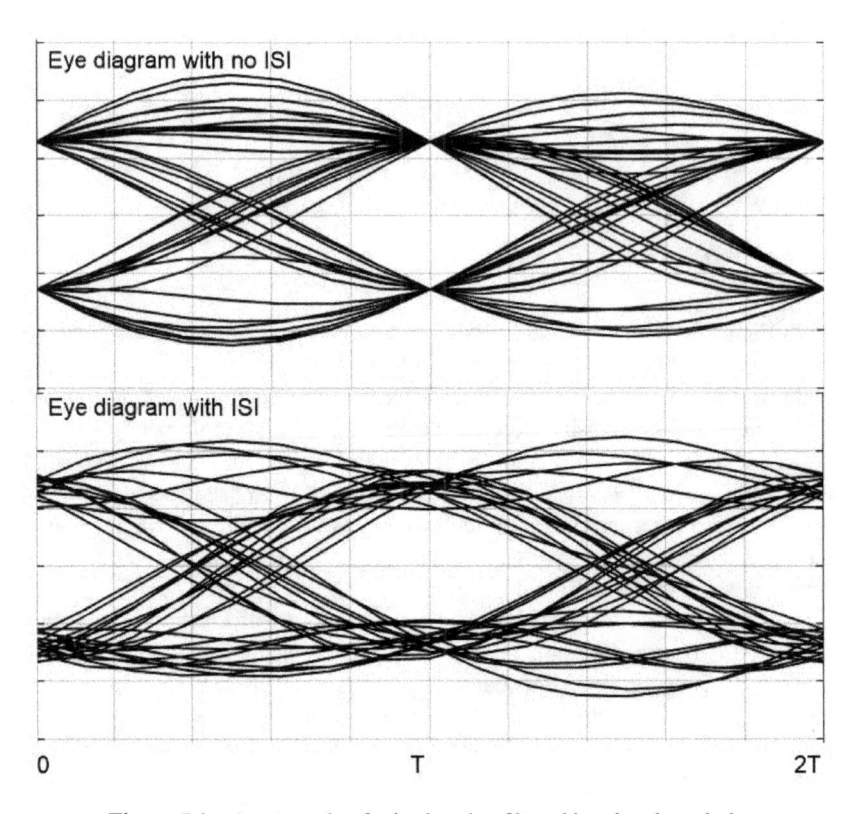

Figure 5.1 An example of raised cosine filtered baseband symbols.

CHAPTER 5. CHANNEL EQUALIZERS

the equation $y_{receive}(n) = 0.5x(n-1) + x(n) + 0.5x(n+1)$. Equalizers discussed in this chapter are for removing ISI such as this using adaptive techniques.

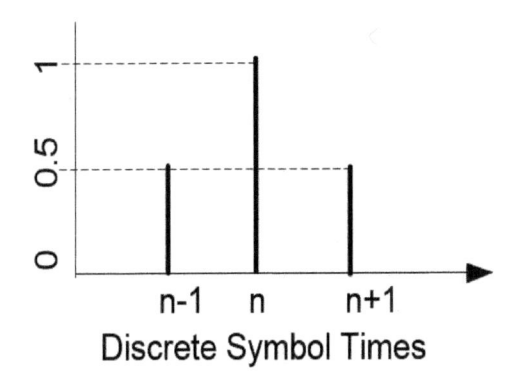

Figure 5.2 An example of an intersymbol interference channel.

Table 5.1 A Simple Example of the Effect of Intersymbol Interference

	Precursor $x(n-1)$	Direct $x(n)$	Postcursor $x(n+1)$	Receive Signal No ISI	ISI
n					
	-1	0	0	0	-0.5
1	-1	-1	0	-1	$-1.5 = -0.5 - 1$
2	-1	-1	-1	-1	$-2 = -0.5 - 1 - 0.5$
3	-1	-1	-1	-1	$-2 = -0.5 - 1 - 0.5$
4	$+1$	-1	-1	-1	$-1 = 0.5 - 1 - 0.5$
5	$+1$	$+1$	-1	$+1$	$1 = 0.5 + 1 - 0.5$
6	$+1$	$+1$	$+1$	$+1$	$2 = 0.5 + 1 + 0.5$
7	$+1$	$+1$	$+1$	$+1$	$2 = 0.5 + 1 + 0.5$

Figure 5.3 illustrates three raised cosine filtered symbols from row 5 of Table 5.1.[2] The detector sees the sum of the boxes at symbol time $n = 5$.

An important observation in Figure 5.3 is that in the "No ISI" trace, the three symbols have traveled the same path to the receiver. Because the pulses have been transmitted at three different times and the channel has no ISI, the three pulses are

---

[2] Note that raised cosine filtering is only for bandwidth control. ISI is caused by channel precursors and postcursors.

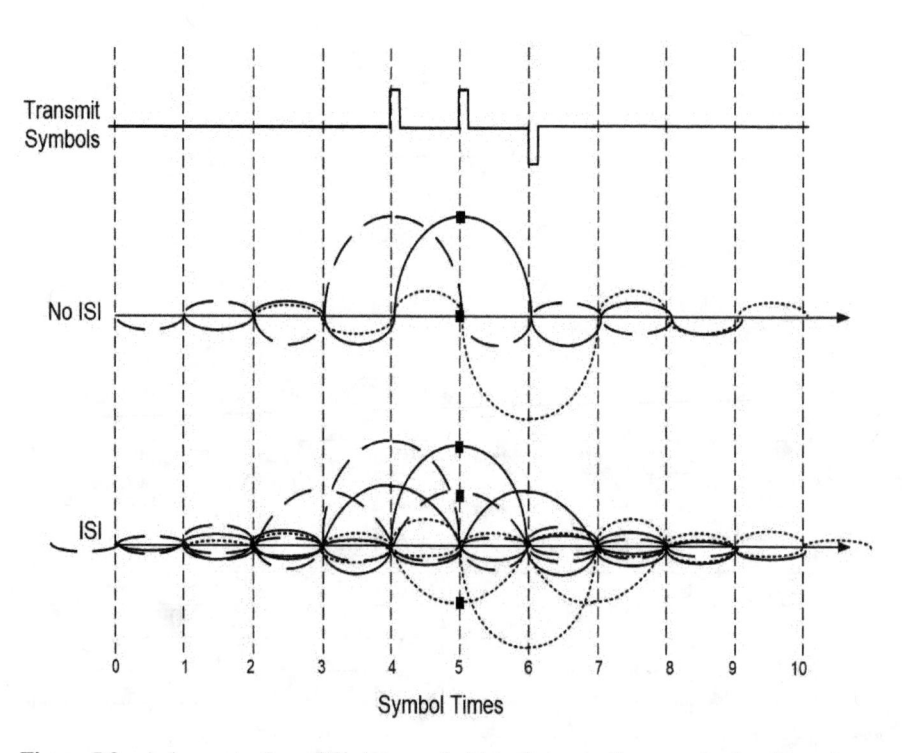

Figure 5.3 A demonstration of the intersymbol interference effect on raised cosine pulses.

CHAPTER 5. CHANNEL EQUALIZERS

properly separated at the receiver. In the ISI case, the receive situation is a little more messy. Because of the three different paths (see Figure 5.2) when a single pulse is transmitted; three are received. The desired pulse: at full amplitude with two adjacent pulses, precursor and postcursor, at half amplitude. Again, in the lower ISI trace, the detector sees the sum of the boxes at symbol time $n = 5$. At any symbol time in Figure 5.3, there is the primary pulse plus two adjacent pulses at half amplitude. The half amplitude pulses are called ISI.

5.2 EQUALIZERS USING LINEAR REGRESSION

Observe again the received symbols in rows two through five in the last column in Table 5.1. The symbol rate finite impulse response (FIR) circuit of Figure 5.4 can be used to remove the ISI from these. Three taps are needed because the channel impulse response has three terms. The received symbols are arranged in matrix H, called a linear regression matrix. The symbol time indexes (same as the rows of Table 5.1) used to construct H are explicitly shown; notice that each column is a set of four sequential symbols offset by 1 from the row above it. In each row, the effect of shifting in a new symbol from the right can clearly be seen.

At the start of Equation (5.1), $HC = D$, represents regression matrix (observed data H) times regression coefficients (to be found) equals desired result (known a priori).

The regression coefficients are simply a linear combination of the observed (ISI corrupted) received symbol values. The ISI on the channel is a linear combination (see Table 5.1) of past, present, and future symbols. Therefore, the channel ISI can be undone by another linear combination of received symbols. Also notice four rows and only three terms (unknowns) in the solution. More equations than unknowns are called an overdetermined system. The solution uses the MATLAB backslash operator.[3] Equation (5.1) illustrates that equalizer taps can be solved for however, as we will see, practical equalizers use other techniques to find the optimum taps.

---

3   The MATLAB backslash operator (type "help mldivide" in the MATLAB workspace for more information) is an implementation of the normal equations of statistics. That is, the solution to $HC = D$ is $C = (H^T H)^{-1} H^T D$. The solution of overdetermined systems is a fascinating and important subject; an excellent reference is [1].

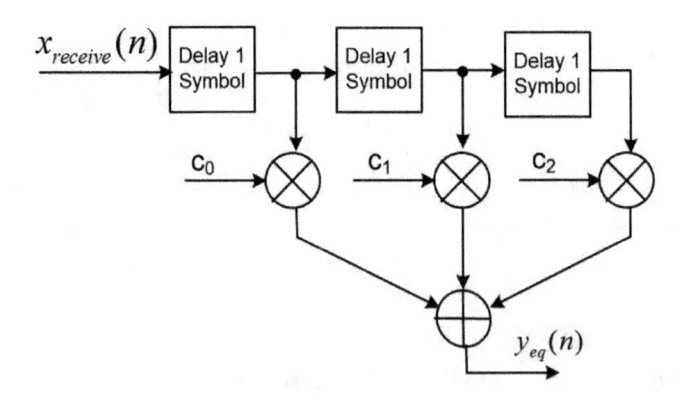

Figure 5.4 A simple equalizer that generates an overdetermined linear system.

$$HC = D$$
$$\begin{bmatrix} x_{receive}(4) & x_{receive}(3) & x_{receive}(2) \\ x_{receive}(5) & x_{receive}(4) & x_{receive}(3) \\ x_{receive}(6) & x_{receive}(5) & x_{receive}(4) \\ x_{receive}(7) & x_{receive}(6) & x_{receive}(5) \end{bmatrix} \begin{bmatrix} c_0 \\ c_1 \\ c_2 \end{bmatrix}$$
$$= \begin{bmatrix} -1 & -2 & -2 \\ 1 & -1 & -2 \\ 2 & 1 & -1 \\ 2 & 2 & 1 \end{bmatrix} \begin{bmatrix} c_0 \\ c_1 \\ c_2 \end{bmatrix} = \begin{bmatrix} -1 \\ -1 \\ 1 \\ 1 \end{bmatrix}$$

(5.1)

$$C = H\backslash D = \begin{bmatrix} -1 & -2 & -2 \\ 1 & -1 & -2 \\ 2 & 1 & -1 \\ 2 & 2 & 1 \end{bmatrix} \backslash \begin{bmatrix} -1 \\ -1 \\ 1 \\ 1 \end{bmatrix} = \begin{bmatrix} -1 \\ 2 \\ -1 \end{bmatrix}$$

Check :
$$HC = \begin{bmatrix} -1 & -2 & -2 \\ 1 & -1 & -2 \\ 2 & 1 & -1 \\ 2 & 2 & 1 \end{bmatrix} \begin{bmatrix} -1 \\ 2 \\ -1 \end{bmatrix} = \begin{bmatrix} -1 \\ -1 \\ 1 \\ 1 \end{bmatrix}$$

5.2.1 LMS Linear Adaptive Equalizer

Up to the mid-1960s, the effect of ISI limited data rates on phone lines to about 1,200 bits/second. On dial-up lines, every new connection would result in a new ISI profile. Dr. Robert W. Lucky of Bell Labs invented a technique for removing the ISI using a set of adjustable taps, what is sometimes called a transversal filter. Each tap provided a one symbol delay. The basic idea is that if the current symbol is distorted by additive interference from adjacent in time symbols (already stored in adjacent taps) then we can remove the ISI by subtracting a proportion of those adjacent taps from the current tap. Dr. Lucky's 1964 adaptive equalizer prototype, Figure 5.5, had 13 taps in a 5-foot-high rack. This invention resulted in maximum data rates on phone lines jumping up from around 1,200 to 9,600 bits/second or more, ushering in a new era in data communications [2].

5.2.1.1 Decision Directed Equalizer

In Figure 5.6 we add some adaptive feedback circuits to continuously adjust the three coefficients of Figure 5.4. This is called a decision directed equalizer because adaptation is based on the slicer estimates of the transmitted symbols. The slicer shown is a simple $sgn(y_{eq}(n))$, which is ideal for BPSK. More complex slicers can be used also. The equalizer error is the difference between the slicer output and the actual equalizer output. Because the error is not 100% reliable, for example, some slicer decisions might be incorrect, the error is scaled down by Δ.

Tap coefficients c_i are updated each symbol time by correlating $e(n)$ with the tap signal $x(i, n)$, where i is the tap number and n is the discrete time index. Notice that $e(n)$ depends on the tap coefficients. The correlation adjusts the tap in the opposite direction of any systematic relationship between the tap signal and the scaled error. The equalizer has achieved convergence when for each tap signal $x(i, n)$: $E\left[x\left(i, n\right) e(n)\right] = 0$ where $E\left[\cdot\right]$ is the expected value, a simple average in this case. The orthogonality principle is behind this and is explained later. For now, take note that the equalizer is converged when the error $e(n)$ is minimized. Also note that for the circuit in Figure 5.6 to work properly, the symbol timing must be correct. In many cases, the symbol timing recovery circuits are ahead of and independent of the equalizer. The carrier frequency offset should also be zero. Sometimes the carrier frequency correction is part of the equalizer design. Note that the small errors in the carrier phase can be corrected by the equalizer itself. Examples are shown in Chapter 12 and 13.

Figure 5.5 R. W. Lucky's first adaptive equalizer for removing ISI.

CHAPTER 5. CHANNEL EQUALIZERS

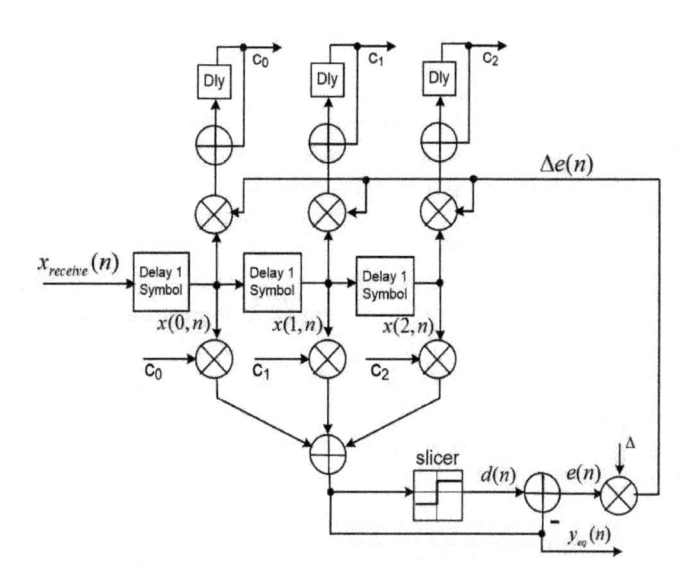

Figure 5.6 A decision directed LMS adaptive equalizer.

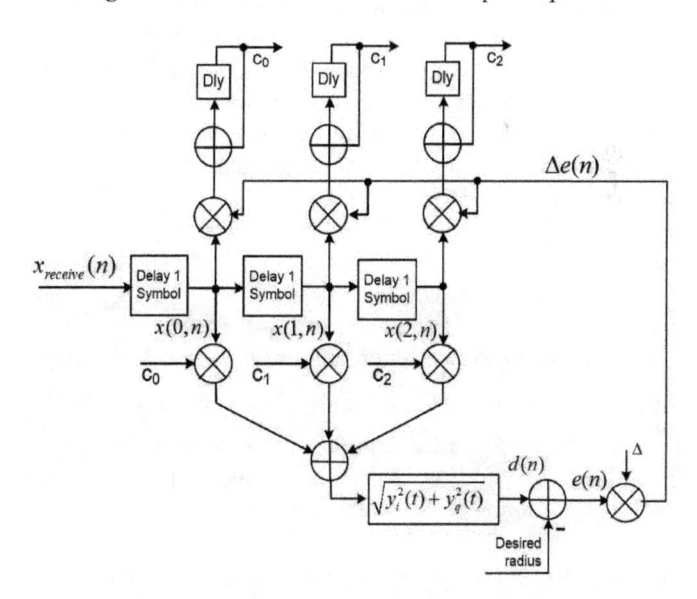

Figure 5.7 A dispersion directed LMS adaptive equalizer.

5.2.1.2 Dispersion Directed Equalizer

For two-dimensional complex symbol sequences, such as QPSK where all constellation points have the same radii, see Figure 10.28, an alternative to decision directed training is called dispersion directed. Dispersion directed bases the feedback error on the difference between the actual symbol vector length at the equalizer output and a desired vector length. The local carrier does not have to be locked to the received signal. Dispersion directed equalization removes ISI that affects symbol amplitude, but has no effect on phase. An example circuit is shown in Figure 5.7. Sometimes a two-mode equalizer is used. The equalizer is in dispersion mode while the carrier tracking is locking and the constellation is still spinning. When the carrier is locked and the constellation stops spinning, the equalizer switches to the decision directed mode to complete the equalization. That way carrier frequency tracking and initial equalization can proceed simultaneously.

5.2.1.3 Fractionally Spaced Equalizer

The first equalizer designs operated at 1 sample per symbol, covering the bandwidth range $[-F_{symbol}/2 \quad F_{symbol}/2]$. The circuits driving the equalizer, for example, the raised cosine receive matched filter, generally operate at two samples per symbol. Thus, the equalizer implies a downsample by 2. As shown in Figure 5.8, the downsample is located prior to the equalizer for the one sample/symbol equalizer. However, for the fractionally spaced equalizer (two samples/symbol), the downsample is after the equalizer.

The top image of Figure 5.8 shows the subsampled by 2 raised cosine spectrum at the equalizer input. This spectrum has the signal power in the range $[-F_{symbol} \quad - F_{symbol}/2]$ folded back (aliased) into the range $[-F_{symbol}/2 \quad 0]$ and likewise for the positive frequency side.[4] An equalizer operating at sampling rate F_{symbol} thus has to work with an aliased spectrum. The ability to separately equalize signals outside the foldover frequencies of $\pm \frac{1}{2T} = \pm \frac{F_{symbol}}{2}$ has been lost. Foldover prior to the equalizer can bring about a spectral null in the equalizer input. This null can cause considerable noise enhancement as the equalizer attempts to correct it. See [3] [4] for more information.

---

4 Note that the raised cosine spectrum has the property that there is no signal power outside the range $[-\frac{1}{T} = -F_{symbol} \quad \frac{1}{T} = F_{symbol}]$. The signal power outside the range $[-F_{symbol}/2 \quad F_{symbol}/2]$ depends upon the excess bandwidth parameter.

CHAPTER 5. CHANNEL EQUALIZERS

The bottom image of Figure 5.8 shows the same signal only now the baseband raised cosine signal spectrum is subsampled by two at the equalizer output. The bottom image shows that an equalizer operating at sampling rate $2F_{symbol}$, has the entire signal spectrum to work with. This situation avoids spectral nulls that can be introduced due to subsampling by two prior to the equalizer.

The sampling rate $2F_{symbol}$ equalizer is widely used in modern designs. A block diagram is shown in Figure 5.9. Notice that, in addition to the sampling rate, the number of taps is also doubled. A subtle aspect of Figure 5.9 is the slicer. The slicer input is two samples per symbol so only one of the slicer outputs per symbol is used as the estimate of that symbol. The equalizer will adapt to the chosen sample.

5.2.1.4 Decision Feedback Equalizer

The linear transversal equalizers that we have examined up until now are trying to remove ISI by linearly inverting the channel response. A drawback to these is increased noise when trying to invert a channel with a spectral null or severe amplitude distortion. These distortions are evidence of channel transfer function zeros. A linear non-recursive equalizer can only provide transfer function zeros (like an FIR filter) and it will attempt to flatten out the null by putting a large gain at the null frequencies and severely attenuating other frequencies. This, in a sense, awkward solution tends to enhance noise at the equalizer output. See [5] and [6].

The nonlinear decision feedback equalizer handles nulls and reduces equalizer noise enhancement. The nonlinear half shifts in a series of past symbol decisions. These symbols are called precursors because their symbol centers have already gone by. However, their "tails" are affecting the current symbol decision; see Figure 5.10. The sum of scaled versions of these precursors are subtracted from the equalizer output to compensate the tail ISI. The nonlinear equalizer is simply learning the ISI instead of inverting the channel response.

As with the linear equalizers discussed above, the error is the difference between the equalizer output $y_{eq}(n)$ and the slicer output. This error goes to separate post-cursor and precursor LMS coefficient adaptation. There is much more detail and a working model in Chapter 12. See also [6]. DFEs are commonly integrated into the receive data pins of high-speed integrated microcircuits. For a high enough switching frequency, a circuit board trace can start to behave like an ISI channel.

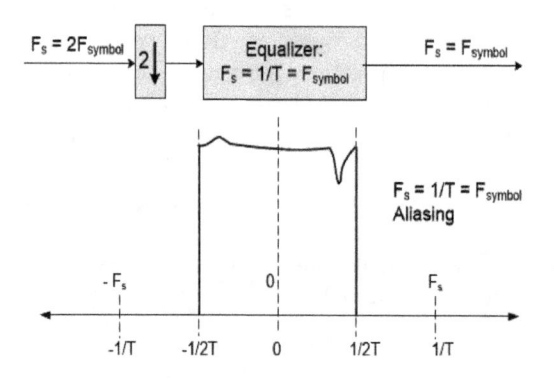

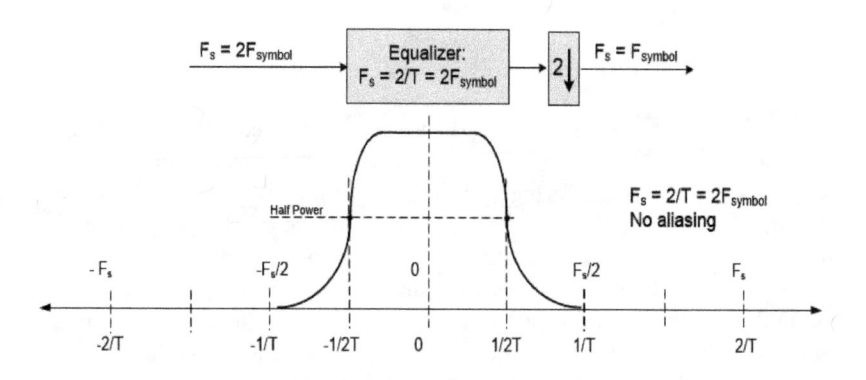

Figure 5.8 Two different equalizer sampling rate options.

5.3 LMS EQUALIZER THEORY

Whereas the previous sections were introductory, this section goes a little deeper into what makes least mean squared adaptation work. Readers who want only a basic introduction to equalization can probably skip this part.

CHAPTER 5. CHANNEL EQUALIZERS

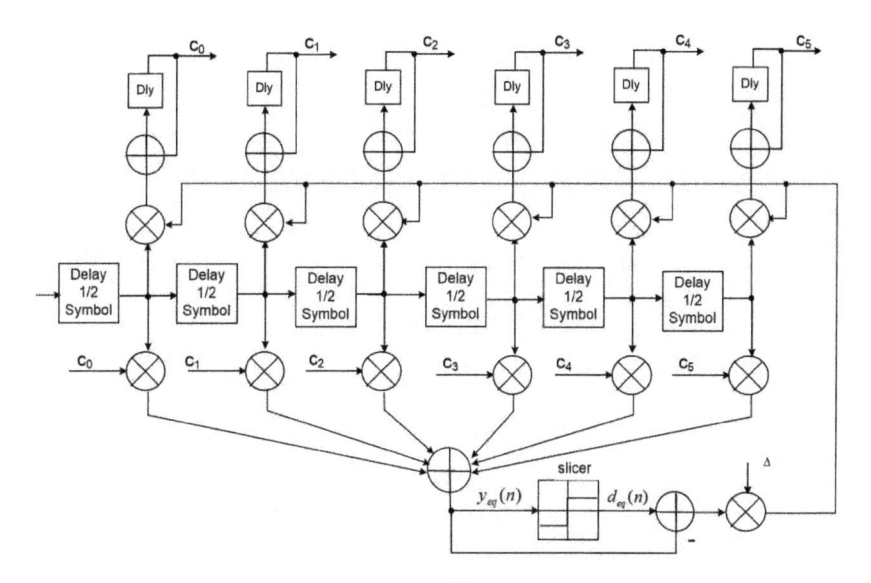

Figure 5.9 A fractional, two samples per symbol, equalizer design.

5.3.1 The Orthogonality Principle

The orthogonality principle states that the error vector of the optimal estimator is at a right angle to the estimation. Consider Equation (5.2), a rearrangement of our original equalizer equation $HC = D$ to show the error. In this example, the 3×2 regression matrix H has two columns, which can be considered two lines in a three-dimensional space. Because two lines can only span (i.e., define) a maximum two-dimension space, we say that H, with its two lines, represents a 2-D column subspace of the 3-D space shown in Figure 5.2.

Inner product HC is only one column, thus, one line, and is the LMS solution to match the desired vector D. However, because HC is simply a linear combination of the two columns of H, HC must exist in the same 2-D subspace. However, the desired vector D can be anywhere in the 3-D space and is thus independent of HC (i.e., not parallel to HC).

The key to finding the optimum solution (smallest error magnitude) is to choose the regression coefficients C so that the $error = D - HC$ is orthogonal to the solution HC. This orthogonality principle is a sort of "The best that can be done under the circumstances" principle. Figure 5.11 shows the geometry of this principle.

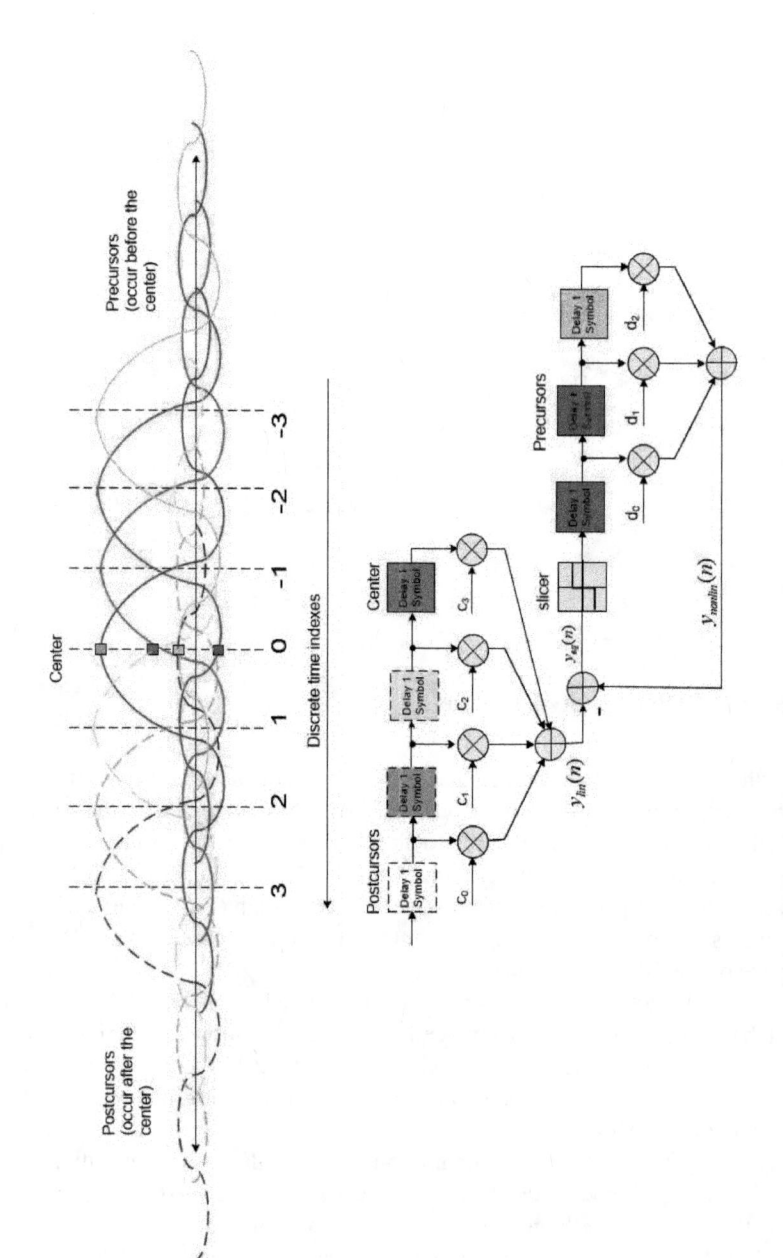

Figure 5.10 A decision feedback equalizer design.

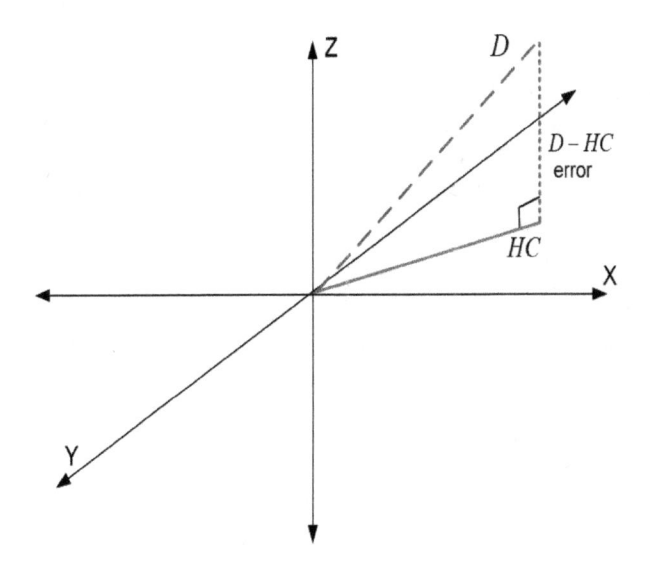

Figure 5.11 A demonstration of the orthogonality principle.

Notice that as the solution HC is moved anywhere else in the 2-D subspace, the orthogonality is lost and the vertical dotted error line gets longer. This is the key principle behind an optimal LMS solution to an overdetermined system. Statistically, the orthogonality principle says that when the equalizer converges to the optimum solution, the error and the chosen tap coefficients are uncorrelated.

$$HC = \begin{bmatrix} x(0,n) & x(1,n) \\ x(0,n) & x(1,n) \\ x(0,n) & x(1,n) \end{bmatrix} \begin{bmatrix} c_1 \\ c_2 \end{bmatrix} = \begin{bmatrix} d_1 \\ d_2 \\ d_3 \end{bmatrix} \text{ is the goal, however :}$$

$$\text{error} = D - HC = \begin{bmatrix} d_1 \\ d_2 \\ d_3 \end{bmatrix} - \begin{bmatrix} x(0,n) & x(1,n) \\ x(0,n) & x(1,n) \\ x(0,n) & x(1,n) \end{bmatrix} \begin{bmatrix} c_1 \\ c_2 \end{bmatrix}$$

(5.2)

5.3.2 Equalizer LMS Adaptation Equations

We have not yet explained how the equalizer circuit of Figure 5.6 adapts. Let's start with some basic definitions:

Three symbol column vector:
$$X(n) = \begin{bmatrix} x_{receive}(n-1) & x_{receive}(n-2) & x_{receive}(n-3) \end{bmatrix}^T$$
Tap coefficient row vector:
$$C(n) = \begin{bmatrix} c_0(n) & c_1(n) & c_2(n) \end{bmatrix} \quad (5.3)$$
Equalizer output:
$$y_{eq}(n) = C(n)X(n)$$

Recall from Equation (5.2) that when the equalizer converges to the optimum solution, the error and the chosen tap coefficients are uncorrelated. That is the only goal of the LMS equalizer. The equalizer output error, Equation (5.4), is the difference between actual transmitted symbol, represented by the slicer output, and actual equalizer output at symbol time index n.

$$e(n) = d(n) - y_{eq}(n) = d(n) - \hat{C}(n)X(n)$$
$$\hat{C}(n) = \min_{C} \|CX(n) - d(n)\| \quad (5.4)$$

Notice in Equation (5.4) that $d(n)$ is simply the slicer output that of course can change every symbol time. The second line of Equation (5.4) defines mathematically the optimum tap coefficients. These change as the equalizer converges. They also dither slightly after convergence.

In Figure 5.6 convergence is brought about by correlating the scaled error, $\Delta e(n)$, with the shift register data. We will explain why this works. We start by defining the equalizer adjustment cost function in Equation (5.5). This is the mean square output error. Like many adaptive systems, we attempt to minimize this cost function.

$$\begin{aligned} J_{MSE}(n) &= E\left\{e^2(n)\right\} = E\left\{(d(n) - y_{eq}(n))(d(n) - y_{eq}(n))\right\} \\ &= E\left\{d^2(n)\right\} - 2E\left\{y_{eq}(n)d(n)\right\} + E\left\{y_{eq}^2(n)\right\} \\ &= E\left\{d^2(n)\right\} - 2E\left\{C(n)X(n)d(n)\right\} + E\left\{C(n)X(n)X^T(n)C^T(n)\right\} \end{aligned} \quad (5.5)$$

As shown in Equation (5.6), for coefficients fixed during the cost function calculation (constant $C(n)$), we can simplify the cost function.

$$\begin{aligned} J_{MSE}(n) &= E\left\{d^2(n)\right\} - 2E\left\{C(n)X(n)d(n)\right\} + E\left\{C(n)X(n)X^T(n)C^T(n)\right\} \\ &= E\left\{d^2(n)\right\} - 2C(n)E\left\{X(n)d(n)\right\} + C(n)E\left\{X(n)X^T(n)\right\}C^T(n) \\ &= E\left\{d^2(n)\right\} - 2C(n)P + C(n)RC^T(n) \end{aligned}$$
$$(5.6)$$

CHAPTER 5. CHANNEL EQUALIZERS

We define P, above, as the vector of average cross-correlation between input samples and desired output. Also, R is the input sample correlation matrix, for example:

$$R = E\left\{\begin{bmatrix} x^2(1) & x(1)x(2) & x(1)x(3) \\ x(2)x(1) & x^2(2) & x(2)x(3) \\ x(3)x(1) & x(3)x(2) & x^2(3) \end{bmatrix}\right\} \quad P = E\left\{\begin{bmatrix} x(1)d(n) \\ x(2)d(n) \\ x(3)d(n) \end{bmatrix}\right\} \quad (5.7)$$

For coefficient adjustment, we need the gradient of the MSE with respect to the coefficients:

$$\nabla_c J_{MSE} = \frac{\partial J_{MSE}}{\partial c} = \frac{\partial}{\partial c}\left(-2C^T P + CRC^T\right) = -2P + 2RC^T \quad (5.8)$$

Note that $E\{d^2(n)\}$ does not depend on coefficients. Dropping the expected value in Equation (5.7), we get sample by sample estimates of $\hat{R}(n) = X(n)X^T(n)$ and $\hat{P}(n) = d(n)X(n)$. Plugging in these estimates, the gradient becomes:

$$\nabla_c J_{MSE} = -2\hat{P} + 2\hat{R}C^T(n) = -2(d(n)X(n)) + 2(X(n)X^T(n))C^T(n)$$
$$= -2X(n)(d(n) - X^T(n)C^T(n)) = -2X(n)e(n) \quad (5.9)$$

Finally, Equation (5.10) shows the practical LMS equalizer tap adjustment procedure. Notice that the last term is a noisy estimate of the correlation between the error and the shift register symbols.

$$C^T(n+1) = C^T(n) - \Delta(\nabla_c J_{MSE}) = C^T(n) + \Delta 2e(n)X(n) \quad (5.10)$$

5.3.2.1 LMS Equalizer Convergence

What affects the convergence of the practical LMS equalizer design?

The power spectrum of the sequence of input symbol values should be flat. This is usually achieved by a random data preamble or by the transmitter exclusive or'ing the payload data with a random sequence, called a scrambler. After the equalizer, an unscrambler is required. Sometimes, even though the transmitted data spectrum is flat, the channel introduces nulls that slow down convergence.

The step size Δ must be chosen carefully, from [4]: "Fastest convergence is obtained when the step size is chosen to be the inverse of the product of the received signal power and the number of taps." If Δ is too high, the taps can wander around their optimum settings causing error power that is in excess of the minimum obtainable (Δ should not be greater than one). If Δ is too low, convergence may take excessively long and the equalizer may not be able to follow channel fading.

5.3.2.2 Complex Valued LMS Equalizer

Many practical LMS equalizers will have a complex baseband input. As shown in Equation (5.11) the equalizer output, desired signal, error, and tap values are all complex.

$$y(n) = C(n)X(n) = \begin{bmatrix} c_1^* & c_2^* & c_3^* \end{bmatrix} \begin{bmatrix} x(n) \\ x(n-1) \\ x(n-2) \end{bmatrix} \quad (5.11)$$

$$d(n) = d_i(n) + jd_q(n)$$
$$e(n) = (d_i(n) - y_i(n)) + j(d_q(n) - y_q(n))$$

Equation (5.12) shows that the final operation, tap update, is a complex computation.

$$\begin{bmatrix} c_1(n+1) \\ c_2(n+1) \\ c_3(n+1) \end{bmatrix} = \begin{bmatrix} c_1(n) \\ c_2(n) \\ c_3(n) \end{bmatrix} + \Delta \left(e_i(n) + je_q(n) \right) conj \left(\begin{bmatrix} x_i(n) + jx_q(n) \\ x_i(n-1) + jx_q(n-1) \\ x_i(n-2) + jx_q(n-2) \end{bmatrix} \right) \quad (5.12)$$

5.3.2.3 Equalizer Delay Problem

The slicer in Figure 5.6 is a decision device whose job is to estimate the original transmitted symbol. In some cases, more complicated decision devices that require multiple symbol time delays may be used. For example, forward error control (FEC). These may produce better decisions; however, their delay can degrade the LMS equalizer performance. Figure 5.12 shows a four-symbol decision delay model.

In Table 5.2, notice how the first equalizer output, $y(0)$ in symbol time 0, is dependent on $x(k)$ for $k = 0, -1, -2, -3, -4, -5, -6$. Although $y(0)$ is computed at symbol time 0, it is not available for computing error $d(0) - y(0)$ until it exits the four-symbol delay at symbol time 4. At that time, the only tap values left that

CHAPTER 5. CHANNEL EQUALIZERS

would have affected $y(0)$ are $x(0)$, $x(-1)$, and $x(-2)$. The gray shaded tap values at symbol time 4 had nothing to do with $y(0)$ and so cannot be used to correlate with error $d(0) - y(0)$. There is no basis in that error for adjusting those taps. A possible solution is to take a low delay sliced symbol decision prior to the more complicated symbol decision circuit.

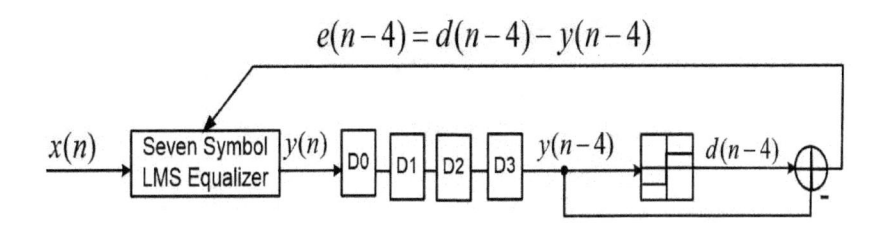

Figure 5.12 A example of the equalizer delay problem.

Table 5.2 An Equalizer Delay Problem Illustration

Tap Values	Symbol Times, n				
	0	1	2	3	4
$T(0)$	$x(0)$	$x(1)$	$x(2)$	$x(3)$	$x(4)$
$T(1)$	$x(-1)$	$x(0)$	$x(1)$	$x(2)$	$x(3)$
$T(2)$	$x(-2)$	$x(-1)$	$x(0)$	$x(1)$	$x(2)$
$T(3)$	$x(-3)$	$x(-2)$	$x(-1)$	$x(0)$	$x(1)$
$T(4)$	$x(-4)$	$x(-3)$	$x(-2)$	$x(-1)$	$x(0)$
$T(5)$	$x(-5)$	$x(-4)$	$x(-3)$	$x(-2)$	$x(-1)$
$T(6)$	$x(-6)$	$x(-5)$	$x(-4)$	$x(-3)$	$x(-2)$
Eq Out	$y(0)$	$y(1)$	$y(2)$	$y(3)$	$y(4)$
D0 Out	0	$y(0)$	$y(1)$	$y(2)$	$y(3)$
D1 Out	0	0	$y(0)$	$y(1)$	$y(2)$
D2 Out	0	0	0	$y(0)$	$y(1)$
D3 Out	0	0	0	0	$y(0)$
Error	0	0	0	0	$d(0) - y(0)$

5.4 FURTHER ADVANCES IN EQUALIZER DESIGN

Dr. Lucky's invention made much higher data rates possible sixty years ago. However, as data rates got higher, the transversal equalizer, such as in Figures 5.6 and 5.9, started to become less practical. For the same RF multipath delay spread profile, high data rates meant transversal equalizers had to include many more received symbols. Enter OFDM, covered in Chapter 3. Now OFDM blocks could be equalized in the frequency domain instead of long serial streams of symbols. Chapter 13 has a complete description of an OFDM equalizer implemented in Simulink as part of an IEEE802.11a receiver.

5.5 QUESTIONS FOR DISCUSSION

1. How is the CDMA rake receiver in the previous chapter similar to the LMS channel equalizer?

2. Discuss how the orthogonality principle applies to the solution of overdetermined sets of simultaneous linear algebraic equations. Are the equalizers discussed here always in an overdetermined system?

3. As discussed, Dr. Lucky's invention of the LMS adaptive equalizer made modern data communications rates possible. Frequency domain based OFDM equalization seems likely to give us another big jump in data rates. Both of these techniques are based on initial channel estimation. Comment on what might be the next big step in channel estimation capabilities.

REFERENCES

[1] G. Strang. *Linear Algebra and its Applications*. Saunders College Publishing, 1988.

[2] R. W. Lucky and H. R. Rudin. "An Automatic Equalizer for General Purpose Communications Channels". *Bell System Technical Journal*, November 1967.

[3] G. Ungerbock. "Fractional Tap Spacing Equalizer and Consequences for Clock Recovery in Data Modems". *IEEE Transactions on Communications*, 24(7), August 1976.

[4] J. A. C. Bingham. *The Theory and Practice of Modem Design*. John Wiley and Sons, 1988.

[5] A. Leclert and P. Vandamme. "Decision Feedback Equalization of Dispersive Radio Channels". *Bell System Technical Journal*, 33(7), July 1985.

[6] J. W. Mark and W. Zhuang. *Wireless Communications and Networking*. Pearson Education, 2003.

Chapter 6

Coding

6.1 INTRODUCTION

Coding has helped make modern data communications possible. From the compact disc to the transmission of images from the far reaches of our solar system, almost every digital message is coded. There are two broad categories of coding:

Source coding. Sources such as voice, text, and pictures generally contain redundancy. Source coding attempts to reduce the amount of data that must be transmitted by removing redundancy in the source data.

Channel coding. RF transmission channels have multiple hazards such as fading, multipath, and noise. These can cause serious damage to the received communications signal, making detection difficult. Channel coding attempts to introduce controlled redundancy to help to protect the data signal.

Hopefully the reader of this chapter can scratch the surface of the subject of coding. However, please understand that this is no deep dive into the subject of coding. For that, you need something like the 1,250 pages of [1].

Figure 6.1 is an example of voice processing used in some cell phones. The input data rate is raw voice samples at 64,000 bits per second. The vocoder (voice coder) removes redundancy in the human voice, reducing the data rate to 9600 bps.[1] The convolutional channel coder adds known redundancy to facilitate error correction at the receiver.

---

1 Vocoders are complicated algorithms that reduce voice samples to a set of model parameters.

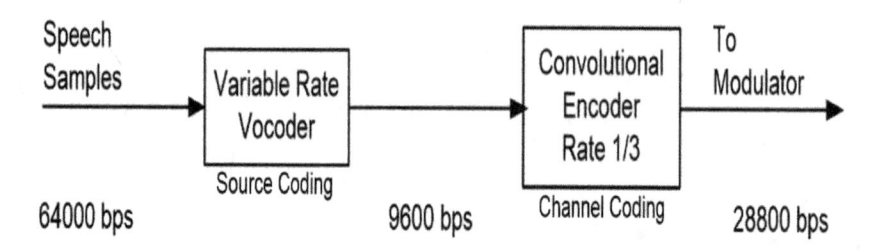

```
     Speech                                         Convolutional      To
     Samples      Variable Rate                       Encoder       Modulator
              →     Vocoder      →                    Rate 1/3    →

                  Source Coding
     64000 bps                    9600 bps         Channel Coding     28800 bps
```

Figure 6.1 A simple source and channel coding example.

6.2 SOURCE CODING

The amount of information in a random event depends upon just how random the event is. Consider a random variable x with sample space $S_x = \{A, B\}$. Outcomes A and B represent the only two messages that a weather station in Death Valley, California can send. The messages and associated probabilities are:

Message A: "It is sunny" $p_x(A) = 0.9$
Message B: "It is raining" $p_x(B) = 0.1$

Self-information in these two messages is:

$$h(A) = -\log_2(p_x(A)) = 0.152$$
$$h(B) = -\log_2(p_x(B)) = 3.322 \qquad (6.1)$$

Notice that the self-information in a rare event such as message B is greater because it is unexpected. Message A has low self information because not much is learned from receiving it. The expected value of self information is called the entropy.[2]

$$H(X) = -\left(p_x(A)\log_2(p_x(A)) - p_x(B)\log_2(p_x(B))\right)$$
$$= -0.9(-0.152) - 0.1(-3.322) = 0.47 \text{ bits} \qquad (6.2)$$

Entropy $H(x)$ is the minimum number of bits required to encode this source. Let's take a simple example. Entropy can be thought of as the average information obtained by observing an outcome. Consider a coin toss: $p_x(heads) = p$, $p_x(tails) = 1 - p$. Random variable x is assigned to represent

---

[2] In thermodynamics, entropy is the amount of disorder in a system. Here entropy is the level of uncertainty or surprise in a probability experiment.

CHAPTER 6. CODING

the outcome of the coin toss experiment. For $p = 0.5$, the entropy of x is shown in Equation (6.3).

$$\begin{aligned} H(x) &= -\left(p_x\left(heads\right)\log_2\left(p_x\left(heads\right)\right) - p_x\left(tails\right)\log_2\left(p_x\left(tails\right)\right)\right) \\ &= -(0.5(-1) - 0.5(-1)) = 1 \text{ bit} \end{aligned} \quad (6.3)$$

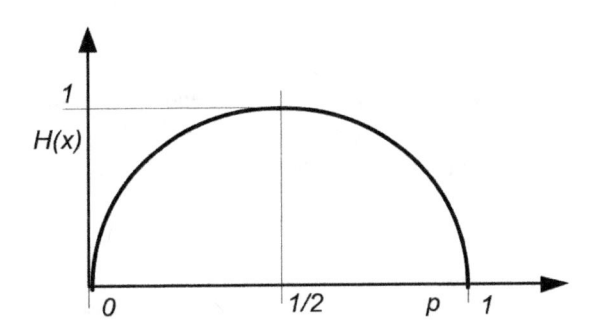

Figure 6.2 The entropy of a coin toss.

A fair coin has a probability of 1 = probability of 0 = 1/2. The entropy of a fair coin, shown in Figure 6.2, equals 1 because its toss outcome is completely surprising. Thus, 1 bit is required to encode a fair coin. Note that for $p = 0$ or $p = 1$ there is no new information, entropy = 0, and no bits are required.

6.2.1 Weather Station Encoding

Getting back to the weather station, we know from Equation (6.2) that a coder can be designed that needs an average of $H(x) = 0.47$ bits. Thus, if the weather station sends independent observations x at rate r per second, then the minimum source rate R_{min} is calculated by Equation (6.4).

$$R_{min}\left(\frac{bits}{second}\right) = r\left(\frac{source\ outputs}{second}\right) H(X) \left(\frac{bits}{source\ output}\right) \quad (6.4)$$

Equation (6.4) shows the minimum theoretical bit rate for this source but does not design the coder for us. Can we get a low source rate by grouping observations? Let's try assigning code words according to joint probability:

From Table 6.1, the expected value of number of bits needed for each observation pair is $1.2 = 0.81(1) + 0.09(4) + 0.01(3)$. Divide this by 2 to get 0.6, the number of bits needed for each observation.

Table 6.1 A Source Coding Example

Pairs of Observations	Joint Probability	Code Word	Bits per Pair
A,A	0.81	0	1
A,B	0.09	10	2
B,A	0.09	11	2
B,B	0.01	111	3

Let's summarize. Say the weather station sends independent observations x at rate $r = 100$ observations per second. We have three ways to code this source: no source coding, suboptimal source coding that we just designed, and optimal source coding based on entropy.

$$R_{noCoding} = r = \left(\frac{100 \text{ source outputs}}{\text{second}}\right)\left(\frac{1 \text{ bit}}{\text{source output}}\right) = \frac{100 \text{ bits}}{\text{second}}$$

$$R_{suboptimumCoding} = rH(X) = \left(\frac{100 \text{ source outputs}}{\text{second}}\right)\left(\frac{0.6 \text{ bits}}{\text{source output}}\right) = \frac{60 \text{ bits}}{\text{second}} \quad (6.5)$$

$$R_{optimumCoding} = rH(X) = \left(\frac{100 \text{ source outputs}}{\text{second}}\right)\left(\frac{0.47 \text{ bits}}{\text{source output}}\right) = \frac{47 \text{ bits}}{\text{second}}$$

So just by guessing a coding scheme, we went from 100 to 60 bits/second. An exercise for the reader is to design a better source coder to get closer to the theoretical optimum rate of 47 bits/second. Shannon's source coding theorem tells us that 0.47 bits per source output is the absolute minimum. So if you get that low, you are done.

6.2.2 Lempel-Ziv Coding

Lempel-Ziv (LZ) is a source coding algorithm for removing redundancy in text and binary. LZ is a variable to fixed coding scheme; code words are always fixed length, but they can represent longer or shorter data segments. Unlike Huffman coding, an older technique that has been widely used, LZ does not require advance knowledge of the source statistical distribution (for example, the probability of occurrence of letters in the alphabet).[3]

---

[3] In some cases, this source distribution is known and Huffman coding may be a better choice. For example, black and white pixels on a newspaper page tend to have a consistent distribution.

CHAPTER 6. CODING

LZ works by parsing the source data stream into segments that are the shortest subsequences not seen previously. We will explain by example. Let's say that we need to LZ encode the input binary test sequence: 000101110010100101. This procedure is shown in Table 6.2. Starting at the left, each step builds up a code book for this data by selecting the shortest subsequence of data not yet seen.

Table 6.2 A Lempel-Ziv Coding Example

Step	Data to be Parsed	Code Book Subsequences
0	000101110010100101	0,1
1	**00**0101110010100101	0,1,00
2	000**10**1110010100101	0,1,00,01
3	000**101**110010100101	0,1,00,01,011
4	000101**11**0010100101	0,1,00,01,011,10
5	000101110**010**100101	0,1,00,01,011,10,010
6	000101110010**100**101	0,1,00,01,011,10,010,100
7	000101110010100**101**	0,1,00,01,011,10,010,100,101

Table 6.3 LZ Code Book Example

Code	1	2	3	4	5	6	7	8	9
Seq	0	1	00	01	011	10	010	100	101
Rep			11	12	42	21	41	61	62
Out			0010	0011	1001	0100	1000	1100	1101

Our example code book is shown in Table 6.3. This is derived from step 7, column 3 in Table 6.2. In Table 6.3, the first line is the code book index, the second line is the subsequence, the third line is the integer representation and the fourth line is the code word output. The bold 0 or 1 is called the innovation bit, the bit that makes a code word unique.

To encode 011, we send 1001. The decoder points to code book 100 or 4. The decoder finds a 01 at codebook position 4 and appends the innovation bit of 1 for a decoded result of 011. To encode 10, send 0100. The decoder points to codebook 010 or 2 The decoder finds a 1 at codebook position 2 and appends a 0 innovation bit for a decoded result of 10. The reader may not find the data compression of these simple examples very impressive. However, let's look at a much longer example.

The MATLAB script file below is an LZ encoder. Input LenAddr is the number of bits in the code book address. Input InputBits is a vector of integer 0 or 1 to be coded. The InputBits are scanned through and concatenated into variable temp until temp is no longer found in the code book. That is the signal to find the latest unused code book address and use it to form the first LenAddr bits of the current output code word. The last bit of the current output code word is the innovation bit, the last bit in temp (i.e., the bit that made temp unique and not in the code book). This goes on until the end of the input data is reached.

```
function output = LZencode(LenAddr,InputBits)
CodeBookMax = power(2,15);
CodeBook = [];
temp = [];
output = "";
as2 = strcat('%0',num2str(addressBit),'d');
as1 = sprintf(as2,0);
for i=1:length(input)
temp = [temp,convertStringsToChars(string(input(i)))];
% Concatenate next input bit to next until it becomes
% unique (not in codeBook)
if (isempty(find(strcmp(CodeBook,temp)))...
||i==length(input))
% See if there is room in CodeBook
if length(CodeBook) < CodeBookMax) CodeBook...
= [CodeBook,string(temp)];
% Special case for single 1 or 0. Send 0 address (as1).
if (length(temp)==1) output=[output strcat(as1,temp)];
else
tempadd = dec2bin(find(strcmp(CodeBook,temp(1:end-1))));
asa = (sprintf(as2,str2num(tempadd)));
%Concatenate address to innovation bit to make next
% fixed length codeword
%Concatenate next fixed length
%codeword to current output
output = [output strcat(asa, temp(end))];
end
temp = [];
end
end
```

```
save CodeBook.mat CodeBook;  % Decoder needs this
```

There are two primary factors affecting the data compression ratio (i.e., number of input bits over number of output bits). The first factor is code book length. The code book contains $2^{LenAddr}$ entries. A long source sequence of bits might use all of these and then coding must stop. Some ad hoc solutions can be tried. When code words are used up, older codes could be purged to make room for new, probably longer, codes. The purging must be coordinated at the coder and decoder. The second factor affecting data compression is the source entropy. A sequence of bits x with $H(x) = 1$ (i.e., a fair coin toss) will result in no compression; see [2]. Since most text and picture data is not completely random, we test with an entropy $H(x) = 0.8$ source, as generated below:

```
% Binary source for testing LZ
LL =50000;
input = zeros(1,LL);
inp = rand(1,LL);
ixh = find(inp>=0.75);
input(ixh) = 1;
% H(input) = -0.25*log2(0.25) - 0.75*log2(0.75) = 0.8
```

6.2.2.1 Lempel-Ziv Coding Results

Table 6.4 shows some results for long sequences of input bits and a 32,768 entry code book. The encoder starts to show data compression for an input sequence of 520,000 bits.[4] However, as the input sequence gets up to 640,000, the code book runs out of space for new patterns. The entropy is fixed at 0.8, as described above.

---

4 Note that every output word is a fixed length code book address of 15 bits plus one additional innovation bit. Thus, to achieve data compression, the code book entries will need to represent patterns much longer than 16 bits.

Table 6.4 LZ Coding Test Results, Code Book Address = 15 Bits

Number Bits In	Number Bits Out	Compression	Errors
280,000	295,632	0.947	0
440,000	444,496	0.990	0
520,000	515,536	1.009	0
600,000	588,208	1.020	0
640,000	623,217	1.030	2,408

6.3 CHANNEL CODING

So far, we have looked at ways to reduce the source rate by removing redundancy. We showed a couple of simple examples of this. After source coding, the next step is to get the data from transmitter to receiver over a generally noisy channel. There are two broad categories of channels and they require slightly different channel coding:

Power-limited Maximum power is fixed but bandwidth expansion can improve performance. Among the many examples is an orbiter to lander space communications channel. There are several coding options for this situation.
Bandwidth-limited Maximum channel bandwidth is fixed, but performance can be improved by increasing power or by trellis coding. Analog telephone lines are generally band-limited.

Channel coding is a process of adding coding bits to detect and/or correct errors in received data. The added channel coding bits have special properties for this critical job. Consider two primary methods of channel coding that add coding bits to power-limited channels. These improve performance by increasing bandwidth but not power.[5]

Block coding Blocks of parity bits are appended to distinct blocks of transmit data.
Convolutional coding A continuous stream of transmit data is converted into a continuous stream of encoded data, bit by bit.

6.3.1 Block Coding

Block coding transforms distinct blocks of k successive payload symbols into longer blocks of $n > k$ coded symbols. We call this an (n, k) block code. As shown in

---

5 Many practical channels have both power and bandwidth limitations.

CHAPTER 6. CODING

Figure 6.3, there are 2^k payload messages and 2^n total code words; see also Table 6.5. These codes have the following properties:

Linear Any sum of code words is also a code word.
Systematic The resulting code words contain the source bits unchanged.
Optimal Minimum distance between code words is maximum for given (n, k).

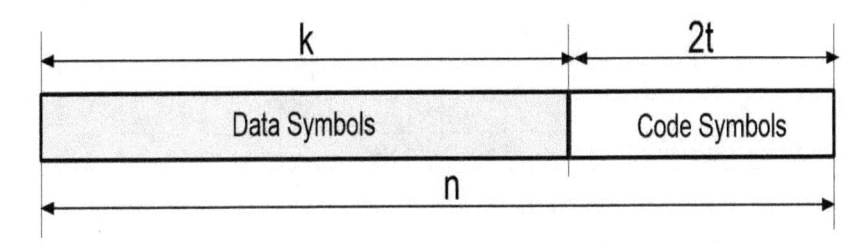

Figure 6.3 The basic structure of block coding.

k = number of payload symbols (message) in a block.

n = number of payload and parity symbols in a block.

n-k = number of parity symbols in a block.

Table 6.5 presents a (6,3) code to illustrate the basic idea: $n = 6$, $k = 3$. In Table 6.5 bold type code words carry messages, as well as provide for error detection and correction. For example, say code word 110100 is transmitted and code word 111100 = 110100+001000 = transmit code word + error is received. A (6,3) block code can correct one error and detect two errors, so this error will be corrected. Let's see how this works.

6.3.1.1 Encoding Block Codes

Block code words can be formed from a set of k linearly independent[6] n element rows, called a generator matrix. Equation (6.6) is an example. The identity matrix in the last three columns makes this a systematic code. Equation (6.7) has three examples of using a generator matrix for our (6,3) code. The 3-bit message is multiplied by the generator matrix and the result is the code word corresponding to that message.

---
6 Linear independent means no row can be written as a linear combination of the others.

Table 6.5 Block Coding, All Possible Code Words for $n = 6$

Message	Code Word	Message	Code Word	Message	Code Word	Message	Code Word
000	**000000**		010000		100000		110000
	000001		010001		100001		110001
	000010		010010		100010		110010
	000011		010011		100011	011	**110011**
	000100		010100		100100	100	**110100**
	000101		010101		100101		110101
	000110		010110		100110		110110
111	**000111**		010111		100111		110111
	001000		011000		101000		111000
	001001		011001	001	**101001**		111001
	001010	010	**011010**		101010		111010
	001011		011011		101011		111011
	001100		011100		101100		111100
	001101	101	**011101**		101101		111101
	001110		011110	110	**101110**		111110
	001111		011111		101111		111111

CHAPTER 6. CODING

$$G = \begin{bmatrix} P & I_k \end{bmatrix} = \begin{bmatrix} 1 & 1 & 0 & 1 & 0 & 0 \\ 0 & 1 & 1 & 0 & 1 & 0 \\ 1 & 0 & 1 & 0 & 0 & 1 \end{bmatrix} \quad (6.6)$$

$$\begin{bmatrix} 0 & 0 & 0 \end{bmatrix} \begin{bmatrix} 1 & 1 & 0 & 1 & 0 & 0 \\ 0 & 1 & 1 & 0 & 1 & 0 \\ 1 & 0 & 1 & 0 & 0 & 1 \end{bmatrix} = \begin{bmatrix} 0 & 0 & 0 & 0 & 0 & 0 \end{bmatrix}$$

$$\begin{bmatrix} 1 & 1 & 0 \end{bmatrix} \begin{bmatrix} 1 & 1 & 0 & 1 & 0 & 0 \\ 0 & 1 & 1 & 0 & 1 & 0 \\ 1 & 0 & 1 & 0 & 0 & 1 \end{bmatrix} = \begin{bmatrix} 1 & 0 & 1 & 1 & 1 & 0 \end{bmatrix} \quad (6.7)$$

$$\begin{bmatrix} 0 & 1 & 0 \end{bmatrix} \begin{bmatrix} 1 & 1 & 0 & 1 & 0 & 0 \\ 0 & 1 & 1 & 0 & 1 & 0 \\ 1 & 0 & 1 & 0 & 0 & 1 \end{bmatrix} = \begin{bmatrix} 0 & 1 & 1 & 0 & 1 & 0 \end{bmatrix}$$

6.3.1.2 Detecting and Correcting Block Code Errors

For each generator matrix G, there is a corresponding parity check matrix, H, used to calculate the received syndrome S. $GH^T = 0_k$ where 0_k refers to a k-by-k matrix of all zeros.

$$H = \begin{bmatrix} I_{n-k} & P^T \end{bmatrix} = \begin{bmatrix} 1 & 0 & 0 & 1 & 0 & 1 \\ 0 & 1 & 0 & 1 & 1 & 0 \\ 0 & 0 & 1 & 0 & 1 & 1 \end{bmatrix} \quad (6.8)$$

To find H, G must first be in systematic form; this means $G = \begin{bmatrix} P & I_k \end{bmatrix}$, where P is a k by $n-k$ full rank matrix that sets the properties of the code and I_k is a k by k identity matrix; for example, Equation (6.6). In this case, $H = \begin{bmatrix} I_{(n-k)} & P^T \end{bmatrix}$.

For received vector r, $S = rH^T$. A syndrome of 0 represents correct reception; we say that correct code word is in the null space of H. Let's try an example. For message bits = [110], Table 6.5 tells us the code vector r = [101110]. Assume these bits are correctly received. Equation (6.9) calculates a zero syndrome indicating no errors.

$$S = rH^T = \begin{bmatrix} 1 & 0 & 1 & 1 & 1 & 0 \end{bmatrix} \begin{bmatrix} 1 & 0 & 0 \\ 0 & 1 & 0 \\ 0 & 0 & 1 \\ 1 & 1 & 0 \\ 0 & 1 & 1 \\ 1 & 0 & 1 \end{bmatrix} = \begin{bmatrix} 0 & 0 & 0 \end{bmatrix} \qquad (6.9)$$

$$S = rH^T = \begin{bmatrix} 0 & 0 & 1 & 1 & 1 & 0 \end{bmatrix} \begin{bmatrix} 1 & 0 & 0 \\ 0 & 1 & 0 \\ 0 & 0 & 1 \\ 1 & 1 & 0 \\ 0 & 1 & 1 \\ 1 & 0 & 1 \end{bmatrix} = \begin{bmatrix} 1 & 0 & 0 \end{bmatrix} \qquad (6.10)$$

Now let's say for the same message bits, the received code vector has an error, r= **0**01110. Equation (6.10) calculates a nonzero syndrome, $\begin{bmatrix} 1 & 0 & 0 \end{bmatrix}$, indicating one or more errors. Table 6.6 matches that syndrome to an error pattern of $\begin{bmatrix} 1 & 0 & 0 & 0 & 0 & 0 \end{bmatrix}$. Now the received code word can be corrected, as shown in Equation (6.11). If the received code word has one error, it will be corrected. For the syndrome = 111, two errors can be corrected; see Table 6.6. This correction process may break down for high error rates.

$$\begin{aligned} r_{corrected} &= \begin{bmatrix} 0 & 0 & 1 & 1 & 1 & 0 \end{bmatrix} \oplus \begin{bmatrix} 1 & 0 & 0 & 0 & 0 & 0 \end{bmatrix} \\ &= \begin{bmatrix} 1 & 0 & 1 & 1 & 1 & 0 \end{bmatrix} \end{aligned} \qquad (6.11)$$

6.3.1.3 Block Code Performance

To analyze performance, we first need d_{min}, the minimum distance of the code. The minimum distance of a linear code equals the minimum weight, or minimum number of ones, of its nonzero code words. For the (6, 3) code, this turns out to be 3. This is easy to see from the bold type code words in Table 6.5.

The guaranteed error correcting capability of a block code is $(d_{min} - 1)/2$ and the error detecting capability is $(d_{min} - 1)$. These are 1 and 2, respectively, for our example code.

CHAPTER 6. CODING

Table 6.6 All Possible Block Coding Error Patterns for $n = 6$

Syndrome	Error Pattern
000	000000
101	000001
011	000010
110	000100
001	001000
010	010000
100	100000
111	010001

6.3.1.4 Alternative Notation

Sometimes discrete time polynomial notation is used to describe coding. This can make multiplication and long division easier. For example, $f(x) = x^5 + x^4 + x^2 + x^0$ is equivalent to 110101 and $g(x) = x^3 + x^0$ is equivalent to 101. In the polynomial, replace x by 2 to get the equivalent decimal number. Equation (6.12) is an example of multiplying two numbers using different representations.

$$\begin{aligned}
f(x) &= x^5 + x^4 + x^2 + x^0 = 110101 = 53 \\
g(x) &= x^3 + x^0 = 1001 = 9 \\
f(x)g(x) &= (x^5 + x^4 + x^2 + x^0)(x^3 + x^0) = 53 * 9 = 477 \\
&= (x^5 + x^4 + x^2 + x^0)x^3 + (x^5 + x^4 + x^2 + x^0)x^0 \\
&= x^8 + x^7 + x^4 + x^3 + x^2 + x^0 \\
&= (110101 << 3) + (110101 << 0) \\
&= 110101000 + 000110101 = 110011101 = 477
\end{aligned} \quad (6.12)$$

6.3.2 Block Coding Example

Let's work through a practical communications link example using block coding. We require: $R = 9{,}600$ bits/second at BER (bit error rate) $< 10\text{e-}7$ at $E_b N_0$ dB $= 11.7$ dB ($E_s N_0$ dB $= 16.5 = 11.72 + 10 log_{10}(3)$). We are given $W = 4{,}000$ Hz bandwidth. This channel is bandwidth-limited and also power-limited because we are not given the option to increase $E_b N_0$ by simply raising the transmit power.

We select 8PSK for modulation; see Figure 6.4. These symbols are Gray coded, meaning each one differs in only 1 bit position from its nearest neighbor. For $M = 8$

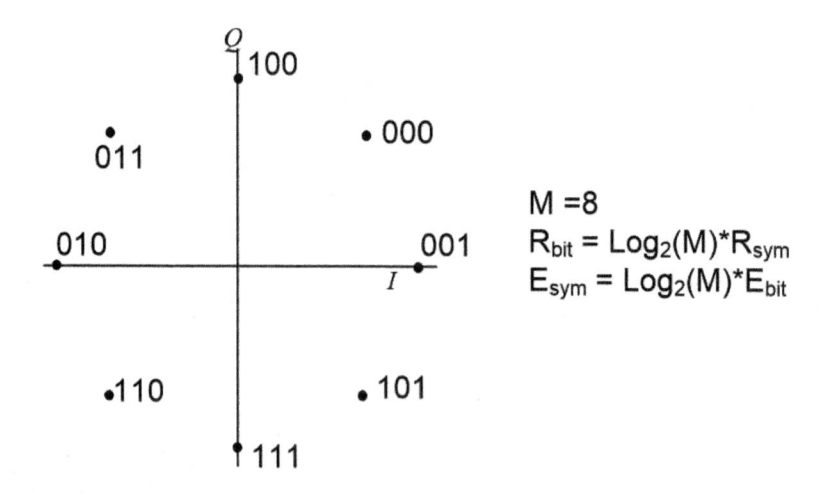

```
            Q
            ↑ 100
        
    •                • 000
    011
                                    M = 8
    010            001              R_bit = Log2(M)*R_sym
    ─────────────┼──•───── I        E_sym = Log2(M)*E_bit


    •110           • 101

            ↑ 111
```

Figure 6.4 An 8PSK Gray coded constellation and parameters.

the symbol rate = $9{,}600/(\log_2 M) = 3{,}200$ symbols/sec. This is less than our channel BW of 4,000 Hz, so the signal fits. However, we still need BER < 10e-7. Let's see where we are now. From [3], for PSK, $M > 2$ we have:

$$P(\text{symbol error}) = 2Q\left[\sqrt{\frac{2E_s}{N_0}}\sin\left(\frac{\pi}{M}\right)\right] \tag{6.13}$$

See Section 3.2.4.1 for the definition of the Q function. For our 8PSK example, $E_s N_0$ dB = 16.5 dB, which is equivalent to $E_s N_0 = 10^{\left(\frac{E_s N_0 (\text{dB})}{10}\right)} = 44.668$, as a ratio (not dB).

Equation (6.13) results in SER = 2.98e-4, BER = SER/3 = 9.93e-05, not good enough. To meet the BER < 10e-7 requirement, we turn to block coding. The uncoded minimum BW requirement is 3,200 Hz. We have 4,000 Hz available. Thus we can utilize a block code that improves BER by increasing signal BW by 25%. Let's take a look at Bose-Chaudhuri-Hocquenghem (BCH) block codes in Table 6.7. Note the coding gain in the last column. Say without coding we are achieving BER

CHAPTER 6. CODING

= X at $E_b N_0$ = Y dB, then with a coding gain[7] of 2 dB, we can achieve BER = X at $E_b N_0$ = (Y-2) dB.

Table 6.7 Some Choices for BCH Block Coding

Total (n)	Info (k)	Code (n-k)	%BW	Correct (t)	Gain G(dB)
31	26	5	19	1	2.0
63	57	6	11	1	2.2
63	51	12	24	2	3.1
127	120	7	6	1	2.2
127	113	14	12	2	3.3
127	106	21	20	3	3.9

From Table 6.7, the (63,51) code looks promising. A 24% bandwidth increase makes our symbol rate 3,200*1.24 = 3,968, which still fits the 4,000 Hz channel (with $\pm 16 Hz$ leftover for Doppler carrier frequency shift). The next step is to calculate the performance obtained. Given the probability of bit error = p, the probability of j errors in any block of n bits is based on the formula for the number of combinations of n items, taken j at a time.[8] We know this (63,51) BCH code can correct $t = 2$ errors in an n bit block. This implies that the code will fail to produce error-free blocks for number of errors = $t+1, t+2, \ldots n$. Thus, the probability of a block error is:

$$P(\text{block error}) = \sum_{j=t+1}^{n} \binom{n}{j} p^j (1-p)^{n-j}$$

$$\binom{n}{j} = \frac{n!}{j!(n-j)!}$$

(6.14)

To understand Equation (6.14), consider an $n = 4$ bit, $t = 1$ block code that can correct 1 error. This code cannot correct two errors, so j starts at 2 in Equation (6.14). Thus, an occurrence of two errors is one way to get a block error. As shown in Table 6.8, there are six ways to get two errors $\frac{4!}{(2!)(2!)} = 6$. The probability of

---

[7] Coding gain is not linear; it generally gets lower for smaller values of $E_b N_0$ and higher BER. The $E_b N_0$ where the coding gain is measured is missing from Table 6.7. Fortunately, we will not need it for this example.

[8] $\binom{n}{j}$ is the number of combinations of n bits taken j at a time. The order of the elements in a combination does not matter. For example, a, b is the same combination as b, a.

Table 6.8 All Combinations of 4, Taken Two at a Time, with Pairs of Possible Errors Shaded

b_1	b_2	b_3	b_4
b_1	b_2	b_3	b_4
b_1	b_2	b_3	b_4
b_1	b_2	b_3	b_4
b_1	b_2	b_3	b_4
b_1	b_2	b_3	b_4

any of the six is $\left(p^2(1-p)^2\right)$. This probability is summed six times in Equation (6.14). The probability for the number of ways to get three and four errors are also summed to get the total probability of block error in this situation.

Given n bits/block, Equation (6.15) expresses the decoder output BER as a function of the channel demodulator BER.

$$P_{\text{decoder}} \text{ (block error)} = \sum_{j=t+1}^{n} \binom{n}{j} p^j (1-p)^{n-j} \qquad (6.15)$$

$$P_{\text{decoder}} \text{ (bit error)} (t) = \frac{1}{n} P_{\text{decoder}} \text{ (block error)}$$

$$\left(\frac{E_s}{N_0}\right)_{\text{Channel}} = \left(\frac{51}{63}\right) 44.668 = 36.160 \qquad (6.16)$$

Coding has increased the channel symbol rate from 3,200 to 3,200(1.24) = 3,968 symbols/sec. Transmitting 51 information bits now requires 63 total bits; increasing the bit rate and lowering E_s, as shown in Equation (6.16). Making use of Equation (6.13), we see that the BCH encoded channel bit error rate has actually increased from the original BER = 9.933e-5 to the demodulator output BER 3.788e-4. This is measured prior to the BCH decoder.

$$P_{\text{demod}} \text{ (symbol error)} = 2Q\left[\sin\left(\frac{\pi}{M}\right)\sqrt{2\left(\frac{E_s}{N_0}\right)_{\text{Channel}}}\right] = 1.1e-3$$

$$P_{\text{demod}} \text{ (bit error)} = \frac{P_{\text{demod}} \text{ (symbol error)}}{\log_2(M)} = 3.788e-4$$

$$(6.17)$$

CHAPTER 6. CODING

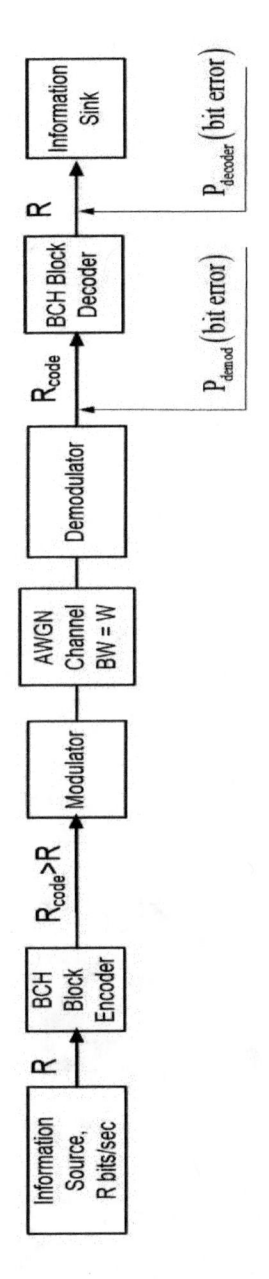

Figure 6.5 Context for a BCH block coding example.

We are most interested in the error rate out of the t error correcting BCH decoder. Using Equation (6.15) and plugging in P_{demod} (bit error) = 3.788e-04, $t = 2$, $n = 63$ we calculate:

$$P_{decoder} \text{ (bit error)} = 3.368\text{e} - 08 \quad (6.18)$$

This is less than our goal of BER = 1e-7, so we are finished. This analysis is summarized in Figure 6.5. The MATLAB code, next, shows the calculations used in this example.

```
% Performance Requirements
EsN0 dB = 16.5;
EbN0 dB = EsN0 dB - 10*log10(3);
EsN0 = 10^(EsN0 dB/10);

% Starting Performance
SER = 2*Qfunc(sqrt(2*EsN0 )*sin(pi/(8)));
BER = SER/3;

% Channel Reduction due to coding
EsN0 Ch = (51/63)*EsN0 ;
EbN0 Ch = EsN0 Ch/3;
SERCh = 2*Qfunc(sqrt(2*EsN0 Ch)*sin(pi/(8)));
BERCh = SERCh/3;

% Performance at BCH decoder output
p = BERCh;
Pblk = 0;
t= 2;
n = 63;
for j = (t+1):n
Pblk = Pblk + (nchoosek(n,j)*(p^j)*( (1-p)^(n-j) ));
end
Pbit = Pblk/n;

% Complementary error function:
function [y]= Qfunc(x)
```

CHAPTER 6. CODING

```
y = 0.5*(erfc(x/sqrt(2)));
```

6.3.3 Convolutional Coding

Convolutional coding inputs a continuous stream of uncoded bits and outputs a continuous stream of channel coded bits. The coder stores some of the most recent input bits. These, together with the new bit, form an internal state. Also, the parameter definitions are slightly different than block codes:

k = number of input information bits per step.

n = number of output coded bits per step.

K = number of bits in encoder state (constraint length).

The convolutional coding procedures described below may be implemented in, for example, C language. However, Simulink can relieve the user from this chore. Note how Figure 13.10 includes a Viterbi encoder and Figure 13.31 includes a Viterbi decoder. These are found in the Simulink Communication System toolbox. The help files for these blocks describe how to set up a particular trellis. The text below describes what a convolutional encoding trellis is and how it works.

6.3.3.1 Encoding Convolutional Codes

As Figure 6.6 shows, convolutional codes are not systematic because the input data is not directly visible in the output coded bit stream. Convolutional codes are considered linear because the encoder obeys the principle of superposition. This means impulse responses to each shifted bit can be generated separately and added together (superimposed) to produce the same result as the impulse response from one continuously shifted input bit.

Figure 6.7 is an example of a $k = 1$, $n = 2$, $K = 3$ encoder state diagram corresponding to Figure 6.6. The state is $\begin{bmatrix} x_i & x_{i-1} \end{bmatrix} = \begin{bmatrix} S_0 & S_1 \end{bmatrix}$. The S_2 helps determine the output bits, however, S_2 is not included in the state because it will be shifted out to the right when the state changes. Figure 6.7 shows the four states enclosed in circles. Starting anywhere in the diagram, an input bit of 1 is shown with a dotted transition line and 0 uses a solid transition line. In addition to a state change, each new input bit results in two output coded bits. These are shown on the transition lines. An alternative to the state diagram is the trellis diagram, shown in Figure 6.8. This diagram shows the progression of states symbol by symbol, from

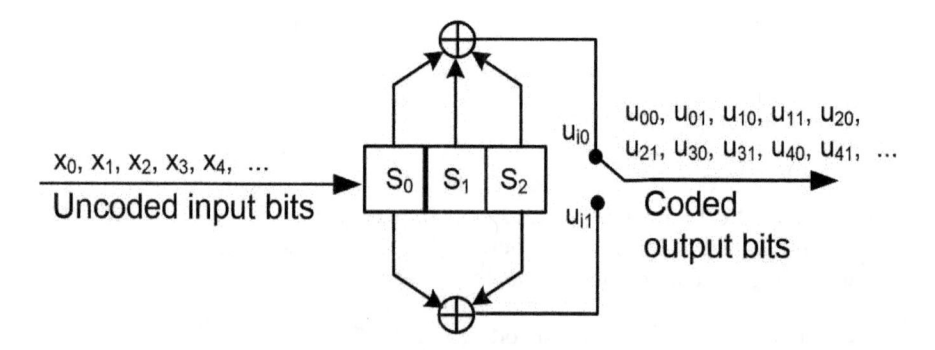

Figure 6.6 Convolutional coding example structure, $k = 1$, $n = 2$, $K = 3$.

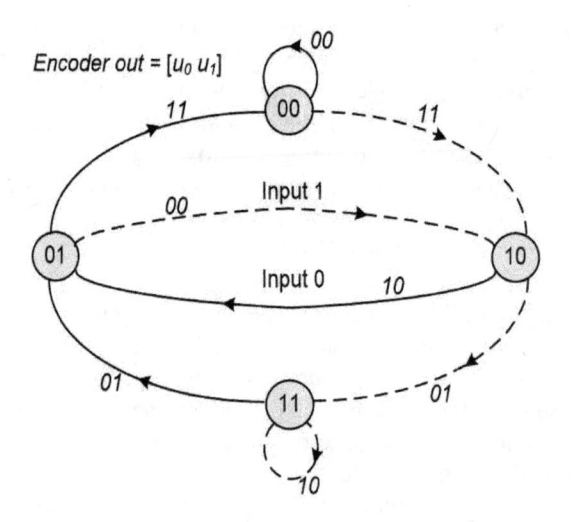

Figure 6.7 Convolutional coding state diagram, $k = 1$, $n = 2$, $K = 3$.

left to right. Notice that not every path through the trellis is allowed. This is part of the reason that convolutional coding is effective. A trellis with all possible paths valid will have zero coding gain.

CHAPTER 6. CODING

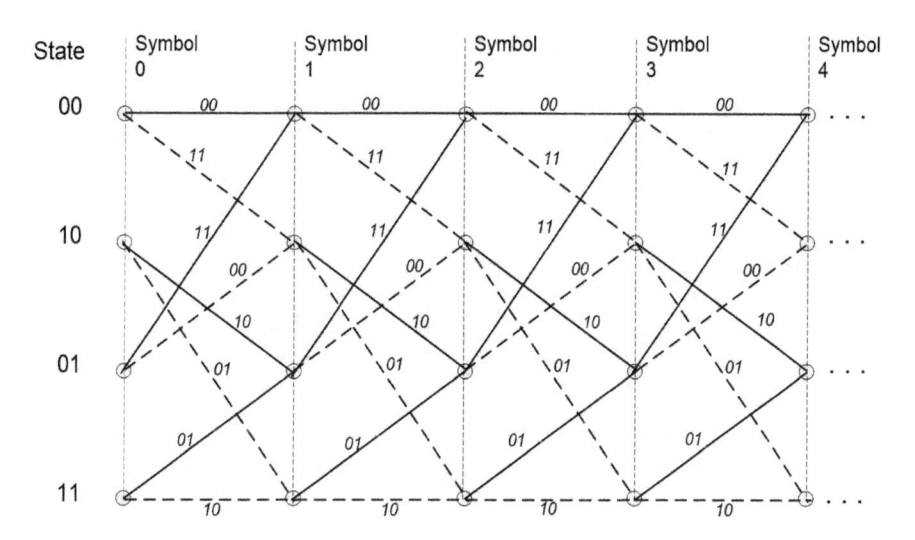

Figure 6.8 Convolutional coding trellis diagram, input 1 dashed, input 0 solid.

6.3.3.2 Decoding Convolutional Codes

Prior to the channel decoder is a filter matched to the transmit symbol shape, which optimizes the SNR. As shown in Figure 6.9, the matched filter output is detected into a hard decision (HD) composed of an absolute plus or minus or a soft decision (SD) which indicates the degree of certainty toward a plus or minus decision. Figure 6.10 generates QPSK symbol decisions by taking one SD from the in-phase rail and one SD from the quadrature rail. HDs are from the set $\begin{bmatrix} 11 & 01 & 00 & 10 \end{bmatrix}$. SDs indicate the same four outcomes but also indicate the confidence levels. For example, in-phase levels of +0.5 and +1.2 are both decided as plus however, +1.2 has a higher confidence level than the +0.5. In other words, +0.5 is more likely to be changed to minus by noise than is +1.2.

Convolutional decoders often input soft decisions for better performance. We will follow an example of rate 1/2 encoding and decoding using two SDs, $\begin{bmatrix} b_I & b_Q \end{bmatrix}$. As shown in Figure 6.6, a single uncoded transmit bit generates a two bit code word. These two transmit bits become 2 received SDs; we can map each received SD pair to the four-state constellation in Figure 6.10.

Comparing the actual received SDs with the four possible transmit pairs, four branch metrics (BM) are calculated using the MATLAB code next. These BMs are

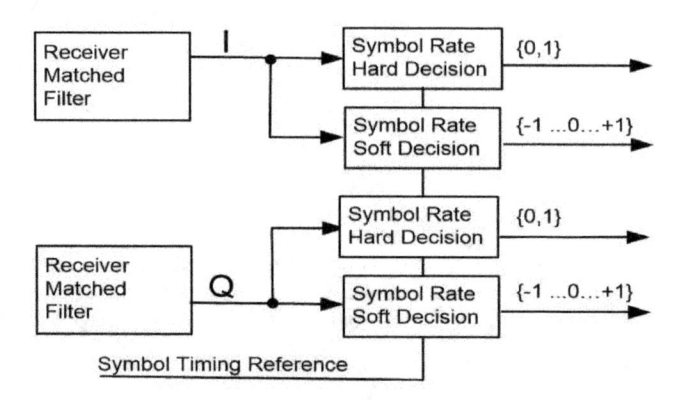

Figure 6.9 The detector output just prior to channel decoding.

the Euclidean distance between received point and each of the four hard decision points in Figure 6.10.

```
a = [0.30 0.24]; % next received symbol
BM11 = FindDist(a(1),a(2), 1,1);
BM01 = FindDist(a(1),a(2), -1,1);
BM10= FindDist(a(1),a(2), 1,-1);
BM00 = FindDist(a(1),a(2), -1,-1);

function dist = FindDist(x0,y0,x1,y1)
% This is Euclidean distance dist = sqrt((x0-x1)*(x0-x1)
+ (y0-y1)*(y0-y1));
```

If the matched filter output is hard decision sliced around zero where HD = 0 implies input less than zero and HD = 1 implies input greater than zero (see Figure 6.10), the Hamming distance is appropriate. Branch metrics are based on 00, 01, 10, 11 and Hamming distance is the number of bit differences (e.g., Hamming distance between 01 and 10 is 2). If the matched filter output is soft decision sliced around zero where -1 implies less than zero and +1 implies greater than zero, the Euclidian distance measure is used (as in this example).

CHAPTER 6. CODING

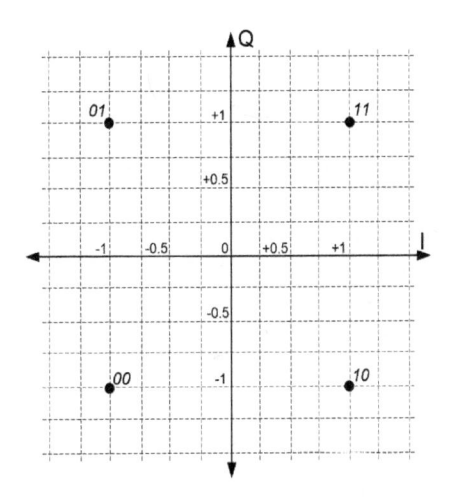

Figure 6.10 Hard decisions versus soft decisions.

6.3.3.3 Viterbi Decoder Example

Table 6.9 is an example of received two bit SDs representing the code words in the third column. Using the MATLAB code above, for each received SD the four calculated BMs are shown across the corresponding row.

Table 6.9 Convolutional Coding Branch Metric Examples

Col.	Tx	Code	Received SD	BM11	BM01	BM10	BM00
1	1	11	[0.30, 0.24]	1.0332	1.5059	1.4239	1.7966
2	0	10	[0.33, -0.28]	1.4447	1.8459	0.9835	1.5124
3	1	01	[-0.30, 0.04]	1.6160	1.1881	1.6648	1.2536
4	1	01	[-0.24, 0.32]	1.4142	1.0198	1.8111	1.5232
5	1	10	[0.24, -0.22]	1.4374	1.7395	1.0890	1.4649

We are now ready to make use of the trellis diagram of Figure 6.8 to implement the Viterbi convolutional decoder algorithm [4]. The states shown in Figures 6.6 and 6.7, gray circles, correspond to the four states on the far left side of Figure 6.8. Notice that, like Figure 6.7, the trellis has 0-bit transitions (solid line) and 1-bit transitions (dashed line). In Figure 6.7 and Figure 6.8; bits were input to the encoder that determined the state transition and the decoder two bit code word output. In the

Viterbi decoder algorithm, we reverse this by starting with the receive matched filter [I, Q] SD code word outputs in Figure 6.9 and estimate the state transition and the encoder input. The [I, Q] SD pair is the received version of the transmit coder output in Figure 6.6. For example, compare the third and fourth columns of Table 6.9.

The BMs in Table 6.9 are not states, they are only Euclidean distances in 2-D space between the received SD [I,Q] pair and the ideal received points in Figure 6.10. Each new received SD pair starts a new trellis column. Each column has a set of four circled input states on the left and four circled output states on the right. Note from Figure 6.8 that each of four output states per column has two branches going into it. There are also path metrics (PM) that accumulate the best possible set of preceding BMs. These are shown in bold type. Here is the trellis column update procedure for one received SD:

1. **Calculate** the BM for each of the four constellation points.
2. **Record** the appropriate BM next to each of the eight trellis column paths.
3. For each of the four output states:
    (a) **Add** upper BM to the upper input state PM to which it branches back to.
    (b) **Add** lower BM to the lower input state PM it which it branches back to.
    (c) **Compare** the two competing sums.
    (d) **Select** the lowest of the upper and lower PMs.
    (e) **Delete** the path with the highest PM.

We say each output state in each column has an add-compare-select (ACS) applied to it. Perhaps the best way to illustrate the Viterbi decoder is to continue with our example.

The left side of Figure 6.11 shows the four trellis states (corresponding directly to the four states in Figure 6.7). Each state has a path metric (PM), initialized to zero. Each of eight (2 per state) trellis transitions has an associated branch metric (BM), italic numbers. We copy the four different BM results from Table 6.9 to corresponding transitions in the trellis column (first row to first trellis column, and so on). Let's look at how the new path metrics on the right side of the first trellis column in Figure 6.11 as well as the survivor branches (bold lines) in the first column were obtained. For example, we choose state 00 on the output side of the first trellis column. We show how the surviving path to this state comes from state 01 on the input side of the first trellis column. The other four first column trellis output states are updated the same way.

1. **Add** upper BM00 = 1.797 to upper input state PM00 = 0; tentative output PM00 is 1.797.

CHAPTER 6. CODING

2. **Add** lower BM11 = 1.033 to lower input state PM01 = 0; tentative output PM00 is 1.033.
3. **Compare** the two tentative PMs, 1.033 and 1.797;
4. **Select** the lowest, output PM00 = 1.033;
5. **Delete** the upper path that would have resulted in output PM00 = 1.797. Surviving path is BM11 and is shown as a bold line.

This exact same process goes on to find all four new PMs on the right side of the second trellis column in Figure 6.11. The four surviving paths are shown as bold lines. Now, for example, let's see how output PM01 for the second trellis column in Figure 6.11 is calculated. Again, the other three output PMs in Figure 6.11 are calculated and the survivors are chosen in the same way. At first this procedure may seem tricky to understand, however, it is very repetitive.

1. **Add** upper BM10 = 0.984 to upper input state PM10 =1.033, tentative output PM01 is 2.017.
2. **Add** lower BM01 = 1.85 to lower input state PM11 =1.424, tentative output PM01 is 3.27.
3. **Compare** the two tentative PMs, 2.017 and 3.27.
4. **Select** the lowest, output PM01 = 2.017.
5. **Delete** the lower path that would have resulted in output PM01 = 3.27. Surviving path is BM10 and is shown as a bold line.

The trellis for processing of all five input SD pairs with all PMs updated and survivors chosen is in Figure 6.12. After all five symbols have been applied to the trellis, it is time for traceback. Starting with path metric 5.33, the lowest output state PM11 on the far right side, we trace back the continuous sequence of branches to the first column input states. There are other continuous trace backs, but they end in a higher PM value on the right side. The lowest accumulated PM indicates where the most likely sequence of transmitted symbols ended. This traceback bit sequence is 1 1 1 0 1. When reversed, we get 1 0 1 1 1, which is the transmit bit sequence. This example has a shorter traceback length than normal; traceback length is normally at least $4K$, where K is number of bits in encoder state; see [2] and [3]. Notice that the path metrics slowly increase in magnitude. In a fixed point system, there will eventually be overflow. There are various techniques for reducing the four PMs to avoid overflow. See [2] and [3] for more tips on Viterbi algorithm implementation.

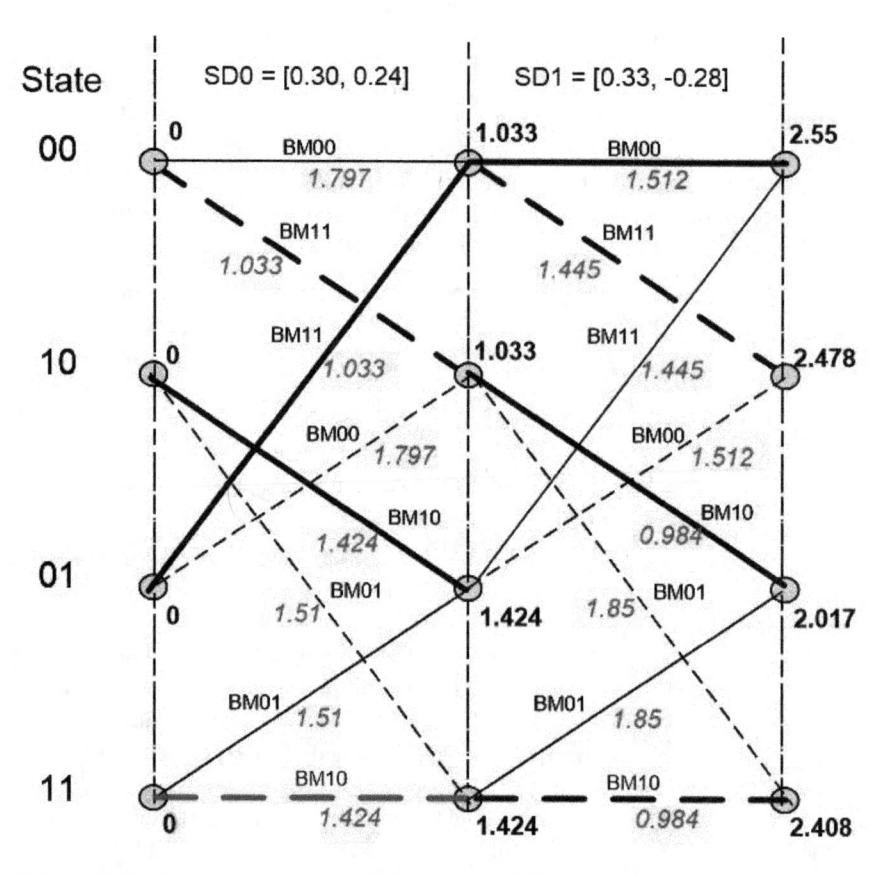

Figure 6.11 Convolutional decoding trellis, SD0 and SD1; path metrics are in bold type, and survivor paths have a bold line.

CHAPTER 6. CODING

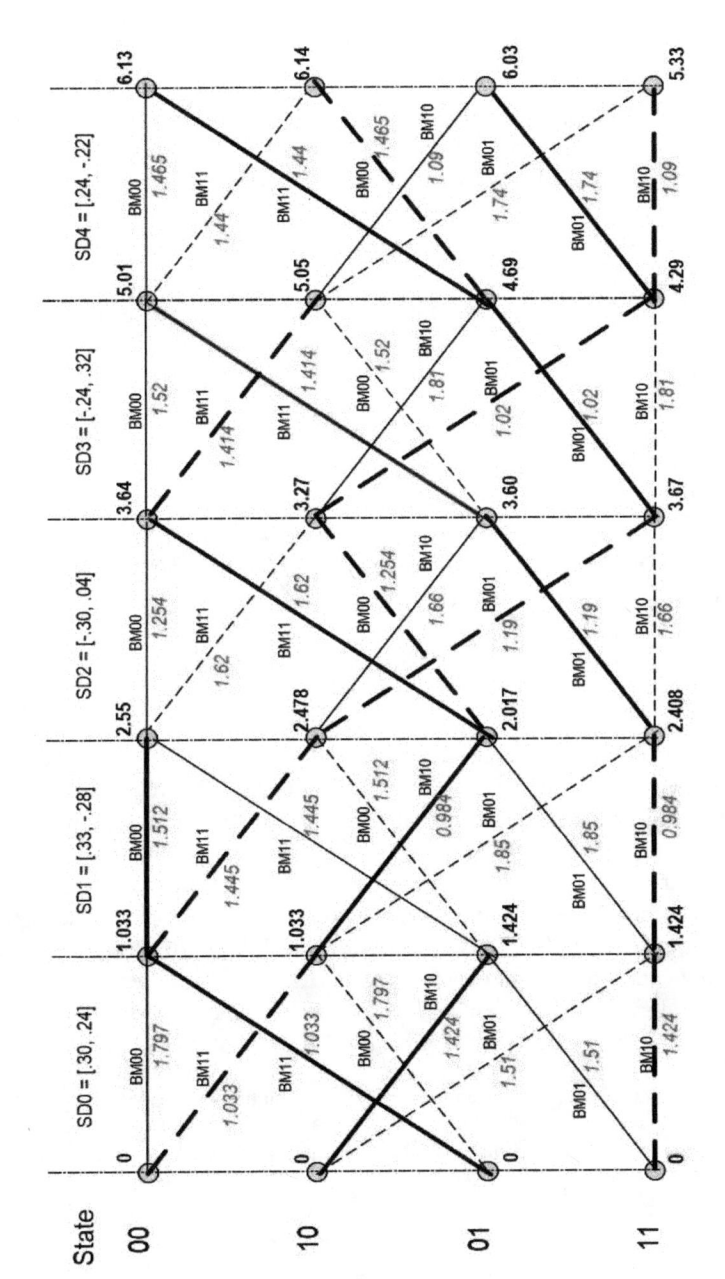

Figure 6.12 Convolutional decoding trellis; path metrics are in bold type, and survivor paths have a bold line.

6.3.3.4 Viterbi Decoder Traceback Techniques

Assume that a long series of M bits has been encoded and transmitted and now must be decoded at the receiver. The traceback length is $N < M$. Now that there are no more received branch metrics, M no longer increases. The lowest PM on the right side starts a traceback of N trellis columns. The one or zero decision from the last column (left side of Figure 6.12) in the traceback is stored and now the traceback length is reduced by 1. This continues until the very last bit to be decided has no traceback. Notice that, as the traceback is reduced, the bit decision reliability is reduced.

For more reliable decoding of the last bit, sometimes a block of N zeros is appended to the series. The extra zeros contain no information and so reduce the effective coding rate. However, the effect of the zeros is to allow all message bits to have the same detection reliability. Another way to accomplish this, called tail biting, is to initialize the encoder so that the start and end states of the Viterbi decoder are identical. This allows the received soft decisions to be processed again, in a circular manner, by the Viterbi decoder to achieve a consistent traceback reliability. Several techniques like this have been developed; see [1].

6.3.3.5 Convolutional Code Performance

Instead of d_{min} of block coding, convolutional coding has a d_{free} measurement. A series of zeros input to the trellis structure will result in a path along the top rail. A nonzero sequence followed by more zeros will cause the trellis path to diverge from and then remerge with the top rail. d_{free} is the minimum total weight of all possible divergent paths. A higher d_{free} means a higher coding gain; see Section 6.3.2. Comparing various coders this way is a complicated undertaking. Fortunately, various researchers have compiled tables of convolution coder performance to help us. Two examples for rate 1/2 coders are shown in Table 6.10. Recall that rate 1/2 means $k = 2$ code bits out for every $n = 1$ bit in. Thus, both encoders have 2 bit code vectors. The shift register connections are shown in Figure 6.13. More complete lists of optimum coders are discussed in [5] and [6] and Chapter 12 of [1].

Table 6.10 Optimum Rate 1/2 Convolutional Encoders

State	g_0	g_1	d_{free}	Coding Gain Upper Bound
$[S_0\ S_1]$	111	101	5	3.98 dB
$[S_0\ S_1\ S_2]$	1011	1111	6	4.77 dB

CHAPTER 6. CODING

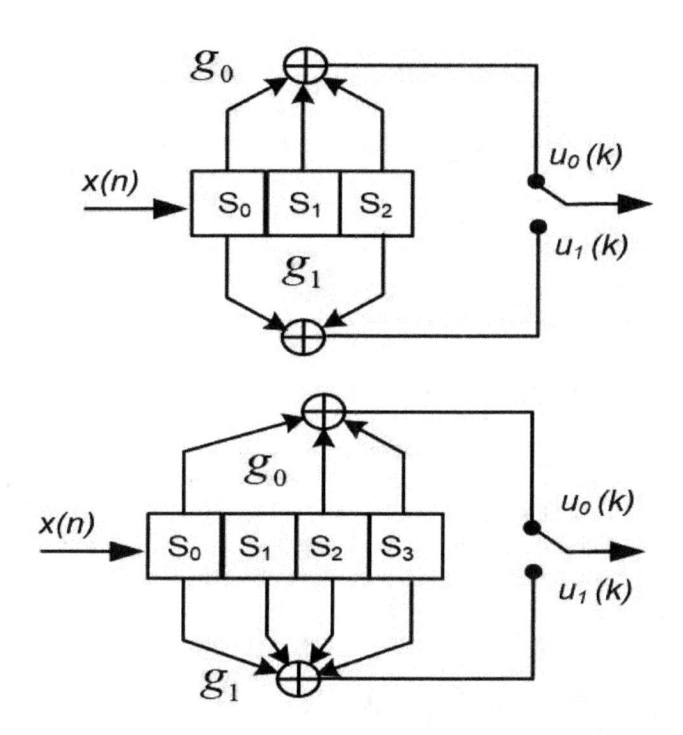

Figure 6.13 Two choices for convolutional coding rate 1/2 (see Table 6.10).

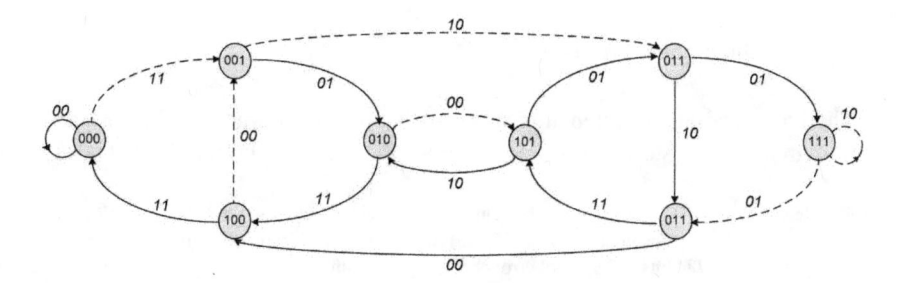

Figure 6.14 Convolutional coding state diagram, $g_0 = 1011$ and $g_1 = 1111$.

226   Software Defined Radio: Theory and Practice

The Figure 6.14 state diagram corresponds to the second coder in Figure 6.13. The state diagram for the first coder is in Figure 6.7.

Note in Figure 6.14 that the states are in the gray circles, written in binary corresponding to $[S_2\ S_1\ S_0]$. The trellis diagram can be constructed by listing the states in two columns and connecting them according to the possible transitions in Figure 6.14.

6.3.4 Concatenated Coding

6.3.4.1 Block-Interleave-Convolutional

Concatenated coding combines block and convolutional coding by sandwiching a convolutional inner coder/decoder between a Reed-Solomon[9] outer coder/decoder. There is an interleaver/deinterleaver between the two error correction systems. In Figure 6.15, if the inner convolutional decoder starts to output burst errors, the deinterleaver spreads these errors out into different outer code words. These outer code words are likely to have fewer errors per code word than if there were no convolutional decoder. This system tends to have lower overall complexity than a single coder of equivalent performance.

An example of concatenated coding used on the Voyager mission to Saturn and Uranus is in Figure 6.15. Voyager telecommunications achieved 10^{-6} BER at $E_b N_0$ = 2.53 dB, 2 Mbits/sec [7]. Power efficiency (see Section 3.2.1) is extremely important for deep-space communications. A coding gain of 6 dB can double the communications range between spacecraft and Earth [8] (as of May 2023, Voyager 1 was 14.6 billion miles from Earth and still communicating). Bandwidth efficiency is not as important because there are not many other users in deep space.

6.3.4.2 Block-Interleave-Block

The block coding concatenation in Figure 6.16 is sometimes used to protect data in computer disc storage systems [9]. For this example, the inner (10,4) code can

---

9   Reed-Solomon (RS) codes have the same overall structure as Figure 6.3. However, RS codes are symbol-based. For example, an RS(255,223) code with s = 8 bit symbols actually has a data payload of 223*8 = 1,784 bits. RS symbol error correcting capability = t. Since $2t = n - k$, t = (255-223)/2 = 16 symbols can be corrected anywhere in the received code word. A symbol error means 1, 2, 3, 4, 5, 6, 7 or all 8 symbol bits are incorrect. Thus, a burst of up to 128 bit errors can be corrected if the errors are confined to 16 symbols within the block of 255 symbols. Because RS works with symbols, it is sometimes called a nonbinary code.

CHAPTER 6. CODING

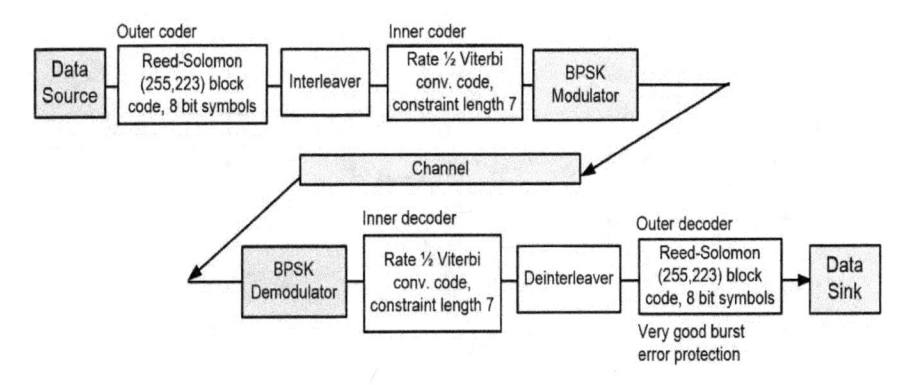

Figure 6.15 A concatenated coding example.

correct 2 errors (d_{min} = 5) and the outer (6,3) code can correct 1 (d_{min} = 3). The overall error correcting capability is:

$$\left\lfloor \frac{d_{min\,row} d_{min\,col} - 1}{2} \right\rfloor = \left\lfloor \frac{3 * 5 - 1}{2} \right\rfloor = 7 \tag{6.19}$$

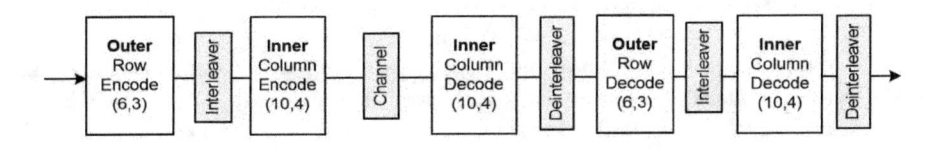

Figure 6.16 Concatenated block coding configuration.

Figure 6.17 is a simple implementation of the receive part of Figure 6.16. We leave out the interleaving for simplicity. The 12 unshaded receive bits in the upper left corner have the receive errors shown. In box A, the inner column decoder (first) can correct 2 errors, so errors remain. In box B, the outer row decoder can only correct 1 error per row, so a_2 and d_1 get fixed. When the inner row decoder has another chance, it can correct the remaining 2 errors per column.

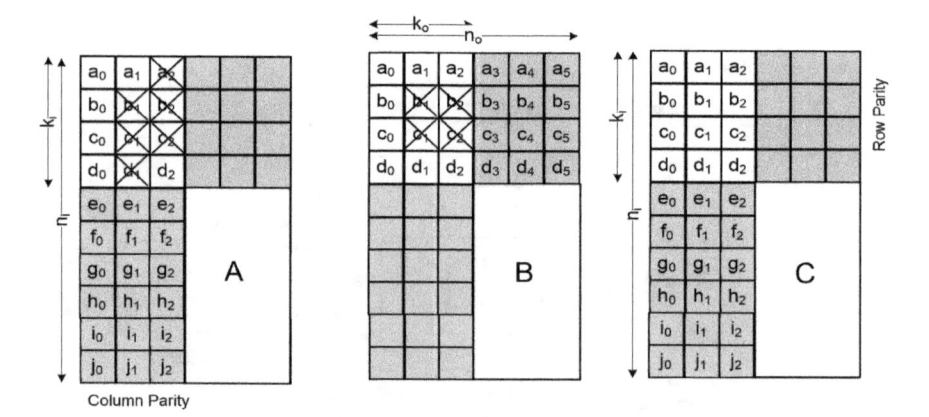

Figure 6.17 A concatenated block coding example.

6.3.5 Trellis-Coded Modulation

In the early 1970s, phone lines were still an important means of data communication for business. Communication engineers believed that bandwidth always had to increase when convolutional coding was employed. Thus, convolutional coding was not considered useful for bandwidth-limited telephone lines. Gottfried Ungerboeck, an engineer at IBM in Zurich, was convinced that an alternative coding scheme existed to protect data without increasing bandwidth. Ungerboeck invented trellis coded modulation (TCM) in the mid-1970s. Dial-up internet was largely made possible by TCM.

Ungerboeck's insight was that not all constellation points need the same coding protection. An example of trellis coding gain is shown in Figure 6.18. Recall from Section 2.7.3 that a Shannon curve assumes the same BER performance for each point plotted. Also, both 8PSK 2/3 and QPSK have the same bandwidth efficiency of 2 bits/symbol. However, trellis-coded 8PSK 2/3 has achieved a 3 dB coding gain. Let's explain this in more detail.

Ungerboeck's invention is for bandwidth-limited channels. In our example, Trellis coding adds one additional bit to the bits in a symbol. This doubles the number of constellation points, however, the symbol rate seen by the user is not changed. Therefore, for the same amount of data and same error rate, lower $E_b N_0$ is required, as shown in Figure 6.18. This is the definition of coding gain.

CHAPTER 6. CODING

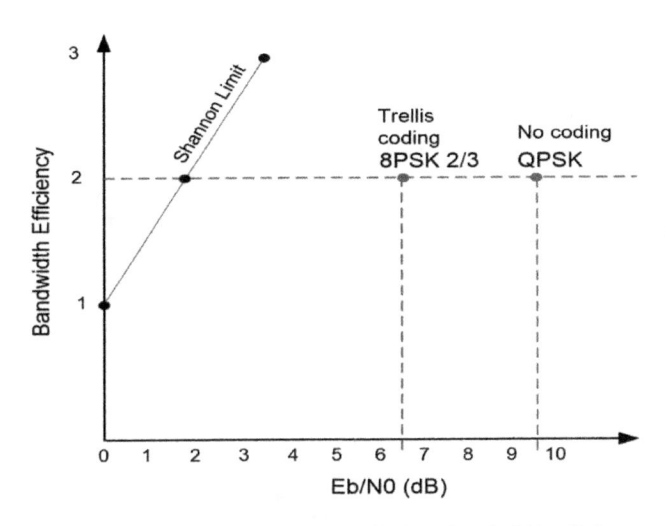

Figure 6.18 QPSK trellis-coding gain, fixed bandwidth efficiency.

Figure 6.19 is the circuit for trellis-coded QPSK. What is actually transmitted is 8PSK (2/3). This means an 8PSK constellation with symbols consisting of three total bits for every two QPSK data bits. Notice in the coding circuit on the left side that only one of the QPSK data bits is convolutionally coded, using the circuit of Figure 6.6. Also notice that the 8PSK constellation points are not Gray-coded but rather mapped around the circle in a natural counting order. This is discussed in [1].

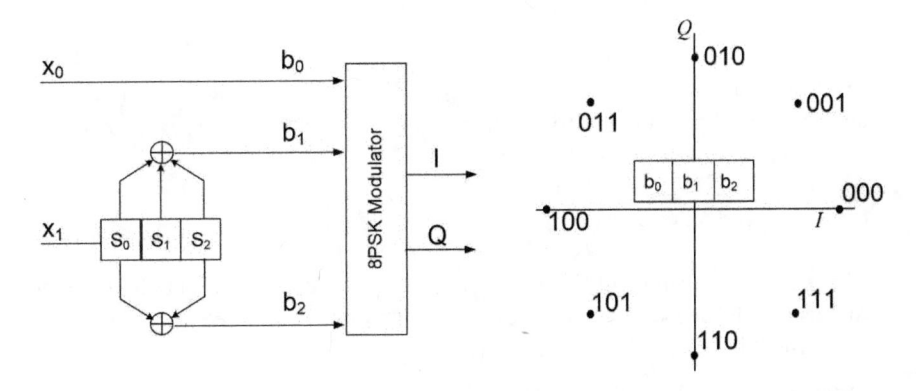

Figure 6.19 A QPSK trellis-coding circuit.

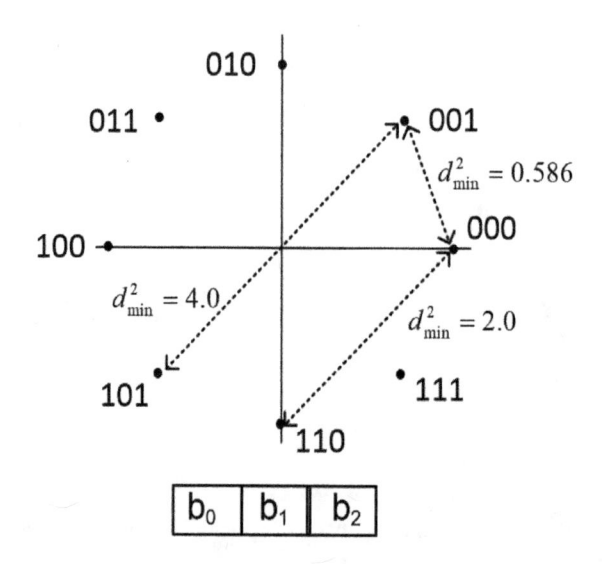

```
                010
       011 •         001
                   \  $d^2_{min} = 0.586$
                    \  000
  100 ←         
     $d^2_{min} = 4.0$   
                    $d^2_{min} = 2.0$
       101          • 111
                110

        | b₀ | b₁ | b₂ |
```

Figure 6.20 Three trellis-coding minimum distance sets.

Figure 6.20 may provide some insight. First, however, we need a review of d^2_{min}. This the squared Euclidean distance between two points in an N dimensional space. Equation (6.20) calculates the squared Euclidean distance between two points, (x_1, y_1) and (x_2, y_2), in $N = 2$ space:

$$d^2_{min} = (x_1 - x_2)^2 + (y_1 - y_2)^2 \tag{6.20}$$

Under AWGN conditions, a higher d^2_{min} results in a lower probability of error. In Figure 6.20, bit b_0 has the highest $d^2_{min} = 4.0$. Because constellation points with a change in $x_0 = b_0$ only are 180° apart, x_0 is not coded at all. x_1 is convolutional encoded and so is represented by b_1, b_2.

6.3.5.1 TCM Viterbi Decoder Steps

The trellis coding circuit of Figure 6.19 is fairly straightforward. Figure 6.21 attempts to illustrate the decoding set partition procedure. Given a received symbol point $r(n)$ somewhere in the 2-D I,Q space:

Step 1 Decide which of the 8 constellation points at the top is closest to $r(n)$; call it $u(n)$.

CHAPTER 6. CODING

Step 2 Decide which 4-point constellations include $u(n)$; this sets the value of b_2.
Step 3 Decide which 2-point constellations include $u(n)$; this sets the value of b_1.
Step 4 From the chosen 2-point constellation, decide b_0.

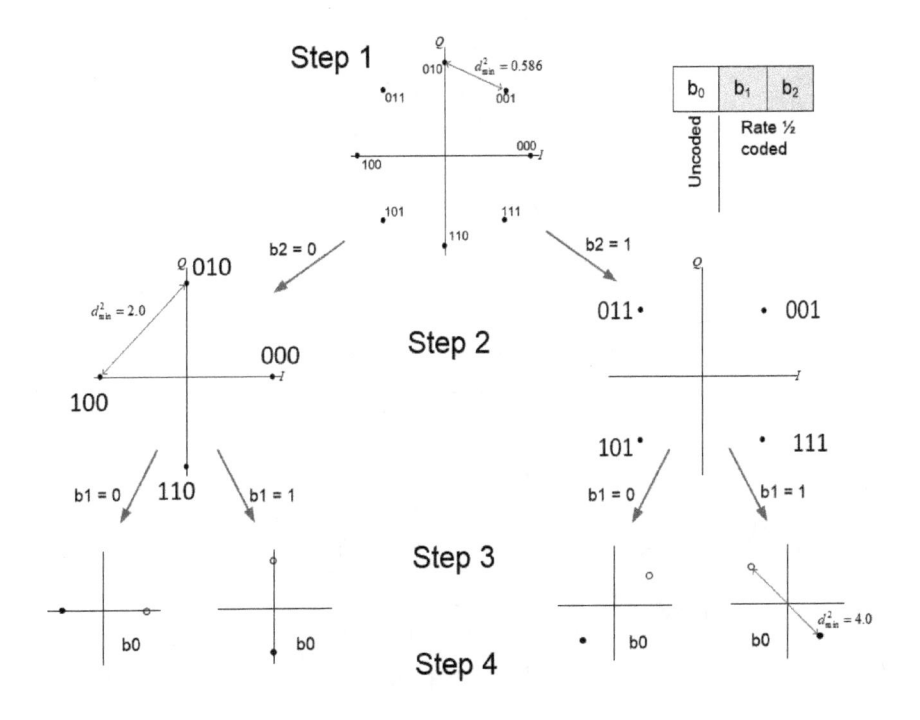

Figure 6.21 Example of trellis decoding set partition.

6.3.5.2 TCM Viterbi Decoder Example; b0 = 1, Solid, b0 = 1, Open

Let's look at a practical TCM decoder example. We use the circuit on the left side of Figure 6.19 to encode X_0 and X_1 from Table 6.11. The received SD is a noisy received version of the encoder output 8PSK constellation point.[10] Finally, the branch metrics are the squared Euclidean distance of the received SD from each of the eight ideal constellation points, using Equation (6.20).

---

10 The received SD is 2-D, I and Q. Compare Table 6.11 with Table 6.9.

Table 6.11 TCM Branch Metric Example

Sym Num	Tx Bits X_0	Tx Bits X_1	8PSK Point	Received SD [I, Q]	Branch Metrics (BM) 000	001	010	011	100	101	110	111
0	0	1	011	[-0.70 0.74]	1.85	1.40	0.74	0.04	0.80	1.45	1.88	2.02
1	1	0	110	[-0.05 -0.98]	1.43	1.85	1.98	1.81	1.37	0.72	0.05	0.80
2	1	1	101	[-0.70 -0.73]	1.85	2.01	1.87	1.44	0.79	0.03	0.75	1.41
3	0	1	001	[-0.70 -0.73]	0.78	0.01	0.75	1.41	1.84	2.00	1.85	1.42
4	1	1	110	[0.07 -0.95]	1.32	1.77	1.95	1.83	1.43	0.81	0.09	0.68

CHAPTER 6. CODING

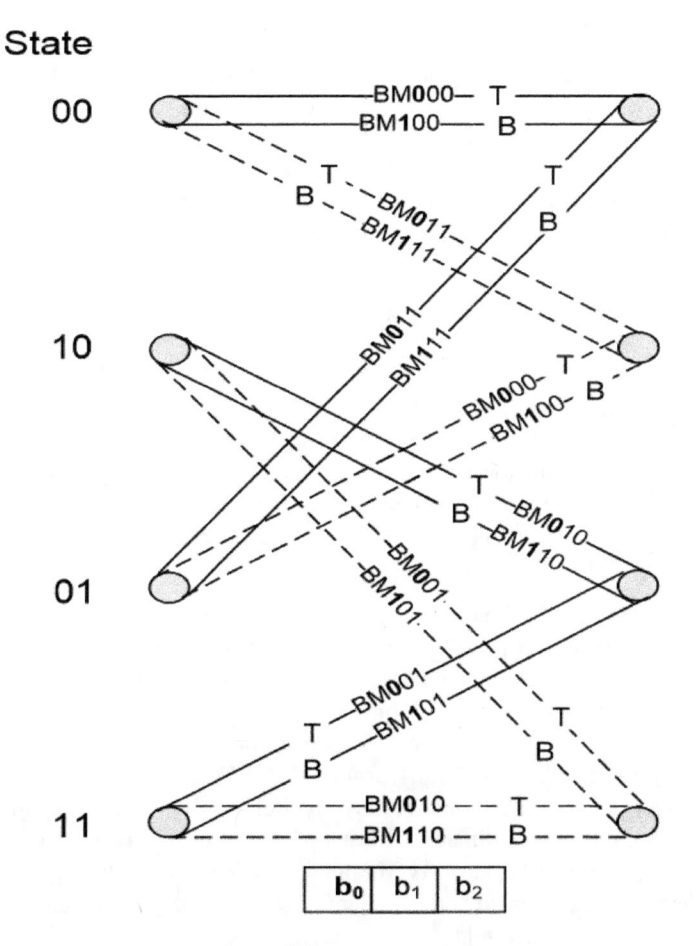

Figure 6.22 A TCM trellis for 8PSK (2/3).

TCM for the circuit in Figure 6.19 corresponds to the Figure 6.22 trellis. The difference between Figure 6.22 and the convolutional coding trellis in Figure 6.8 is the double branches marked T for top and B for bottom. These correspond to a decision for b0 when b1 and b2 are unchanged. Here is the trellis column update procedure for one received symbol (i.e., one row of Table 6.11).

1. **Calculate** the BM for each of the eight constellation points; see Table 6.11.
2. **Label** the new symbol's trellis branches with the appropriate BM.
3. For each of the eight branches:
   (a) **Compare** T-BM and B-BM branches.
   (b) **Select** the lowest BM (this is the uncoded bit).
   (c) **Delete** the branch with the highest BM.
4. For each of the four output states:
   (a) **Add** upper BM to upper input state PM to which it branches back (left side).
   (b) **Add** lower BM to lower input state PM to which it branches back (left side).
   (c) **Compare** the two competing sums.
   (d) **Select** the lowest of the upper and lower input PM + BM.
   (e) **Delete** the path with the highest input PM + BM.
   (f) **Update**, on the right side of the trellis column, the accumulated output PM for that state.

Figures 6.23 and 6.24 show a trellis update example for Symbol 0 and Symbol 1 (i.e., received SD0 and SD1 from Table 6.11). Figure 6.25 is the column update procedure applied to all five symbols in Table 6.11. The traceback bit sequence (for a 0 bit indicated by a solid line and a 1 bit indicated by a dotted line) is 1,1,1,0,1. When this is reversed, we obtain the correct X_1 bit pattern is 1,0,1,1,1. The X_0 bit pattern is simply B,T,B,B,T for 0,1,1,0,1, from Table 6.11. Recall that X_0 = b0 decisions are memoryless (i.e., they involve only one trellis column), but they must be made first to obtain the best estimate of b1 and b2. (i.e., best trellis branch as revealed by the trellis pruning procedure). TCM implementations and analysis go far beyond the simple example presented here. Readers who want to dig deeper are encouraged to study the TCM chapter in [1].

CHAPTER 6. CODING   235

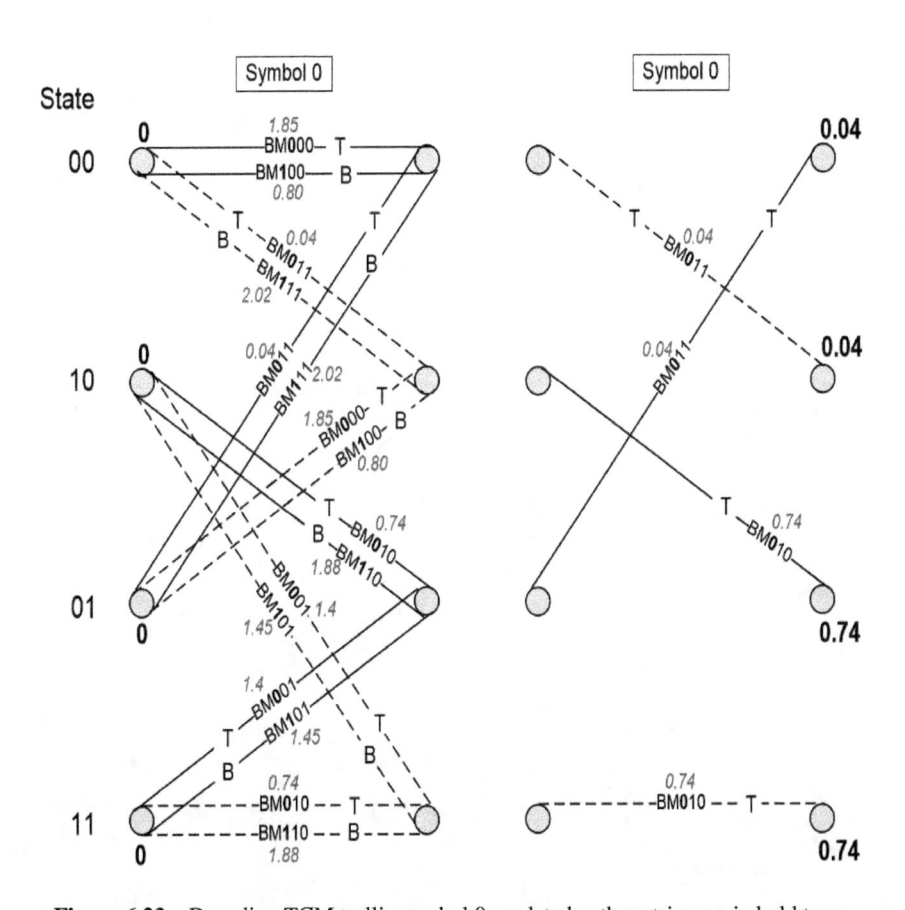

Figure 6.23 Decoding TCM trellis symbol 0; updated path metrics are in bold type.

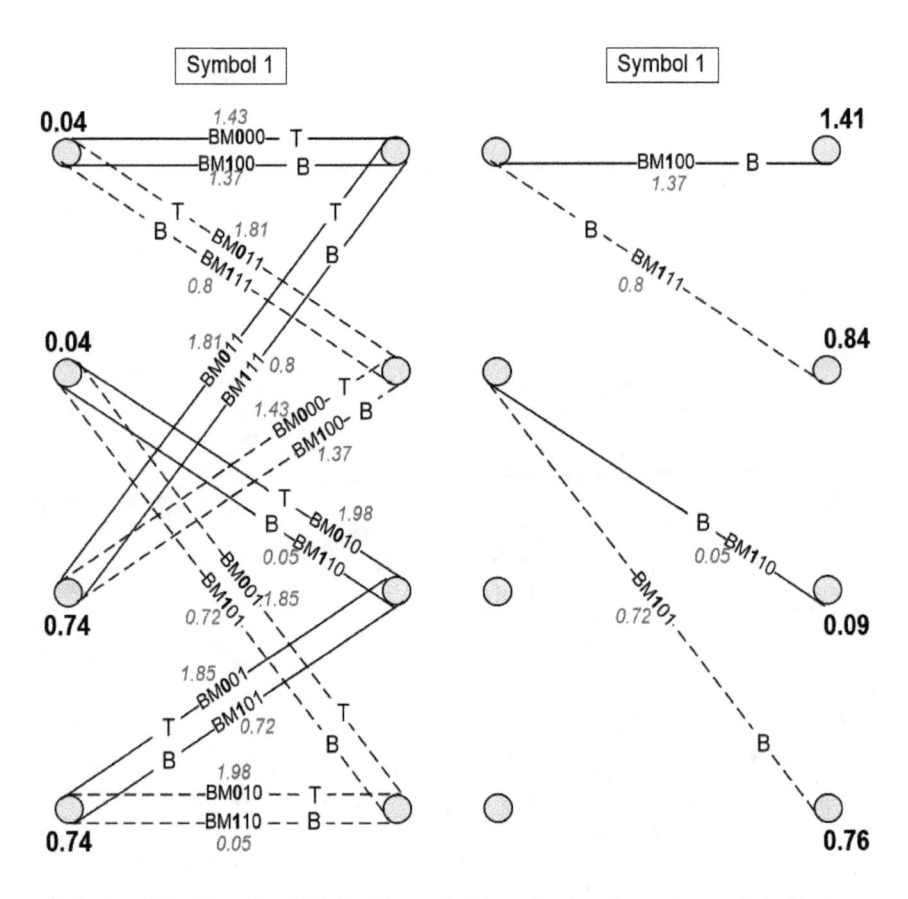

Figure 6.24 Decoding TCM trellis symbol 1; updated path metrics are in bold type.

CHAPTER 6. CODING

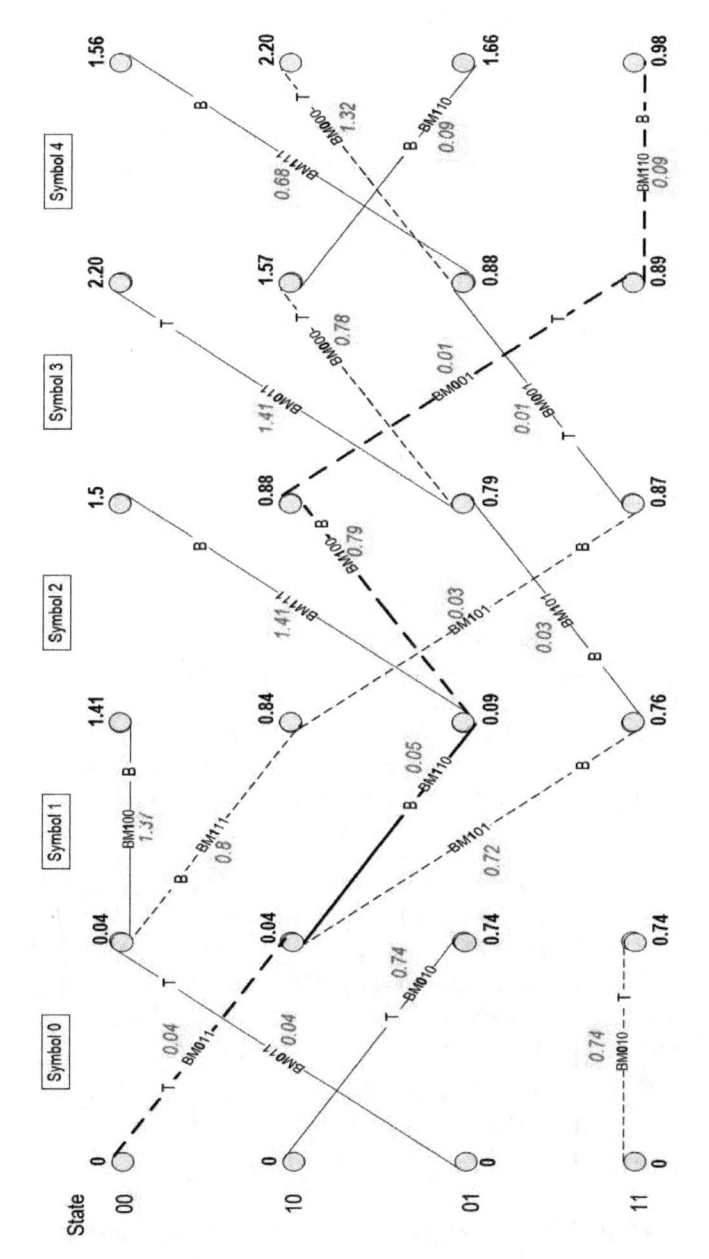

Figure 6.25 TCM trellis, five symbol traceback; path metrics are in bold type, and survivor paths have a bold line.

6.4 QUESTIONS FOR DISCUSSION

1. A linear block code is described by the generator matrix in Equation (6.21). Determine n, k, r. Write out all the code words. What is the minimum Hamming distance? How many errors can be detected?

$$G = \begin{bmatrix} 1 & 1 & 1 & 0 & 0 \\ 0 & 1 & 1 & 1 & 1 \\ 1 & 0 & 1 & 0 & 1 \end{bmatrix} \quad (6.21)$$

2. Draw the circuit diagram for the convolutional encoder specified by the generator matrix G in Equation (6.22). Determine the coding rate.

$$\begin{bmatrix} v_1 & v_2 & v_3 & v_4 & v_5 \end{bmatrix} = \begin{bmatrix} u_1 & u_2 \end{bmatrix} \begin{bmatrix} 1+D & D & 0 & 1+D & D^2 \\ D & 1 & D+D^2 & 0 & 1 \end{bmatrix} \quad (6.22)$$

3. Figure 6.26 shows a small sample of the IBM 9-track, 0.5-inch-wide, magnetic tape used in the 1960s. Eight bit bytes and an even parity bit are written across the tape on the nine tracks. This tape was able to store about 800 bytes per inch. A parity bit is a very simple way to detect bit errors. An even parity bit is 1 to indicates that an even number of 1s was written to the tape. What different error situations can a parity bit detect? Are there any bit error situations that a parity bit cannot correct?

bit 7	(0,128)	(1,128)	(2,128)	...
bit 6	(0,64)	(1,64)	(2,64)	...
bit 5	(0,32)	(1,32)	(2,32)	...
bit 4	(0,16)	(1,16)	(2,16)	...
bit 3	(0,8)	(1,8)	(2,8)	...
bit 2	(0,4)	(1,4)	(2,4)	...
bit 1	(0,2)	(1,2)	(2,2)	...
bit 0	(0,1)	(1,1)	(2,1)	...
parity	(0,p)	(1,p)	(2,p)	...

Figure 6.26 Small sample of IBM 9-track data storage tape.

CHAPTER 6. CODING

4. Draw the encoder diagram (form of Figure 6.13) for the rate 1/2 convolutional encoder defined by $g_0 = 1101$ and $g_1 = 1011$.

5. Draw the state transition diagram for the encoder in problem 4.

6. Draw a column of the Viterbi decoder trellis for the encoder in problem 4.

7. For a hard decision received vectors $r(n) = 10$, show all the branch metrics on the trellis of problem 6. Use a Hamming distance instead of the Euclidean distance for calculating branch metrics. Note how some of the tracebacks must be randomly chosen.

8. Trellis coding:

a. Increases the symbol rate by addition of code bits;
b. Expands the constellation to add code bits without increasing the symbol rate;
c. Is a new computer programming language;
d. None of the above.

REFERENCES

[1] S. Lin and D. J. Costello. *Error Control Coding.* Prentice Hall, second edition, 2004.

[2] J. G. Proakis. *Communication System Engineering.* Prentice-Hall, 2002.

[3] B. Sklar. *Digital Communications.* Prentice Hall, 1988.

[4] G. D. Forney. "The Viterbi Algorithm". *Proceedings of the IEEE*, 61(3):109–116, March 1973.

[5] J. P. Odenwalder. *Error Control Coding Handbook.* Linkabit Corp., 1976.

[6] D. Bhargava. *Digital Communications by Satellite.* John Wiley and Sons, 1981.

[7] S. A. Butman. "Performance of Concatenated Codes for Deep Space Missions". *TDA Progress Report*, April 1981.

[8] A. Burr. *Modulation and Coding for Wireless Communications.* Prentice Hall, 2001.

[9] Gibson G. A. Patterson, D. A. and R. Katz. "A Case for Redundant Arrays of Inexpensive Disks". *Proceedings of the SIGMOD International Conference on Data Management*, 1988.

Chapter 7

Analog Signal Processing

7.1 INTRODUCTION

For SDR, analog signal processing for a receiver includes the radio frequency (RF) front-end from the antenna to the analog-to-digital converter (ADC) input. Transmitter analog signal processing starts with a digital-to-analog converter (DAC) and ends with the transmit antenna. RF front end implementations will vary from individual components wired on a circuit board to an advanced highly integrated part such as the AD9361.

7.2 COMPONENTS

We start our review of analog signal processing for SDR by studying the characteristics of the various analog components involved. These are amplifiers, mixers, oscillators, and filters.

7.2.1 RF Amplifiers

Amplifiers increase the signal amplitude linearly for inputs below a certain power. Unfortunately, they can distort the signal as the input gets too high and degrade the signal-to-noise ratio as the signal gets too low.

7.2.1.1 Amplifier Linearity

For an arbitrary signal $x(t)$, we would like the output to be $y(t) = Gx(t)$ where G is the gain, a proportional scaling factor that is constant at any amplitude or frequency. Unfortunately, what we typically get is a nonlinear function based on weighted powers of the input signal envelope:

$$y(t) = x(t) \left(g_1 + g_3 x(t)^2 + g_5 x(t)^4 + g_6 x(t)^6 + \cdots \right) \tag{7.1}$$

This is the nonlinear AM to AM gain; that is, amplitude out versus amplitude in. In Equation (7.1) notice that even orders of the $x(t)$ input are proportional to input power (i.e., voltage squared). There are other amplifier transfer function equations such as phase out versus amplitude in (AM to PM gain).

For SDR use, we are mainly concerned about an amplifier's intermodulation distortion (IMD). A perfectly linear amplifier can never produce an output at a frequency not present on the input. However, a practical amplifier, such as in Equation (7.1), certainly can. We call these extra outputs IMDs. A two-tone test is a simple way to characterize them. The input tones, at f_1 and f_2, should be specified. The choice of frequencies is not critical, however, they should not be further apart than the signal bandwidth planned for normal operation. For a worst-case test, the peak power of the two-tone composite should be near the power limit of the amplifier input. The combiner should have good isolation between inputs.

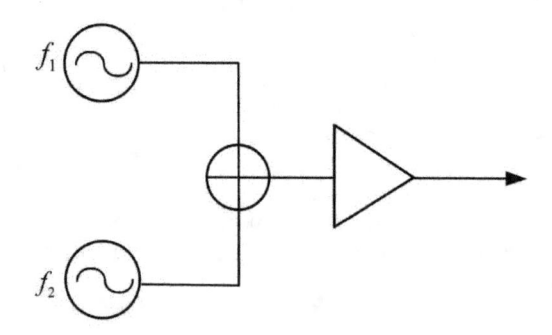

Figure 7.1 A two-tone amplifier test setup.

The IMD frequencies present on the output are predicted by Equation (7.2). Here J and K are integers and $H = |J| + |K|$ is the IMD order.

$$f_{IMD} = J(f_1) + K(f_2) \tag{7.2}$$

CHAPTER 7. ANALOG SIGNAL PROCESSING

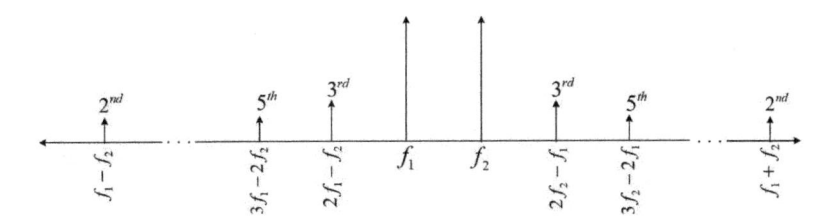

Figure 7.2 A typical amplifier IMD frequency spectrum.

As shown in Figure 7.2 third and fifth-order IMD products can fall near the desired signal (assumed to be between f_1 and f_2). This can result in undesired spectral regrowth and/or adjacent channel interference on, for example, RF power amplifiers (see Figure 7.50). The amplitude of IMDs on the amplifier output depends on the amplifier nonlinear characteristic; for example, Equation (7.1). Due to the nonlinearity, different IMDs increase at different rates relative to the input. For example, for input $x(t)$, third-order products increase proportional to $x^3(t)$; that is, the third-order products increase faster than the input increases.

Figure 7.3 shows an amplifier first-order gain as a solid diagonal line. The more rapidly rising output power of the third-order IMD product intersects the first-order output power at the OP3 (output intercept point third order). This is a commonly used figure of merit for power amplifiers; higher OP3 indicates more linear operation. Receiver amplifiers, such as the front end LNA, generally refer to IP3 (input intercept point third order). OP3 is simply IP3 plus the amplifier gain.

A cascade of amplifiers has a composite intercept point. Figure 7.4 is an example where the composite IP3 is calculated as:

$$IP3_{composite} = \frac{1}{\frac{1}{IP3_1} + \frac{G_1}{IP3_2} + \frac{G_1 G_2}{IP3_3}} \qquad (7.3)$$

Notice that for Equation (7.3) input intercept points and gains in this equation are not in dB. The result on the left side is simply a power level and so can be converted to dBm. This is a worst-case calculation that assumes IMD products add in phase.

7.2.1.2 Cross-Modulation

Receiver cross-modulation is directly related to RF amplifier nonlinearity. Cross modulation occurs when a strong amplitude modulated off-channel and unwanted

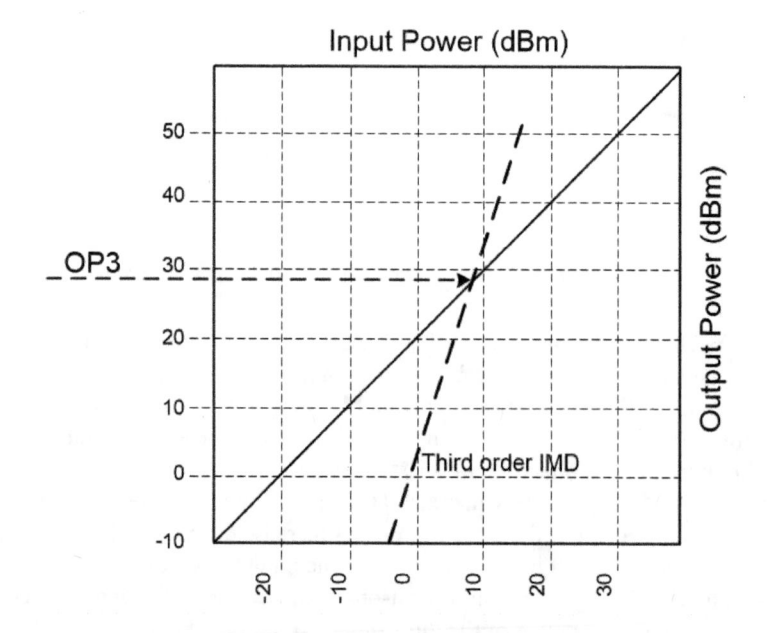

Figure 7.3 Amplifier linear gain and typical output intercept.

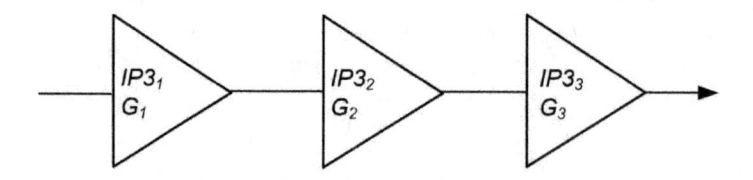

Figure 7.4 A three-amplifier cascade.

signal causes variable compression of the receiver front end gain. The resulting variable gain transfers undesired amplitude modulation to the desired on-channel signal.

An example of cross-modulation is shown in Figure 7.5. The ideal linear gain is the dotted diagonal line; the actual nonlinear receiver gain is the solid line. A large undesired signal (shaded) with random envelope power variation between 7 and 10 dBm causes the output to compress and the gain to vary randomly between g_1 and

CHAPTER 7. ANALOG SIGNAL PROCESSING   245

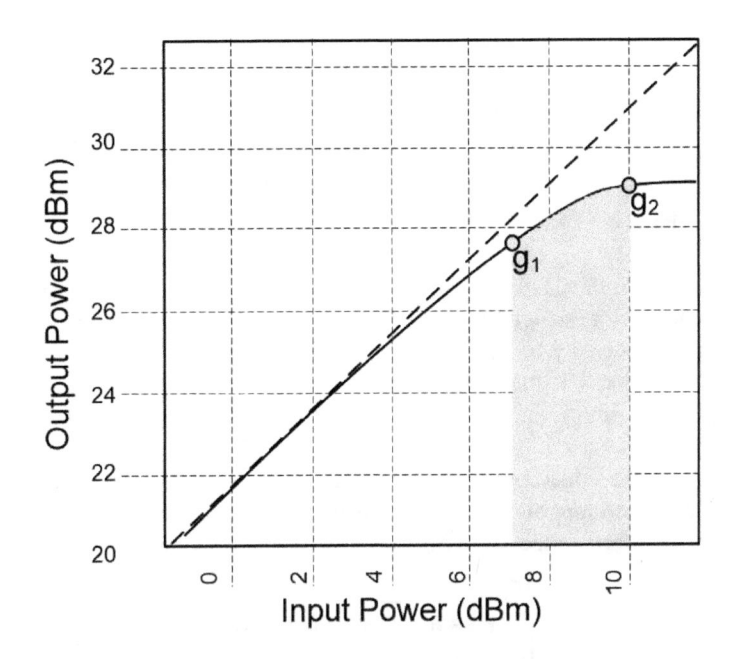

Figure 7.5 Example of cross-modulation through an input RF amplifier.

g_2. This gain variation transfers undesired amplitude modulation to the lower power desired signal.

7.2.1.3 Noise Factor

The noise factor is a measure of the signal-to-noise power change through a two-port device. For example, an amplifier typically has transistors that generate noise along with their primary job of increasing the input signal. From Equation (7.4), for an amplifier the noise factor (F) is always greater than one and the noise figure (NF) is always greater than zero.

The noise figure of LNAs can be dependent on frequency and physical temperature. Noise figure is generally measured with a noise figure analyzer, such as the Agilent 8970B. For a receiver front end noise figure measurement, the LO should be replaced by an extremely low phase noise bench oscillator. This will keep LO phase noise out of the measurement.

$$F = \frac{SNR_{in}}{SNR_{out}}$$

$$NF = 10 \log_{10} F \tag{7.4}$$

7.2.1.4 Passive Circuit Noise

A special case is the passive circuit such as a filter or waveguide or other transmission line. If the circuit of Figure 7.6 is terminated and in thermal equilibrium, then the thermal noise density of $N_0 = kT_{Kelvin}$ watts per Hertz is present at the input and output. Note that k is the Boltzmann constant. The signal through the device may have its bandwidth reduced but the thermal noise power density at the input and output is not affected. The composite signal and noise presented at the passive device input are reduced together, but the noise power density, $N_0 = kT_{Kelvin}$ is still the same at the input and output. The resistors in Figure 7.6 are like independent noise sources forcing the same N_0.

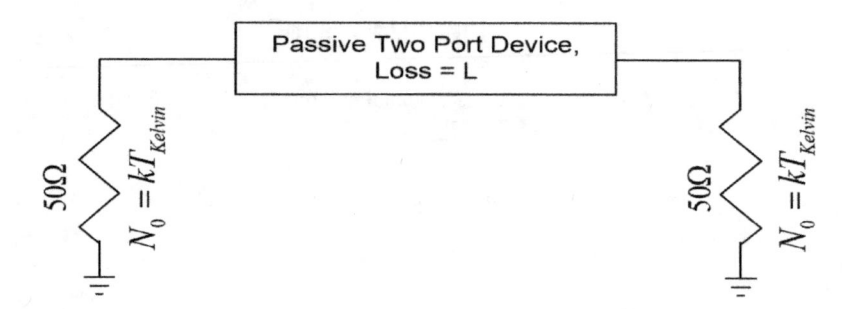

Figure 7.6 Passive two port network noise calculation.

To show this a little more concretely, consider Equation (7.5), where L is the passive device loss (for example, the loss of a transmission line or waveguide). The resulting noise factor is the inverse of L, or $F = 1/L$. P_x is the power into the two-port passive device and B is the bandwidth.

$$F_{passive} = \frac{SNR_{in}}{SNR_{out}} = \frac{\frac{P_x}{BkT_{Kelvin}}}{\frac{LP_x}{BkT_{Kelvin}}} = \frac{1}{L} \tag{7.5}$$

CHAPTER 7. ANALOG SIGNAL PROCESSING

7.2.1.5 Amplifier Noise

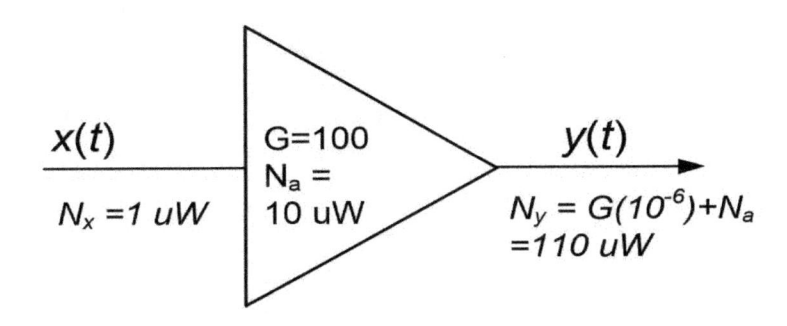

Figure 7.7 Example of amplifier noise calculation.

Amplifier noise is another important performance factor, especially for receiver front end circuits. In Figure 7.7, P_x is input signal power. Also, N_x and N_y are input and output noise power, respectively. Finally, N_a is the noise power generated internally by the amplifier. This example is from [1].

For the amplifier in Figure 7.7 the noise factor is calculated as:

$$F = \frac{SNR_x}{SNR_y} = \frac{P_x/N_x}{GP_x/(GN_x + N_a)} = 1.1 \qquad (7.6)$$

The noise figure in dB is calculated from the noise factor as:

$$NF = 10 log_{10}(F) = 0.4 dB \qquad (7.7)$$

7.2.1.6 Cascade Component Noise Factor

A cascade of amplifiers and other two-port devices has a composite noise factor. Figure 7.8 is an example where the composite noise factor is calculated as shown in Equation (7.8).

$$F_{total} = F_1 + \frac{F_2 - 1}{G_1} = 3 + \frac{9-1}{24} = 3.33 \qquad (7.8)$$

In Equation (7.8), we must use noise factors (F) and gains that are not in dB (noise figures (NF) are in dB). The noise factor of first amplifier dominates the total; first amplifiers dominate, and additional amplifiers have less effect on cascade noise

```
        ┌─────────┐           ┌─────────┐
        │ F₁ = 3  │           │ F₂ = 9  │
   ─────│ G₁=24   │───────────│ G₂=10   │──────▶
        └─────────┘           └─────────┘
```

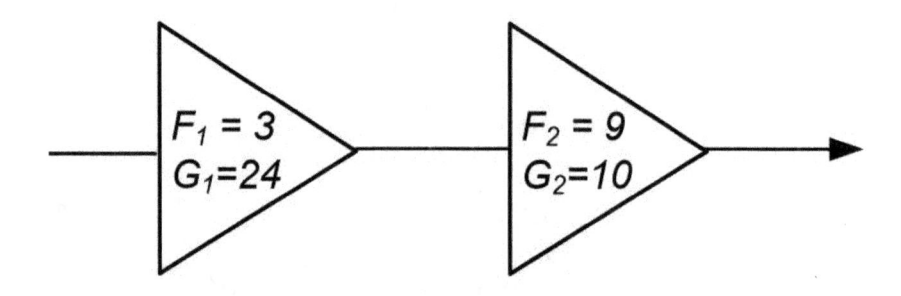

Figure 7.8 A two-amplifier cascade.

factor. That is why the first amplifier in a cascade should be low noise and high gain (e.g., low noise amplifier (LNA) mounted in a satellite dish). That is also why the noise factor of the downstream amplifier driving the receiver ADC is not as critical.

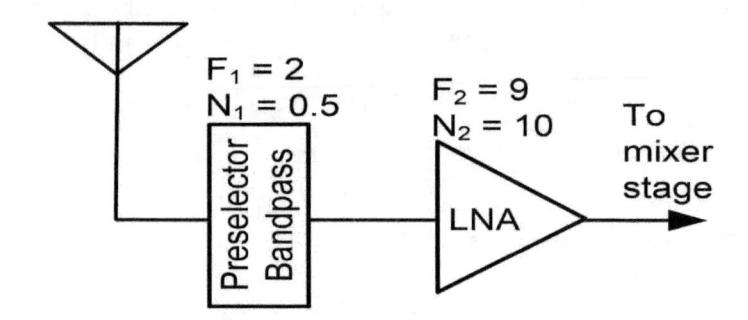

Figure 7.9 A filter and amplifier cascade.

(with $F_1 = 2$, $N_1 = 0.5$ for the Preselector Bandpass, and $F_2 = 9$, $N_2 = 10$ for the LNA, to mixer stage.)

Let's look at another example of how an LNA affects cascade noise factor. Consider Figure 7.9, a typical SDR front end. Equation (7.10) is the composite noise factor of this circuit. The antenna is not a two-port network so it has no noise factor; accounting for antenna noise was discussed in Section 2.6.1.

$$F_{total} = F_1 + \frac{F_2 - 1}{G_1} = 2 + \frac{9 - 1}{0.5} = 18 \quad (7.9)$$

In Figure 7.10, let's put an LNA in front of the previous circuit. The composite noise factor of the circuit in Figure 7.10 is shown in Equation (7.10).

CHAPTER 7. ANALOG SIGNAL PROCESSING

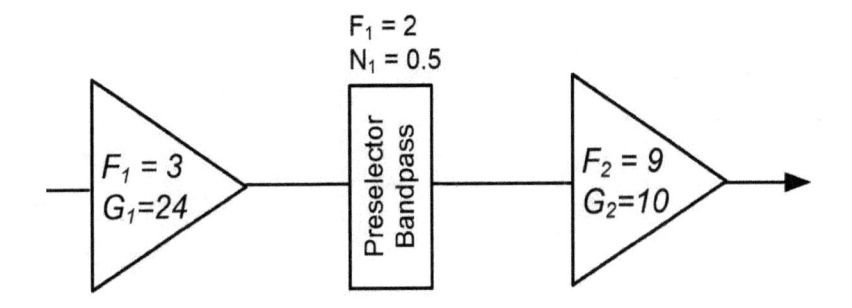

Figure 7.10 A low noise amplifier, filter, and amplifier cascade.

$$F_{Total} = F_1 + \frac{F_2 - 1}{G_1} + \frac{F_3 - 1}{G_1 G_2} = 3 + \frac{2-1}{24} + \frac{9-1}{12} = 3.71 \qquad (7.10)$$

7.2.2 RF Mixers

Let's review RF mixer characteristics.

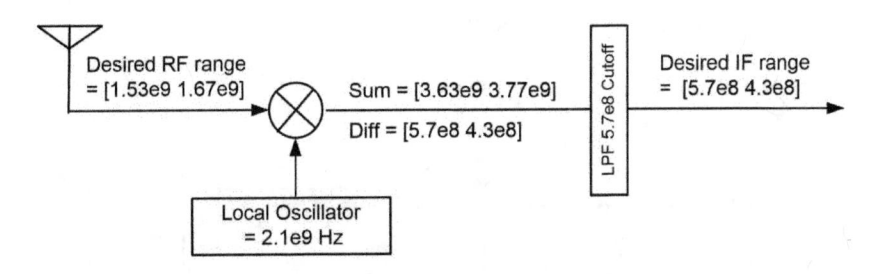

Figure 7.11 A basic high side mixer circuit.

Electronic RF mixers output the sum and difference of their two input frequencies. For example, in Figure 7.11 our desired RF range spans the frequencies we want our receiver to process. At the mixer output the difference between the LO (local oscillator) and the desired RF range becomes the desired IF range (intermediate frequencies, a range of frequencies that is better suited for subsequent electronic processing). The sum of the LO and the desired RF range produces an unneeded set of frequencies that we filter out with the lowpass filter (LPF). This circuit is called high side injection because the LO frequency is higher than the RF range.

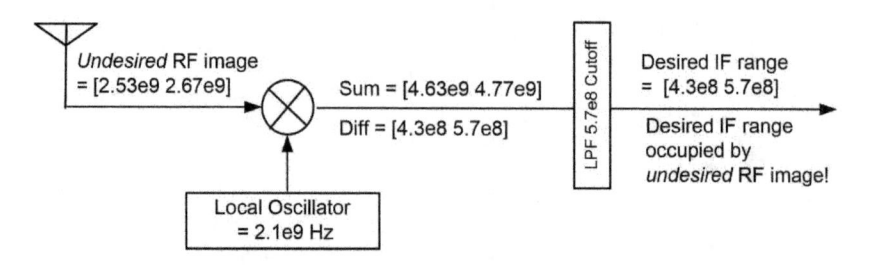

Figure 7.12 A mixer circuit demonstrating the image frequency problem.

7.2.2.1 Image Filter

A separate problem related to what we just discussed but not quite the same is the RF image frequency. Figure 7.12 illustrates how the desired IF range can be produced by an undesired range of received frequencies at the antenna called the image frequencies. The desired IF in Figure 7.11 and undesired IF in Figure 7.12 are right on top of each other. The undesired RF spectrum may contain an undesired signal that will corrupt the desired IF output.

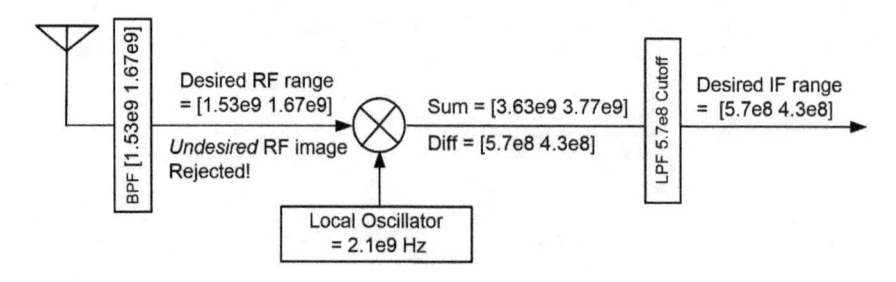

Figure 7.13 A mixer circuit with an image rejection bandpass filter (preselector).

A bandpass filter, also called a preselector, is often used to keep out the undesired image frequencies. This is shown in Figure 7.13. A preselector also limits the receiver tuning range. Later on in this chapter, we will see that a zero IF receiver also eliminates the image frequency problem.

CHAPTER 7. ANALOG SIGNAL PROCESSING

7.2.2.2 Image Rejection Mixer

The Hartley image rejection mixer in Figure 7.14 is another way to keep out the undesired image frequency.

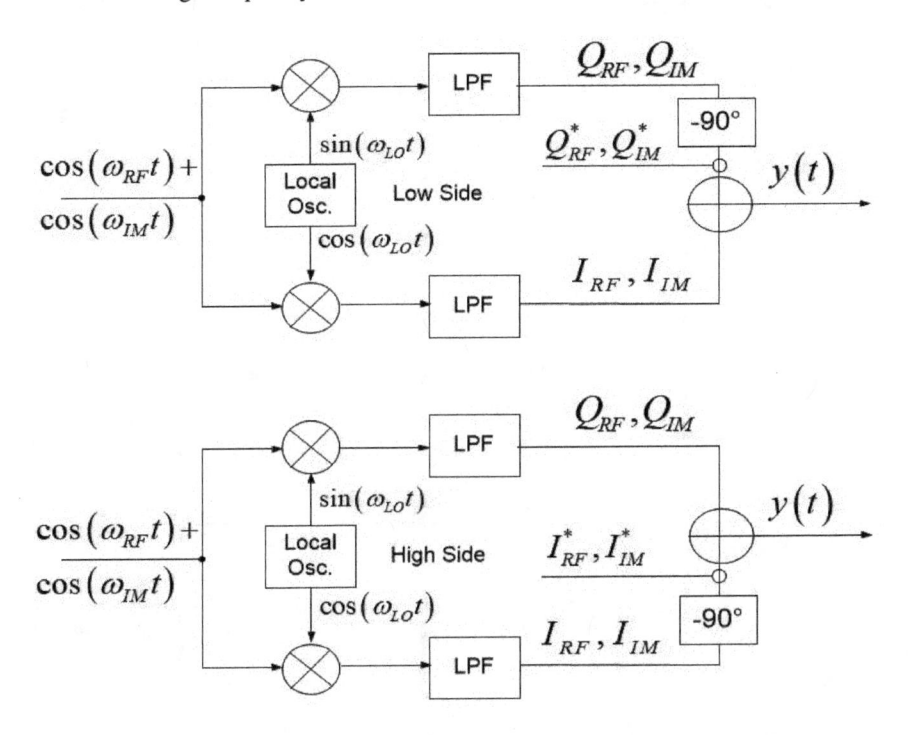

Figure 7.14 Low and high side rejection Hartley image rejection mixers.

Figure 7.15 illustrates low side injection; the desired ω_{RF} is labeled ω_{IF}^{good} and the undesired ω_{IM} is labeled ω_{IF}^{bad}. The top circuit of Figure 7.14 passes desired ω_{RF} and rejects undesired ω_{IM}.

To accomplish high side injection, we use the bottom circuit of Figure 7.14. Now we pass ω_{IM} and reject ω_{RF}. The frequencies do not change, we are simply redefining what we want to pass and what we want to reject.

Here are some detailed examples of the math behind the Hartley image rejection mixer. The 90° phase shifts are left out because they are simply trigonometry identities.

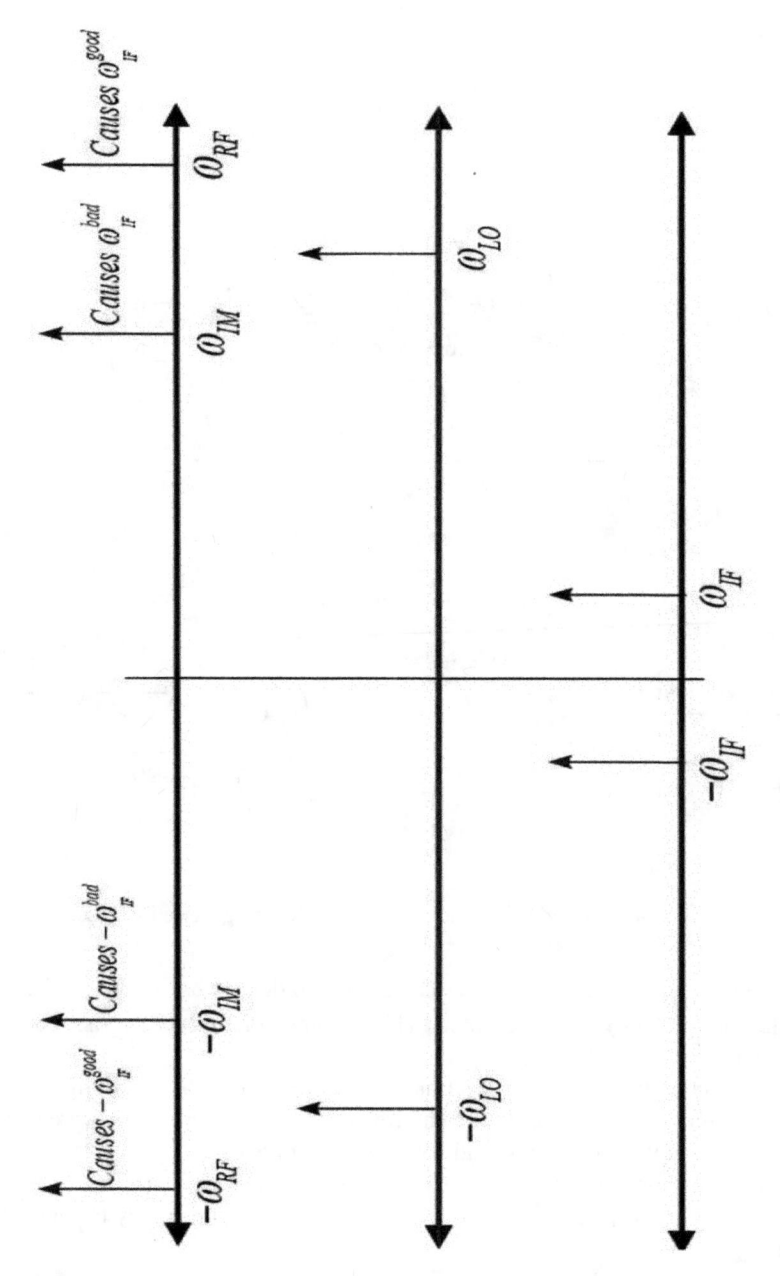

Figure 7.15 A frequency domain review of desired and undesired low side mixing.

CHAPTER 7. ANALOG SIGNAL PROCESSING

$$\omega_{IF}^{good} = \omega_{RF} - \omega_{LO}$$
$$\omega_{IF}^{bad} = \omega_{LO} - \omega_{IM}$$
(7.11)

Equation (7.12) lists some definitions we will need:

$$\omega_{LO} = \omega_{RF} - \omega_{IF}^{good}$$
$$\omega_{LO} = \omega_{IF}^{bad} + \omega_{IM}$$

$$\cos(\omega_{RF}t) = \frac{1}{2}\left(e^{j\omega_{RF}t} + e^{-j\omega_{RF}t}\right)$$
$$\cos(\omega_{IM}t) = \frac{1}{2}\left(e^{j\omega_{IM}t} + e^{-j\omega_{IM}t}\right)$$
$$\cos(\omega_{LO}t) = \frac{1}{2}\left(e^{j\omega_{LO}t} + e^{-j\omega_{LO}t}\right)$$
$$\sin(\omega_{LO}t) = \frac{1}{2j}\left(e^{j\omega_{LO}t} - e^{-j\omega_{LO}t}\right)$$
(7.12)

Equation (7.13) calculates I_{RF}, the in-phase IF signal in Figure 7.14.

$$\begin{aligned} I_{RF}(t) &= \cos(\omega_{LO}t)\cos(\omega_{RF}t) \\ &= \frac{1}{4}\left(e^{-j\omega_{LO}t} + e^{+j\omega_{LO}t}\right)\left(e^{-j\omega_{RF}t} + e^{j\omega_{RF}t}\right) \\ &= \frac{1}{4}\left(e^{-j\omega_{IF}^{good}t} + e^{+j\omega_{IF}^{good}t}\right) = \frac{1}{2}\cos\left(\omega_{IF}^{good}t\right) \end{aligned}$$
(7.13)

Equation (7.14) calculates Q_{RF}, the quadrature IF signal in Figure 7.14.

$$\begin{aligned} Q_{RF}(t) &= \sin(\omega_{LO}t)\cos(\omega_{RF}t) \\ &= \frac{j}{4}\left(e^{-j\omega_{LO}t} - e^{+j\omega_{LO}t}\right)\left(e^{-j\omega_{RF}t} + e^{j\omega_{RF}t}\right) \\ &= \frac{j}{4}\left(e^{j\omega_{IF}^{good}t} - e^{-j\omega_{IF}^{good}t}\right) = -\frac{1}{2}\sin\left(\omega_{IF}^{good}t\right) \end{aligned}$$
(7.14)

Equation (7.15) calculates I_{IM}, the IF signal at the in-phase rail in Figure 7.14.

$$\begin{aligned} I_{IM}(t) &= \cos(\omega_{LO}t)\cos(\omega_{IM}t) \\ &= \frac{1}{4}\left(e^{-j\omega_{LO}t} + e^{+j\omega_{LO}t}\right)\left(e^{-j\omega_{IM}t} + e^{j\omega_{IM}t}\right) \\ &= \frac{1}{4}\left(e^{-j\omega_{IF}^{bad}t} + e^{+j\omega_{IF}^{bad}t}\right) = \frac{1}{2}\cos(\omega_{IF}^{bad}t) \end{aligned}$$
(7.15)

Finally, Equation (7.16) calculates Q_{IM}, the IF signal at the quadrature rail in Figure 7.14.

$$Q_{IM}(t) = \sin(\omega_{LO}t)\cos(\omega_{IM}t)$$
$$= \frac{j}{4}\left(e^{-j\omega_{LO}t} - e^{+j\omega_{LO}t}\right)\left(e^{-j\omega_{IM}t} + e^{+j\omega_{IM}t}\right) \quad (7.16)$$
$$= -\frac{j}{4}\left(e^{j\omega_{IF}^{bad}t} - e^{-j\omega_{IF}^{bad}t}\right) = \frac{1}{2}\sin\left(\omega_{IF}^{bad}t\right)$$

Table 7.1 summarizes all the calculations that apply to Figure 7.14.

Table 7.1 A Summary of Hartley Image Rejection Mixer Signals, Low Side Use Bold

Q_{RF}	Q_{IM}	Q_{RF}^*	Q_{IM}^*
$-0.5\sin\left(\omega_{IF}^{good}t\right)$	$0.5\sin\left(\omega_{IF}^{bad}t\right)$	$0.5\cos\left(\omega_{IF}^{good}t\right)$	$-0.5\cos\left(\omega_{IF}^{bad}t\right)$
I_{RF}	I_{IM}	I_{RF}^*	I_{IM}^*
$0.5\cos\left(\omega_{IF}^{good}t\right)$	$0.5\cos\left(\omega_{IF}^{bad}t\right)$	$0.5\sin\left(\omega_{IF}^{good}t\right)$	$0.5\sin\left(\omega_{IF}^{bad}t\right)$

For the low-side rejection, we want to pass ω_{IF}^{good} (corresponding to the desired RF) and reject ω_{IF}^{bad} (corresponding to the undesired IM). We use the top circuit of Figure 7.14. Results are shown in Equation (7.17).

$$Q_{RF}^* + I_{RF} = 0.5\cos\left(\omega_{IF}^{good}t\right) + 0.5\cos\left(\omega_{IF}^{good}t\right) = \cos\left(\omega_{IF}^{good}t\right)$$
$$Q_{IM}^* + I_{IM} = -0.5\cos\left(\omega_{IF}^{bad}t\right) + 0.5\cos\left(\omega_{IF}^{bad}t\right) = 0 \quad (7.17)$$

For the high side rejection, we want to pass ω_{IF}^{bad} (now corresponding to the desired IM) and reject ω_{IF}^{good} (now corresponding to the undesired RF). The terminology redefinition may take some getting used to. We use the bottom circuit of Figure 7.14. Results are shown in Equation (7.18).

$$Q_{RF} + I_{RF}^* = -0.5\sin\left(\omega_{IF}^{good}t\right) + 0.5\sin\left(\omega_{IF}^{good}t\right) = 0$$
$$Q_{IM} + I_{IM}^* = 0.5\sin\left(\omega_{IF}^{bad}t\right) + 0.5\sin\left(\omega_{IF}^{bad}t\right) = \sin\left(\omega_{IF}^{bad}t\right) \quad (7.18)$$

Perhaps because of size constraints, commercially available image rejection mixers tend to be in the microwave frequency bands above 1,000 MHz. An example is the Miteq® IRM0208LC2A. Let's say we connect a 3,000 MHz sine wave to a

CHAPTER 7. ANALOG SIGNAL PROCESSING 255

low side image reject mixer LO input, and inject a 2,900 MHz sine wave at the RF input. This is the desired RF. We should see a 100 MHz sine wave on the IF output. Now change the RF input to the 3,100 MHz image frequency. We should see zero signal at the IF output. We rejected the image frequency without needing any image rejection filter. This might be a big advantage compared with a high Q image rejection filter required at the RF input of a standard mixer. The range of RF, LO, and IF frequencies and bandwidths over which the mixer is designed to work must be carefully adhered to. The cost of an integrated Hartley circuit versus an image rejection filter must also be considered.

7.2.2.3 Mixer Spurs

So far, we have illustrated the operation of an ideal mixer. Real electronic mixers have additional undesired outputs called spurs. For N and M integers, spur frequencies are characterized by: $f_{IF} = N f_{RF} + M f_{LO}$. Using the LO, RF, and IF frequencies from the example we started with, $(N, M) = (-1, 1)$ represents a correct high side injection mixer operation. This is shown in Figure 7.16.

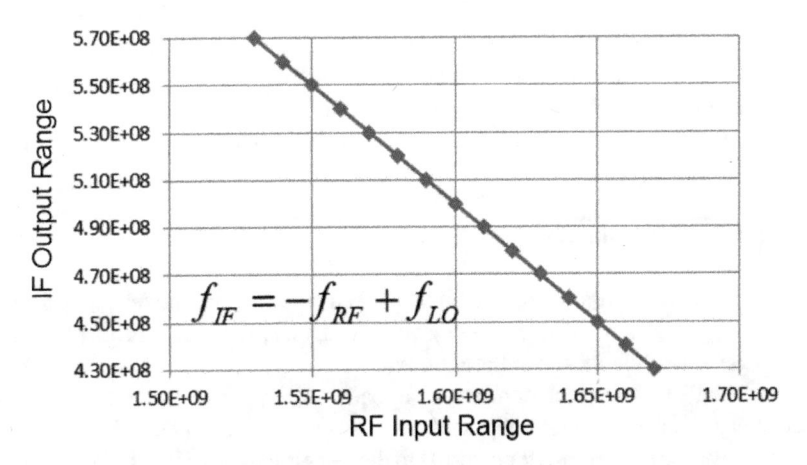

$$f_{IF} = -f_{RF} + f_{LO}$$

Figure 7.16 The mixer desired characteristic, where $(N, M) = (-1, 1)$.

Our primary concern is to keep mixer spurs out of the IF passband (IF output range). For a given combination of RF and LO frequencies, many combinations of N and M do not end up in the IF passband. However, in this case, an undesired fifth-order product, $(N, M) = (3, -2)$, does. Note that the spur output frequencies,

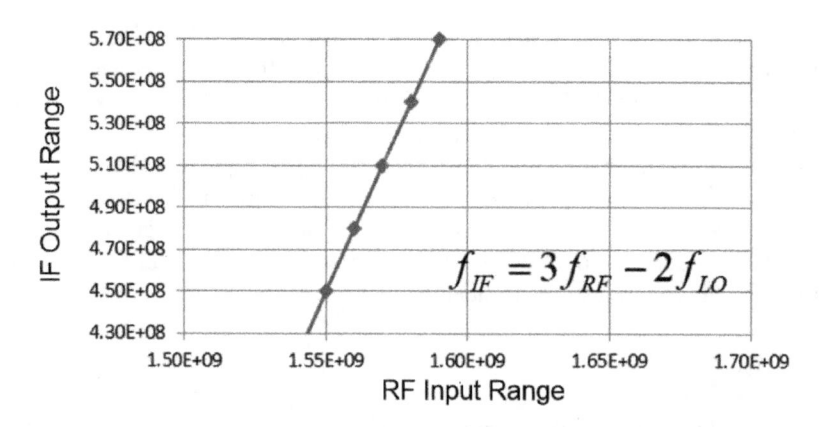

Figure 7.17 Example of mixer undesired characteristic, $(N, M) = (3, -2)$.

shown in Figure 7.17, land right on top of the desired mixer output. Depending on what is in the RF passband, the spur output may be a tone or distribution of tones. In any case, it will be different from the desired $(N, M) = (-1, 1)$ output and it will probably degrade performance. Detailed analysis of a particular mixer configuration is generally performed using simulation software. Manufacturers often provide simulation models that help the designer see what are the worst-case spur orders that will appear in the output spectrum. Mixers are also rated for their power handling capability.

7.2.3 Local Oscillators

The term local oscillator traditionally referred to a receiver mixer input for down-converting the RF to IF. Either a crystal (XTAL) oscillator or a PLL (phase locked loop) generated the local oscillator output.

The PLLs and clock dividers are already designed and integrated into modern receiver and/or transmitter integrated circuits, such as the AD9361. However, a fixed frequency reference oscillator external to the integrated circuit must still be supplied. For the AD9361, this reference oscillator has a frequency specified to be between 10 and 80 MHz, with an output level of 1.3Vpp; see [2]. Typical specifications for the reference oscillator are phase noise, frequency accuracy, and frequency drift with temperature.

Phase noise is a measure of the noise spectrum around the fixed frequency reference oscillator output. A typical reference oscillator phase noise requirement is

CHAPTER 7. ANALOG SIGNAL PROCESSING

shown in Figure 1 of [2]. Offset frequency means how far from the nominal output frequency is the phase noise measured. At each frequency offset on the horizontal scale, the plot shows the ratio of power at the rated output frequency to the power contained within a 1 Hz bandwidth. Units are dBc/Hz; dBc means with respect to the carrier power.

Frequency accuracy is usually measured in parts per million (ppm) frequency error. For example, a 10 MHz oscillator with a maximum error of 30 ppm will be within the frequency limits of $\pm \left(30/10^6\right)\left(10^7\right) = \pm 300 Hz$

Modern TCXOs (temperature-controlled crystal oscillators) achieve very low frequency drift with temperature, for example, ±0.3 ppm[1] over the industrial temperature range of -40 to 85°C where 25°C is the reference point of 0 frequency error. TCXOs are especially important for SDRs designed for continuous monitoring in an outdoor environment with wide temperature swings.

7.2.3.1 Reciprocal Mixing

Excessive LO phase noise is sometimes responsible for reciprocal mixing in a receiver. Figure 7.18 shows how a high-power interferer is mixed down out of the receiver IF passband. However, the interferer's high phase noise skirt, which came from the receiver LO, is partly mixed into the receiver passband. Thus, the passband noise level is increased by phase noise imparted to the interfering signal. An easy way to understand this is to imagine that the interferer mixed down the LO signal, with its phase noise skirt, instead of the other way around. A low phase noise LO or a well designed preselector filter might have prevented this situation. This has to do with design quality. A receiver for a commercial aircraft will have reciprocal mixing prevented. A low-cost receiver for hobby use may not.

7.2.4 RF Filters

A typical SDR receiver front end makes use of two filter functions, a preselector to reject unwanted RF signals from the antenna and an antialias filter to restrict the frequency spectrum at the ADC input. A few analog RF filter definitions may be in order here.

Quality factor (Q) For a bandpass filter, Q is the ratio of the center frequency to the -3 dB bandwidth. In Figure 7.19 the -3 dB BW extends from 45 MHz to 55 MHz and the filter is centered at 50 MHz. Thus, quality factor = 50/10 = 5.

1   ppm = parts per million, actual frequency drift in Hz is ±0.3/1000000.

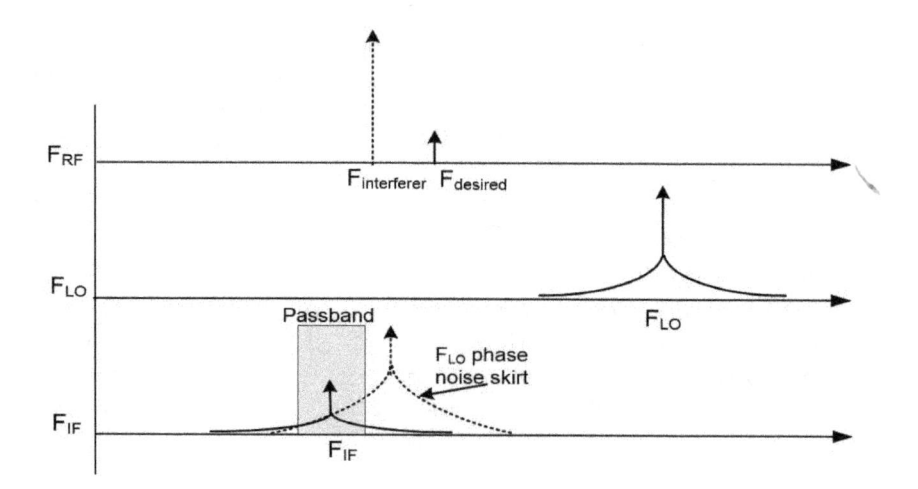

258   *Software Defined Radio: Theory and Practice*

Figure 7.18 An example of receiver reciprocal mixing.

Shape factor (SF) For any filter, SF is the ratio of the -60 dB stop bandwidth to the -3 dB bandwidth. SF measures the filter skirt steepness In Figure 7.19, shape factor = 67.5/10 = 6.75.

Insertion loss (IL) Attenuation at the center frequency, for a bandpass filter, or at 0 Hz for a lowpass filter. In Figure 7.19, insertion loss = -7dB. Low IL can be important for a preselector; see Section 7.2.1.5.

Group delay Group delay is defined as the negative of the derivative of the phase shift versus frequency; see Figure 9.26.

7.2.4.1 Lumped Element RLC Preselector

Preselectors can improve receiver performance by rejecting RF image frequencies and noise in the receive bandwidth. Many SDRs are zero IF designs that do not have image frequencies; see Section 7.3.2.2, so often a preselector is not provided; see Table 1.1 for some examples. Without any filtering prior to the first mixer, the RF components will need enough linearity to handle the extra power from the unfiltered, undesired receive signals; see Section 7.3.1.3. Nonzero IF SDRs may be subject to reciprocal mixing for unfiltered close in signals, see Figure 7.18.

Preselectors for SDR are often built with discrete resistors, capacitors, and inductors (RLC). These are economical and work well at frequencies below about 1 GHz, about 30 cm wavelength. Above 1 GHz, bandpass filters will have to reject

CHAPTER 7. ANALOG SIGNAL PROCESSING

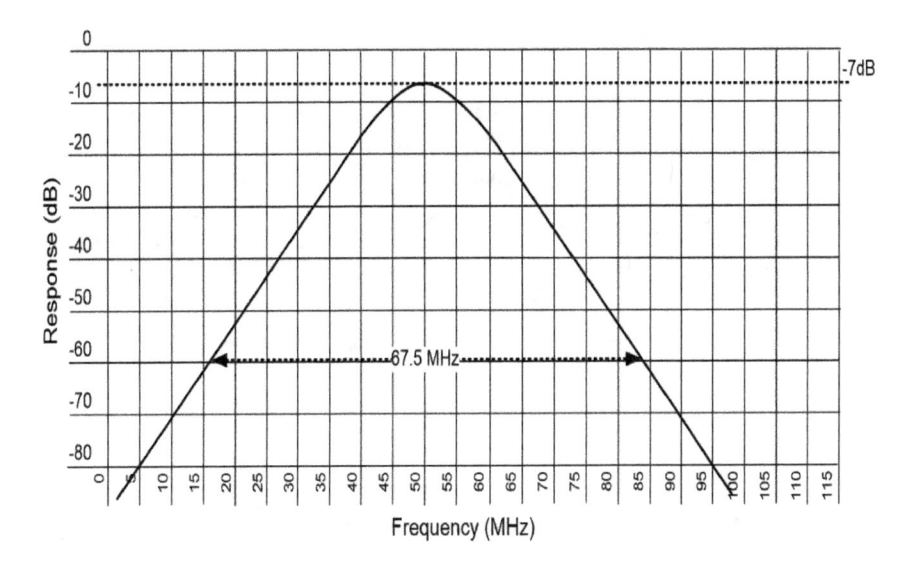

Figure 7.19 Example microwave filter response.

higher frequencies with shorter wavelengths. Unintended transmission paths across a filter circuit layout may degrade performance. For RLC filters, Q factors > 100 are difficult, requiring a large number of poles, thus making the circuit board layout large. Some other filter technologies, to be discussed, may be a better choice because they will provide higher Q in a smaller space. There are many free RLC circuit design tools available; see, for example, www.electroschematics.com or www.rf-tools.com.

7.2.4.2 Active Filters

Antialias lowpass filters driving the ADC are often active filters. Active filters are built around op-amps (operational amplifiers). These op-amps can also provide a low impedance output suitable for driving an ADC input. Figure 7.31 is an example of an active ADC drive circuit that uses an impedance matching, DC blocking, transformer. Some of the major op-amp manufacturers provide excellent free software tools to design active filters around the op-amps that they sell. See, for example, www.tools.analog.com or www.ti.com.

7.2.4.3 Advanced Technology Filters

More advanced technology filters are smaller and work at higher frequencies and have higher Q for a smaller space. They are often used in radio hardware.

The film bulk acoustic resonator (FBAR) filter shown in Figure 7.20 has evolved from the decades-old surface acoustic wave filter. FBAR filters are commonly used in cell phones due to their compact size (for example 1.1 × 1.4 × 0.8 mm). Also, the low insertion loss causes minimal degradation of receiver cascade noise factor.

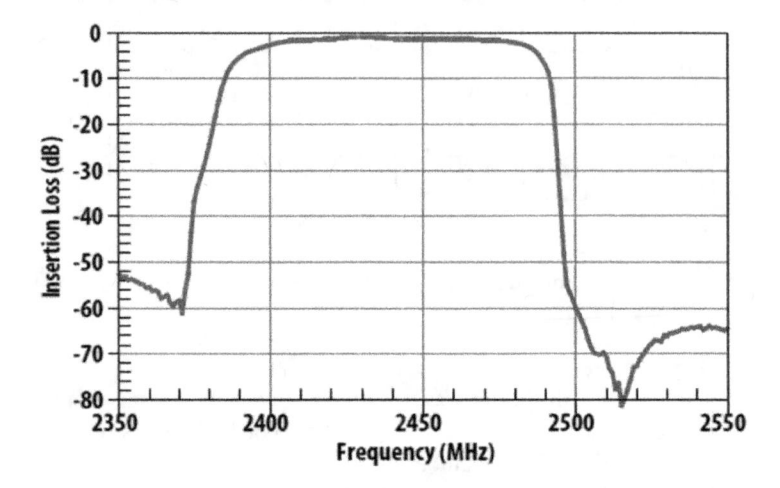

Figure 7.20 Film bulk acoustic resonator response.

Dielectric filters use ceramic dielectric material as a resonator for microwave frequencies. A ceramic filter passband is shown below in Figure 7.21.

7.3 RECEIVER CONFIGURATIONS

The traditional double conversion receiver configuration (two mixer stages) is not commonly used for SDR. The more modern single conversion receiver has only one mixer and one IF (intermediate frequency). The single IF can be nonzero or zero frequency. We will examine both designs and focus on frequency planning, gain planning, noise, and AGC (automatic gain control).

CHAPTER 7. ANALOG SIGNAL PROCESSING

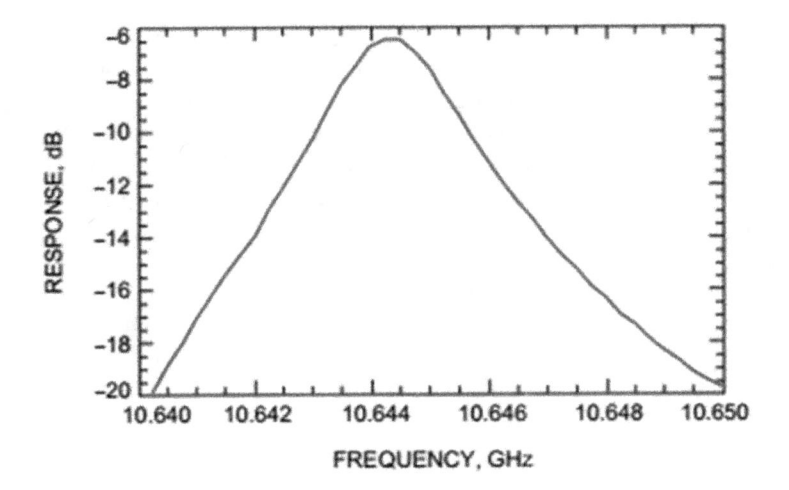

Figure 7.21 Ceramic resonator bandpass filter response.

7.3.1 Nonzero IF Receiver

7.3.1.1 Nonzero IF Receiver Frequency Planning

Figure 7.22 shows our starting point for understanding the nonzero intermediate frequency (NZIF) radio for use as an SDR RF front end. Our first consideration is frequency planning.

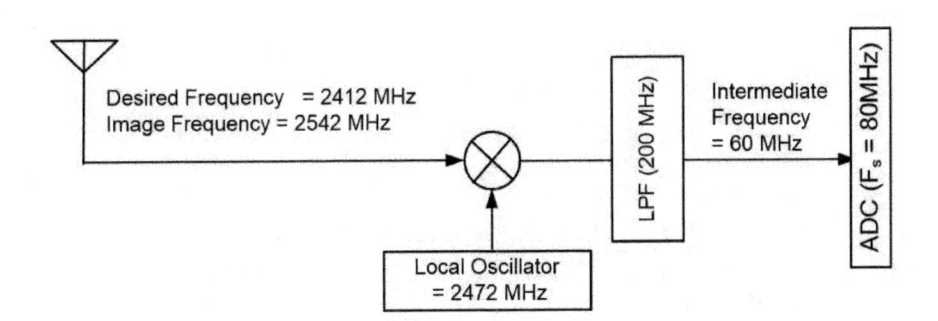

Figure 7.22 Nonzero IF SDR high side receiver topology.

To illustrate the NZIF concept, let's design a single conversion NZIF receiver for an 802.11 wireless LAN: center frequency = 2,412 MHz, bandwidth = 20 MHz. This signal is in the 2,400 to 2,500 MHz ISM (Industrial, Scientific, Medical) band that is set aside for unlicensed low-power communications. We start by downconverting the RF channel at 2,412 MHz to 60 MHz using local oscillator (LO) = 2,472 MHz. The use of an LO frequency above the desired RF is called high side injection.

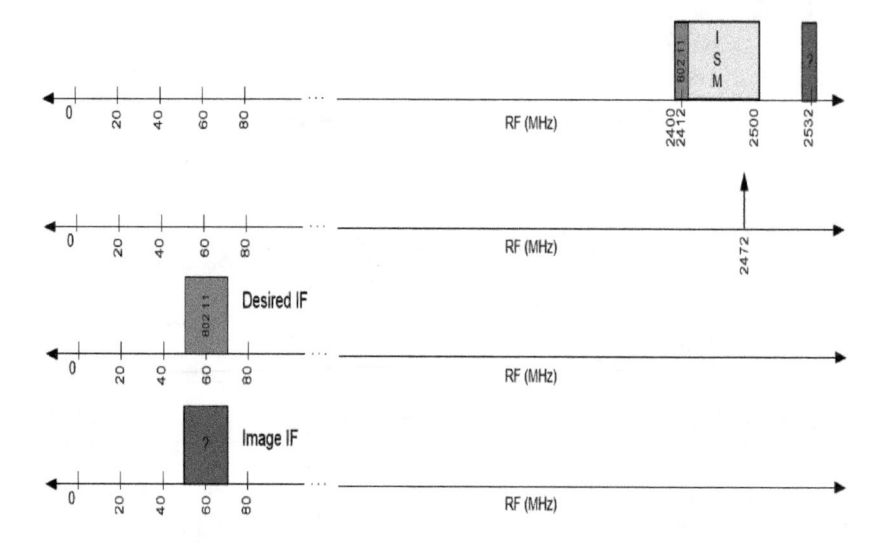

Figure 7.23 Example of nonzero IF high side injection.

Downconverting the RF channel at 2,412 MHz results in four mixer products (only the first two are shown in Figure 7.23):

1. Difference frequency: LO - Desired = $|2,472 - 2,412| = 60$ MHz
2. Difference frequency: LO - Image = $|2,472 - 2,532| = 60$ MHz
3. Sum frequency: LO + Desired = $|2,472 + 2,412| = 4884$ MHz
4. Sum frequency: LO + Image = $|2,472 + 2,532| = 5004$ MHz

We need to get rid of the sum frequencies so we add a lowpass filter (LPF) to remove them at the mixer output. The lowpass cutoff of 200 MHz in Figure 7.22 will easily remove the sum frequencies and pass the difference frequencies. Figure 7.24 shows the frequency and phase response of a typical analog filter for this

CHAPTER 7. ANALOG SIGNAL PROCESSING

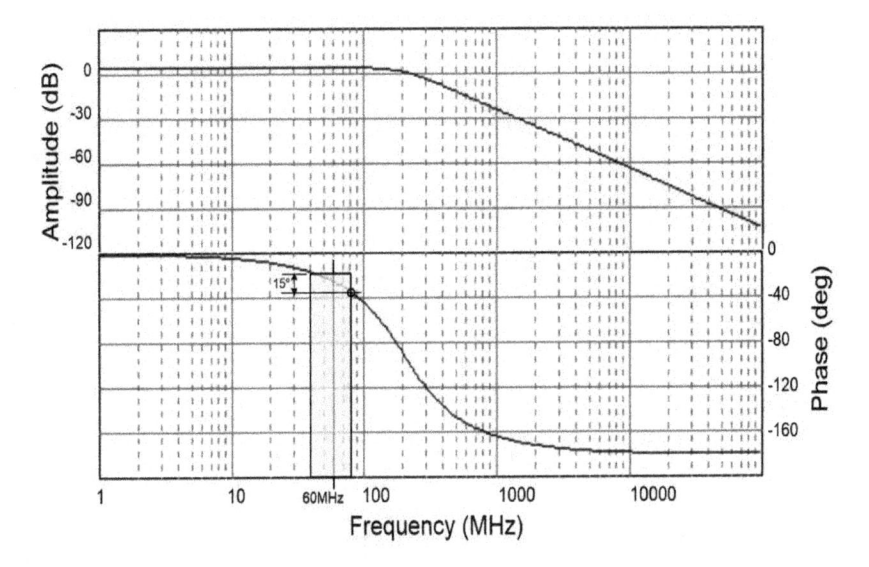

Figure 7.24 A second-order analog filter for removing mixer sum products.

application. The phase plot response shows why the 40 MHz wide sampling band centered at 60 MHz is not closer to the 200 MHz cutoff frequency. Moving back from the cutoff frequency reduces the phase change across the sampling bandwidth. The Butterworth lowpass chosen here has a flat passband and smooth monotonic phase rolloff. The filter also has a very simple design using an op-amp and a few resistors and capacitors. Butterworth filters also have a maximally flat passband. Figure 9.25 compares Butterworth with some other standard filter shapes. The phase rolloff characteristics of these will all be quite different.

Notice from Figure 7.23 that we still have the undesired image frequency downconverted to the 60 MHz IF. Figure 7.25 solves this problem by adding a preselector filter before the mixer. Blocking the image signal ensures that the only signal at 60 MHz will be the desired 802.11 signal, first item in the list above. This preselector disadvantage may be that it places an absolute limit on tuned bandwidth.

The preselector in Figure 7.26 has bandwidth = 45 MHz, insertion loss < 1dB and high Q = (2,410 MHz center frequency) / (45 MHz BW) = 47.5 (Q is a dimensionless indicator of bandpass filter sharpness). A preselector is generally required. For example, with no preselector, a signal at 1,266 MHz may reach the mixer shown in Figure 7.22. Mixer nonlinearity can result in a third-order IF product at $|2F_{RF} - F_{LO}| = 60$ MHz, right on top of our desired signal.

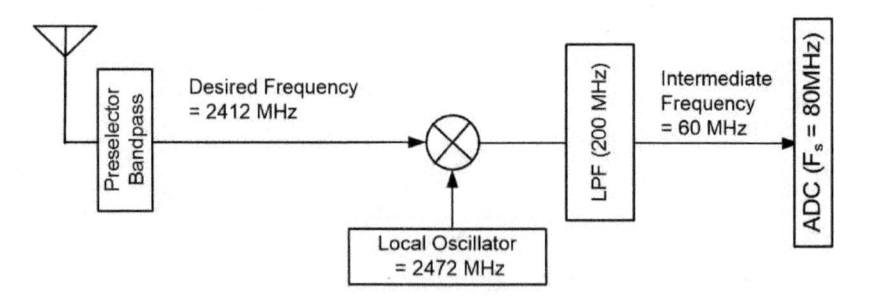

Figure 7.25 A nonzero IF high side injection with a preselector.

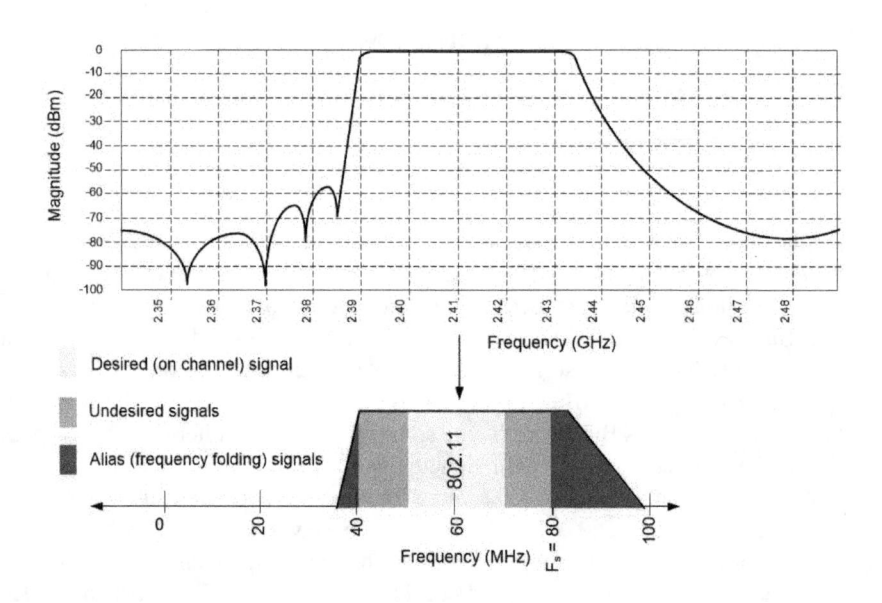

Figure 7.26 An ISM band channelizing filter.

Figure 7.26, lower part, shows the spectrum at the output of the 200 MHz lowpass filter (LPF) in Figure 7.25. Notice the shape similarity after downconverting to 60 MHz IF. The 200 MHz LPF, discussed above, removes the sum frequencies

CHAPTER 7. ANALOG SIGNAL PROCESSING

from the mixer output with very little effect on the desired signal spectrum at the mixer output.

The choice of 60 MHz for our intermediate frequency is the result of a trade-off. As shown in Figure 7.23, the frequency difference between the desired and image signals is 120 MHz, twice the IF. This frequency difference is fixed for any RF tuning frequency. Thus, as the tuning frequency goes up, the preselector must be more selective. For example at 3,000 MHz tuning frequency, percent bandwidth for a 120 MHz preselector is 4%. This requires a Q of more than 5,000. The only way around this requirement is to raise the NZIF frequency IF, thus increasing the frequency distance between desired RF and image signals. Then the ADC sampling rate will need to be increased, a possible disadvantage.

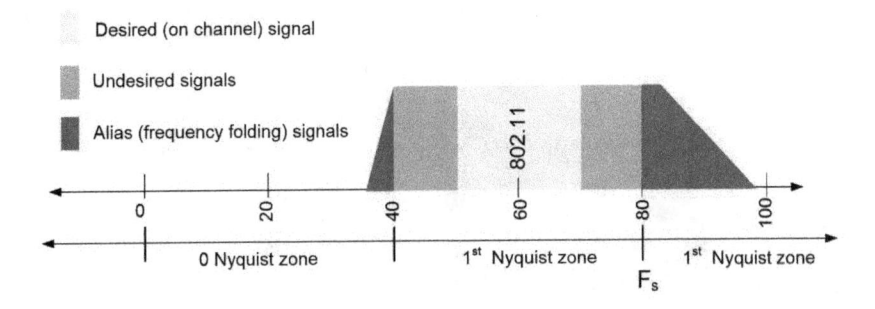

Figure 7.27 Receiver analog IF spectrum at the ADC input.

The IF spectrum at the ADC input in Figure 7.25 is shown in Figure 7.27. This figure shows the 80 MHz sampling frequency and the 60 MHz IF in N1, all prior to sampling. A band-limited IF signal in any Nyquist zone will be folded into N0. Spectral inversion (flipping low and high frequencies around the center IF) occurs for folding from odd-numbered Nyquist zones (for example, N1) into Nyquist zone zero. Spectral inversion does not occur for folding from even numbered Nyquist zones into Nyquist zero. Spectral inversion can easily be corrected in the digital processing of the sampled signal. As a DSP topic, this will be described in Section 9.3.1.

7.3.1.2 4/3 Sampling

Figure 7.28 illustrates a technique sometimes called 4/3 sampling, $F_s = (4/3)F_{IF} = 60(4/3) = 80$ MHz. Figure 7.28 illustrates how the sampling rate of 80 MHz results in a pivot around 40 MHz causing the first Nyquist zone to fold into Nyquist zone zero. This pivot can be thought of as a hinge between the unsampled and sampled spectrums. As shown in Figure 7.28, aliasing occurs in the sampled signal when the IF signal to be sampled (top spectrum in Figure 7.28) is not confined to the first Nyquist zone. The triangle-shaped alias spectra fold back and can damage the desired 802.11 signal, as shown in the lower spectrum of Figure 7.28.

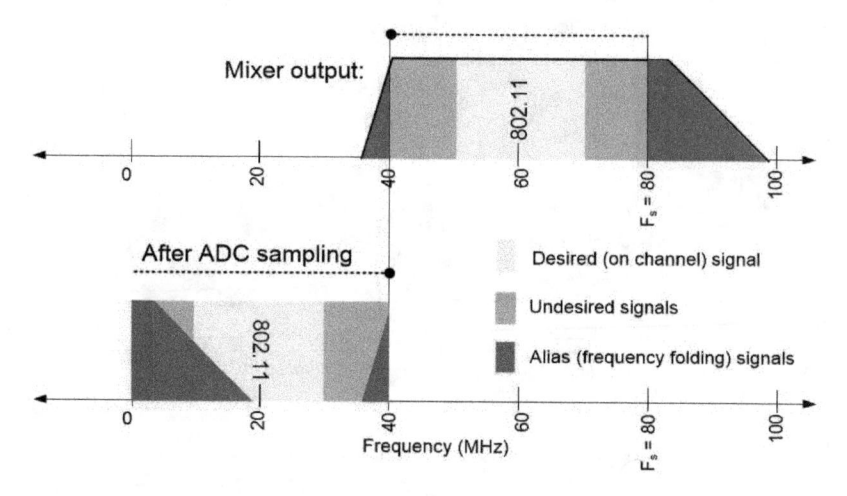

Figure 7.28 ADC sampled IF spectrum with aliasing through the ADC.

The solution to aliasing is to add one more filter to tighten up the spectrum in the first Nyquist zone prior to sampling. The antialias filtered spectrum to be sampled is in the center part of Figure 7.29. The antialias BPF frequency response is shown as a black line overlaid on this spectrum. As mentioned above, the sampled spectrum in the lower part of Figure 7.29 will be inverted because it folds into Nyquist zone 0 from zone 1.

So far, we have our desired receive signal frequency translated to the correct IF with minimal interference from the image frequency. As shown in Chapter 9, the signal centered at 40 MHz will be converted, after sampling, to complex baseband centered at 0 Hz. Our next step is receiver gain planning.

CHAPTER 7. ANALOG SIGNAL PROCESSING 267

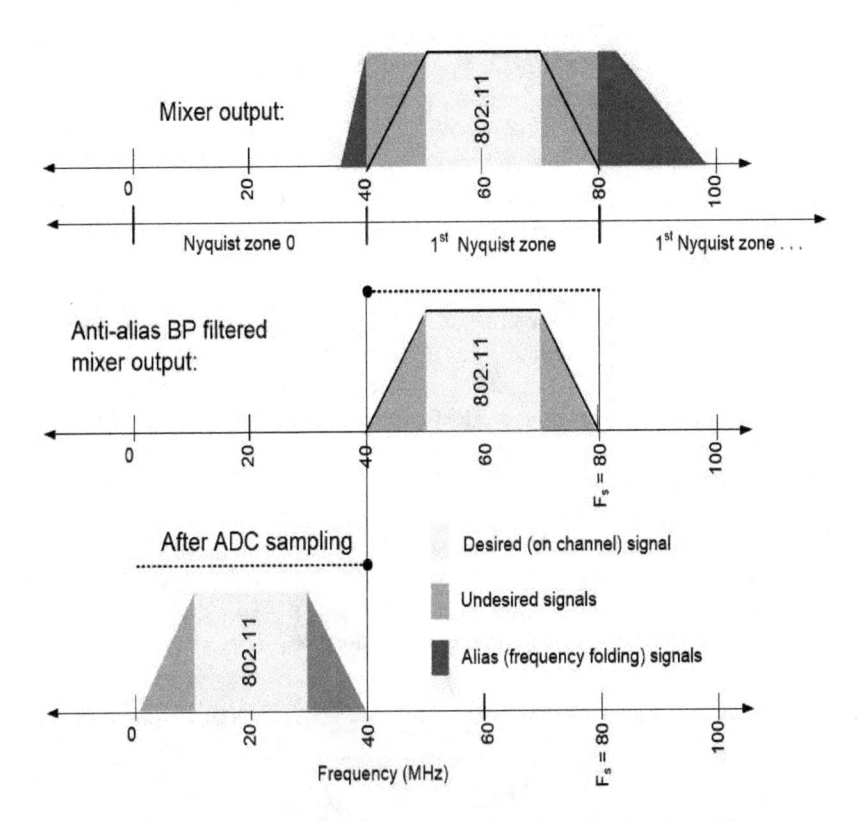

Figure 7.29 ADC sampled IF spectrum with antialias bandpass filter.

7.3.1.3 Nonzero IF Receiver Gain Planning

Quantizing error (see Section 9.4) is minimized by controlling the analog voltage at the ADC driver output to make full use of the ADC input range.[2]

As stated, the AGC task is to maintain a constant signal level at the ADC input. The AGC lowers the receiver gain for high-power signals and raises gain for low-power signals. As shown in Figure 7.30, the receiver has to cope with contradicting requirements. At the ADC input, only one optimum power level results in best performance, whereas the antenna input needs to handle a much wider range, for example, from -100 dBm or less to -20 dBm or more. To implement AGC in our receiver, we include two types of amplifiers.

At the receiver front end we add an low noise amplifier (LNA). Typical LNA specifications are gain = 26 dB, NF = 0.5 dB, IP3 = 10 dBm, OP3 = 36 dBm. Recall that high intercept points result in the LNA staying linear for a wide range of input power. This means, for example, that a high-power undesired signal that gets through the preselector has less chance of generating IMD (intermodulation distortion) in the receiver IF band. Sometimes the AGC controls a gain reduction switch circuit prior to the LNA. This is to attenuate very strong in-band signals so the downstream AGC components will not be overloaded. Prior to the AGC, we include variable gain amplifiers VGA1 and VGA2. These have a digital (or analog) control input that varies the gain from 0 to 38 dB.

We choose the ANALOG DEVICES AD6645 as our example ADC. From the data sheet, the maximum voltage swing at the ADC balanced input pins is $2.2V_{PP}$. With reference to the ADC input circuit of Figure 7.31, we convert $2.2V_{PP}$ to RMS and then divide it by 2 to get a practical ADC input voltage setpoint. There are other ADC driver circuits that do not require a transformer [3].

$$V_{FS,RMS} = \frac{0.5V_{PP}}{\sqrt{2}} = \frac{1.1}{\sqrt{2}} = 0.778 V_{RMS}$$

$$V_{ADC} = \frac{V_{FS,RMS}}{2} = 0.389 V_{RMS}$$

(7.19)

The 1:2 turns ratio transformer has a 1:4 resistance ratio (transformer resistance ratio is the turns ratio squared). Thus, the transformer input and 60.4 ohm output resistor match the 50 ohm output impedance of VGA1 to the 1,000 + 2(25) = 1,050 ohm input impedance of the ADC as shown in Equation (7.20).

---

2 Even though the quantizing error is minimized by using the full ADC range, in practice the ADC input signal level may be set to only half the ADC range to account for signal peaks and/or AGC (automatic gain control) inaccuracy.

CHAPTER 7. ANALOG SIGNAL PROCESSING

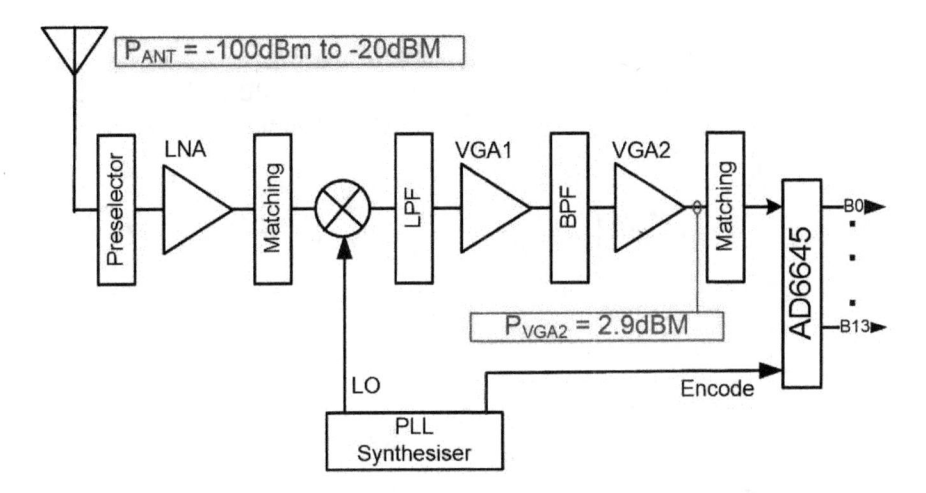

Figure 7.30 Outline of receiver automatic gain control task.

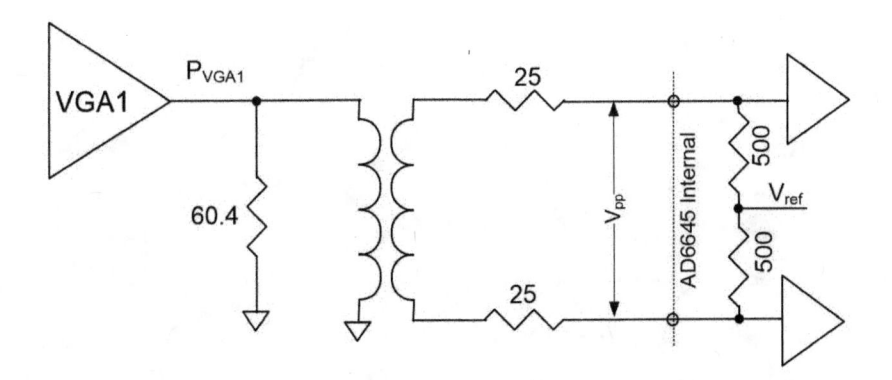

Figure 7.31 Transformer-based receiver ADC interface circuit.

$$R_{VGA1} = 50 \approx (60.4) \parallel \left(\frac{1050}{4}\right) = 49.1 \; ohms \qquad (7.20)$$

Now calculate the VGA1 output power that produces full-scale ADC input voltage using Equation (7.21). Finally, reduce this by 2 dB to further reduce the occurrence of overload. $P_{VGA1} = 4.9$ dBm - 2 dB = 2.9 dBm.

$$P_{VGA1\_MAX} = 10 log_{10}\left(\frac{V^2}{0.001 R_{VGA1}}\right) = 10 log_{10}\left(\frac{(0.389)^2}{0.0491}\right) = 4.9 dBm \qquad (7.21)$$

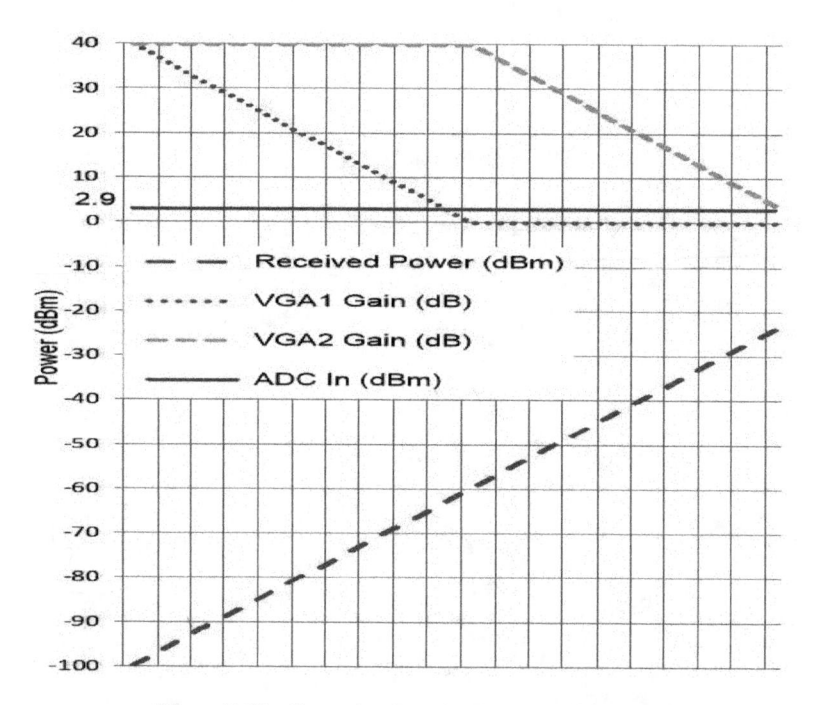

Figure 7.32 Example of receiver overall AGC control.

Figure 7.32 shows the VGA control voltages required to hold the 2.9 dBm set point as the input power goes up. These were calculated on a spread sheet. Notice how the gain closest to the ADC is reduced first. Except for fast-acting RF protection front-end gain reduction, gain nearer to the antenna is reduced last. This is because

CHAPTER 7. ANALOG SIGNAL PROCESSING

the input composite noise figure will worsen as input RF gain is reduced, but is not affected much by reducing ADC driver gain.

7.3.1.4 Nonzero IF Receiver AGC Details

This is a summary of important AGC principles:

1. The desired range or set point for AGC control should be set to about one-half of the ADC input range. Only one bit of ADC range is lost and the feedback control has a large linear range in which to operate.
2. As input power goes up, gain closest to the ADC (VGA1) is reduced first. Gain near the antenna is reduced last. This is because the RF input composite noise figure will worsen as input gain nearest the antenna is reduced.
3. A switchable RF input attenuator can be added (PIN diode, MEMS) to control receive power closer to antenna (i.e., the first line of defense for strong signals). However, AGC closed-loop dynamics may limit the response time to high-power transients; therefore, front-end components may have to handle short high power inputs without damage.

7.3.1.5 AGC in the Digital Domain

Figure 7.33 shows a possible configuration for AGC DSP processing inside the FPGA or DSP that interfaces to the radio ADC. Assuming that the ADC is linear, signal power can be measured at complex baseband as $p_{rec}(k) = E\left(I^2(k) + Q^2(k)\right)$. E is the expected value or, in this case, simple sliding average. There will be an averaging time that needs to be considered carefully.

$log\left(p_{rec}(k)\right)$ linearizes the received power with respect to the input signal. Power is the square of input voltage. The log function is close to the inverse square so the power measurement becomes near linear in dB. Thus, the AGC response will be similar for small and large input signals. The linearized power then goes to an exponential lowpass average (see Section 9.2.3). This LPF uses a single parameter u to determine how much of the next output is due to the current input or to the previous output. Parameter u should be low enough so that the AGC does not try to follow the AM signal peaks (constant envelope or FM will not have this problem).

The threshold comparison circuit controls the feedback to gain setting elements such as VGA1 and VGA2. If this is a receiver for a constant envelope signal, such as MSK, the comparison will be very simple. However, for a signal with unpredictable peaks, such as voice, the situation is more complicated. Consider Figure 7.34. For

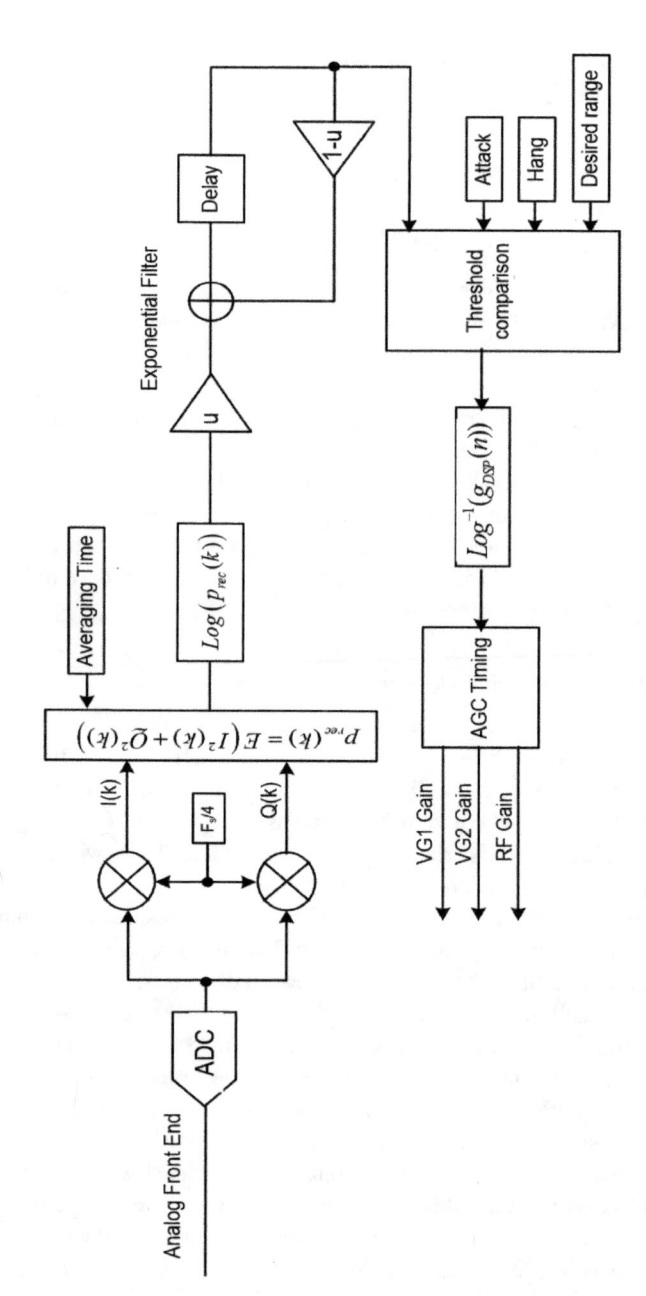

Figure 7.33 Example of a receiver AGC circuit contained in DSP.

CHAPTER 7. ANALOG SIGNAL PROCESSING

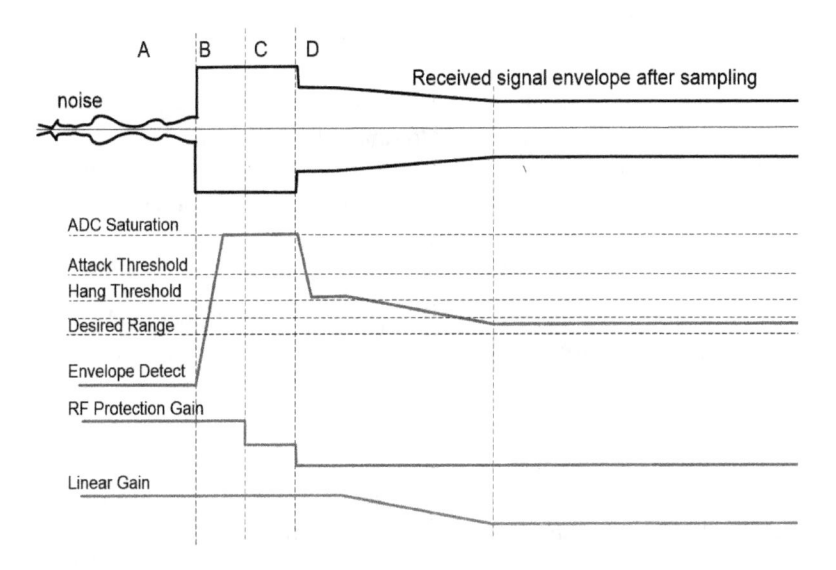

Figure 7.34 Receiver DSP AGC control timing.

a high power AM voice signal,[3] the series of AGC steps are listed next. A good reference on voice AGC is [4].

A: The high-power signal has not arrived yet. The receiver gain is maximum.
B: The AM signal overwhelms the receiver causing the ADC input to clip.
C: The AGC control enables two fast acting gain reduction steps at the antenna input.
D: The input signal power at the AGC threshold comparison circuit drops down to the hang threshold. The hang threshold stabilizes the AGC by inserting a fixed gain delay to prevent the AGC from reacting to pauses in voice power or large voice peaks. After the hang time, the gain is allowed to change and the threshold compare can fine adjust linear gain setting elements like VG1 and VGA2. Measuring average voice power is difficult because of the peaks and pauses.

---

3 For a linear modulation like AM, the voice signal will be directly represented in the RF envelope.

7.3.1.6 Nonzero IF Receiver Noise Considerations

For a digital receiver, receiver sensitivity is the minimum received signal power needed for the maximum error rate allowable on the receiver final output. First, we need to calculate the composite ADC noise floor. This consists of receiver noise plus ADC noise. Noise power density (i.e., noise power in 1 Hz) at the antenna input is:

$$10 log_{10} k_B T_a = -174 dBm/Hz \quad (7.22)$$

k_B = Boltzmann constant = 1.38e-23 (Watts/Hz)/Kelvin and T_a = antenna noise temperature = 290° Kelvin. The total noise power in receiver tuned bandwidth B is:

$$P_{AntNoisedBm} = -174 dBm/Hz + 10 log_{10} B \quad (7.23)$$

An example of a cascade receiver noise factor analysis is shown below in Figure 7.35. Note that three passive circuits are combined for a single passive noise factor[4] and there is only one VGA. All the gains and noise factors are linear, not in dB. See Section 2.6.2 for additional background.

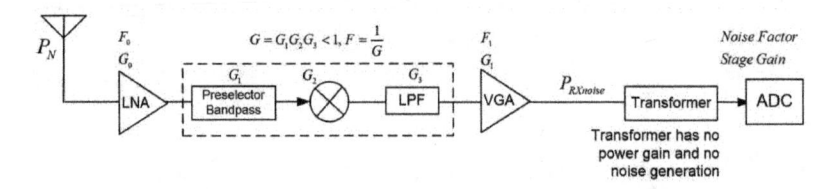

Figure 7.35 Model of receiver noise factors and gains.

The composite gain is $G_{RXdB} = 10 log_{10} [G_0 G G_1]$. The composite noise figure up to the VGA input is shown in Equation (7.24). Note that the VGA NF does not affect the overall NF very much; see Section 7.2.1.

$$NF_{RXdB} = 10 log_{10} \left[F_0 + \frac{F-1}{G_0} + \frac{F_1-1}{G_0 G} \right] \quad (7.24)$$

Finally, the noise at the ADC input due to the antenna and receiver is:

$$P_{RxADCnoisedBm} = P_{AntNoisedBm} + NF_{RXdB} + G_{RXdB} \quad (7.25)$$

---
4 To convert from noise factor to noise figure, we take the log; see Equation (7.7).

CHAPTER 7. ANALOG SIGNAL PROCESSING

To complete our composite ADC noise floor calculation, we convert receiver noise at the ADC input to an RMS voltage:

$$V_{RxADCnoise} = 2\sqrt{R_{ADC}\left(10^{\left(\frac{P_{RxADCnoisedBm}}{10}-3\right)}\right)} \qquad (7.26)$$

Given the ADC data sheet SNR rating, calculate the ADC-generated input noise RMS voltage:

$$V_{ADCnoise} = V_{FS,RMS}\left(10^{-SNR/20}\right) \qquad (7.27)$$

Note that the ADC data sheet SNR assumes a full-scale sine wave input. We assume that $V_{ADCnoise}$ and $V_{RxADCnoise}$ are normally distributed random variables, therefore, we can sum the variance and take the square root to get back to a composite ADC noise floor voltage; see Equation (7.28).

$$V_{NoiseFloorADC} = \sqrt{V_{ADCnoise}^2 + V_{RxADCnoise}^2} \qquad (7.28)$$

Figure 7.36 shows ADC dynamic range details for one in-band signal. The received signal power is shown on the left, however, because an ADC is generally a voltage-driven device, the ADC input is measured in volts. At the bottom of the range, the signal must be above $V_{NoiseFloorADC}$ to provide the signal-to-noise ratio required to attain the specified error performance. This performance is what the receiver delivers on its output and could be measured in bit error rate (BER) or packet error rate. At the top of the ADC input range, the signal must be backed off to accommodate the signal peaks. For practical designs, there is usually a couple dB more backoff at the top to account for AGC inaccuracies.

7.3.2 Zero IF Receiver

This section describes a receiver configuration with an analog complex IF signal centered at 0 Hz. This is a common configuration for SDR receivers. All the required analog signal processing is available in an integrated circuit, for example, the Analog Devices AD9361. Figure 7.37 shows our starting point for understanding the zero intermediate frequency (ZIF) radio for use as an SDR RF front end.

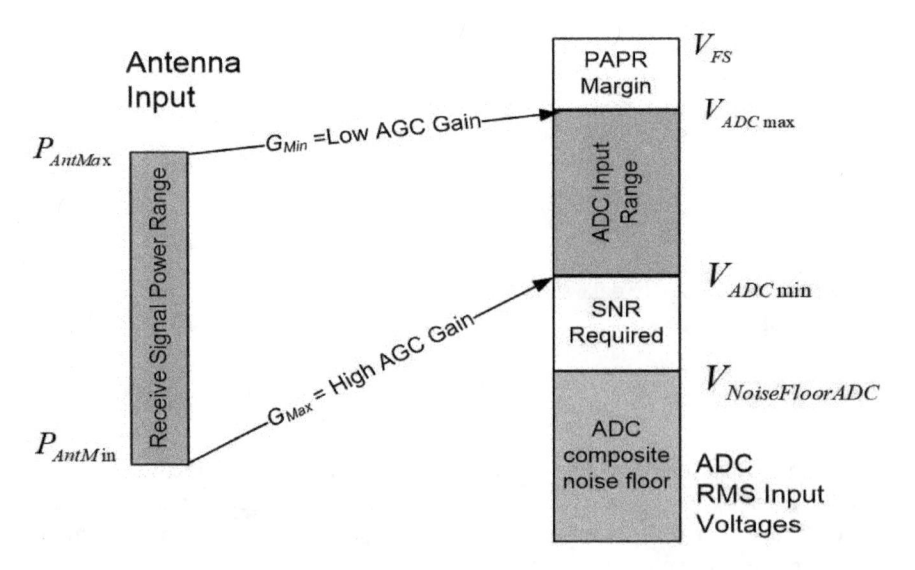

Figure 7.36 Typical ADC dynamic range details.

7.3.2.1 Zero IF Receiver Gain Planning

Figure 7.37 shows the AGC components for the ZIF receiver. Note that these are substantially the same as for the NZIF receiver. As we saw for the NZIF receiver, the interface between VGA2 and the ADC is carefully designed for impedance matching. In addition, the VGA2 output circuit may be able to supply the current needs of the ADC input better than a LPF circuit could. Thus, we put the antialias filter prior to VGA2.

7.3.2.2 Zero IF Receiver Frequency Planning

To illustrate the ZIF concept, we will design a ZIF receiver for an 802.11 wireless LAN: center frequency = 2,412 MHz, bandwidth = 20 MHz. We start by downconverting the RF channel at 2,412 MHz to 0 MHz using LO frequency = 2,412 MHz. The LO is a complex signal at the same frequency as the desired RF. Notice from Figure 7.37 that a complex LO results in real and imaginary IF signals. These must be separately sampled.

CHAPTER 7. ANALOG SIGNAL PROCESSING

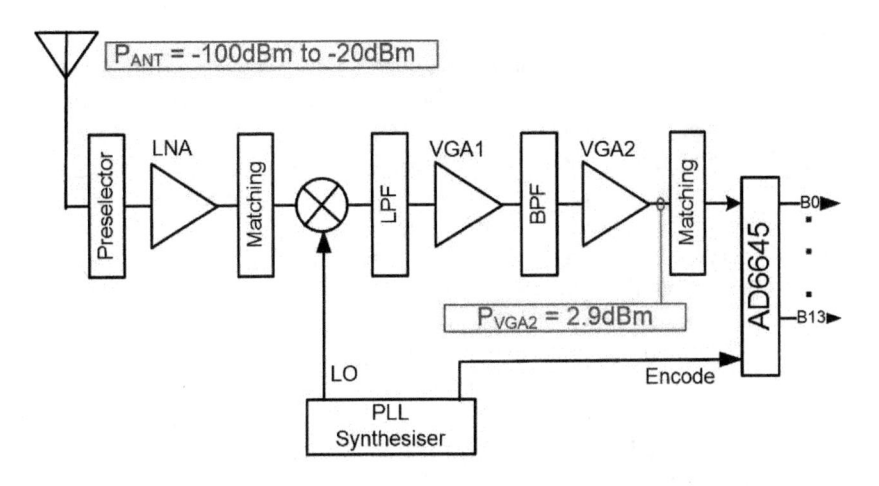

Figure 7.37 Typical zero IF topology and gain planning.

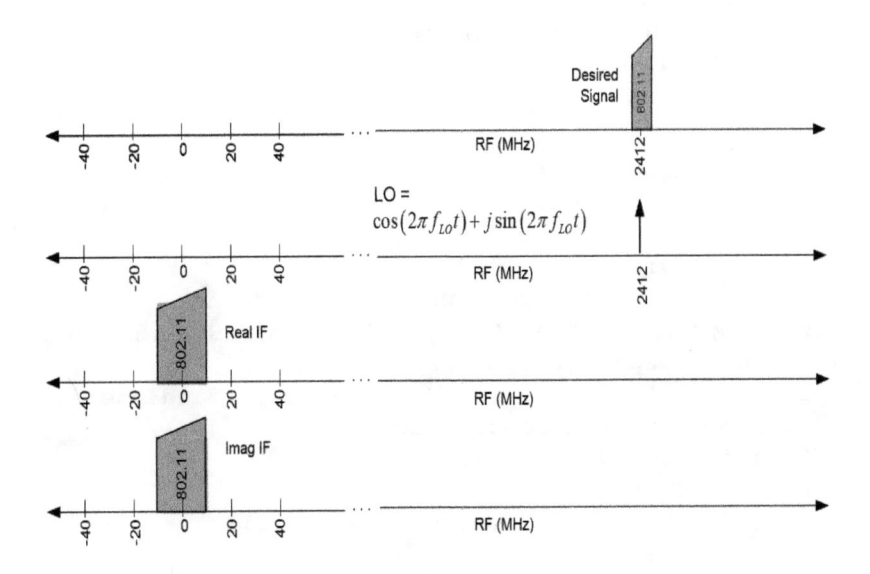

Figure 7.38 Example of zero IF SDR local oscillator tuning.

As shown in the frequency plan of Figure 7.38, the ZIF receiver has no separate image frequency.

The ZIF tuning range can be wider because there is no image frequency to reject. Also, because the ZIF IF is 0 Hz, only the signal bandwidth must be considered for ADC sampling. For this example, the ZIF only needs a 40 MHz sampling rate instead of 80 MHz required for NZIF in Section 7.3.1

Without a preselector, a very wide frequency range may be present at the mixer outputs. Depending on the mixer linearity, there may be a need to limit this bandwidth prior to the mixers. However, for this example, we assume no preselector.

The ADC antialias lowpass filter is a critical component. As shown in Figure 7.39, for an ADC sampling rate of 40 MHz, any signals above 20 MHz will be aliased into the -20 to +20MHz passband. An analog lowpass filter with a -3dB cutoff of +12 MHz attenuates this aliasing as shown in Figure 7.40. Both the Chebyshev 2 and the Butterworth have flat passbands. The roll-off rate and ultimate attenuation of the Chebyshev 2 is larger. Note that these are not digital filters so their response does not repeat at multiples of the sample rate.

Also important is group delay.[5] For a digitally modulated signal, the maximum group delay should generally be no more than 10% of the symbol rate. As shown in the lower plot of Figure 7.40, the Chebyshev 2 group delay is only slightly higher than the Butterworth. This is primarily due to the way the frequency responses have been aligned. Some of the MATLAB code used to generate Figure 7.40 is included.

```
% Set up analog filter parameters
PassBandEdge = 10e6;
N = 6; % Filter order
% Calc transfer function,
% note scaling to force cutoffs to line up
(zbb,pbb,kbb)= butter(N, 2.5*pi*PassBandEdge,'low','s');
(bbb,abb) = zp2tf(zbb,pbb,kbb);
(z2c,p2c,k2c) = cheby2(N,80,6.75*pi*PassBandEdge,'s');
(b2c,a2c) = zp2tf(z2c,p2c,k2c);
% Calc frequency response
(hButter,wButter)= freqs(bbb,abb,4096);
(hCheby2,wCheby2) = freqs(b2c,a2c,4096);
% Calc Grpdelay
```

---

5 Group delay is the negative of the derivative of filter phase response with respect to frequency.

CHAPTER 7. ANALOG SIGNAL PROCESSING

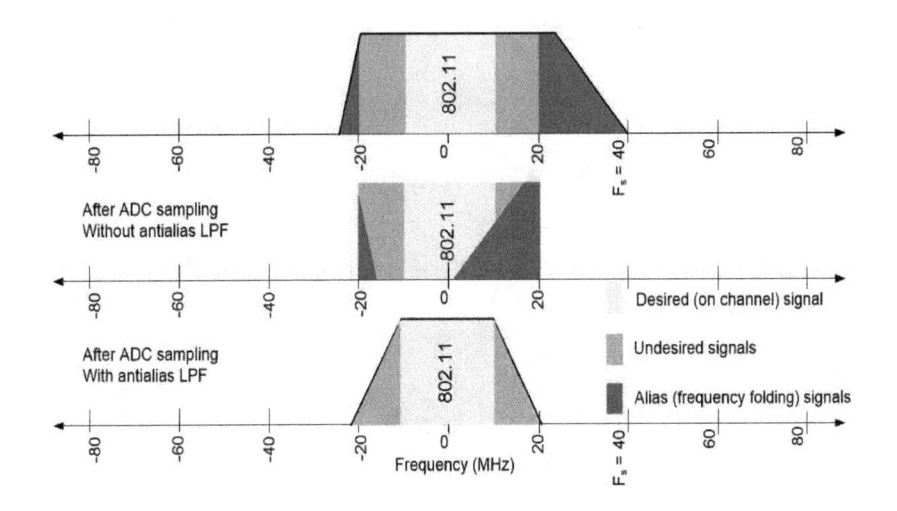

Figure 7.39 A typical zero IF SDR ADC input spectrum.

```
aButter=(-diff(unwrap(angle(hButter)))) ./diff(wButter);
aCheby2=(-diff(unwrap(angle(hCheby2)))) ./diff(wCheby2);
```

7.3.2.3 Zero IF Receiver Noise Considerations

Analysis of the ZIF ADC noise floor is that same as for the NZIF receiver. The only difference is now we have two ADCs instead of one.

7.3.2.4 Zero IF Receiver Pros and Cons

We have seen that ZIF receivers have some important advantages:

1. Wide tuning range;
2. No image frequency;
3. Low ADC sampling rate.

However, ZIF also introduces new challenges:

1. Static DC offset;
2. Dynamic DC offset;

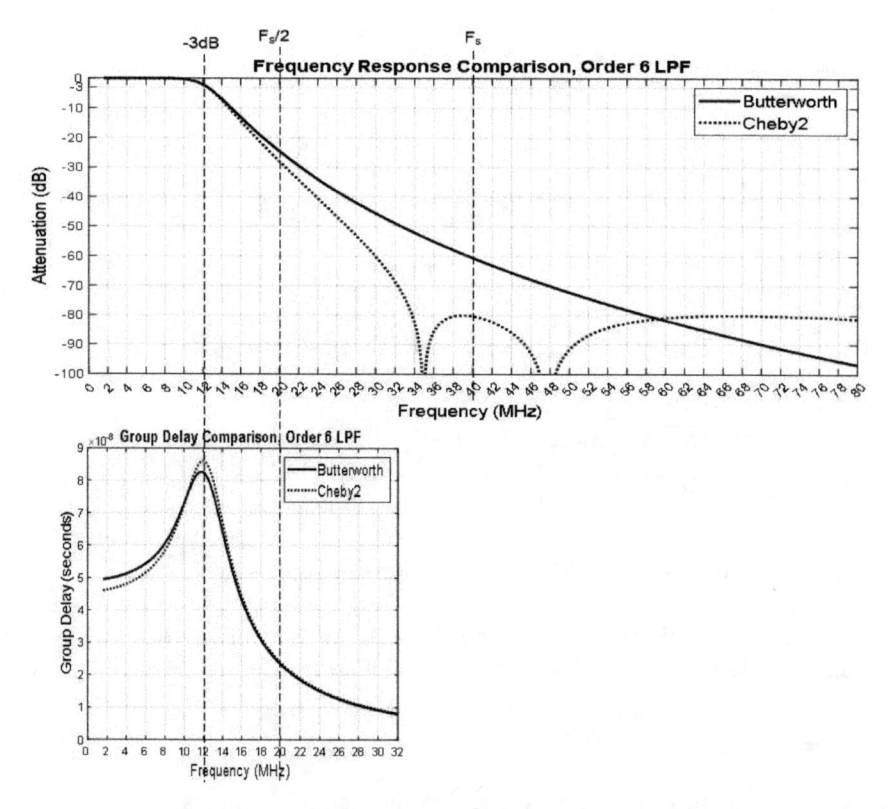

Figure 7.40 Zero IF SDR analog antialias LPF examples.

CHAPTER 7. ANALOG SIGNAL PROCESSING

3. 1/f noise;
4. IQ imbalance.

Let's consider each challenge separately.

1. Static DC offset due to the on-channel LO feeding back to RF input. Because RF and LO frequencies are the same, there is no filtering to prevent this feedback into receiver through typical paths shown in Figure 7.41. Consider LO frequency f_{LO} feeding back to the receiver antenna input and mixing with a phase-shifted version of itself, Equation (7.29).

$$f_{IF}(t) = \cos(2\pi f_{RF}(t)) \cos(2\pi f_{RF}(t) + \theta) = \frac{1}{2}\cos(\theta) + \frac{1}{2}\cos(4\pi f_{RF}(t) + \theta) \tag{7.29}$$

In Equation (7.29, the first term on the right side, $0.5\cos(\theta)$, is a static (i.e., constant) DC offset. Because the desired signal is centered at 0 Hz, this offset gets sampled by the ADCs. This DC offset can result in increased errors in QAM and PSK digital radio.

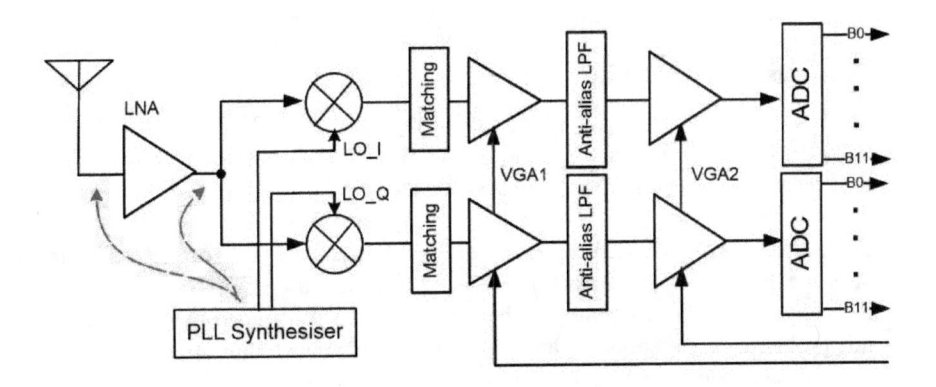

Figure 7.41 A zero IF LO feedback potential problem.

Integrated ZIF receiver parts, like the ANALOG DEVICES AD9361, measure and compensate for RF DC offset. This tracking only runs when the received energy goes below some threshold or the receiver gain changes. DC offset tracking can be very effective because it acts like a highpass filter, rejecting DC offset and passing the desired receive signal.

The ZIF LO signal leaking back through the receive antenna input leads to another possible problem: ZIF receivers can be more difficult to hide than receivers where LO feedback is rejected by the preselector. This means that the LO emitting from the receive antenna can be subject to detection by, for example, a radar detector or a direction finding system.

2. Dynamic DC offset due to second-order harmonic distortion. The LNA is not a strictly linear device, it can be modeled as in Equation (7.30).

$$x(t) = g_1 y(t) + g_2 y(t)^2 + \cdots \qquad (7.30)$$

The first-order term is the primary gain, even-order terms cause DC offset, and odd-order terms cause clipping and intermodulation distortion. Consider the amplitude-modulated blocker shown in Equation (7.31).

$$S_{AM}(t) = (1 + m(t)) \sin(2\pi f_{RF} t) \qquad (7.31)$$

The second-order gain term of the nonlinear LNA gain causes a dynamic DC offset by demodulating the AM envelope ($m(t)$) to baseband. Equation (7.32) shows this.

$$(S_{AM}(t))^2 = (1 + 2m(t) + m^2(t)) \sin^2(\omega_{RF} t) \qquad (7.32)$$

This is a dynamic DC offset, difficult to predict and correct for; see [5]. An LNA with a high IP2 as well as some kind of input bandwidth limitation can help.

3. 1/f noise at IF amplifier output. The typical IF amplifier test result in Figure 7.42 clearly shows an increase in noise figure near 0 Hz. This could be a problem in a ZIF receiver, however, there are other considerations; for example, a high gain, low noise figure LNA will dominate the composite receiver NF and may help to alleviate this situation; see Section 7.2.1. Also, widely used OFDM signals, such as LTE and IEEE 802.11a, generally have their center frequency bin zeroed to avoid this problem. When choosing an amplifier for this application, make sure the data sheet specs are tested down to zero frequency.

4. IQ imbalance of quadrature downconverting mixer. IQ imbalance consists of amplitude and phase differences in the I,Q outputs of the quadrature LO. Assume that the amplitude error is α and the phase error is φ. The effect of these errors in the frequency domain is seen in Figure 7.43. This figure shows a downconversion to nonzero IF using a quadrature (i.e., complex) LO. This downconversion is supposed to result in a positive frequency only complex IF. However, due to IQ imbalance, a negative image frequency artifact appears. Also shown is an artifact popping up at 0

CHAPTER 7. ANALOG SIGNAL PROCESSING

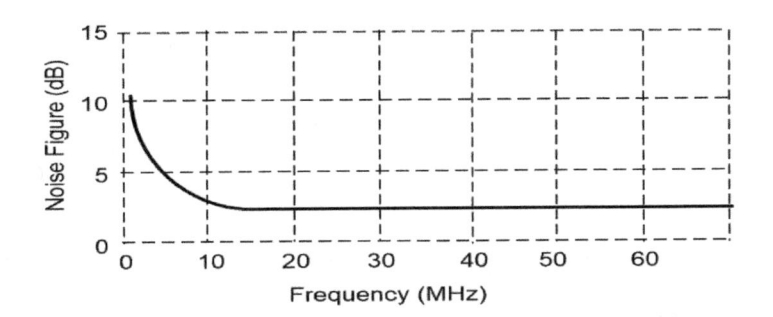

Figure 7.42 IF amplifier typical measured 1/f performance.

frequency due to the DC bias error discussed previously. For an additional example, see Section 9.3.4.

Here is a practical example. OFDM, a multichannel digital modulation discussed at length in Chapter 3, is susceptible to I,Q imbalance. This is shown in simplified form in Figure 7.44. Here we see a 9-channel OFDM signal that is supposed to be downconverted to 0 Hz center frequency. The signal has center bin zeroed to protect against zero frequency offset. However, due to phase and amplitude imbalance, four of the positive frequency channels have interfering images on the negative frequency side and vice versa for four of the positive frequency images.

Figure 7.45 shows a receive RF at the LO frequency ($\omega_{RF} = \omega_{LO}$). We assume phase coherence between the received tone and the LO so that the received IF, after the lowpass filter, should be $1 + j0$, a 0 Hz vector. However, without correction of the LO amplitude imbalance error α and phase imbalance error φ, the output is a vector $\alpha + j\sin(\varphi)$. Figure 7.45 shows the corrector that we would like to design.

For $\omega = \omega_{RF} - \omega_{LO}$, the ideal lowpass filter output in Figure 7.45 is shown in Equation (7.33).

$$\begin{bmatrix} x_I(t) \\ x_Q(t) \end{bmatrix} = \begin{bmatrix} \cos(\omega t) \\ \sin(\omega t) \end{bmatrix} \tag{7.33}$$

With phase and amplitude imbalance and no correction we have Equation (7.34) where: ($k_{Ia} = 1$, $k_{Ib} = 0$, $k_{Qa} = 1$, $k_{Qb} = 0$).

$$\begin{bmatrix} y_I(t) \\ y_Q(t) \end{bmatrix} = \begin{bmatrix} x_I(t) \\ x_Q(t) \end{bmatrix} = \begin{bmatrix} \alpha \cos(\omega t) \\ \sin(\omega t + \varphi) \end{bmatrix} \tag{7.34}$$

Following part of the development in [6], we have Equation (7.35).

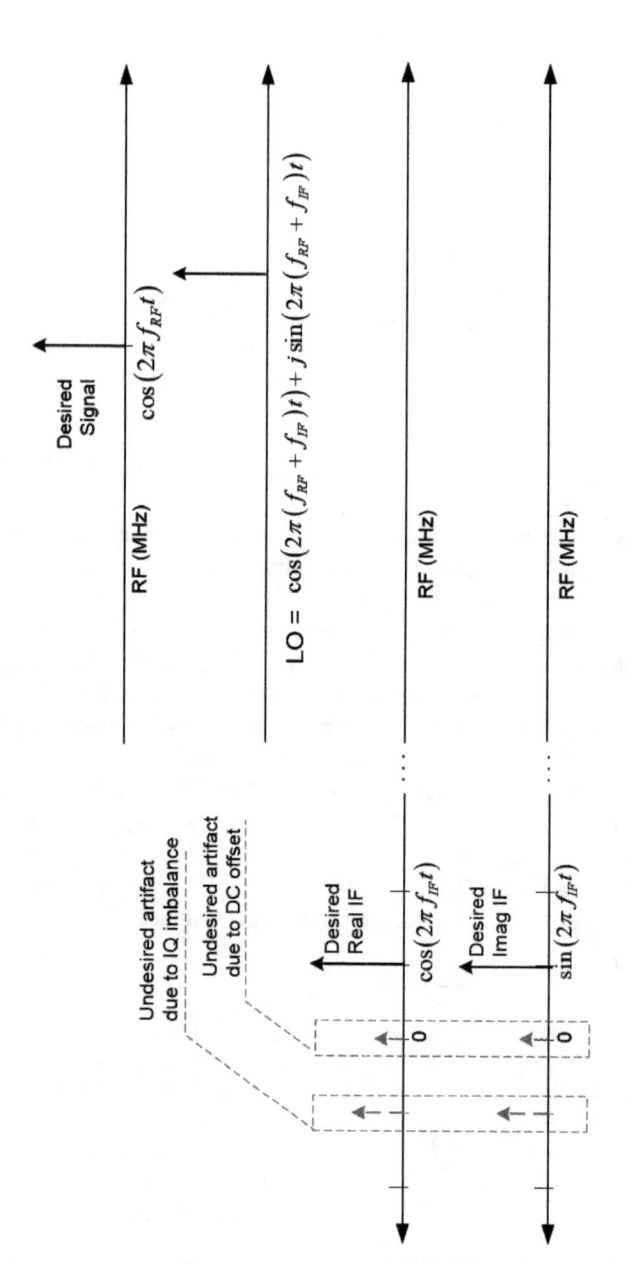

Figure 7.43 Effect of quadrature IF amplitude and phase differences.

CHAPTER 7. ANALOG SIGNAL PROCESSING

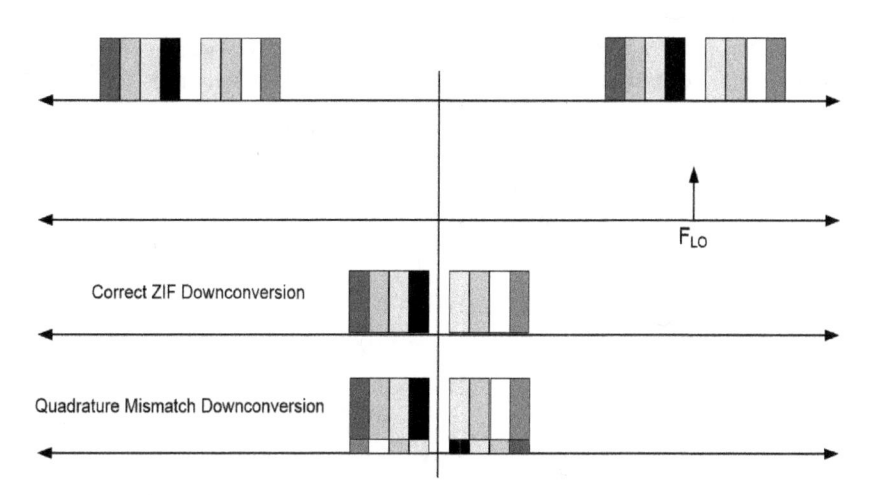

Figure 7.44 Zero IF amplitude and phase difference effect on OFDM.

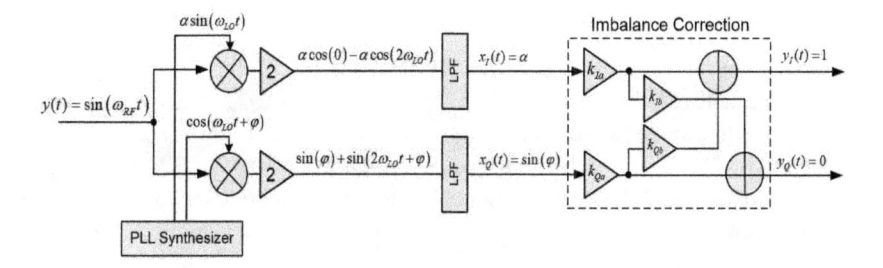

Figure 7.45 Correcting for IF amplitude and phase imbalance.

$$\begin{bmatrix} y_I(t) \\ y_Q(t) \end{bmatrix} = \begin{bmatrix} \alpha & 0 \\ \sin(\varphi) & \cos(\varphi) \end{bmatrix} \begin{bmatrix} x_I(t) \\ x_Q(t) \end{bmatrix} \qquad (7.35)$$

Where, as in [6], we have used: $\sin(\omega t + \varphi) = \sin(\omega t)\cos(\varphi) + \cos(\omega t)\sin(\varphi)$. We have also set $\omega = \omega_{RF} - \omega_{LO} = 0$. The next step is shown in Equation (7.36).

$$\begin{bmatrix} x_I(t) \\ x_Q(t) \end{bmatrix} = \begin{bmatrix} \alpha & 0 \\ \sin(\varphi) & \cos(\varphi) \end{bmatrix}^{-1} \begin{bmatrix} y_I(t) \\ y_Q(t) \end{bmatrix} \qquad (7.36)$$

After calculating the above matrix inverse, we can calculate the coefficients for the imbalance correction in Figure 7.45; see Equation (7.37).

$$\begin{bmatrix} 1/\alpha & 0 \\ (1/\alpha)\tan(\varphi) & \sec(\varphi) \end{bmatrix} = \begin{bmatrix} k_{Ia} & k_{Qa}k_{Qb} \\ k_{Ia}k_{Ib} & k_{Qa} \end{bmatrix}$$

$$\begin{aligned} k_{Ia} &= 1/\alpha \\ k_{Ib} &= \tan(\varphi) \\ k_{Qa} &= \sec(\varphi) \\ k_{Qb} &= 0 \end{aligned} \qquad (7.37)$$

We still need to find α and ϕ. If only one set, α and ϕ, results in minimum imbalance, then we say the solution is unique (also called convex) and we can search to find it. Equation (7.38) is an objective function that is minimum when the solution is found. For a calibration input, as in Figure 7.45, the imbalance is minimum when $L_2(t)$ is minimum.[6] For actual hardware, this solution for α and ϕ will be valid only at the temperature at which the test is performed. For operation over a wide temperature range, multiple calibrations may be required.

$$L_2(t) = \left\| (y_I(t) + jy_Q(t)) - (1 + j0) \right\|_2 \qquad (7.38)$$

The entire 2-D space for $\alpha_{low} < \alpha < \alpha_{high}$ and $\varphi_{low} < \varphi < \varphi_{high}$ must be searched either in an exhaustive way or by using some algorithm such as simplex or genetic. Table 7.2 shows the results of an exhaustive search of 100 values of each quantity or 10,000 total trials. Finally, the MATLAB code to perform the search is listed below.

```
%Two Dimensional Search to estimate α and φ
ComplexIn = complex(1.01,0.01); % sample errored input
```

---

6 $L_2(t)$ is also known as the Euclidean distance between two vectors.

CHAPTER 7. ANALOG SIGNAL PROCESSING

Table 7.2 Example of Results of an Exhaustive 2-D Search for α and φ

	α	φ **(degree)**
True	1.01005	0.5672664
Estimate	1.01515	0.5644014

```
Iin = real(ComplexIn); Qin = imag(ComplexIn);
alpha = abs(ComplexIn);
phid = (180/pi)*angle(ComplexIn); %degrees
M = 100;
% Center search space around known solution
% Normally we don't know the solution so space
% will be bigger. However, normaly we use a
% more complicated search algorithm that does
% not have to search the entire 2D space
alp = linspace(alpha-(alpha/2),alpha+(alpha/2),M);
phd = linspace(phid-(phid/2),phid+(phid/2),M);
error = zeros(M,M);
for j = 1:M
for k = 1:M
temp = complex(Iin*(1/alp(j)),...
Qin*secd(phd(k))-Iin*(1/alp(j))*tand(phd(k)));
temp2 = abs(temp - complex(1,0));
if(temp2>0.08)
error(j,k) = 0.08;
else
error(j,k) = temp2;
end
end
```

7.4 TRANSMITTER CONFIGURATIONS

7.4.1 SDR Transmitters

This section describes typical SDR transmitter configurations; for example, Figure 7.46.

7.4.1.1 Transmitter Example

As shown in the TX DAC block on the left of Figure 7.46, modern DAC integrated circuits tend to incorporate reconfigurable upsampling and frequency upconversion.

The complex baseband modulated input consists of band-limited signal samples, real and imaginary, centered at 0 Hz. In this example, we assume baseband digital modulation at 50 MHz symbol rate with a 200 MHz sample rate. As shown in the first two traces of the frequency domain in Figure 7.47, the DAC first upsamples this to 800 MHz. After upsampling, the signal is frequency-shifted up from 0 Hz to $F_s/4$ using an LO generated digitally, as in Figure 8.8. This $F_s/4$ DAC output avoids DC offsets on the DAC outputs. Note that the outputs of the two upconverters, DACs and LPFs, are real signals when considered separately, as on the quadrature modulator input. Thus, they have double- sided spectra as shown in the last three traces of Figure 7.47.

The DAC analog outputs, shown in the fourth trace in Figure 7.47, reveal the multiple sampling images that must be removed by the LPF. These have a sinc function amplitude reduction with frequency nulls at multiples of the sampling frequency. Because the sinc function amplitude is always changing, the DAC output imparts a slight droop to the signal amplitude at frequencies in Nyquist band 0. The upsampling helps to alleviate this. Also, at the cost of using more DAC output range, this droop can be corrected by imparting an opposite shape droop to the signal amplitude prior to the conversion. This is a form of predistortion.

The last trace in Figure 7.47 is the input to the quadrature modulator in Figure 7.46. The use of an analog quadrature modulator allows for transmission at higher frequencies than the selected DAC might support. The modulator output is a real signal centered at $\pm f_{RF} + F_s/4$, in Figure 7.48. There is an undesired remnant at f_{RF}. This is due to DC offset on the modulator inputs. There is also an image frequency that is due to phase and amplitude imbalance in the modulator local oscillator and mixers. Note that moving the signal center frequency to $F_s/4$ helps move the image away from the desired RF making the BPF (bandpass filter) easier to design.

The bandpass filter prior to the PA (power amplifier) cleans up any remaining frequency artifacts at the expense of limiting the transmit output frequency.

This example can be designed for any RF and any transmit power level. If only a few milliwatts of transmit power are needed and the RF is less than 6 GHz, the AD9361 implements the entire transmitter in one integrated circuit. This is shown in Figure 7.49, which was derived from ANALOG DEVICES, UG-570, "AD9361 Reference Manual" and [2]. A higher power design will require separate DAC and upconverter components.

CHAPTER 7. ANALOG SIGNAL PROCESSING

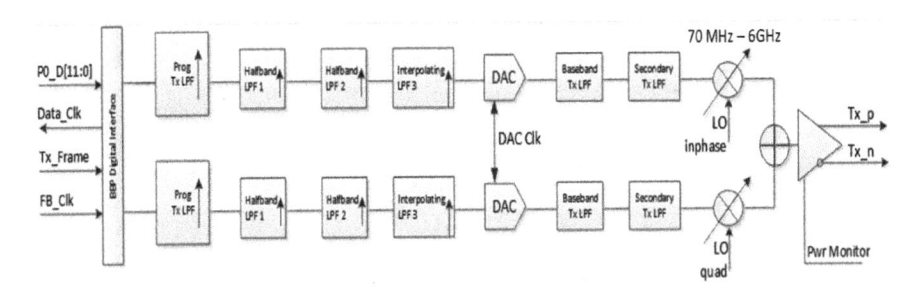

Figure 7.49 ANALOG DEVICES AD9361 transmitter simplified block diagram.

7.4.1.2 Transmitter Considerations

Several transmitter parameters are important.

1. **Power efficiency**: How much power do we have to transmit to achieve a certain bit error rate? Power efficiency is an important factor in the link budget of power-limited systems, such as satellite communications. The choice of digital modulation can have a big effect on power efficiency.
2. **Energy efficiency**: How much energy do we have to provide to obtain a highly linear PA output power? Energy efficiency has to do with required PA headroom and PA design. Note that power and energy efficiency are sometimes confused. For clarification and more information about efficiency, see [7].
3. **Peak to Average Power Ratio (PAPR)**: Maximum power of signal peaks that occur with some minimum probability (10^{-4} is typical) divided by average power. The PAPR of a constant envelope signal, such as MSK, is 0 dB. The CCDF signal characteristic is a useful tool for ensuring the PA has the headroom needed for the signals for which it is intended. See Section 14.5 for a complete description.
4. **Spectral regrowth**: Intermodulation distortion due to PA nonlinearity (i.e., clipping) can cause the PA to dump intermodulation distortion into adjacent channels. Figure 7.50, top, shows a PA output that meets the spectral mask requirement with a high PAPR signal output. The lower plot shows a PA output, of the same signal, but not in compliance with the required spectral mask. The spectral regrowth in the lower plot is mostly due to odd-order intermodulation distortion (IMD). There are two primary ways to correct this problem. One is adding headroom and the other is predistortion of the signal to counteract the PA nonlinear gain curve. The are also advanced PA designs,

such as Doherty, that have a separate peaking amplifier to preserve peak power integrity and prevent IMD.

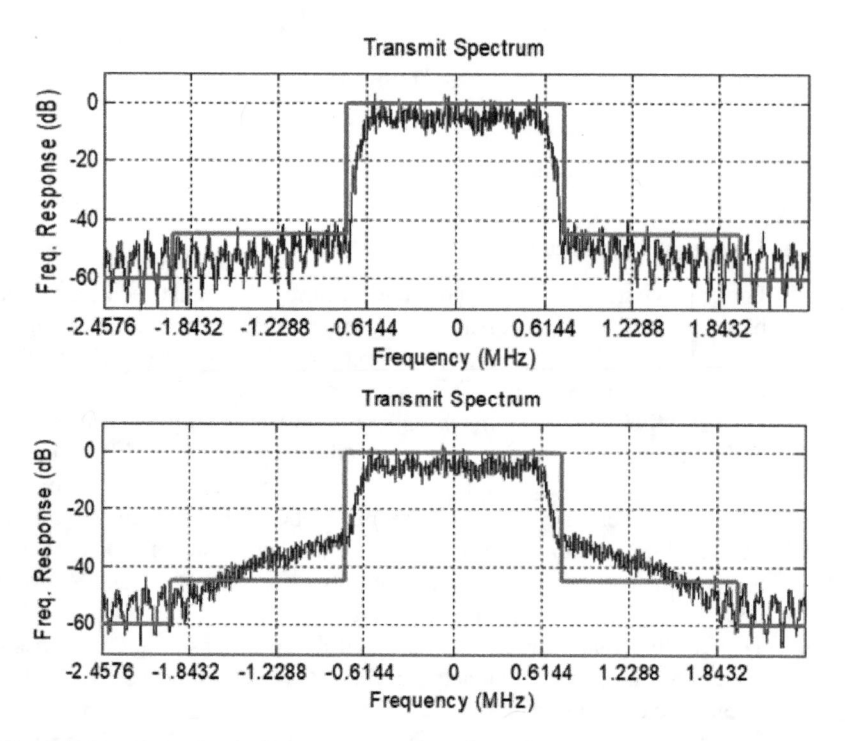

Figure 7.50 Example of cellular system power amplifier outputs with and without spectral regrowth.

7.5 QUESTIONS FOR DISCUSSION

1. An alternative ADC interface circuit to that in Figure 7.31 is shown in Figure 7.51. The ADC driver converts input voltage V_{IN}, referenced to ground, into a differential ADC input voltage centered around V_{CM} where CM stands for common mode. The ADC driver amplifier has very low output impedance. Compare these two ADC drive techniques (i.e., Figure 7.31 and Figure 7.51) with the two coupling techniques from Section 2.4.1.

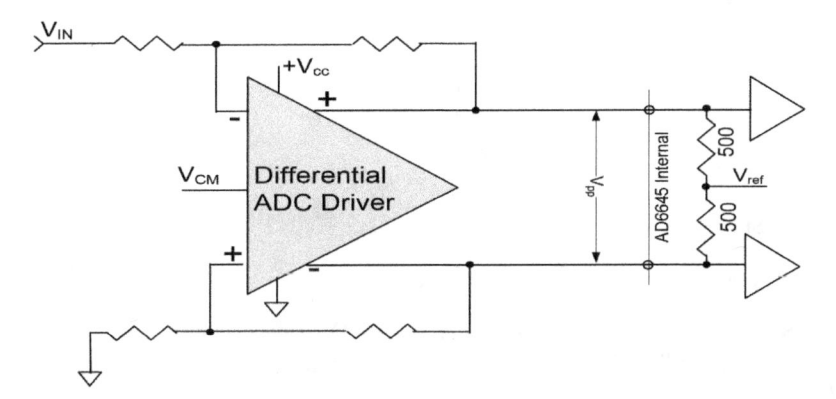

Figure 7.51 Alternative ADC driver circuit.

2. For the nonzero IF receiver, low side injection, discuss the pros and cons of a higher or lower IF at the ADC input. This discussion should include the LO frequency and the ADC sampling rate.

3. Consider a 30 MHz bandwidth signal centered at 90 MHz analog IF. Draw a frequency-domain diagram showing downconverting this signal from the Nyquist zone three to Nyquist zone 0. What is the required sampling rate? Will the signal in Nyquist 0 be inverted? What will the ADC bandwidth need to be?

4. Like any closed-loop control system, AGC has a response time. Generally fast AGC response is an advantage. However, what might set a lower limit on desired

AGC response time? Hint: consider how AGC might respond to an AM versus an FM signal.

5. Note from Figure 7.33 that the input signal power is measured close to the ADC output. Why is it important not to have any filtering between the ADC output and the point where the signal power if measured for the purpose of AGC?

6. Output third-order intercept is the output power where third-order intermodulation = first-order signal output. True or false?

REFERENCES

[1] B. Sklar. *Digital Communications*. Prentice Hall, 1988.

[2] Analog Devices. AD9363 Data Sheet. *www.analog.com*, 2016.

[3] S. Pithadia. "Smart Selection of ADC/DAC Enables Better Design of Software Defined Radio". *Texas Instruments Application Report SLAA407*, April 2009.

[4] W.E. Sabin and E. O. Schoenike. *Single-Sideband Systems & Circuits*. McGraw-Hill, 1987.

[5] P. Kenington. *RF and Baseband Techniques for Software Defined Radio*. Artech House, 2005.

[6] S. W. Ellingson. "Correcting I-Q Imbalance in Direct Conversion Receivers". *Electroscience Lab, Ohio State University*, February 2003.

[7] E. McCune. *Practical Digital Wireless Signals*. Cambridge University Press, 2010.

Chapter 8

ADC and DAC Technology

This chapter describes some specification and design techniques for ADCs (analog-to-digital converters) and DACs (digital-to-analog converters). We start with a brief review of ADC sampling theory.

8.1 ADC SAMPLING THEORY

8.1.1 Time-Domain

Figure 8.1, first trace, is an unsampled continuous-time waveform. To understand the second trace, let's first establish a definition for the Dirac delta function:

$$\delta(t) = \begin{cases} +\infty & t = 0 \\ 0 & t \neq 0 \end{cases} \qquad (8.1)$$

The second trace shows the ADC sampling time instances: $\delta(t - nT)$. Given constant sampling time interval T and integer sample number n, the argument $(t - nT)$ is only nonzero at the desired sampling times. Thus, for each sample n, as t increases from 0 to ∞, only one sample $x_s(n)$ results. This is shown in the third trace of Figure 8.1.

$$x_s(n) = x(t) \sum_n \delta(t - nT) = x(nT) \qquad (8.2)$$

Finally, the last trace shows how continuous-time, *quantized* samples look when processed by a zero-order hold, Equation (8.3). A DAC output looks like the

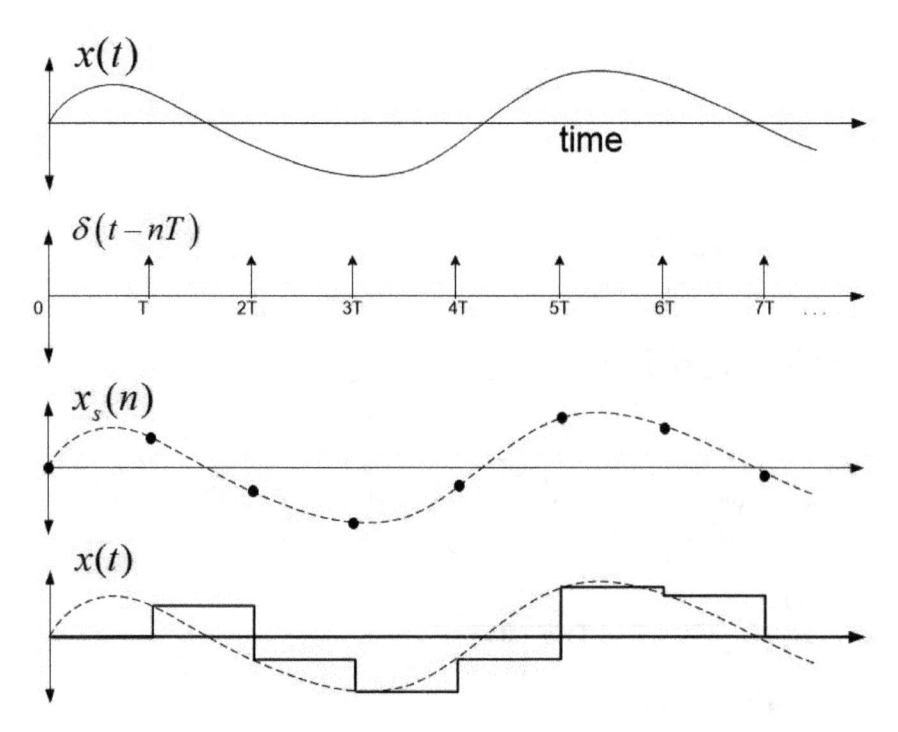

Figure 8.1 ADC sampling in the time-domain, no aliasing.

fourth trace, however, the third trace shows what ADC samples used inside the DSP look like.

$$x(t) = \sum_{n=-\infty}^{\infty} x_s(n) \prod \left(\frac{t - nT}{T} - \frac{1}{2} \right)$$

$$\prod(t) = \begin{cases} 0 & |t| > 0.5 \\ 0.5 & |t| = 0.5 \\ 1 & |t| < 0.5 \end{cases}$$

(8.3)

8.1.2 Frequency-Domain

In the frequency-domain, the multiplication of the input continuous time waveform by the impulse train (traces one and two in Figure 8.1) becomes a convolution

CHAPTER 8. ADC AND DAC TECHNOLOGY

between the input signal and the impulse train spectrum. In the convolution[1] of Equation (8.4) next, ω_s is the sampling frequency in radians per second (sampling rate is $T = \frac{2\pi}{\omega_s}$) and $*$ indicates convolution.

$$X_s(\omega) = \omega_s X(\omega) * \sum_m \delta(\omega - m\omega_s) = \omega_s \sum_m X(\omega - m\omega_s) \qquad (8.4)$$

Interestingly, the frequency-domain impulse train spectrum looks the same as the time-domain impulse spectrum except the time-domain impulses are spaced by multiples of the sampling time and the frequency-domain impulses are spaced by multiples of the sampling rate. The lowest trace in Figure 8.2 shows the complete sampling spectrum on the ADC digital output. The convolution between the first two traces shifts the input signal spectrum in the first trace to the center frequency of each impulse. An important observation is that the frequency difference between each impulse corresponds to one trip around the unit circle[2]. It is always true that in the sampled frequency-domain, the spectrum centered around zero frequency is replicated exactly at multiples of the sampling frequency. There are no unique frequency elements outside of Nyquist zone zero, N0; see Figure 9.4.

Aliasing results when the signal spectrum to be sampled exceeds the N0 frequency limit, that is, $[\ -F_{ssample}/2\ \ +F_{sample}/2\ ]$. Aliasing is also called frequency folding because the excess spectrum folds back on top of the desired spectrum. The top trace of Figure 8.3 shows frequency regions (shaded) that will alias when the signal is sampled in the third trace. Note again that the alias spectrum is repeated in all Nyquist zones. Aliasing generally causes distortion; however, a useful aspect of aliasing is discussed in Section 7.3.1.

8.2 ADC SPECIFICATIONS

Here we review some commonly used specifications that can be handy for comparing ADCs. Additional information is at [1].

---

1 Convolution is defined as: $(f*g)(t) = \int_{-\infty}^{\infty} f(\tau)g(t-\tau)d\tau$. Note that function $f(\tau)$ goes forward with variable of integration, τ, and $g(t-\tau)$ goes backward with τ.
2 This is easy to see in Nyquist zone zero where the circle starts at $F_{sample}/2$ at 180°, goes counterclockwise through 0°, and ends at $-F_{sample}/2$ at 180°; see Figure 9.3.

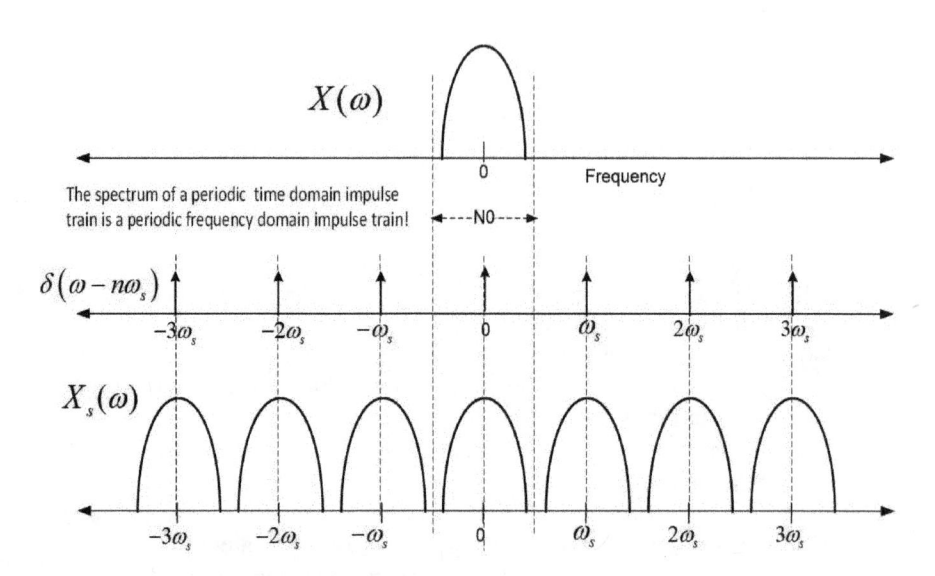

Figure 8.2 ADC sampling in the frequency-domain, no aliasing.

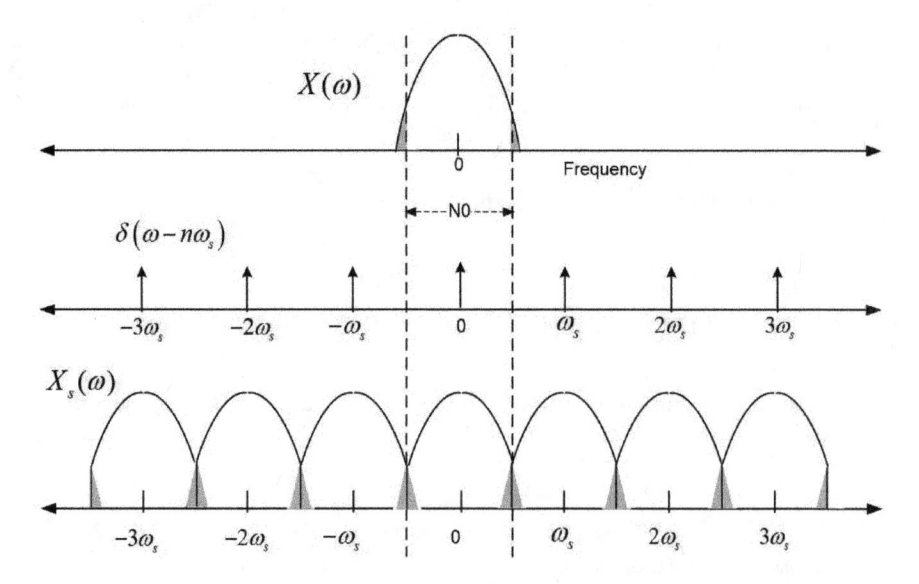

Figure 8.3 ADC sampling in the frequency-domain, aliasing shaded gray.

8.2.1 Signal-to-Noise Ratio

For an ideal N-bit ADC, signal-to-noise-ratio (SNR) is all about quantizing uncertainty. From the quantizing noise analysis in Section 9.4, we can say, the RMS quantizing noise power for quantizing step voltage Δ is:

$$\sqrt{E[\sigma^2]} = \frac{\Delta}{\sqrt{12}} \qquad (8.5)$$

The $E[\ ]$ above means expected value or simply average in this case. Assuming stationary white[3] quantizing noise that is uncorrelated to the desired input tone, the ratio between the RMS value of a sine wave with peak-to-peak amplitude equal to the ADC input voltage range to the RMS value of the quantizing noise is shown in Equation (8.6); see [1] and [2]. This equation is only valid under the conditions just mentioned.

$$SNR = (6.02N + 1.76)\ dB \qquad (8.6)$$

Sampling a complex baseband signal (as in a ZIF receiver; see Section 7.3.2) requires two ADCs, one for in-phase and one for quadrature signals. Unlike a real signal, the noise on either side of zero frequency is unique and uncorrelated; thus the noise power doubles.

The bandwidth of the sampled communications signal of interest may be considerably less than the Nyquist frequency range. Thus, sometimes we can filter the signal inside the DSP to discard unneeded sampling bandwidth and increase the SNR:

$$SNR = \left(6.02N + 1.76 + 10\log\left(\frac{F_{sampling}}{2B_{signal}}\right)\right) dB \qquad (8.7)$$

This is called processing gain and will be further explored in Section 9.3.2.5. Rearranging Equation (8.6) to make the number of ADC bits N depend on the SNR, we can write an equivalent expression for N called $ENOB$ or (Effective Number of Bits). A 4× reduction in sampling bandwidth reduces SNR and results in a 1-bit ENOB increase.

$$ENOB = \frac{SNR_{measured\_dB} - 1.76}{6.02} \qquad (8.8)$$

For an ADC, SNR is equivalent to dynamic range. This is because a signal level below the noise floor cannot be converted with much accuracy. To see this, consider

---

3 Evenly distributed in frequency.

a digital voltmeter with 4 fractional digits. Say the right most digit is bouncing around and cannot be read reliably. Loosely speaking, the voltmeter's noise floor is around 0.000x. An ADC with a noisy least significant bit is similar.

8.2.2 ADC Nonlinearity

ADCs are voltage-driven devices; they do not have the predictable spurious responses (e.g., multiple order intercept points) of amplifiers, as described in Section 7.2.1.

For signals between the ADC input voltage range minimum and maximum, ADCs can be considered linear with some undesired low-level extra tones (called spurs) on the output. At any input voltage level, the ADC transfer curve has overall quantizing nonlinearity called integral nonlinearity (INL) and also may contain small random quantizing errors called differential nonlinearity (DNL).

In Figure 8.4, the solid 45° line in both graphs is the slope of the ideal ADC transfer curve. Along this line are solid line ideal ADC quantizing steps. The actual ADC quantizing steps are shown dotted. Note that for INL a bow-shaped dotted line shows the integral slope distortion. The DNL transfer curve overall has no integral slope distortion. However, there are small differential distortions due to the unequal step sizes.

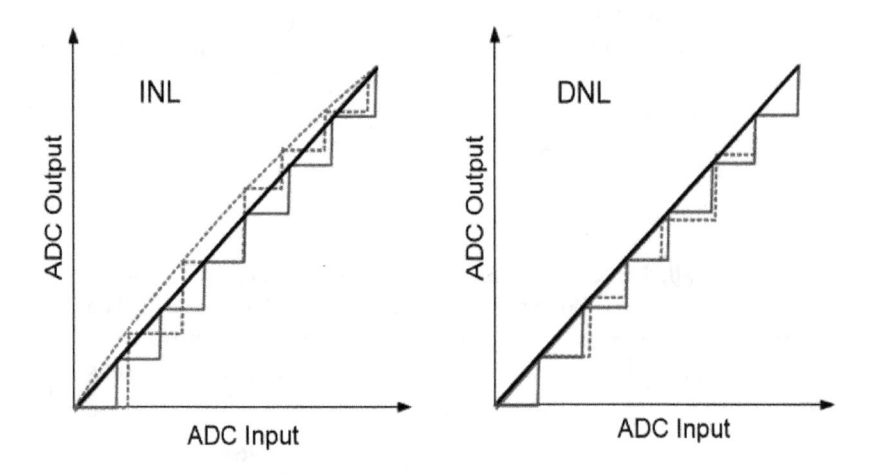

Figure 8.4 INL and DNL are ADC primary ADC nonlinearities.

DNL is fixed relative to the converter's full-scale range. Good ADC design can spread out the DNL effect over the entire input range. In this case, INL generates

CHAPTER 8. ADC AND DAC TECHNOLOGY

most of the distortion. Note that the DNL curve in Figure 8.4 is not random, a tone input will produce a predictable pattern of output distortions that result in a fixed spur. ADC designers can randomize the DNL steps by adding a small amount of jitter, also called dither, to the sampling clock. This reduces the spurs by spreading out their power over the sampling frequency band. Many modern ADCs are designed with dither circuits that are always on. See [2] for an excellent discussion of ADC dither.

8.2.3 ADC Measurements

ADC are usually tested in the frequency-domain using input tones or pairs of tones. Figure 8.5 shows the FFT output for an $N = 16$ bit ADC with a 124 MHz full-scale sine wave input.

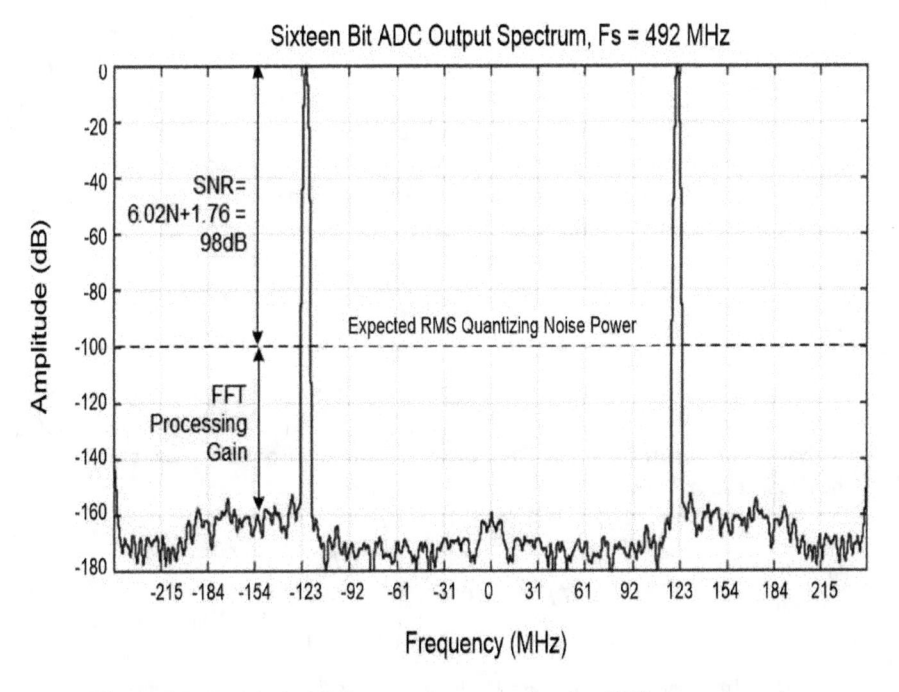

Figure 8.5 Example ADC output spectrum, showing FFT processing gain.

8.2.3.1 ADC Quantizing Noise

Direct measurement of quantizing noise can be misleading. The dotted line in Figure 8.5 shows the expected level of quantizing noise, as shown by Equation (8.6). However, the noise floor in the graph is about 60 dB below this. This is due to FFT processing gain. FFT processing gain is a combination of the FFT number of bins[4] and multiple FFT averaging. As discussed in [3], doubling the number of FFT bins results in an average 3 dB reduction in quantizing noise for a sine wave centered in an FFT bin. For an FFT of length $N = 2,097,152$, the processing gain is $60.2db = 10\log_{10}(N/2)$.

8.2.3.2 ADC Spur-Free Dynamic Range

Spur-free dynamic range (SFDR) is the ratio of the RMS value of a full-scale sine wave at the ADC input[5] to the RMS value of the peak spurious spectral content. SFDR is the power difference (in dB) between the desired sine wave ADC output and the largest spur. SFDR is often referred to as dBc instead of just dB (see Chapter 2). Figure 8.6 shows an example of an unwanted ADC output spur and the SFDR. This particular spur will probably be removed by subsequent filtering. Unlike amplifiers, ADC spur production is difficult to predict because it is caused by a systematic correlation between an ADC nonlinearity, such as DNL, and a frequency component of the ADC input. Instead of trying to characterize something that complicated, ADCs generally use dither to knock down the spur energy.

8.2.3.3 ADC Total Harmonic Distortion

For a full-scale sine wave, total harmonic distortion (THD) is the ratio of the sum of the power in the harmonics ($H_1, H_2, ...$) to the power in the desired signal (P_0), as shown in Equation (8.9). THD is simply a sum of harmonic power normalized by the desired signal power.

$$THD = 10\log_{10}\left(\frac{\sum_{n=1}^{\infty} H_n}{P_0}\right) dB \qquad (8.9)$$

---

[4] Each FFT bin can be thought of as a bandwidth reducing filter with a bandwidth equal to the FFT size divided by the number of bins.

[5] RMS voltage of a sine wave is $\frac{1}{\sqrt{2}}$ times the peak voltage of the sine wave.

CHAPTER 8. ADC AND DAC TECHNOLOGY

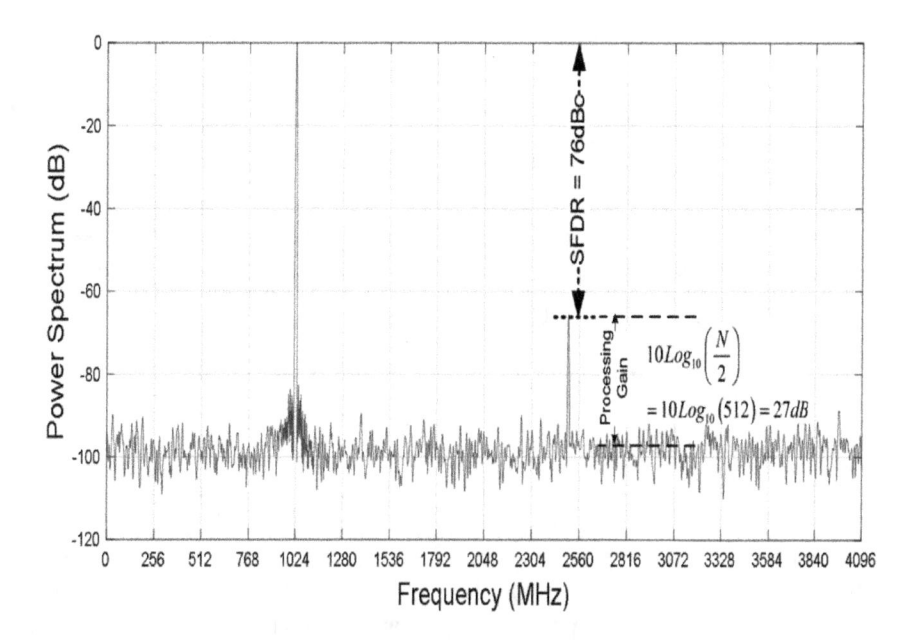

Figure 8.6 Example ADC output spectrum, spur free dynamic range, $F_{sample} = 8,192$ MHz.

8.2.3.4 ADC Signal to Noise and Distortion

Signal to noise and distortion (SINAD) is the ratio between signal power and the power of everything else (except DC) that appears in the ADC output spectrum. SINAD is similar to SNR but includes the power in the harmonics, all other spurs, and the noise floor.

$$SINAD = 10\log_{10}\left(\frac{P_0}{\sum_{n=1}^{\infty} H_n + \sum_{n=1}^{\infty} S_n + N}\right) dB \qquad (8.10)$$

For undersampling applications, it is important to understand if SINAD degrades at the frequency intended for the received IF. Both bandwidth and SINAD must hold up at this frequency. For a high-quality ADC, this information can usually be found on the data sheet.

8.2.3.5 ADC Sampling Jitter

The process of converting an analog input to a digital output must start with a series of absolutely stable unchanging time epochs, i.e., the sampling intervals. Sampling jitter, also called aperture jitter, is undesired random variation in this timing, [4]. Aperture jitter is due to external ADC sample clock jitter and internal ADC sample and hold uncertainty. In either case, the result is SNR degradation at the ADC output. For an input signal frequency of f_{input} and an RMS jitter amplitude of t_a, SNR degradation has been modeled as:

$$SNR_{jitter} = -20 Log_{10}\left(2\pi t_a f_{\text{input}}\right) \quad (8.11)$$

8.2.3.6 ADC Bandwidth

A sometimes overlooked ADC specification is input bandwidth. Figure 8.7 shows this for two different ADCs. They both perform well over the Nyquist bandwidth from 0 to 20 MHz. However, consider the undersampling technique discussed in Section 7.3.1. In that example, because the sampled input is at 60 MHz, the dotted line bandwidth ADC will function better than the solid line bandwidth. Check the bandwidth carefully of an ADC intended for undersampling.

8.2.4 ADC Designs

8.2.4.1 1-Bit ADC

There are sometimes serious power constraints on an ADC that is part of a digital radio system. Here are some examples:

1. Satellite. The solar cells have size constraints that limit power supplied to charge batteries. Batteries have size and weight constraints.
2. Implantable heart pacemaker. If changing the pacemaker battery requires a new surgery, then the internal digital radio circuits should be very power-conserving.
3. Portable radios carried by soldiers. Battery life may be critical to mission success.
4. Moon rover system. The solar cells have size constraints that limit power supplied to charge batteries. Batteries have size and weight constraints.

CHAPTER 8. ADC AND DAC TECHNOLOGY

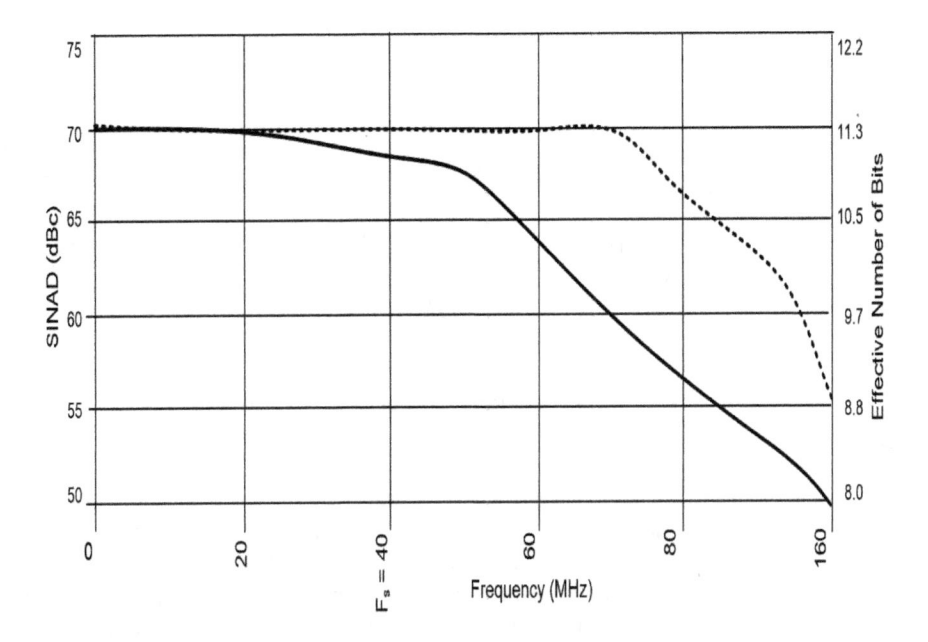

Figure 8.7 ADC input bandwidth, a critical parameter for undersampling.

A solution may be a very low power 1-bit ADC. However, the quantizing noise of one bit is very high. The key to making the 1-bit ADC work is processing gain and the use of simple communications signals, such as BPSK. Let's start our investigation by reviewing the $F_{sample} = 4F_{IF}$ local oscillator.[6]

Figure 8.8 and Equation (8.12) show that because $F_{sample} = 4F_{IF}$, the LO must advance in phase by $\pi/2$ every sample. The other useful feature about this LO is that, if the number of samples per symbol is a multiple of 4, the sequence of four LO samples is the same for every new symbol.

$$\left(\frac{2\pi \text{ radians}}{\text{cycle}}\right)\left(\frac{F_{IF} \text{ cycles}}{\text{second}}\right)\left(\frac{\text{second}}{4F_{IF} \text{ samples}}\right) = \left(\frac{\pi}{2}\right)\frac{\text{radians}}{\text{sample}} \quad (8.12)$$

The complete 1-bit ADC is shown in Figure 8.9. The antenna and RF to IF downconversion on the left is shown even though these are not part of the ADC.

[6] In this example we assume a local oscillator (LO) that frequency shifts the ADC input IF down to complex baseband; see Chapter 9.

$\sin\left(2\pi n\left(\dfrac{f_{LO}}{F_s}\right)\right)$

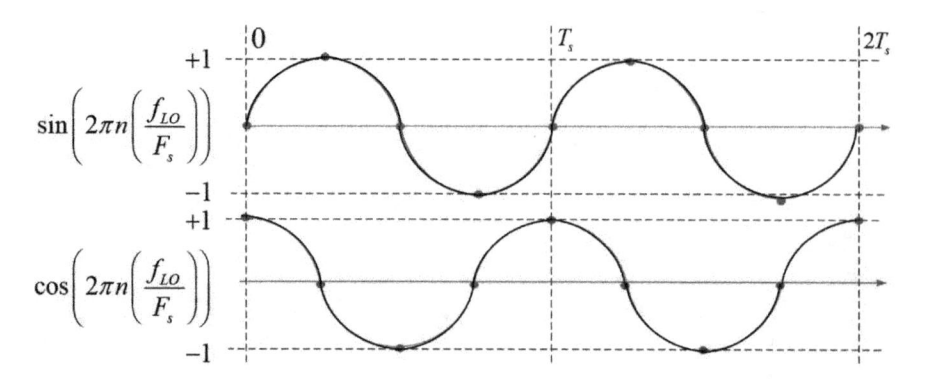

$\cos\left(2\pi n\left(\dfrac{f_{LO}}{F_s}\right)\right)$

Figure 8.8 Four sample per cycle local oscillator.

The ADC starts at the sample and hold just prior to the sgn(.) function. The sgn(.) function has the standard mathematical definition and produces a 1-bit polarity indicator. After that, the complex LO is applied as a multiplication by -1, 0, or +1 (this multiplication could be implemented with a switch). After the even/odd switching, and a one sample delay in the Q rail, the integrators start averaging to produce a symbol estimate. The averaging length is the same as the number of bits per symbol, so this design tends to have a high bits per symbol to reduce noise.

Figure 8.10 shows the timing of the important signals in Figure 8.9. To aid understanding, notice the vertical dotted lines (input signal polarity indicators) on the even numbered input samples on the top trace. These are multiplied by the in-phase LO to produce $I_{T_{sample}}$. The odd numbered input samples are multiplied by the quadrature phase LO to produce $Q_{T_{sample}}$. $Q_{T_{sample}}$ is delayed by one sample to line up with $I_{T_{sample}}$. Notice that if the analog input were to flip 180°, that I_k would invert and Q_k would continue to dither randomly.

For processing gain, $I_{T_{sample}}$ and $Q_{T_{sample}}$ are averaged by the number of samples per symbol. It is easy to see that this system can reliably detect both BPSK and QPSK. The key to making this work is the coordination of the sample rate, LO frequency, and symbol rate. Beyond that, a large number of samples per symbol can result in a very robust system.

8.2.4.2 ΔΣ ADC

A variation on the 1-bit ADC is the ΔΣ ADC. Power consumption is not quite as good as the 1-bit design just discussed but is better than older designs such as

CHAPTER 8. ADC AND DAC TECHNOLOGY

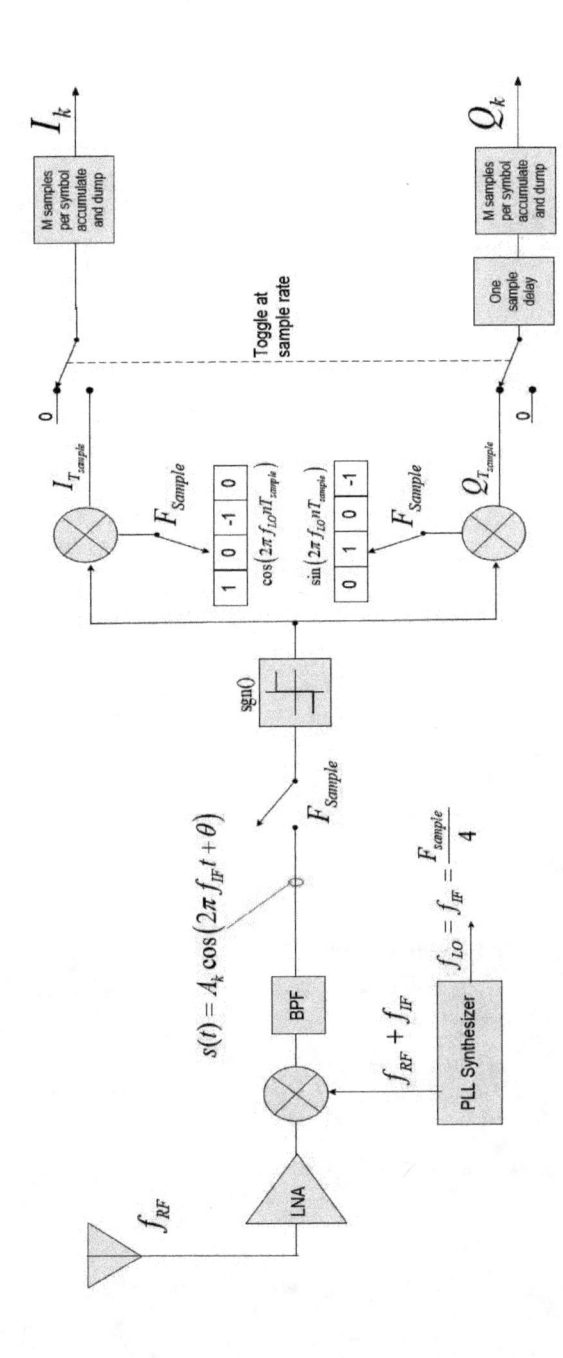

Figure 8.9 A 1-bit ADC for BPSK and QPSK.

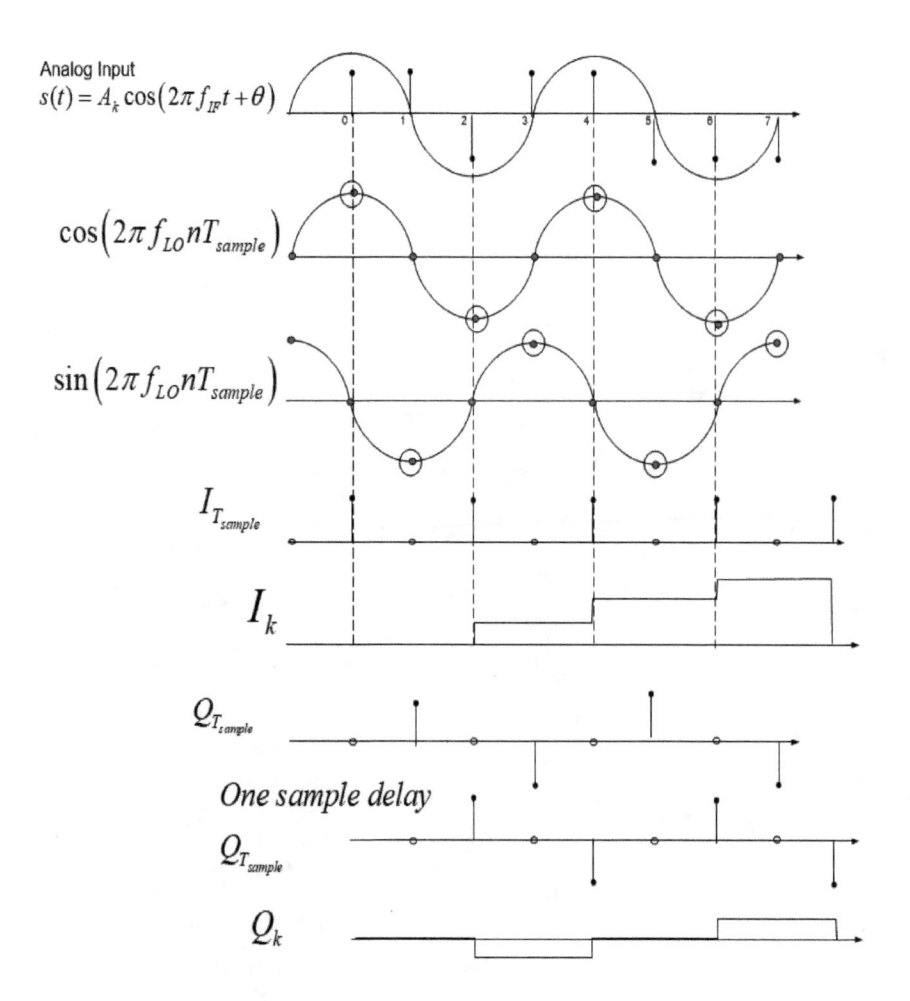

Figure 8.10 A 1-bit ADC timing for BPSK and QPSK.

CHAPTER 8. ADC AND DAC TECHNOLOGY

successive approximation (to be discussed). We will also show that the ΔΣ ADC has inherently low noise for the same number of bits. Figure 8.11 is the basic circuit. The timing is shown in Figure 8.12. ΔΣ ADCs use an oversampled one-bit ADC that has a variable ones-density[7] output. The 1-bit ADC output is connected to a digital filter, decode and downsample circuit that converts the high-speed 1 bit stream into a multibit sample at the sample rate.

Referring to Figure 8.11, the input samples can change at index n but the ΔΣ circuit operates at a much higher indexing rate $k \gg n$. Thus, although the 1 bit DAC only has two output values, +4 and -4 in this example, it switches so fast that the average DAC output can be made equal to $x(n)$. In Figure 8.12, n is constant during the time shown.

Two examples are shown in Figure 8.12; top is for a constant input of +2 and the bottom is for -2 constant input. These are fixed by the sample and hold for the duration of one set of output sample k indices. Notice how $V_{Diff}(k)$ changes the integrator ramp direction. The integrator output is sliced around 0 to form $y(k)$ = binary 0 or 1. As shown in Figure 8.11, V_{DAC} changes with $y(k)$ and determines the integrator direction. The integrator rate is controlled by the amplitude of the subtractor output, V_{Diff}. This amplitude controls how many k steps are needed to flip $y(k)$. As shown in Figure 8.12, the distribution of $y(k)$ represents the converted input voltage.

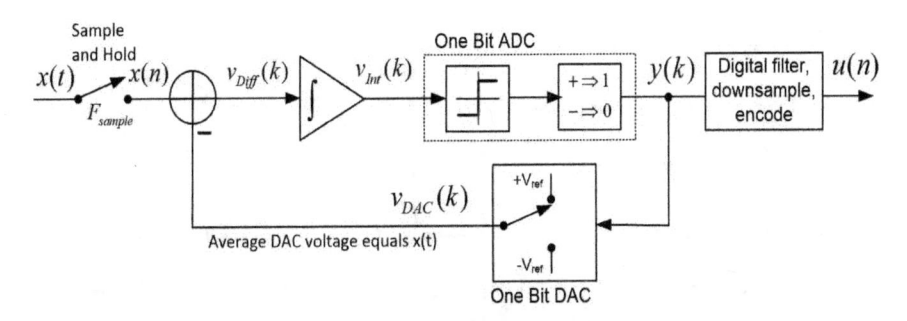

Figure 8.11 Simplified block diagram of a ΔΣ ADC.

It is important to understand that the switching shown in Figure 8.12 covers only part of a sample. There may be hundreds of integrator cycles required to convert to digital an incoming voltage that the sample and hold grabs. At one time, ΔΣ ADCs were used for slow-speed applications, such as temperature monitoring.

---

7 A high ones-density simply means there are a lot more ones than zeros

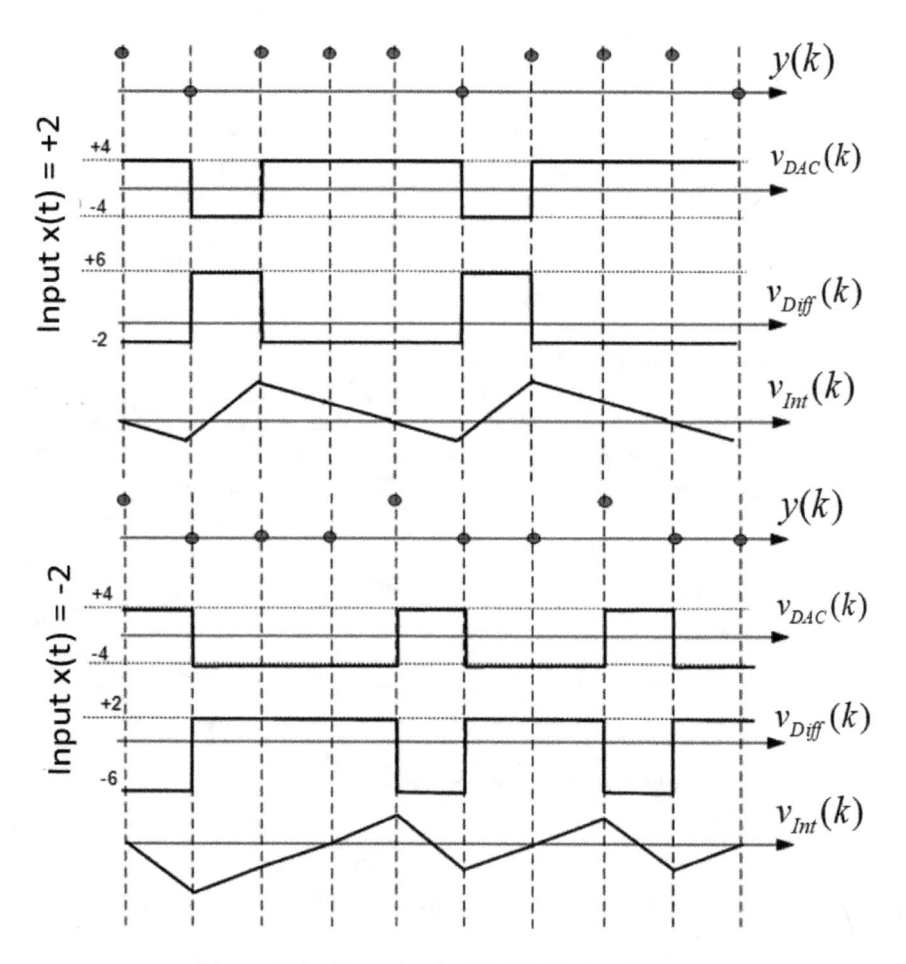

Figure 8.12 Example of a ΔΣ ADC timing diagram.

CHAPTER 8. ADC AND DAC TECHNOLOGY

Current technology has enabled higher-speed usages, such as audio processing and communications signals.

To appreciate the unique noise shaping advantage of the $\Delta\Sigma$ ADC, we need a frequency-domain analysis. Figure 8.13 is the model we will use.

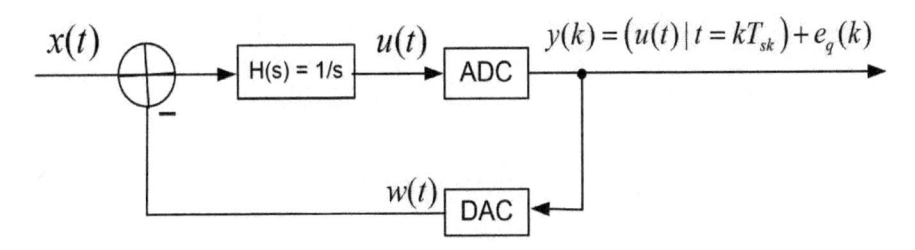

Figure 8.13 A $\Delta\Sigma$ ADC frequency-domain model.

$$U(s) = H(s)(X(s) - W(s))$$
$$Y(s) = U(s) + E_q(s) = H(s)(X(s) - W(s)) + E_q(s)$$

$$\text{Assume}: Y(s) \approx W(s)$$
$$Y(s) = H(s)(X(s) - Y(s)) + E_q(s) \quad (8.13)$$
$$Y(s)(1 + H(s)) = H(s)X(s) + E_q(s)$$
$$Y(s) = \left(\frac{H(s)}{1 + H(s)}\right)X(s) + \left(\frac{1}{1 + H(s)}\right)E_q(s)$$
$$Y(s) = S(s)X(s) + N(s)E_q(s)$$

From Equation (8.13), the transfer functions for the signal and the quantizing noise through the $\Delta\Sigma$ ADC are $S(s)$ and $N(s)$, respectively. These are shown in Equations (8.14) and (8.15).

$$S(s) = \left(\frac{H(s)}{1 + H(s)}\right) \quad \text{Input} = H(s) = \frac{1}{s} = \text{Frequency response of unit step}$$

$$S(s) = \frac{\frac{1}{s}}{1 + \frac{1}{s}} = \frac{1}{1+s} = \frac{1}{1+j\omega} = \text{Complex frequency response}$$

$$(8.14)$$

$$N(s) = \left(\frac{1}{1+H(s)}\right) \quad H(s) = \frac{1}{s}$$
$$N(s) = \frac{1}{1+1/s} = \frac{s}{1+s} = \frac{j\omega}{1+j\omega} \tag{8.15}$$

Equation (8.16) shows that the signal transfer function in Equation (8.14) is lowpass and the noise transfer function in Equation (8.15) is highpass. This is shown using limits as frequency goes to 0 and ∞.

$$|S(\omega)|^2 = \left(\frac{1}{1+j\omega}\right)\left(\frac{1}{1-j\omega}\right) = \frac{1}{1+\omega^2}$$
$$\lim_{\omega \to 0} |S(\omega)|^2 = \lim_{\omega \to 0} \left(\frac{1}{1+\omega^2}\right) = 1$$
$$\lim_{\omega \to \infty} |S(\omega)|^2 = \lim_{\omega \to \infty} \left(\frac{1}{1+\omega^2}\right) = 0$$
$$|N(\omega)|^2 = \left(\frac{j\omega}{1+j\omega}\right)\left(\frac{-j\omega}{1-j\omega}\right) = \frac{\omega^2}{1+\omega^2}$$
$$\lim_{\omega \to 0} |N(\omega)|^2 = \lim_{\omega \to 0} \left(\frac{\omega^2}{1+\omega^2}\right) = 0$$
$$\lim_{\omega \to \infty} |N(\omega)|^2 = \lim_{\omega \to \infty} \left(\frac{\omega^2}{1+\omega^2}\right) = 1$$
$$\tag{8.16}$$

Figure 8.14 shows the desired complex baseband signal as a tall box centered around 0. The noise floor is shown below this over the entire sampling frequency range. Note how for the standard ADC the noise floor is flat; that is, white. For the ΔΣ ADC design, the noise floor is lower in the center because of Equation (8.15). Thus, digital filtering the desired signal after sampling, as shown in the second and fourth traces, will improve the ΔΣ ADC SNR.

Interestingly, because the ΔΣ ADC is always integrating a difference between the input and the internal DAC, the circuit works the same at any input voltage and the INL and DNL measurements do not apply. Also, note that Equation (8.6) for SNR due to quantization noise does not apply to ΔΣ ADCs. See [5] for a better approximation.

Several commercially available parts start with a high-speed ΔΣ ADC and then downsample to take advantage of the noise shaping around 0 Hz. For an ADC sampling rate of 12.8 MHz, the output sample rate can be as low as 3.125 kHz with an OSR (oversample ratio or downsampling amount) of 4096. The downsampling filters are built in and the processing gain they provide results in a 16 bit output.

CHAPTER 8. ADC AND DAC TECHNOLOGY

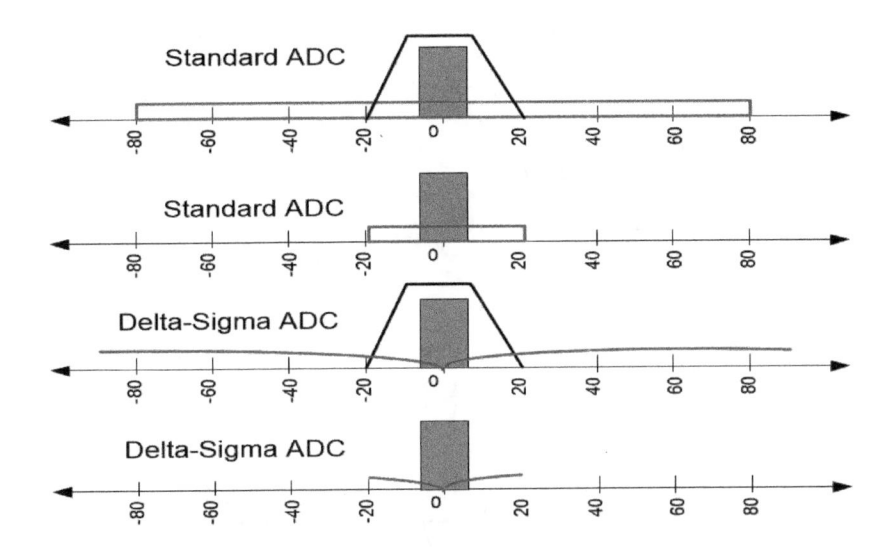

Figure 8.14 Standard and $\Delta\Sigma$ ADC noise shaping comparison.

The ANALOG DEVICES® AD9363 used in the ADLAM-Pluto SDR provides a receiver output of 12 bits using a smaller number of sampled buts from a $\Delta\Sigma$ front end ADC. In this case, the downsampling filters are a series of halfband, subsample by two designs (see Section 9.2.8).

The second-order (and higher) $\Delta\Sigma$ ADC is another very effective technique used by commercially available $\Delta\Sigma$ ADCs. The noise shaping integrator in Equation (8.15) becomes second-order instead of first-order.

8.2.4.3 Resolution Improvement

Let's try an experiment. Figure 8.15 shows an 8 kHz source sine wave. The intended sample rate is the audio standard of 48 kHz. We start with a sample rate of 3.072 MHz, oversampling by 64. The center plot is the spectrum of x(k). Now we pass this oversampled signal through a 24 kHz ideal low pass filter and downsample by 64 to get to the desired 48 kHz sample rate. The last plot is the spectrum of $y(n)$.

For the first spectral plot, we have $N = 1$ bit so the SNR is 6.02N+1.76 = 7.78 dB. For the second spectral plot we have processing gain due to downsampling of $(3dB)(log_2 64) = 18dB$ added to the one bit ADC SNR results in 18+7.78 = 25.78

dB SNR at the downsampler output. The SNR is 25.78 = 6.02N+1.76, resulting in $N = 4$ bits. So our circuit has converted 1 bit resolution into 4 bit resolution. With its highpass noise shaping, the $\Delta\Sigma$ ADC can improve this situation significantly.

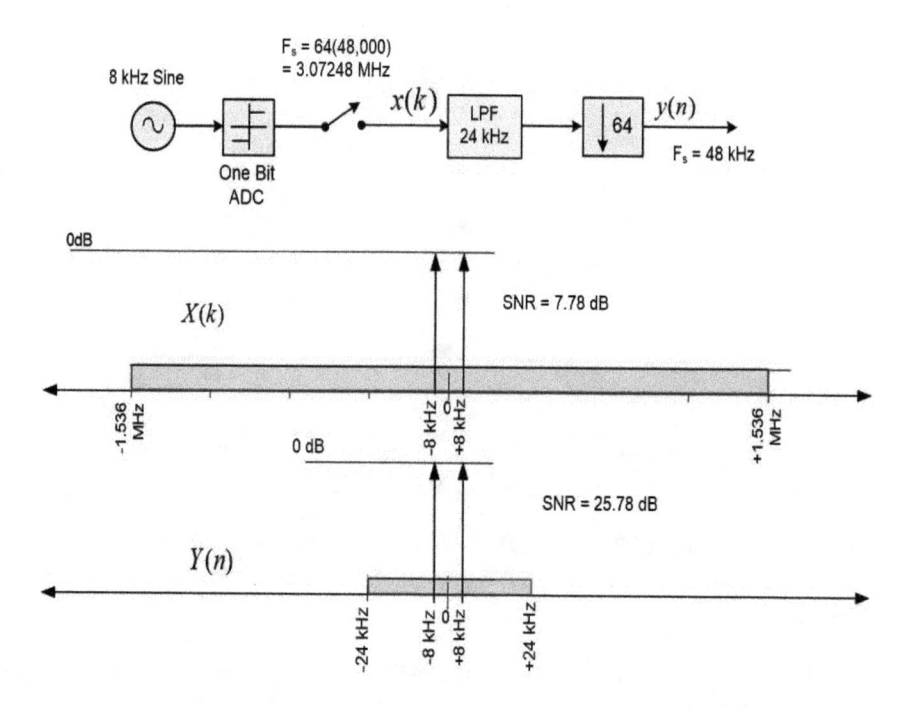

Figure 8.15 ADC noise after lowpass filtering and downsampling.

8.2.4.4 Successive Approximation ADC

Shown in Figure 8.17 is a simplified diagram of the Successive Approximation ADC invented in the 1940s at Bell Labs. To understand the operation, consider Figure 8.18. The input voltage of 0.46 is the dotted line labeled V_{in}. In this simple example, the input V_{Min} is equivalent to binary 00000 and the input V_{Max} is equivalent to binary 11111.

The SAR proceeds in steps 0 through 5, shown at the bottom of Figure 8.18 and in Table 8.1. In this example, the SAR ADC input $V_{in} = 0.46$ Vdc. After six

CHAPTER 8. ADC AND DAC TECHNOLOGY

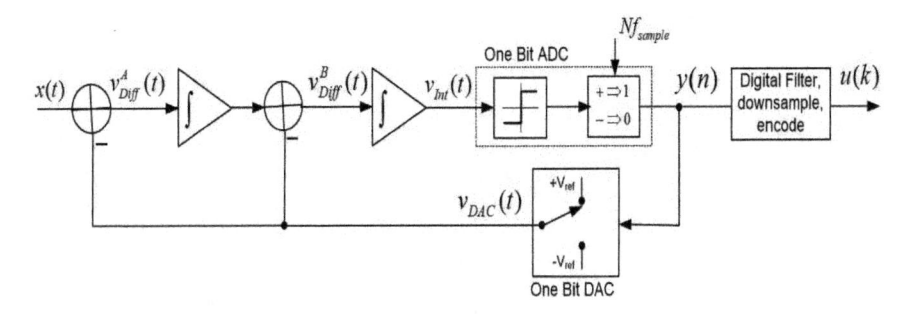

Figure 8.16 Second order $\Delta\Sigma$ ADC.

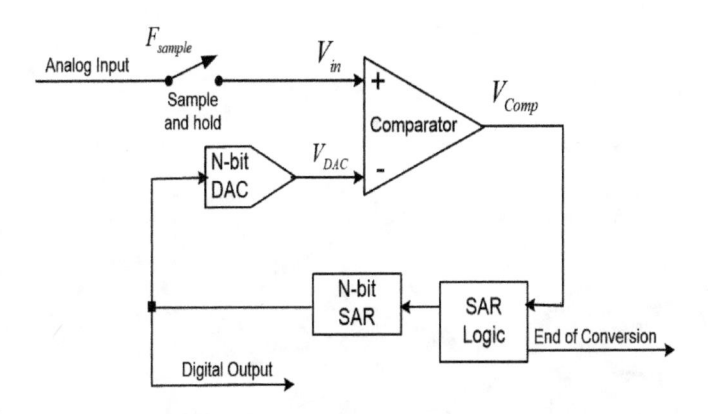

Figure 8.17 Simplified successive approximation ADC.

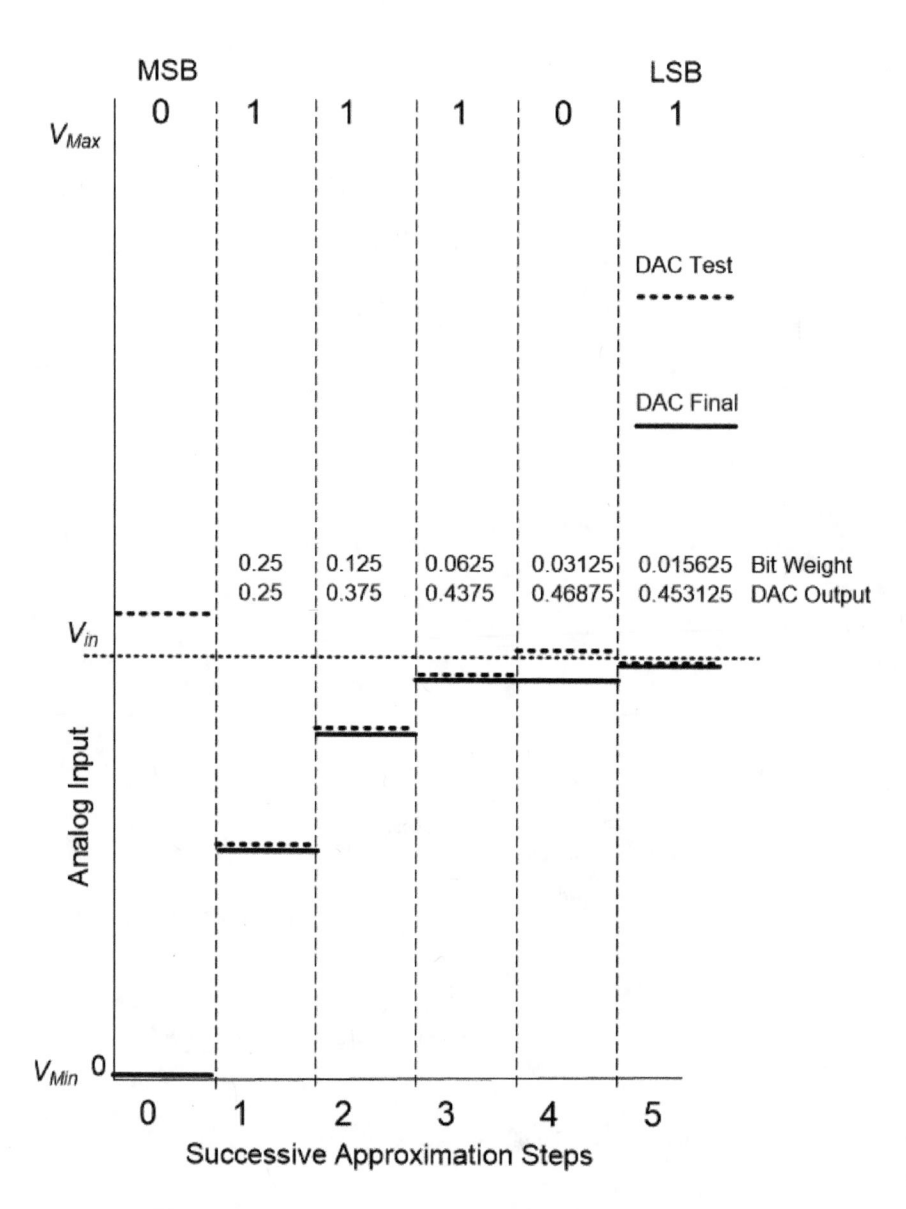

Figure 8.18 Successive approximation detailed ADC operation.

CHAPTER 8. ADC AND DAC TECHNOLOGY

Table 8.1 Successive Approximation ADC Steps for Input 0.46Vdc

Step	Bit	DAC Test	DAC Output	$V_{in} - V_{DAC}$	DAC Final
0	5	**1**00000	0.5	−	**0**00000
1	4	0**1**0000	0.25	+	0**1**0000
2	3	01**1**000	0.375	+	01**1**000
3	2	011**1**00	0.4375	+	011**1**00
4	1	0111**1**0	0.468755	−	0111**0**0
5	0	01110**1**	0.453125	+	01110**1**

steps (for a 6-bit ADC) are completed, the output, shown at the top of Figure 8.18, is 011101.

Each step is for the decision, 0 or 1, of the current bit, starting at the MSB (5) and moving to the LSB (0). The SAR logic sets the current bit to 1 to move the DAC output to DAC test. If the input voltage is greater than this level the current bit stays at one and DAC test becomes DAC final. If the input voltage is less than this level, the DAC test returns to its previous value and DAC final bit is set back to zero. Each line of Table 8.1 shows one new DAC final bit. DAC test and DAC final are shown in bold type.

For any ADC, we would like the transfer function from analog to digital to be strictly monotonic.[8] Because the internal DAC quality affects operation, some SAR ADCs may have missing output codes for a continuous ramp increase, or may even step backwards. A high-quality SAR ADC will be tested for missing codes and results will be on the data sheet.

8.2.4.5 Flash ADC

The flash ADC basic circuit is shown in Figure 8.19. The eight-resistor voltage divider sets up a series of voltage steps. The comparators output 0 for V_{in} less than their voltage step input and 1 for V_{in} greater than their voltage step input.

The priority encoder outputs a 3-bit representation ($\log_2(8) = 3$) of the largest n (i.e., D_n) that is 1. See Table 8.2.

The lowest voltage, V_0, is equal to one LSB (least significant bit) of output. Each resistor going up the stack adds another LSB worth of voltage. As the input voltage goes up and the next comparator output flips from 0 to 1, an LSB bit is added to the output $[Y_2; Y_1; Y_0]$. When the input voltage goes down, one LSB is subtracted

---

[8] A monotonic function either always increases or always decreases, never both.

from the output as each comparator output flips from 1 to 0. Table 8.2 attempts to illustrate this.

Flash ADCs have one big advantage: they are fast. The disadvantage is that, although the speed results from the structure, the internal parts result in high-power consumption. For example, for a high sample rate, the comparators must have high slew rate (i.e., high bandwidth) and that requires power. The amount of hardware is another drawback. As the number of output bits goes up the number of comparators goes up geometrically by 2. Eight output bits resolution requires 255 comparators. Although flash ADCs similar to Figure 8.19 have been used for high-speed video processing, the more practical use today is for subranging ADCs. This will be discussed in the next section.

Table 8.2 Three-Bit Flash ADC Operation

D_0	D_1	D_2	D_3	D_4	D_5	D_6	Y_2	Y_1	Y_0
0	0	0	0	0	0	0	0	0	0
1	0	0	0	0	0	0	0	0	1
x	1	0	0	0	0	0	0	1	0
x	x	1	0	0	0	0	0	1	1
x	x	x	1	0	0	0	1	0	0
x	x	x	x	1	0	0	1	0	1
x	x	x	x	x	1	0	1	1	0
x	x	x	x	x	x	1	1	1	1

8.2.4.6 Subranging ADC

Referring to Table 8.2, an n bit flash ADC will require $2^n - 1$ comparators. So the number of comparators, and the current they need, increases rapidly with the number of bits. To avoid that situation, the subranging ADC was invented in the 1950s. The six bit subranging ADC in Figure 8.21 only needs a total of 12 comparators instead of 63.

Figure 8.21 is best understood by comparison with the example in Figure 8.20. Sampled input voltage V_0 is flash converted to a 2-bit resolution V_{0D}. This coarse conversion is stored as the first two MSBs $[b_5 b_4] = [1\ 0]$ and then subtracted from V_0. The residue from the first stage, V_1, undergoes the same process and generates the next two bits $[b_3 b_2] = [1\ 0]$. Then the final residue, V_2 is converted to the two LSBs $[b_1 b_0] = [1\ 1]$. The ADC output is now 101011. Notice the pipeline delays that

CHAPTER 8. ADC AND DAC TECHNOLOGY					319

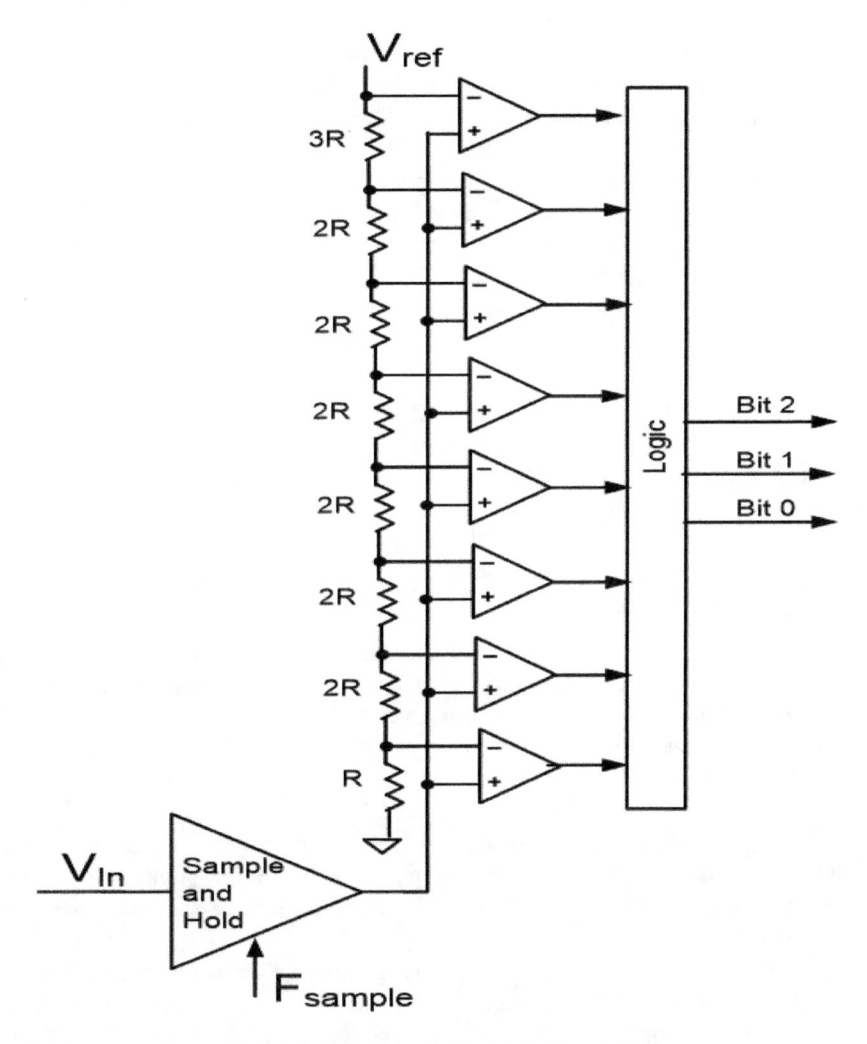

Figure 8.19 Three-bit flash ADC basic idea.

ensure that the three calculations are aligned in time in the output register. Also note that the flash ADC reference voltages are not the same: $V_{Ref0} = 4V_{Ref1} = 16V_{Ref2}$.

Consider an analogy. Start with a big jar of quarters, dimes, and nickels, with total monetary value unknown. Remove and count the quarters to obtain sum $Q. Because the quarters are removed, $Q has been subtracted out and recorded. Now remove and count the dimes to obtain sum $D. Because the dimes are removed, $D has been subtracted out and recorded. Finally, add up the remaining nickels to get $N. The total monetary value can be expressed as [$Q $D $N], where each of the three coin counts contributes its assigned monetary weight. The 6-bit subranging ADC in Figure 8.21 makes the same series of three steps if we assume the total value of quarters is the measurement of V_{0D}, the value of dimes is the measurement of V_{1D} and the remaining nickels sum up to V_2.

8.3 DIGITAL TO ANALOG CONVERTERS

Here we review some commonly used characteristics and specifications for digital to analog converters (DAC).

8.3.1 DAC Comparison with ADCs

ADCs are analog in, digital out. Testing requires sampling a high spectral purity analog sine wave (or combination of sine waves). The digital output is stored and analyzed in the frequency-domain using various MATLAB functions. The ADC input signal is generally confined to one Nyquist zone; see Figure 9.4. ADC linearity is defined by the change in output binary codes produced for each carefully calibrated input voltage step.

As shown in Figure 8.1, the sampled ADC signal consists of an infinite number of frequency-domain replicates centered on multiples of the sampling frequency, one per Nyquist zone. These are all equal amplitude and the unused replicates can simply be ignored.

DACs are digital in, analog out. This time, the input test signal must be generated digitally and the analog output is tested for nonlinearities, distortion and spurs on a laboratory spectrum analyzer. DAC linearity is defined by the variation in output analog voltage for each binary code increase.

As shown in Figure 8.22, the DAC output signal consists of an infinite number of frequency-domain replicates centered on multiples of the sampling frequency (called Nyquist zones, see Figure 9.4) and decreasing in amplitude. Unused replicates

CHAPTER 8. ADC AND DAC TECHNOLOGY

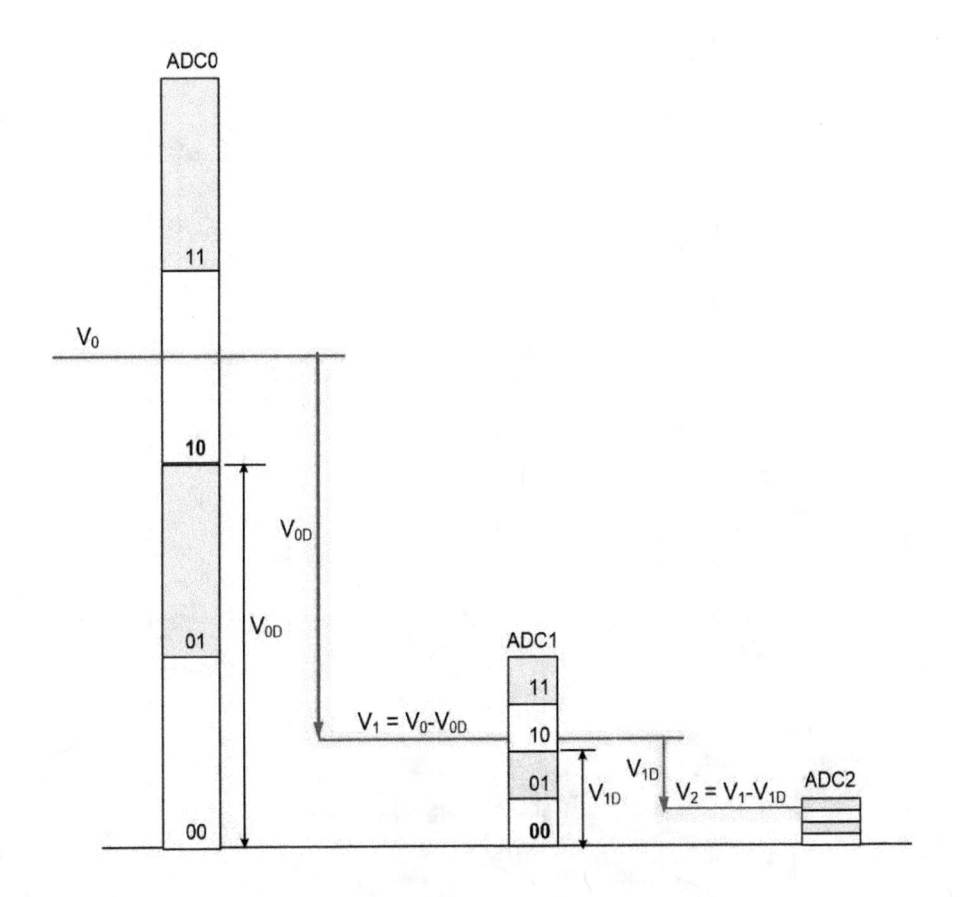

Figure 8.20 Six-bit subranging ADC conversion steps.

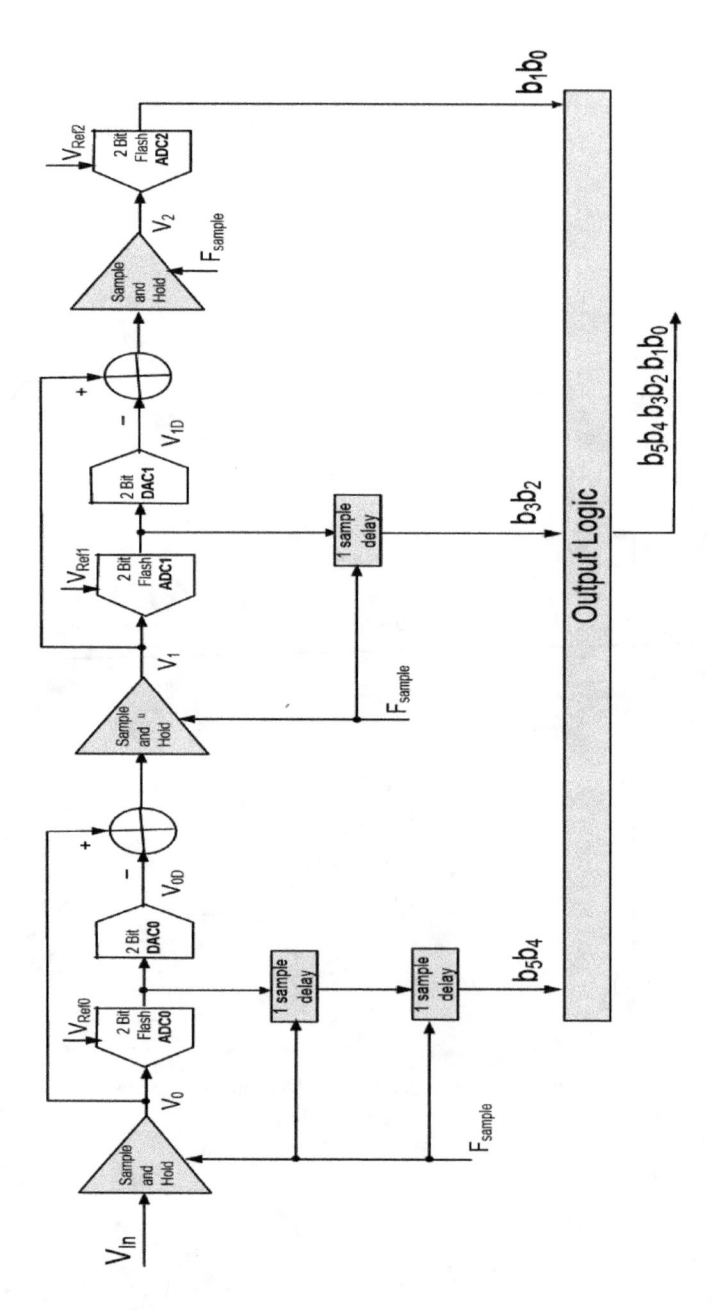

Figure 8.21 Six-bit subranging ADC block diagram.

CHAPTER 8. ADC AND DAC TECHNOLOGY

must be removed with an analog LPF for the DAC signal to match the expected narrow band smooth signal. This is sometimes called signal reconstruction and the analog LPF is then called a reconstruction filter. Figure 8.23 shows the difference between square sample pulses at the DAC output and narrow-band reconstructed signal at the LPF output. The zero and first-order holds are shown for completeness, although generally DACs only output the zero-order hold.

In Figure 8.22, the amplitude reduction over multiple Nyquist zones is due to the square pulses corresponding to each unfiltered DAC sample. These have a sinc() function frequency response with nulls at multiples of the sample rate.

$$A = \left| \frac{\sin\left(\frac{\pi f}{F_s}\right)}{\left(\frac{\pi f}{F_s}\right)} \right| \tag{8.17}$$

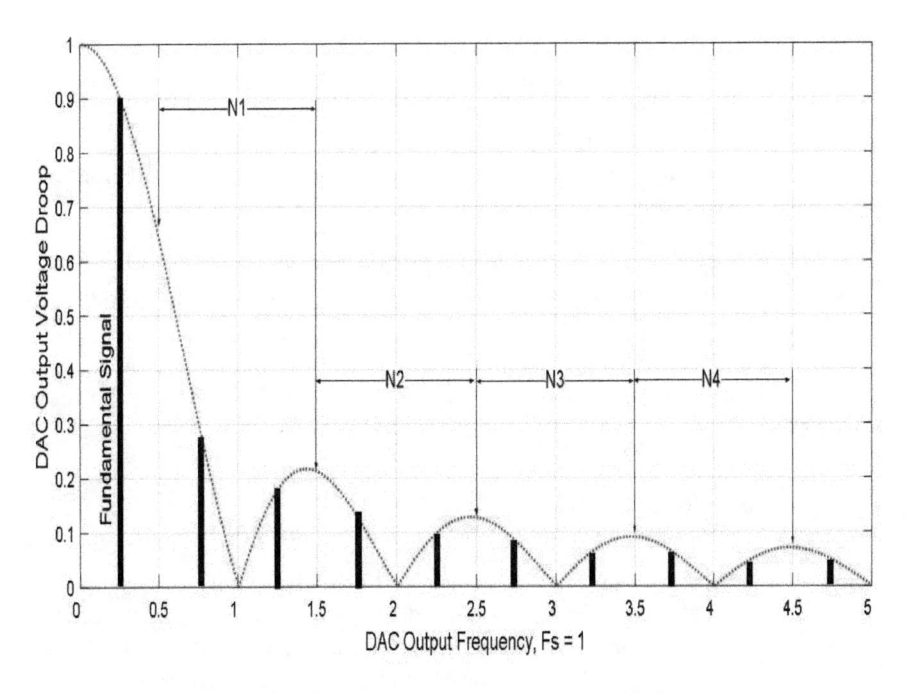

Figure 8.22 DAC output spectrum showing $sin(x)/x$ shaping.

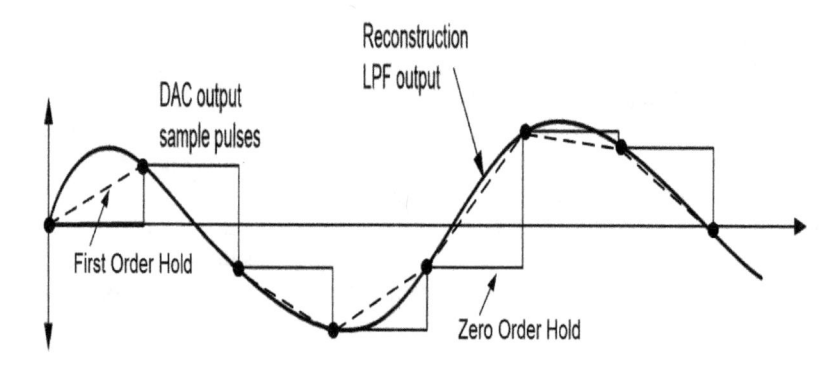

Figure 8.23 DAC and reconstruction LPF output, time-domain.

8.3.2 DAC Specifications

Up until the 1990s, DAC were primarily specified according to linearity, for example, INL and DNL (see Figure 8.4). Linearity is critical in, for example, motor control feedback systems. An extreme case of nonlinearity is lack of monotonicity. This means, for example, the DAC digital input decreases and the DAC analog output increases. This can cause serious stability problems.

As cell phone networks started to build out, DAC linearity specifications measured in the frequency-domain, such as SFDR and SINAD, became more important. Cellular base station power amplifiers typically output adjacent channel interference consisting of third and fifth order IMDs (see Figure 7.50). To reduce this undesired PA behavior, base stations DACs need to produce high data rate, high peak to average power and wideband signals, such as LTE OFDM, with excellent spectral purity.

8.3.2.1 DAC SFDR and IMD

Spur Free Dynamic Range (SFDR) is the ratio of desired signal power to the largest undesired spur, at full-scale output amplitude. Both harmonics and intermodulation products can cause spurs; see Section 7.2.1.1 for a complete discussion. The two-tone test shown in Figure 8.24 can reveal both SFDR and intermodulation (IMD) performance. In both cases, the DAC output has spectral components that were not on the input; that is a definition of nonlinearity.

CHAPTER 8. ADC AND DAC TECHNOLOGY

Consider that the analog DAC output is generally tested on a laboratory spectrum analyzer. The spectrum analyzer analog front end must not cause confusion by introducing its own nonlinearities. Critical laboratory setups to test DACs are discussed in [6].

Notice that the spurs shown in Figure 8.24 are somewhat predictable IMDs. These are directly caused by DAC nonlinearity, for example, DAC output compression. SFDR is probably the most important DAC specification. Many DACs are used to generate transmit signals and close-in spurs can render them useless.

8.3.3 Interpolating DAC

Figure 8.25 shows the unfiltered frequency-domain output of a typical DAC, the top two graphs are at sample frequency $F_s = 1$. The lower two graphs are upsampled to sample frequency $F_s = 2$ (upsampling is also called interpolation). To reconstruct the desired smooth signal from the DAC output, we must lowpass filter all the replicate signals shown except for the first signal (shaded black). Unlike the ADC output, these extra images cannot simply be ignored.

For example, say we require the first replicate, shaded gray in Figure 8.25, top plot, to be attenuated by 25 dB. As shown in Figure 8.25, second plot, a sixteenth-order Butterworth lowpass filter will meet this requirement. Notice how in Figure 8.25, third plot, the shaded gray replicate is moved out further in frequency by the upsampling. This makes it easier to filter. Only a tenth-order Butterworth LPF is needed.

These Butterworth filters[9] are analog at the DAC output; a reduction in order from 16 to 10 will save circuit board space and power. Another advantage is that the filter -3 dB cutoff frequency is moved out further on the lower plot of Figure 8.25. Since most of the filter group delay distortion occurs near the cutoff frequency, the effect of group delay on the fundamental signal is reduced.

The interpolating DAC technique is widely used to reduce the size of the DAC reconstruction filters. For example, Figure 7.49 is a block diagram of the AD9363 transmitter used in the ADALM-Pluto SDR. The inphase and quadrature DAC outputs are filtered by a cascade of two third-order, on-chip, Butterworth filters, see ANALOG DEVICES UG-1040 for details on these filters. The DAC inputs are upsampled by eight to extend in frequency the signal replicates starting in N1 (Nyquist Zone 1), making them easier to remove.

---

9 Butterworth filters are also called maximally flat, meaning the passband ripple is zero.

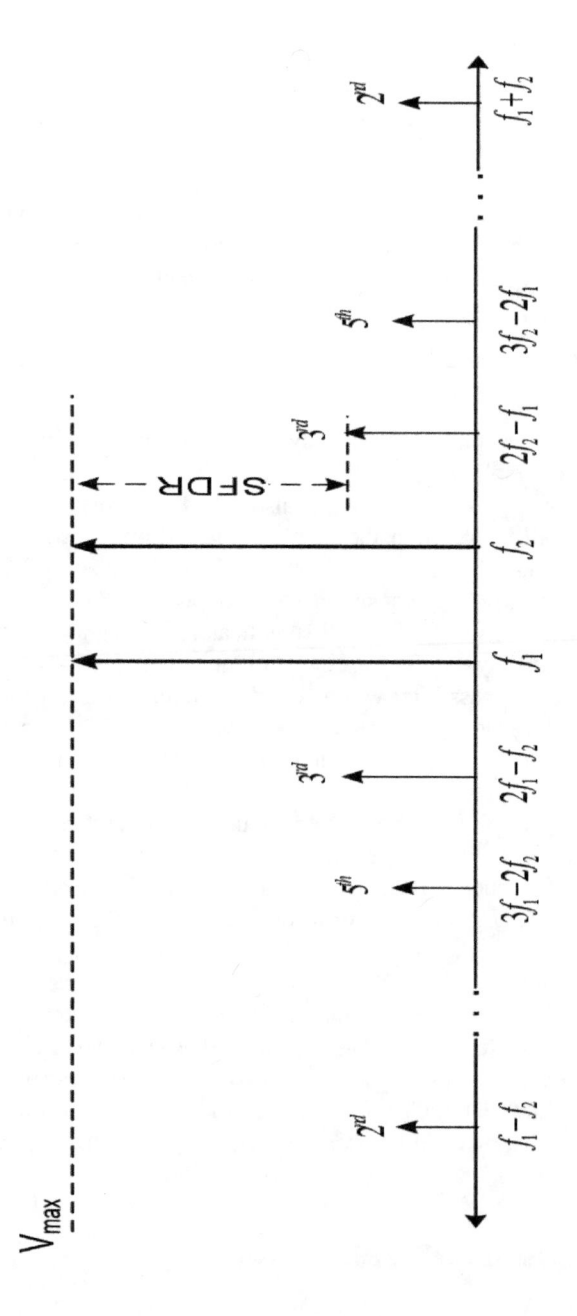

Figure 8.24 Full-scale two-tone DAC output showing IMD and harmonics.

CHAPTER 8. ADC AND DAC TECHNOLOGY

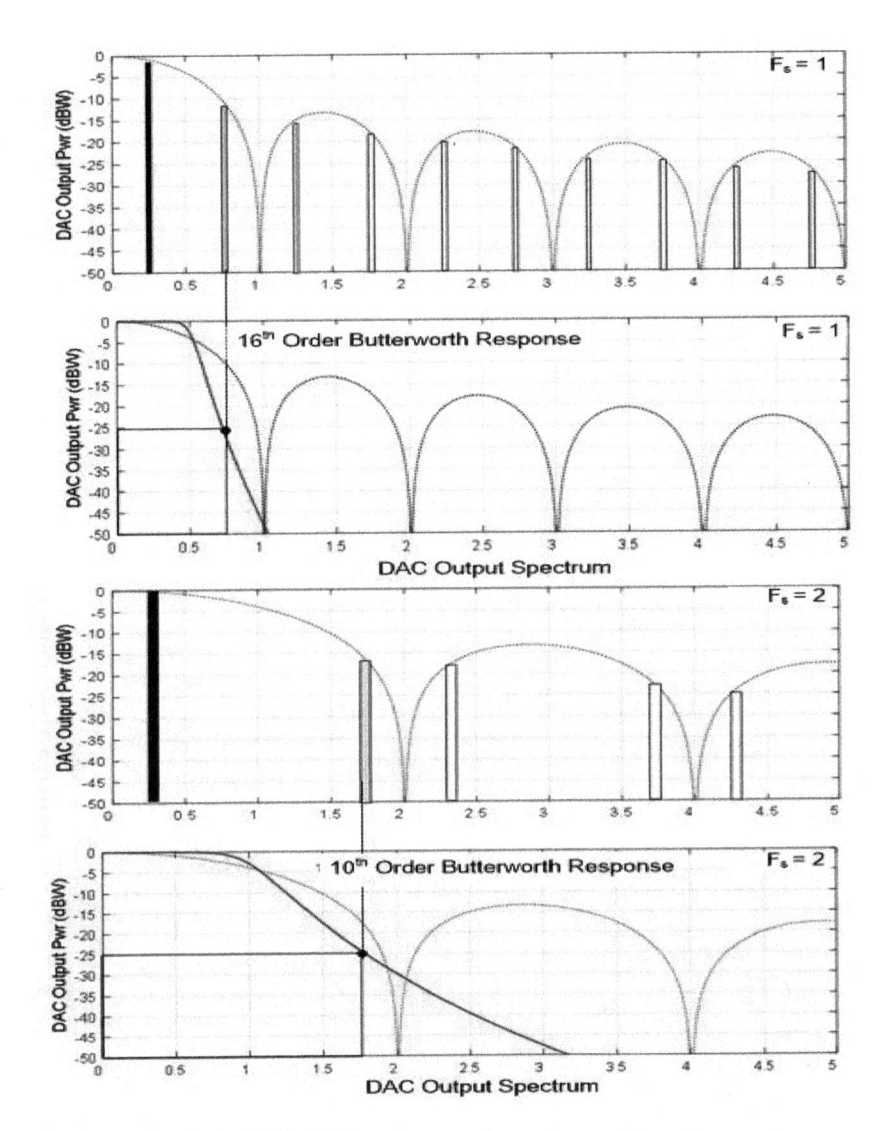

Figure 8.25 DAC output spectrum at two sampling rates.

8.4 QUESTIONS FOR DISCUSSION

1. In Figure 8.9, we mentioned that samples belonging to each symbol (BPSK, QPSK) are summed up in what amounts to a downsample operation. That sum is reset to start summing up the next symbol samples. This will result in processing gain as described in Section 8.2.4.3. Assuming the transmitter and receiver symbol rate are generally the same, what is a clever way to line up the transmitter and receiver symbol boundaries?

2. Consider the frequency-domain DAC output in Figure 8.22. Unlike Figure 8.2, which shows the ADC sampled output in the frequency-domain, the DAC series of images really exist and any of them can be bandpass filtered and used to drive follow-on circuits. What would be the advantages and disadvantages of doing that?

3. An ideal 12 bit ADC has a bipolar sine wave input range extending from $-V_{FS}/2$ to $+V_{FS}/2$. What is the total RMS power of the ADC input (assume a resistance of 1)? What is the approximate SNR at the ADC output?

4. For the example ADC frequency-domain output in Figure 8.26, find the SNR, SFDR, and SINAD.

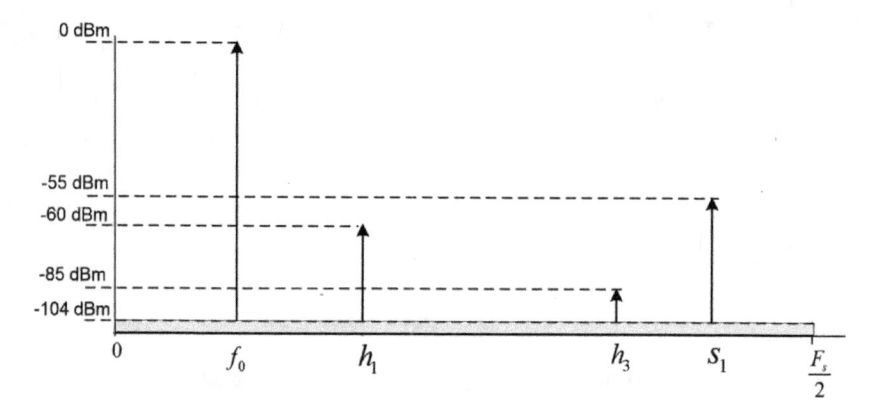

Figure 8.26 ADC output example.

REFERENCES

[1] W. Kester. "Understanding SINAD, ENOB, SNR, THD, THD+N and SFDR so You Don't Get Lost in the Noise Floor". *Analog Dialogue, MT-003 Tutorial.*

[2] W. Kester. "The Good, the Bad, and the Ugly Aspects of ADC Input Noise—Is No Noise Good Noise?". *Analog Dialogue*, February 2006.

[3] R. G. Lyons. *Understanding Digital Signal Processing.* Prentice Hall, 2011.

[4] J. Miller and T. Pate. "Driving High-Speed Analog-to-Digital Converters. *Texas Instruments, Inc.*, 2010.

[5] T. Hentschel and G. Fettweis. *CDMA Techniques for Third Generation Mobile Systems.* Kluwer Academic Publishers, 1998.

[6] J. Munson. "Understanding High Speed DAC Testing and Evaluation". *Analog Devices Application Note, AN-928*, 2006.

Chapter 9

Digital Signal Processing

This chapter provides detailed information about some of the filters and other circuits used in digital communications design. For a more comprehensive treatment of digital signal processing, refer to a book like [1] or [2].

9.1 FUNDAMENTAL DSP CONCEPTS

9.1.1 Unit Delay

Consider the unit delay block, shown in Figure 9.1 and commonly seen on signal processing diagrams of sampled systems:

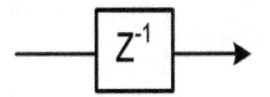

Figure 9.1 The unit delay block.

The letter z is defined as a complex phasor. If f_{signal} is the frequency of a sine wave at the unit delay block input and F_{sample} is the system sampling rate in Equation (9.1).

$$z^{-1} = \left(e^{-j2\pi \frac{f_{signal}}{F_{sample}}}\right) = \cos\left(2\pi \frac{f_{signal}}{F_{sample}}\right) - j\sin\left(2\pi \frac{f_{signal}}{F_{sample}}\right) \qquad (9.1)$$

For a given sample rate and sine wave input frequency, the delay operator phase shift defines a point on a unit circle (a circle with a circumference of 2π and a radius of 1, hence, the name unit circle). For a sine wave input, the ratio f_{signal}/F_{sample} defines the point location and associated phase shift.

Take a simple example with constant frequency $f_{signal} = F_{sample}/4$. From Equation (9.1) we have $z^{-1} = (e^{-j\pi/2})$, just a single complex phase shift at -90°. An input sine wave at frequency f_{signal} will have each new input sample undergoing a CW[1] phase shift of $\pi/2$ radians. Every sample in this sampled sine wave will have this phase shift. Thus, a phase shift equivalent to backing up one sample ($e^{-j\pi/2}$) looks like a one sample delay. Figure 9.2 attempts to illustrate this. The frequency-dependent phase shift z^{-1} is sometimes called a phase shift operator instead of a unit delay because, as we have seen, sample phase shift and sample delay are closely tied together.

Note that the unit delay brought about a unit delay in the sampled signal because it produced the correct amount of phase at $f_{signal} = F_{sample}/4$. For a sine wave at any chosen frequency, the unit delay phase shift and the phase shift brought about by a one sample time delay are the same thing; see Figure 9.2. Change the input frequency and the unit delay will function the same, only a different unit delay phase shift will be implied.

9.1.2 Z-Transform

To analyze circuits built up from unit delays, we first need to understand the nature of the Z plane and the z-transform.

$$X(\omega) = \int_{-\infty}^{\infty} x(t)\, e^{-j\omega t}\, dt$$

$$X(z) = \sum_{n=-\infty}^{\infty} x(n)\, z^{-n} \qquad (9.2)$$

In Equation (9.2), top, the Fourier transform (FT) for continuous time signals is a correlation of $x(t)$ with $e^{j\omega t} = e^{j2\pi f_{basis} t}$, a continuous time complex sine wave

---

1   In this chapter CW = clockwise and CCW = counterclockwise.

CHAPTER 9. DIGITAL SIGNAL PROCESSING

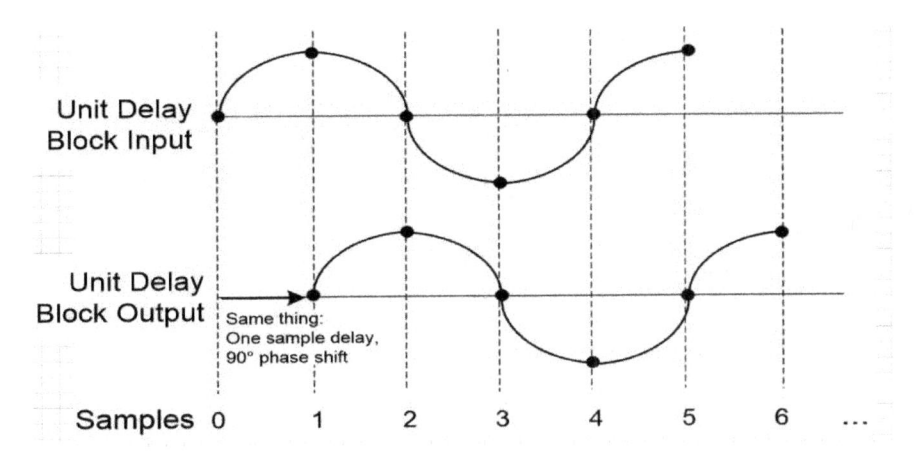

Figure 9.2 Unit delay due to phase shift at frequency $\dfrac{F_{sample}}{4}$.

of infinite duration at frequency f_{basis}. Since the FT is a correlation, we say the FT of time series $x(t)$ at $\omega = 2\pi f_{basis}$ is a complex number the magnitude of which indicates, loosely speaking, the f_{basis} content of $x(t)$. Change the f_{basis} frequency and you will get a different correlation result and hence content. A spectrum is computed by making this calculation over a regular series of f_{basis} frequencies. For example, N spectrum points are based on the progression of Equation (9.3).

$$f_{basis} = \left(-\frac{F_{sample}}{2}\right) \text{ to } \left(\frac{F_{sample}}{2} - \frac{F_{sample}}{N}\right) \text{ in steps of } \frac{F_{sample}}{N} \qquad (9.3)$$

Equation (9.2), top, is continuous time where $\omega = 2\pi f_{basis}$, below that is discrete time $z = e^{j2\pi(f_{basis}/F_{Sample})t}$. For continuous time, any basis frequency can be used; for discrete time, the basis frequency is expressed at a fraction of the sampling frequency and restricted to the range $\pm F_{sample}/2$, see Equation (9.3). A set of z-transforms calculated at uniformly spaced points around the unit circle is also called a spectrum. By contrast, the unit sample delay is simply a delay. However, as described above, we can understand it as a phase shift operator.

Table 9.1 A Few Examples of Z-Transform Basis Functions, $F_{sample} = 1$

f_{basis}	Complex Samples z^n					
	n=0	n=1	n=2	n=3	n=4	...
$F_{sample}/32$	e^{-j0}	$e^{-j\frac{\pi}{16}}$	$e^{-j\frac{\pi}{8}}$	$e^{-j\frac{3\pi}{16}}$	$e^{-j\frac{\pi}{4}}$	...
$F_{sample}/16$	e^{-j0}	$e^{-j\frac{\pi}{8}}$	$e^{-j\frac{\pi}{4}}$	$e^{-j\frac{3\pi}{8}}$	$e^{-j\frac{\pi}{2}}$	...
$F_{sample}/8$	e^{-j0}	$e^{-j\frac{\pi}{4}}$	$e^{-j\frac{\pi}{2}}$	$e^{-j\frac{3\pi}{4}}$	$e^{-j\pi}$	...
$F_{sample}/4$	e^{-j0}	$e^{-j\frac{\pi}{2}}$	$e^{-j\pi}$	$e^{-j\frac{3\pi}{2}}$	$e^{-j2\pi}$	...
⋮	⋮	⋮	⋮	⋮	⋮	⋱

9.1.3 Unit Circle

Referring to Figures 9.3 and 9.4, a fundamental property of sampled systems is that the unique spectrum is confined to the Nyquist range $[-F_{sample}/2, F_{sample}/2]$. This range is centered at 0 Hz in Nyquist zone $N0$ and must be repeated in the range $[F_{sample}/2, 3F_{sample}/2]$ centered at F_{sample} in $N1$, and so on.

In the Figure 9.3, z-plane, $N0$ corresponds to exactly one trip around the circle, starting at $-1 + j0$ (equivalent to $-F_{sample}/2$) and going CCW one complete rotation back to $-1 + j0$ (equivalent to $F_{sample}/2$). Additional trips around the unit circle correspond to higher Nyquist zones and simply repeat the same spectrum, just like the Nyquist theorem requires (see [2] for another good discussion of the Nyquist Theorem).

Figure 9.3 is a closer look at the z-plane unit circle. The numbers on the outside are in units of frequency as a fraction of the sampling rate. The corresponding numbers on the inside are in units of phase advance per sample. Keep in mind that the z-plane is a frequency-domain representation based on a fixed sample rate.

For example, given tone frequency = $F_{sample}/8$, the corresponding phase advance is $e^{j2\pi(f_{signal}/F_{sample})} = e^{j(\pi/4)}$, a complex number at 45° on the unit circle. All the samples of that complex sine wave discrete time series will advance by a 45° phase.

CHAPTER 9. DIGITAL SIGNAL PROCESSING

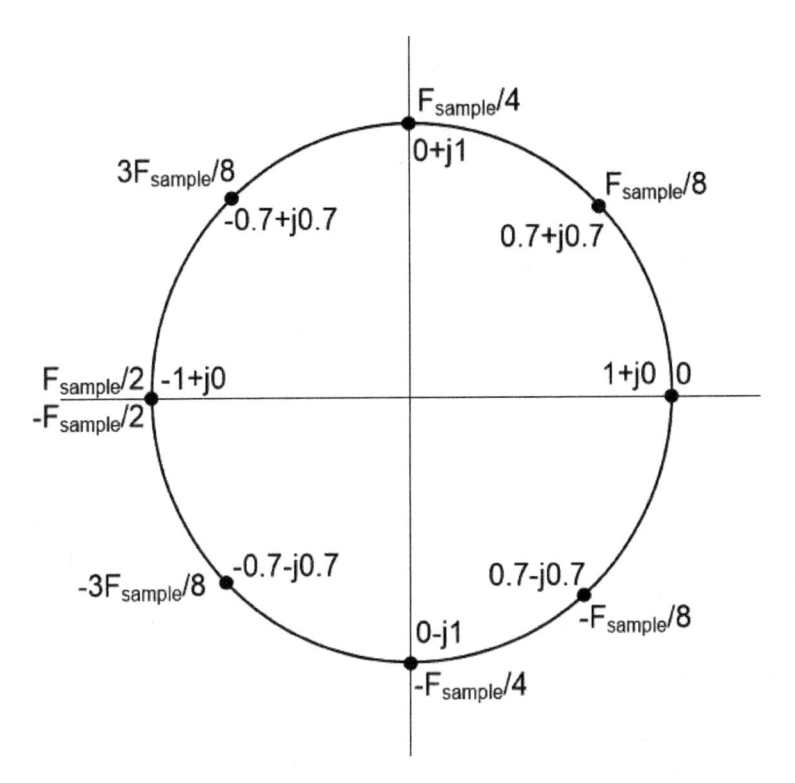

Figure 9.3 Z-Plane example of frequency points on the unit circle for N0.

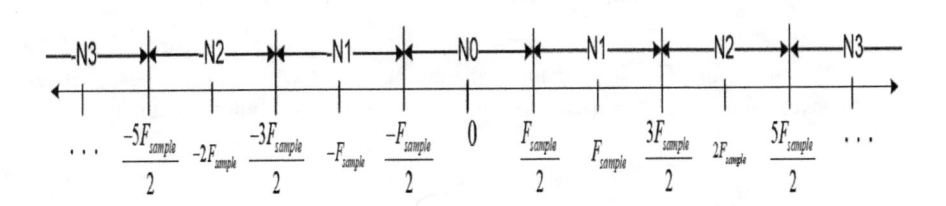

Figure 9.4 Nyquist zone locations for a sampled system.

9.1.4 Poles and Zeros

So far we have discussed:

1. Fourier transform for continuous time signals;
2. z-transform for discrete time signals;
3. Plotting the phase advance of a single-frequency discrete-time sequence on the z-plane unit circle.

We understand the phase, but how do we plot the discrete-time complex sine wave magnitude? In the plane of the page, the z-plane of Figure 9.3 can only show phase. Thus, generally, we do not plot magnitude, except in two cases. When the magnitude goes to infinity we plot an × on the corresponding frequency point and refer to that as a pole. When the magnitude goes to zero we plot a ○ on the corresponding frequency point and refer to that as a zero. Poles and zeros are sometimes called singularities or extremities. They can be useful for predicting the frequency-domain characteristics and stability of a digital circuit. We can find the extremities of a discrete time circuit without computing the z-transform at every frequency in $N0$. Let's study some examples.

9.1.5 Digital Filter Frequency Response

There are at least four ways to generate the frequency response of a digital filter:

1. MATLAB freqz() function;
2. Taking the FFT of the filter impulse response;
3. Substituting $e^{-j2\pi \frac{f_{signal}}{f_{sample}} n}$ for each transfer function term z^{-n}. Then solve for the magnitude and/or phase of the result;
4. A graphical technique based on only the pole and zero locations.

The first two are shown below for the halfband filter. Notice in Figure 9.5 that the result is the same. The second method may be useful when some other software language, such as C, must be used and there is an fft() function but no freqz() function.

```
% Calculate impulse response
ImpulseResponseHB =firhalfband('minorder', PassBandEdge,
StopBandRipple, 'kaiser');
% Check frequency response
```

CHAPTER 9. DIGITAL SIGNAL PROCESSING

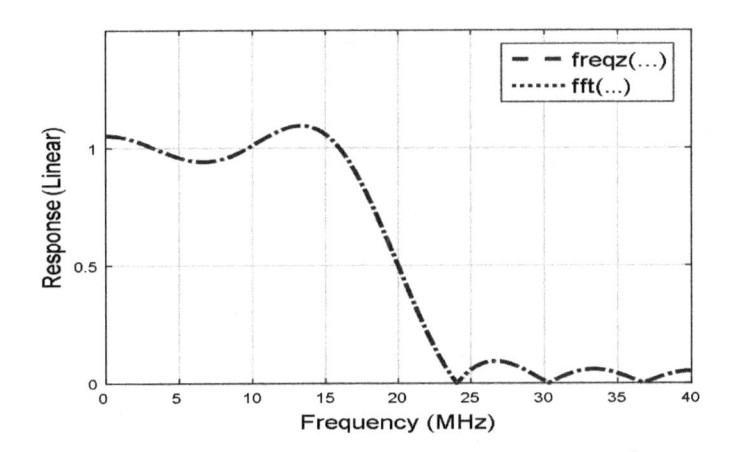

Figure 9.5 Halfband frequency response calculated two ways.

```
NFFT = 2048;
(hb,wb) = freqz(ImpulseResponseHB,1,NFFT);
% Alternative to freqz:
hbi = fft(ImpulseResponseHB,2*NFFT);
```

Method three can be tedious work with a paper and pencil. For a simple example, based on [1], take the two-point averaging circuit of Figure 9.6. Equation (9.4) is the system function, written by inspection. The fourth method is also described in [1].

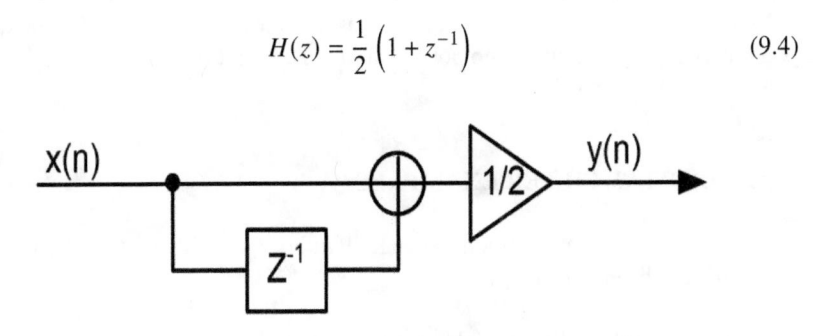

$$H(z) = \frac{1}{2}\left(1 + z^{-1}\right) \tag{9.4}$$

Figure 9.6 A two-point averaging circuit.

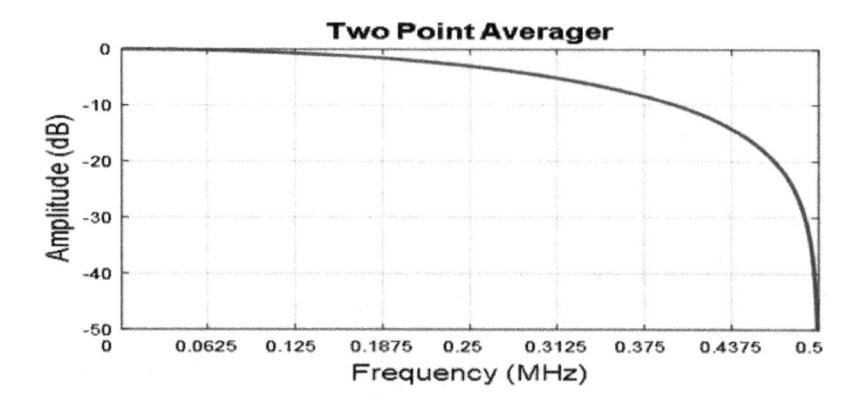

Figure 9.7 A two-point averaging frequency response. $F_S = 1$.

Let's set $F_{sample} = 1$ for simplicity and substitute Equation (9.5) for z^{-1} to get Equation (9.6).

$$e^{-j2\pi \frac{f_{signal}}{F_{sample}} n} = e^{-j2\pi f_{signal} n} \tag{9.5}$$

$$H(f_{signal}) = \frac{1}{2}\left(1 + e^{-j2\pi f_{signal}}\right) = e^{-j\pi f_{signal}} \cos\left(\pi f_{signal}\right) \tag{9.6}$$

The magnitude and phase of the frequency response is shown in Figure 9.7. Note that the 2-point averaging circuit has no poles and one zero. In general, an M point averaging circuit has no poles and $M - 1$ zeros. For example, for $M = 4$, zeros are at +j, -1+j0 and -j.

9.2 DIGITAL FILTER EXAMPLES

This is a collection of commonly seen DSP filters. One of these, the M point average, has already been covered.

9.2.1 Example 1: Single Pole on Unit Circle

Our first example is shown in Figure 9.8.

CHAPTER 9. DIGITAL SIGNAL PROCESSING

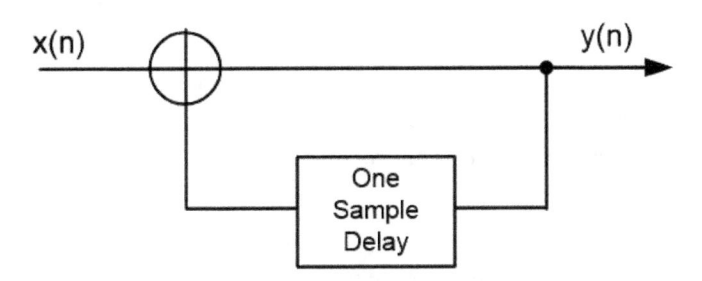

Figure 9.8 Simple DSP circuit with single pole at $z = 1$.

$$y(n) = x(n) + y(n-1)$$
$$Y(z) = X(z) + z^{-1}Y(z)$$
$$Y(z)\left(1 - z^{-1}\right) = X(z) \qquad (9.7)$$
$$\frac{Y(z)}{X(z)} = \frac{1}{1 - z^{-1}}$$

The transfer function in Equation (9.7) is used to compute poles and zeros. Notice how in the second line the z-transform distributes into the sum on the right side of the first line. The last line of Equation (9.7) is the z-transform of the time-domain recursion in the first line. In this case, the pole location is easy to see; solve $\left(1 - z^{-1}\right) = 0$ to see that the system function goes to infinity when $z = 1$.

The impulse response is the output in response to an input impulse. Make input $x(n) = 1, 0, 0, 0, 0, 0, \ldots$ in Figure 9.8. In response to this, the output goes to 1 and stays there, $y(n) = 1, 1, 1, 1, 1, 1, \ldots$.

DSP circuits with poles on or beyond the unit circle do not have the stability that we generally require. We can apply the bounded input, bounded output (BIBO) stability criteria. The BIBO criteria says that an impulse response that outputs an infinite sum (i.e., an infinite sum of ones is infinite and not bounded) indicates the system is not stable. Note that DSP circuits can have zeros (values that make the numerator go to zero) outside the unit circle.[2]

---

[2] A DSP circuit with all zeros inside the unit circle is called minimum phase and, likewise, if all zeros are outside the unit circle we call that circuit maximum phase. A minimum phase circuit has the minimum group delay of any circuit with the same magnitude response.

Finally, what about frequency response? An unstable system is also not linear (i.e., outputs are generated that are not scaled version of the input). So, frequency response is not defined.

9.2.2 Example 2: Single Pole Inside Unit Circle

We can make our simple circuit only slightly more complicated as shown in Figure 9.9. Equation (9.8) is used to compute poles and zeros: $1 - az^{-1} = 0$ implies a pole at $z = a$. For a practical lowpass filter, a is usually a real number between zero and less than +1 on the unit circle horizontal axis.

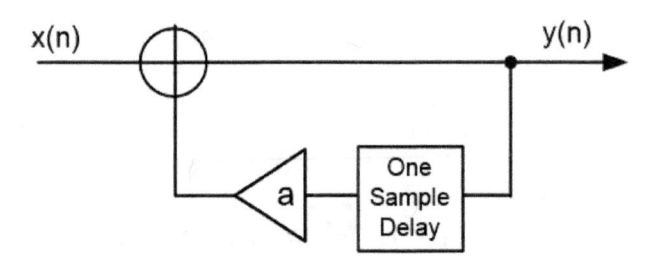

Figure 9.9 Simple DSP circuit with single pole at $z = a$.

$$\begin{aligned} y(n) &= x(n) + ay(n-1) \\ Y(z) &= X(z) + az^{-1}Y(z) \\ Y(z)\left(1 - az^{-1}\right) &= X(z) \\ \frac{Y(z)}{X(z)} &= \frac{1}{1 - az^{-1}} \end{aligned} \quad (9.8)$$

The impulse response is not quite as obvious as in the first example; however, we can calculate the basic equation recursion from the circuit diagram to obtain it. See Equation (9.9) where the first column shows the input, output indexing and the second column shows the resulting impulse response.

CHAPTER 9. DIGITAL SIGNAL PROCESSING

$$
\begin{array}{ll}
y(0) = x(0) & y(0) = 1 \\
y(1) = x(1) + ay(0) & y(1) = 0 + a \\
y(2) = x(2) + ay(1) & y(2) = 0 + a^2 \\
y(3) = x(3) + ay(2) & y(3) = 0 + a^3 \\
y(4) = x(4) + ay(3) & y(4) = 0 + a^4 \\
\vdots & \vdots
\end{array}
\tag{9.9}
$$

The resulting impulse response is, $h(n) = a^n u(n)$. BIBO stability is shown in Equation (9.10).

$$\sum_{n=-\infty}^{\infty} h(n) = \sum_{n=0}^{\infty} a^n < \infty \quad \text{for } |a| < 1 \tag{9.10}$$

Frequency response of this circuit is lowpass, with passband starting at zero, and bandwidth-dependent on pole location a on the real axis in the range $[0, 1)$. We obtain the frequency response by calculating the fast Fourier transform (FFT) of the impulse response. MATLAB code to accomplish this is shown next. We should note that MATLAB also provides a fvtool() built-in function for a quicker look at digital filter responses (type "help fvtool" in the MATLAB workspace).

```
Fs = 2; % sampling rate
L = 256; % number of points on unit circle
a = [0.5 0.9]; % pole locations
h = zeros(length(a),L);
for k = 1:2
% Compute impulse response
h(k,:) = impz(1,[1 -a(k)],L);
% Convert time domain impulse response to
%frequency domain spectrum
h(k,:) = fft(h(k,:));
% normalize peak to 1
h(k,:) = h(k,:)/(max(abs(h(k,:))));
% Change [0 Fs] to [-Fs/2 Fs/2] and take log magnitude
h(k,:) = 10*log(abs(fftshift(h(k,:))));
end
ph = plot(h(1,:),'k');
ph = plot(h(2,:),'k:');
```

```
set(gca,'XLim',[0 256]);
set(gca,'XTick',[0:32:256]);
set(gca,'XTickLabel',[-Fs/2 : Fs/8 : Fs/2]);
yh = ylabel('Response (dB)');
xh = xlabel('Unit Circle Frequency');
legend('a = 0.5','a = 0.9');
```

Note in Figure 9.19, top plot, that the lowpass response rolls off faster as parameter a gets closer one (i.e., closer to the unit circle). This parameter cannot equal 1, however, because then it would be on the unit circle and the circuit would lose stability (like the one pole example above). Note also the horizontal frequency axis tick marks are equivalent to $\left(\dfrac{-F_s}{2}, \dfrac{-3F_s}{8}, \dfrac{-F_s}{4}, \dfrac{-F_s}{8}, 0, \dfrac{F_s}{8}, \dfrac{F_s}{4}, \dfrac{3F_s}{8}, \dfrac{F_s}{2}\right)$. Compare this with the unit circle frequency points in Figure 9.3.

9.2.3 Example 3: Exponential Averaging Filter

A variation on the filter of Example 2 is the EAF (exponential averaging filter) shown in Figure 9.10. The frequency response in Figure 9.11 is parameterized by one number a. Coefficients a and $1 - a$ are a complementary pair whose sum is 1, thus leading to the result that the gain at 0 frequency is always one. EAFs are very useful for constant gain, variable smoothing (averaging) of signals from external sensors. They can also be used for smoothing internal tracking signals, such as the control feedback in a carrier tracking loop. In Figure 9.11 we plot the frequency response difference between $a = 0.1$ and $a = 0.6$.

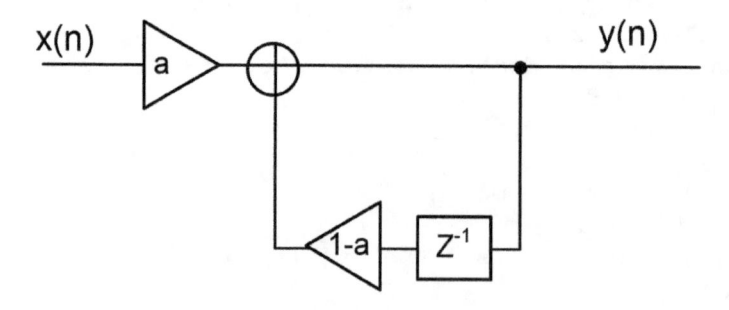

Figure 9.10 An exponential averaging filter circuit.

CHAPTER 9. DIGITAL SIGNAL PROCESSING 343

$$y[n] = y[n-1] \cdot (1 - \alpha) + x[n] \cdot \alpha$$

$$\frac{Y(z)}{X(z)} = \frac{a}{1 - z^{-1}(1-a)} \tag{9.11}$$

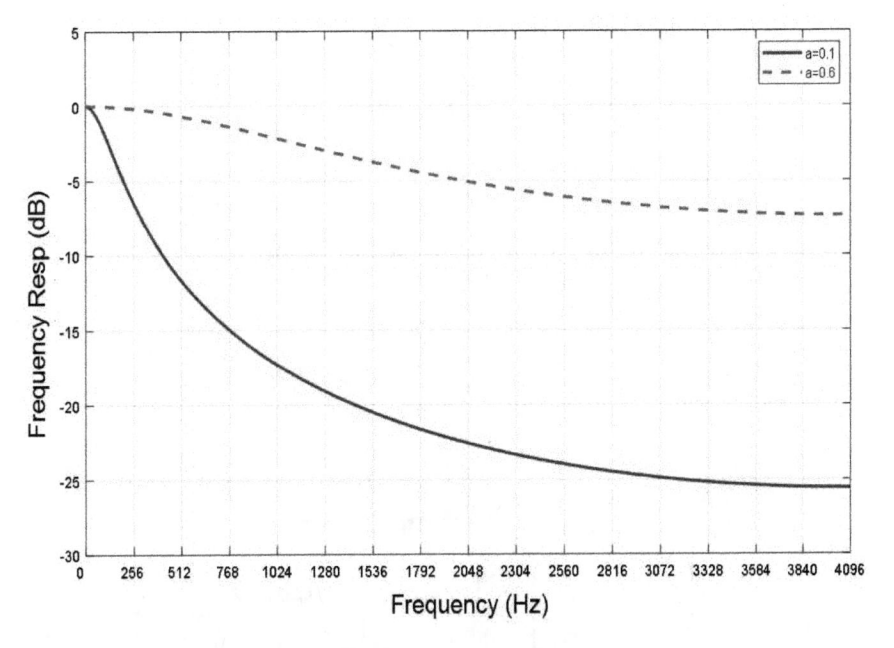

Figure 9.11 Exponential averaging filter frequency response examples.

9.2.4 Example 4: Cascade Integrator Comb

Our next example is the popular cascade integrator comb (CIC) filter, shown in Figure 9.12. To derive the transfer function, we separate the circuit around the center, as shown in Equation (9.12).

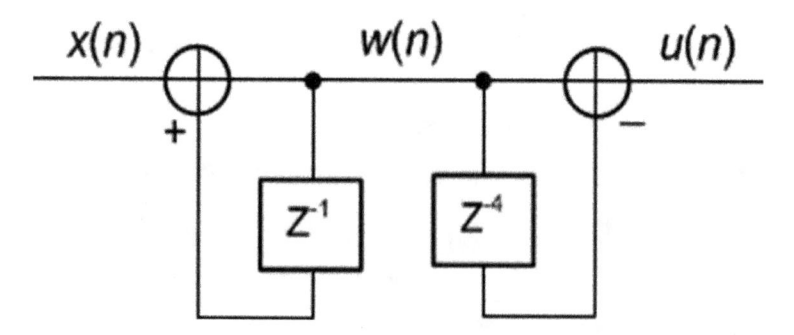

Figure 9.12 A cascaded integrator comb filter.

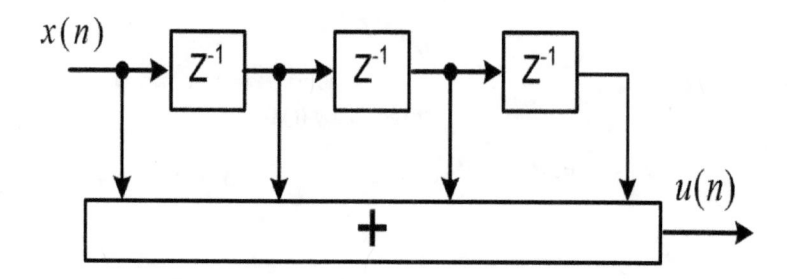

Figure 9.13 Cascaded integrator comb filter, showing boxcar equivalent.

CHAPTER 9. DIGITAL SIGNAL PROCESSING

$$w(n) = x(n) + w(n-1)$$
$$w(n) = x(n) + w(n)z^{-1}$$
$$w(n)(1 - z^{-1}) = x(n)$$
$$\frac{w(n)}{x(n)} = \frac{1}{(1 - z^{-1})}$$

$$u(n) = w(n) - w(n-4)$$
$$u(n) = w(n) - w(n)z^{-4} \quad (9.12)$$
$$u(n) = w(n)(1 - z^{-4})$$
$$\frac{u(n)}{w(n)} = (1 - z^{-4})$$

$$\frac{w(z)}{x(z)} \frac{u(z)}{w(z)} = \frac{u(z)}{x(z)} = \frac{1 - z^{-4}}{1 - z^{-1}}$$

Continuing the transfer function analysis in Equation (9.13), we see that the CIC filter is actually a 4-tap boxcar filter (i.e., a simple average of N signal delays), see Figure 9.13. Thus, an N delay CIC is the same as an N sample average. Figure 9.12 circuit avoids an N input adder, as these can be expensive to construct for large N.

$$\frac{u(z)}{x(z)} = \frac{1 - z^{-4}}{1 - z^{-1}} = \frac{(1 - z^{-2})(1 + z^{-2})}{1 - z^{-1}}$$
$$= \frac{(1 - z^{-1})(1 + z^{-1})(1 - jz^{-1})(1 + jz^{-1})}{1 - z^{-1}} \quad (9.13)$$
$$= (1 + z^{-1})(1 - jz^{-1})(1 + jz^{-1})$$
$$= 1 + z^{-1} + z^{-2} + z^{-3}$$

The impulse response for the averaging circuit in Figure 9.13 will be $h(n) = [1, 1, 1, 1, 0, 0, 0, 0 \ldots]$. The number of ones simply equals the averaging length. This is BIBO stable because the sum of the impulse response is the sum of the averaging length, in this case, 4.

The poles and zeros are calculated in Equation (9.14) from the second to last line in Equation (9.13). These are plotted in Figure 9.14. As a general rule, boxcar averages (or equivalent CIC filters) with N taps will start with N zeros spaced evenly around the unit circle. To complete the diagram, just delete the zero at $1 + j0$ to get final total of $N - 1$ zeros.

$$\begin{aligned}(1+z^{-1}) &\Rightarrow \text{zero}: z = -1+j0\\ (1-jz^{-1}) &\Rightarrow \text{zero}: z = 0+j1\\ (1+jz^{-1}) &\Rightarrow \text{zero}: z = 0-j1\end{aligned} \quad (9.14)$$

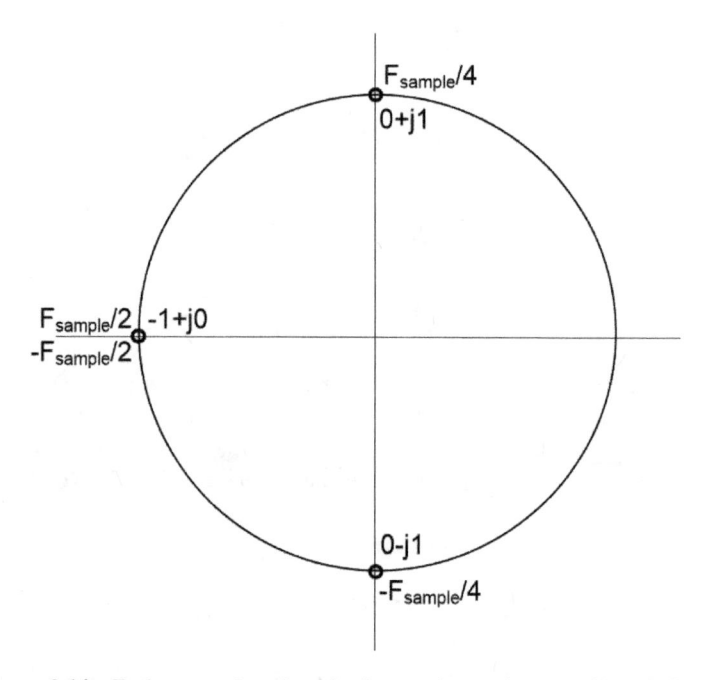

Figure 9.14 Z-plane zero locations for four-tap boxcar lowpass filter (CIC).

As shown in Figure 9.15, CIC frequency response has $N-1$ nulls. Notice also the droop in the response between 0 and $\frac{F_{sample}}{4}$ frequency. It is important to understand if and how this droop affects the communications signal. In some systems, for example, filtering the output of a slowly changing temperature monitor, the value of N could be very large and the droop inconsequential. There is much more information about CIC filters and DSP theory in [2]. The invention of the CIC filter is described in [3]. CIC filters are often used for narrowband communications signals centered on zero frequency.

The CIC filter is sometimes used to lower the sampling rate. The upper CIC circuit in Figure 9.16 shows the obvious way to subsample by 4. The lower CIC circuit in Figure 9.16 shows a simplified but exactly equivalent CIC subsampler. As

CHAPTER 9. DIGITAL SIGNAL PROCESSING

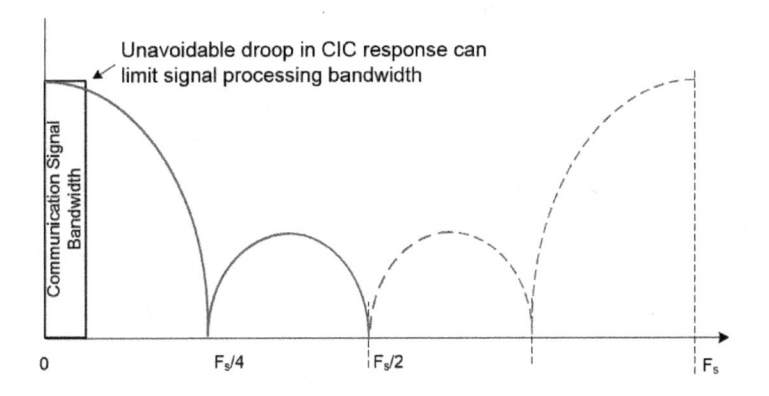

Figure 9.15 Cascaded integrator comb filter, frequency response for $N = 4$.

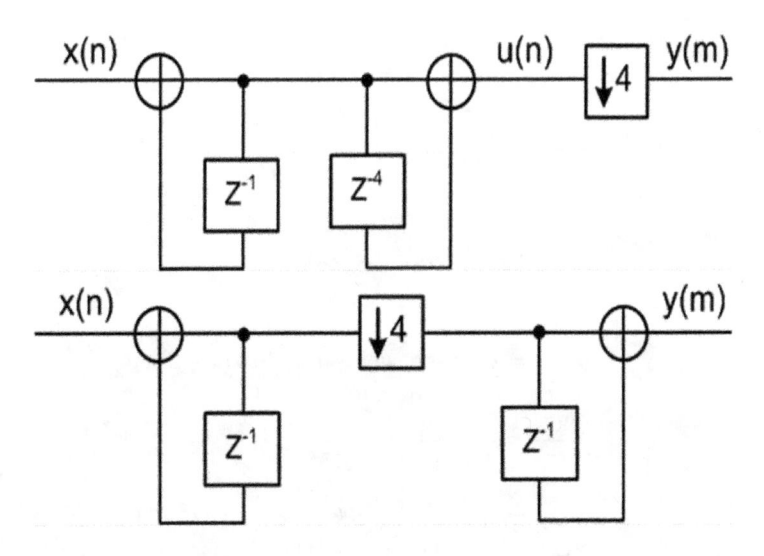

Figure 9.16 Two CIC subsampling filters with the same transfer function.

described in Section 9.2.8, it is important to understand how noise is being aliased when using subsamplers.

9.2.5 Example 5: Resonator

The DSP resonator is a second-order (two-pole) narrow bandpass filter (BPF) that can be used for symbol timing recovery and tone detection. The DSP block diagram is shown in Figure 9.17. In Equation (9.15) the transfer function is calculated directly from this block diagram, using $w(n)$ to separate around the center.

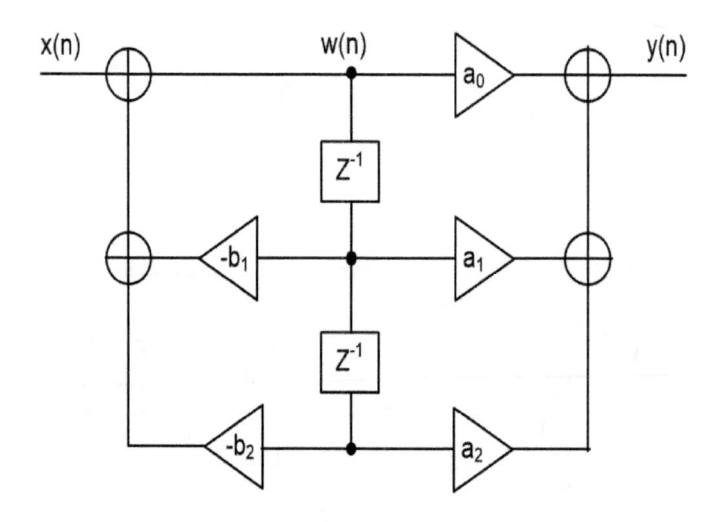

Figure 9.17 A DSP resonator bandpass filter.

$$w(k) = x(k) - b_1 w(k-1) - b_2 w(k-2)$$
$$w(z) = x(z) - b_1 z^{-1} w(z) - b_2 z^{-2} w(z)$$
$$\frac{w(z)}{x(z)} = \frac{1}{1 + b_1 z^{-1} + b_2 z^{-2}}$$

$$y(k) = a_0 w(k) + a_1 w(k-1) + a_2 w(k-2)$$
$$y(z) = a_0 w(z) + a_1 z^{-1} w(z) + a_2 z^{-2} w(z) \qquad (9.15)$$
$$\frac{y(z)}{w(z)} = a_0 + a_1 z^{-1} + a_2 z^{-2}$$

$$\frac{y(z)}{x(z)} = \frac{w(z)}{x(z)} \frac{y(z)}{w(z)} = \frac{a_0 + a_1 z^{-1} + a_2 z^{-2}}{1 + b_1 z^{-1} + b_2 z^{-2}}$$

CHAPTER 9. DIGITAL SIGNAL PROCESSING

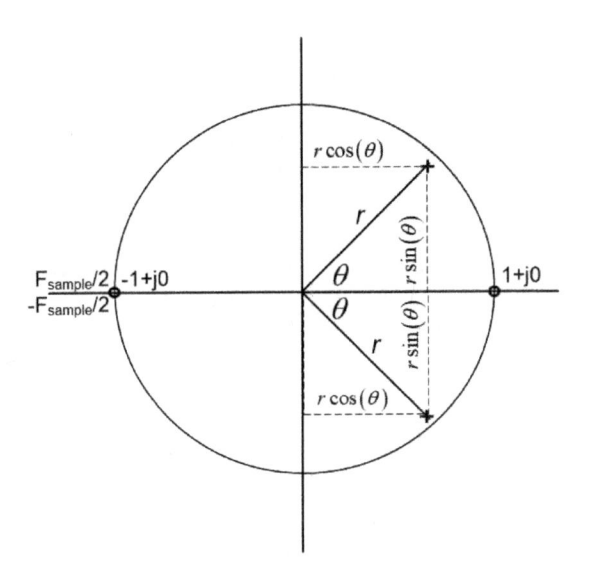

Figure 9.18 DSP resonator typical pole zero plot.

The next step is to set up the second-order resonator transfer function in a form that allows us to choose denominator parameters r and θ. The denominator has two complex conjugate poles at radius r and angles $\pm\theta$; see Figure 9.18. Parameters r and θ affect the rolloff and center frequency, respectively. The numerator is set up with a zero at both $1 + j0$ and $-1 + j0$. This affects rolloff from the resonator peak.

Equation (9.16) is a standard way to parameterize this second-order resonator (see [1]). The resonator poles and zeros are shown in Figure 9.18. The resulting frequency response is in Figure 9.19, lower plot. Note that the zeros shown at -1, 0 and +1 on the horizontal axis.

9.2.5.1 Quality Factor

Quality factor $Q = (BPF\,center\,frequency)/(-3dB\,bandwidth)$. High Q means a sharp narrowband filter; low Q is the opposite. For the resonator filter, pole radius r is generally chosen very close to the unit circle for high Q. However, stability requires keeping $r < 1$. Angle θ is the bandpass frequency on the unit circle, θ is usually chosen by application. Here $\theta = \pi/2$ is equivalent to $F_{sample}/8$ for no particular application.

Note from Figure 9.19, lower plot, how the choice of r affects the resonator bandpass sharpness. The -3dB bandwidth clearly tightens up for $r = 0.99$ versus $r = 0.9$. However, examination of Equation (9.16) shows that the relationship between -3dB bandwidth and r is not linear. Finally, a fixed point, high Q, resonator implementation may require long word lengths to prevent overflow.

Figure 9.19 is set up to compare the frequency response of the one-pole circuit of Section 9.2.2 with the two-pole resonator circuit of this section. First, we notice that both responses are about -45 dB at ±0.25 normalized frequency. However, the ultimate attenuation of the two-pole design is much higher due to the zeros (shown on the lower plot, horizontal frequency axis).

$$\frac{y(z)}{x(z)} = \frac{a_0 + a_1 z^{-1} + a_2 z^{-2}}{1 + b_1 z^{-1} + b_2 z^{-2}} = \frac{\left(z^{-1} + 1\right)\left(z^{-1} - 1\right)}{1 - \left(1 + r^2\right) \cos\left(\theta\right) z^{-1} + z^{-2} r^2}$$

$$= \frac{1 - z^{-2}}{1 - \left(1 + r^2\right) \cos\left(\theta\right) z^{-1} + z^{-2} r^2} \quad (9.16)$$

9.2.6 Example 6: Halfband Filters

The halfband lowpass filter is our next example. Examining the ideal halfband filter frequency response on the DSP only unit circle diagram in Figure 9.20 may be helpful. Starting on the circle on the left side at $-F_{sample}/2$ and moving CCW, we see that the frequency response is 0 (dotted line) until $-F_{sample}/4$ where it becomes 1 (solid line). The frequency response stays at 1 until $F_{sample}/4$ where it goes back to 0 until we get back to the starting point. Then the frequency response cycle just described repeats. The repeat cycle is centered at F_{sample}, instead of 0. The cycle repeats forever, with images centered at integer multiples n of nF_{sample}. In Figure 9.21, five rotations are spread out on a line. Corresponding Nyquist zones are also indicated.

All these images are exactly alike in frequency and power; the only difference is the center frequency. At sampling rate = F_{sample}, only the image in $N0$ is manifest; the others are latent (hidden). Changes to the $N0$ image will be seen on all the latent images because the spectral response in any Nyquist zone must be exactly the same as the $N0$ spectrum.

So far, we have only looked at the ideal halfband lowpass. An ideal filter is difficult to build. Let's find the tap values of a practical finite impulse response halfband filter. Start by defining the filter mathematically in the frequency domain,

CHAPTER 9. DIGITAL SIGNAL PROCESSING

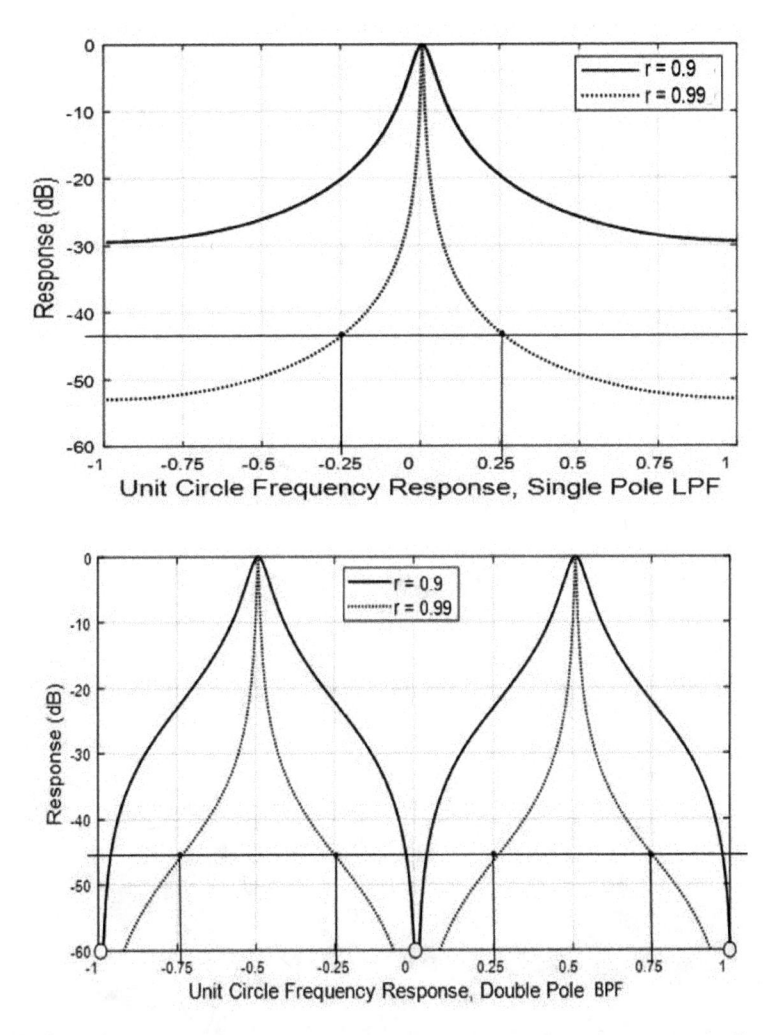

Figure 9.19 Frequency response of 1 and 2 pole resonator filters; $\pm 1 \Rightarrow \pm \frac{F_s}{2}$.

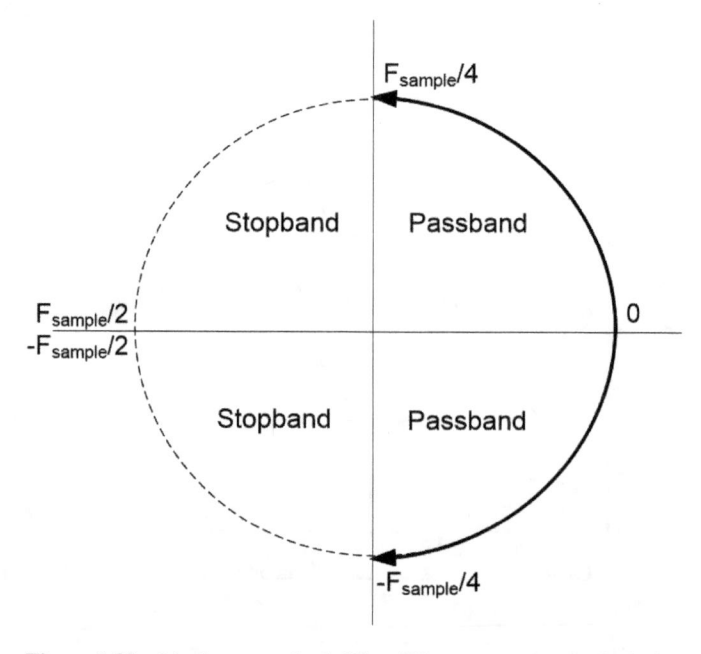

Figure 9.20 Ideal rectangular halfband filter response on unit circle.

Equation (9.17). The ideal halfband LPF in Figure 9.20 acts like a frequency switch so the gate function, Π, represents it correctly.

$$H(f) = \Pi\left(\frac{2f}{F_{sample}}\right) = \begin{cases} 0 & \text{if } |f| > \frac{F_{sample}}{4} \\ 0.5 & \text{if } |f| = \frac{F_{sample}}{4} \\ 1 & \text{if } |f| < \frac{F_{sample}}{4} \end{cases} \quad (9.17)$$

$$h(t) = F^{-1}(H(f)) = \int_{-\frac{F_{sample}}{4}}^{+\frac{F_{sample}}{4}} e^{(2\pi jft)} df = \left(\frac{F_{sample}}{2}\right) \text{sinc}\left(\left(\frac{F_{sample}}{2}\right)t\right)$$

$$(9.18)$$

CHAPTER 9. DIGITAL SIGNAL PROCESSING 353

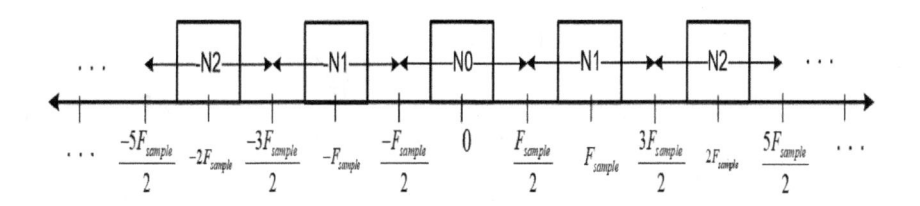

Figure 9.21 Ideal halfband filter response, multiple Nyquist zones.

In Equation (9.18) we calculate the time-domain impulse response using the inverse Fourier transform.[3] However, we cannot use this impulse response directly as FIR filter taps because it is continuous time for $-\infty < t < \infty$. To make this a practical FIR filter, we need to truncate the time interval and show discrete sampling points at $t = nT_{sample}$. Extending the result in Equation (9.18) we get Equation (9.19). For an N tap FIR, we have $-\frac{N}{2} < n < \frac{N}{2}$.

However, we still have a problem. Equation (9.19), top, is noncausal and not practical. To see this, note that the top summation of Equation (9.19) is noncausal because, for $n = 0$, negative values of k result in positive indices of $h(n - k)$, which make the summation dependent on future input samples. To fix this problem, we simply rewrite the convolution as shown in the lower summation of Equation (9.19).

$$y(n) = \sum_{k=-\frac{N}{2}}^{\frac{N}{2}} x(k)h(n-k)$$

$$y(n) = \sum_{k=0}^{N} x(k)h(n-k) \qquad (9.19)$$

---

[3] Notice that this calculation illustrates the fundamental principle that time domain $p(t) = \text{sinc}\left(\frac{t}{T_{sample}}\right)$ transforms to frequency domain $P(f) = \left(\Pi\left(\frac{f}{F_{sample}}\right)\right)$ and vice versa. Π is a gate function, defined as $\Pi(x) = \begin{cases} 0 & |x| > 0.5 \\ 0.5 & |x| = 0.5 \\ 1 & |x| < 0.5 \end{cases}$.

$$h(n) = h(t)_{for\ t=nT_{sample}} = \left(\frac{1}{2T_{sample}}\right) \frac{\sin\left(\frac{\pi t}{2T_{sample}}\right)}{\left(\frac{\pi t}{2T_{sample}}\right)} = \frac{\sin\left(\frac{\pi n}{2}\right)}{\left(\frac{\pi n}{2}\right)} \qquad (9.20)$$

In trace A of Figure 9.22, we plot 64 points of Equation (9.20), centered at 0. Except for the point at index 0, which equals 1, all even numbered points are 0. This is easy to see in Equation (9.20), numerator term $\sin\left(\frac{\pi n}{2}\right)$. There even values of n result in an integer multiple of π for the argument of the sin( ) function.

Truncating the sinc function in trace A of Figure 9.22 is the same as multiplying it by a gate function (a different gate function than Equation (9.17). Thus the spectrum in trace B of Figure 9.22 is the convolution of the sinc spectrum and the gate function spectrum. This produces the familiar Gibbs ringing at the edge of the passband (see [1]). Note also the horizontal frequency axis tick marks line up with $\left(\frac{-F_s}{2}, \frac{-3F_s}{8}, \frac{-F_s}{4}, \frac{-F_s}{8}, 0, \frac{F_s}{8}, \frac{F_s}{4}, \frac{3F_s}{8}, \frac{F_s}{2}\right)$. The halfband points are at $\frac{-F_s}{4}$ and $\frac{+F_s}{4}$ corresponding to -0.5 and +0.5.

To get rid of the Gibbs ringing, we can multiply the impulse response in trace A of Figure 9.22 point by point with a window function, such as the Kaiser window, that tapers to zero at the end points. We chose a Kaiser window because it has an extra parameter beta to trade off stopband attenuation versus transition bandwidth. The result is shown in trace C of Figure 9.22. Note that the zero coefficients are marked with a dot. The final spectral result is in trace D of Figure 9.22. Note that the frequency response plots in Figure 9.22 are not in dB; this shows the halfband characteristic clearly. Thus the dots at the plus and minus halfbands in trace D are where the magnitude is reduced by one-half. Notice the slope on either side of the halfband lines. This is another halfband filter characteristic. For $|f| < 0.5$, it is true that if $|H(0.5 + f)| = w$ then $|H(0.5 - f)| = 1 - w$.

For some filter design tools, the coefficients that are supposed to be zero may actually be some small number compared to the nonzero coefficients. Setting these to zero should have little or no effect on the frequency response. Finally, although the cutoff frequency must be halfband, the number of taps N and the Kaiser window parameter beta both control the stopband rolloff rate.

Because the halfband filter has no recursions like an IIR filter would, it has no poles and is inherently stable and linear phase. Some of the MATLAB signal processing steps used for Figure 9.22 are also shown here.

CHAPTER 9. DIGITAL SIGNAL PROCESSING

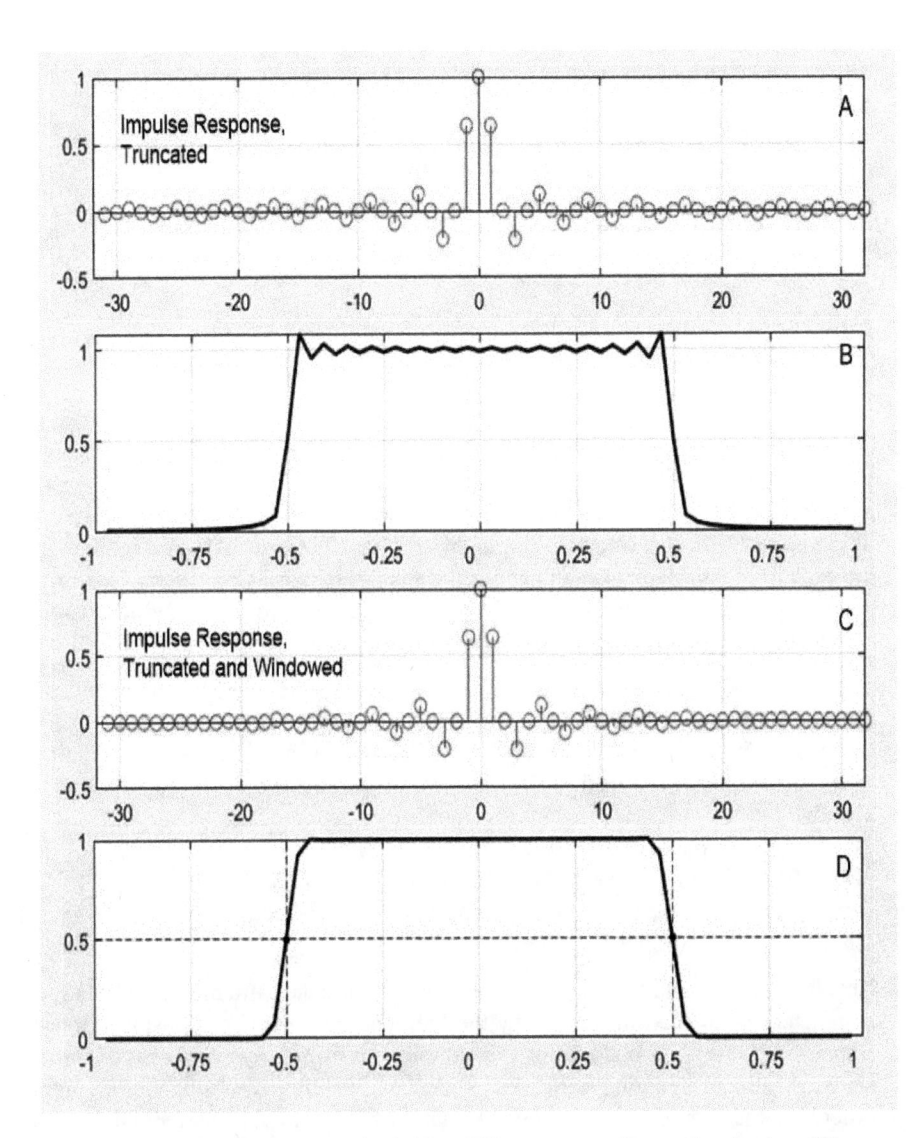

Figure 9.22 Ideal rectangular halfband filter response, linear diagram (not dB).

```
Fs = 2; % sampling rate
N = 64; % number of points on unit circle
n = (1-N/2):N/2; % non-causal range of n
hh = sinc(n/2); % halfband impulse response, truncated
win = kaiser(N,5);
hw = hh.*win'; %Kaiser window for halfband coefficients
hhf = fft(hw); % convert to frequency domain
hhfa = abs(fftshift(hhf));
```

9.2.7 Example 7: Upsampling Filters

Figure 9.23 shows the process of increasing the sample rate by 2. The first block doubles the sample rate by inserting a 0 between each input sample. This upsample block is very simple and is well documented in Simulink. The second block is generally a halfband lowpass filter (HBF). Figure 9.23 also shows the process of upsampling by in the frequency domain. The A, B, C, and D spectrums match the time-domain circuit. Details about the four steps are listed below. Due to the simplicity of halfband filters, larger upsampling ratios are generally achieved by cascading multiple upsample by two circuits; see, for example, Figure 7.49.

A Narrowband signal centered at 0 Hz, sample rate F_s. Frequency domain replicates at multiples of sample rate are shown in a dotted line.
B Double the sample rate $F_s^* = 2F_s$ by adding a zero sample between every time-domain sample. This puts the left side of the first replicate in trace A in the Nyquist bandwidth. We need to filter out this frequency-domain image.
C Shows halfband filter response at the higher sample rate.
D Halfband filter used to remove unneeded replicate at $F_s^*/2$. EThe end result is an upsampled narrowband signal matching the original spectrum in trace A, except that the gain is multiplied by $1/L$.

Note that the halfband filter is close to optimal because the extra image can be almost completely eliminated in trace C by the halfband monotonic decreasing stopband response. Figure 9.23 is doubling the sample rate of a fairly narrowband lowpass signal. A more challenging signal (i.e., wider bandwidth) that would require a higher order HBF that would have a bandwidth extending closer to one-quarter of the sample rate in trace C.

  A naive approach to upsampling by 2, for example, might use a two tap boxcar filter (i.e., two taps = [1,1]). Filtering the upsampled signal in trace B above with this

CHAPTER 9. DIGITAL SIGNAL PROCESSING

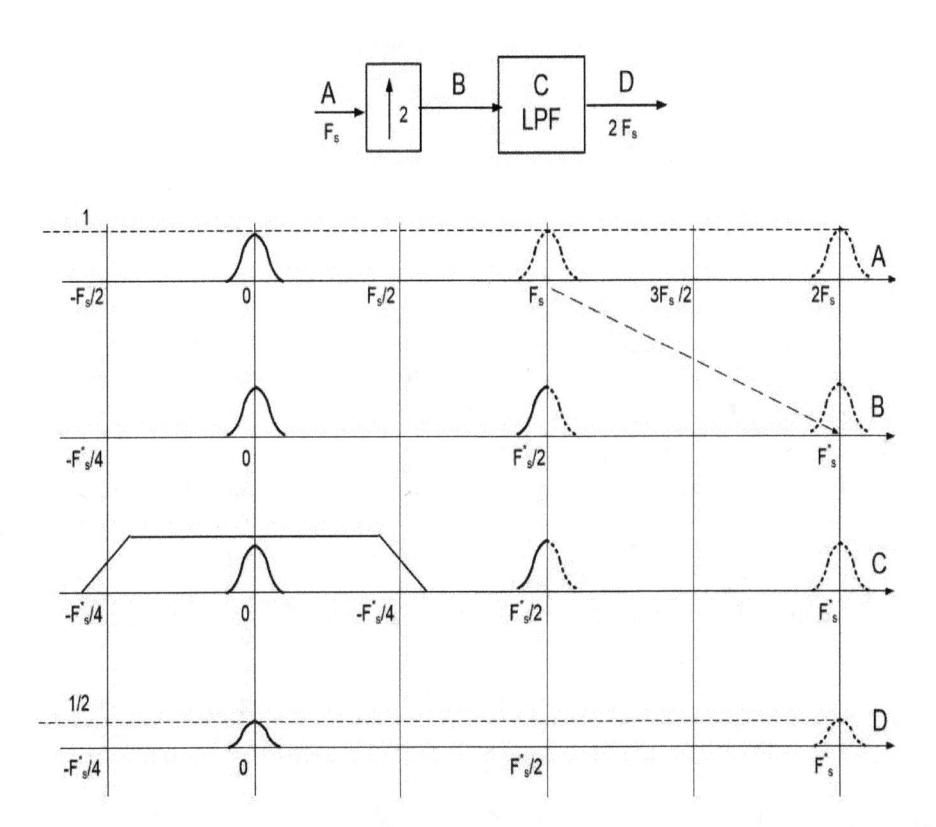

Figure 9.23 Upsample by two circuit and spectrum.

filter would simply repeat each of the original samples. Sample repeating may look like upsampling but it is very suboptimal. A simple way to see this is to consider that replacing the halfband filter by a 2-tap boxcar filter will significantly degrade performance.

Notice that the upsampling process results in a loss of $1/L$ amplitude. In a fixed point system, this can be fixed by a simple shift left by 1. Interestingly, the downsampling process has no such gain change. This is shown with mathematical rigor in [4].

9.2.8 Example 8: Down Sampling Filters

Figure 9.24 shows the process of decreasing the sample rate by 2. The downsample by 2 circuit simply discards every other sample, thus reducing the sample rate by 2. The halfband filter (LPF) with a cutoff at one-fourth the input sample rate prevents aliasing. This downsample block is well documented in Simulink. Figure 9.24 also shows the process of downsampling by 2 in the frequency domain. Details about the four steps are listed below. Cascades of halfband downsamplers are often used for downsample ratios greater than 2; see Figure 11.2.

A Narrowband signal centered at 0 Hz, sample rate F_s^*. Frequency domain replicates at multiples of sample rate are shown dotted. There is an unwanted signal (little triangle) between $F_s^*/4$ and $F_s^*/2$ that will alias near the desired signal if we directly downsample.
B Halfband filter response to get rid of unwanted signal and noise.
C Half the sample rate $F_s^*/2 = F_s$ by deleting every other time-domain sample.
D End result is a downsampled narrowband signal matching the original spectrum in trace A.

9.2.9 Example 9: Standard Filters

These are widely used standard filter designs that were invented as analog filters many years ago. MathWorks has provided digital versions of these with similar characteristics to the original analog filters. Figure 9.25 has examples of four of these. The MATLAB code used to calculate the IIR transfer function for each is shown below also. Note that the end of the passband starts when the response is 3 dB down and the start of the stopband starts when the response is 40 dB down. This is only one of many ways to define limits for these bands. To compare filters these limits must be the same. In Figures 9.25 and 9.26, the specification sixth-order

CHAPTER 9. DIGITAL SIGNAL PROCESSING

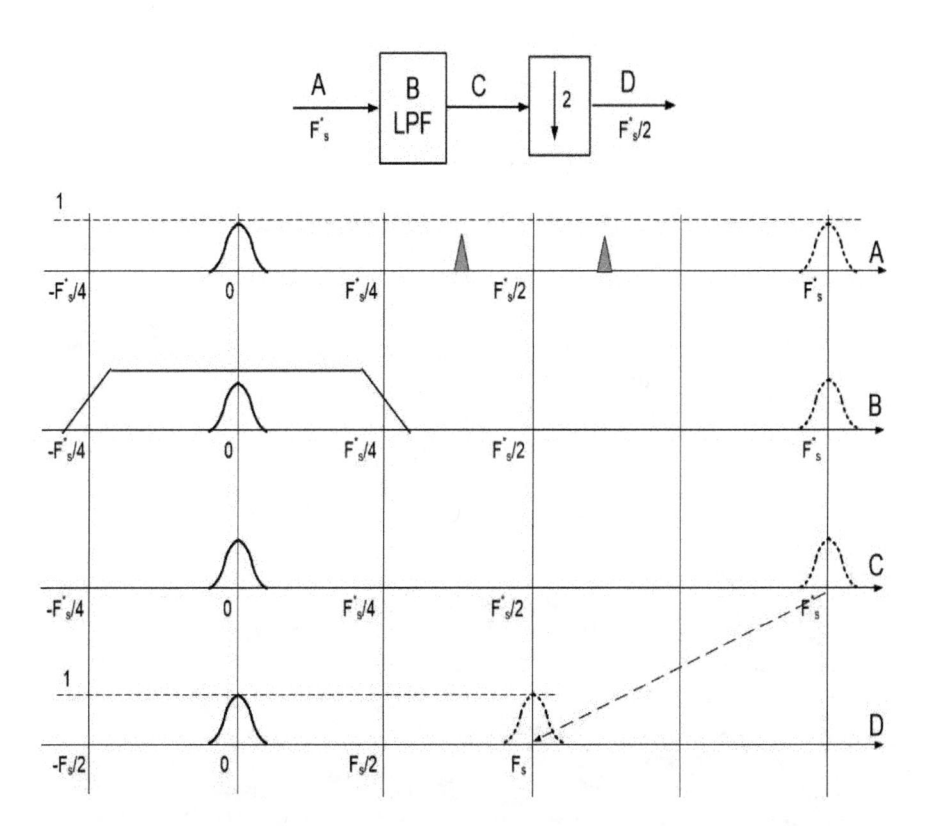

Figure 9.24 Downsample by two circuit and spectrum.

means the filter transfer function has 3 pairs of complex conjugate poles, inside the unit circle.

1. **Butterworth filters** are also know as maximally flat. They have a flat passband and a monotonically decreasing smooth stopband that rolls off at $-6N$ dB per octave, where N is the filter order (number of poles). Butterworth filters also have a smooth phase response.
2. **Chebyshev 1 filters** have a monotonically decreasing smooth stopband that is steeper than Butterworth filters. Passband ripple inherent in this filter can be selected to fit a particular application; see the code below. These filters may be very useful in audio applications were phase response is not critical. For a digital communication application, the group delay (negative derivative of phase response) will have to be checked against the symbol rate.
3. **Chebyshev 2 filters** have a flat passband and short transition band compared to Butterworth and Chebyshev 1. Their stopband rejection is not monotonic and does not continue downward. Like their cousin Chebyshev 1, the phase response causes a group delay spike near the cutoff frequency. This must be considered for digital radio applications.
4. **Elliptic filters** have equal amounts of ripple in both the passband and the stopband. Elliptic filters also have the narrowest possible transition band.

```
Fss = 80e6; % sampling rate
PassBandEdge = 1/4; % Lowpass cutoff around 10e6
StopBandRipple = 40; % Stopband attenuation
PassBandRipple = 5; % Passband variations
N = 6; % Filter order
% Calculate impulse response
(nButter dButter) ...
=butter(N, PassBandEdge,'low');
(nCheby1,dCheby1)...
=cheby1(N,PassBandRipple,PassBandEdge,'low');
(nCheby2,dCheby2)...
=cheby2(N,StopBandRipple,PassBandEdge,'low');
(nElip,dElip)...
=ellip(N,PassBandRipple,StopBandRipple,...
PassBandEdge ,'low');
% Check frequency response of completed design
```

CHAPTER 9. DIGITAL SIGNAL PROCESSING

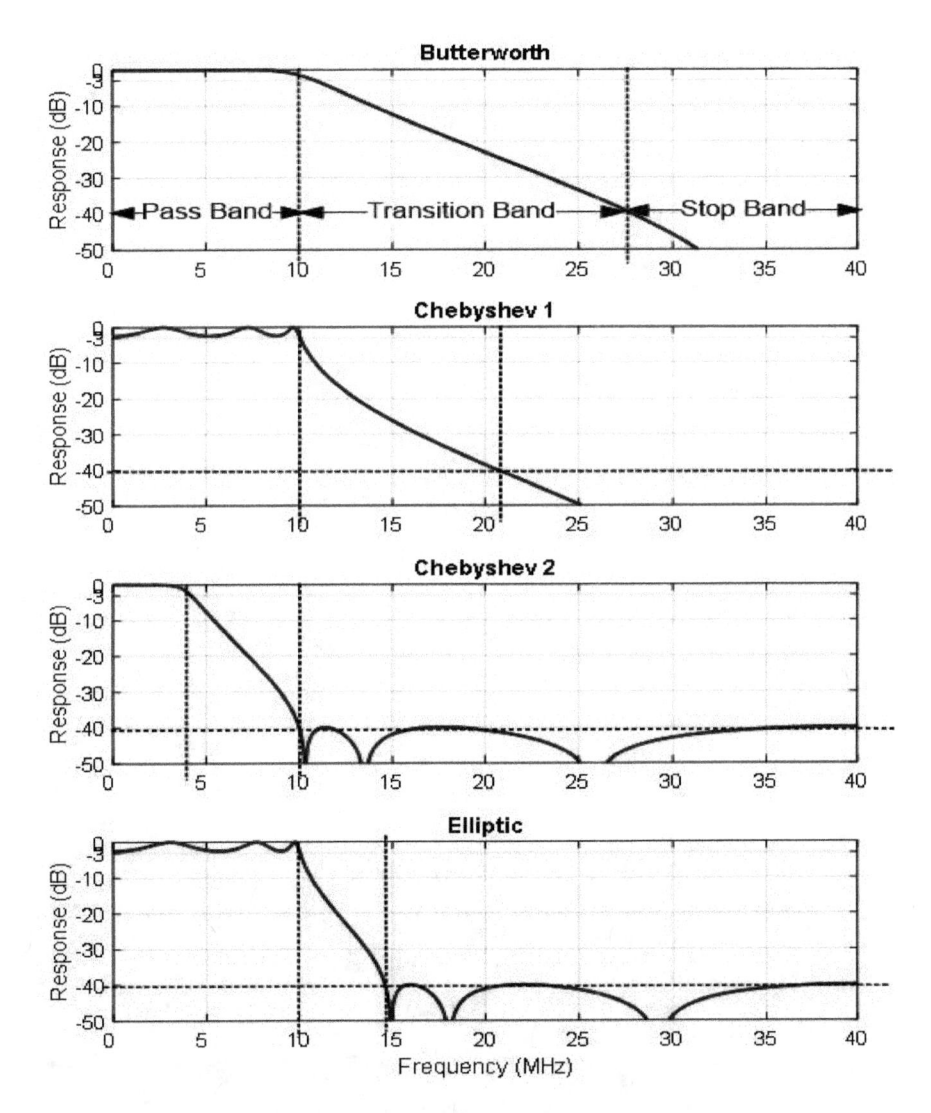

Figure 9.25 Digital filters, sixth-order, F_{sample} = 80 MHz and cutoff = 10 MHz.

```
(hButter,wButter) = freqz(nButter, dButter, 2048);
(hCheby1,wCheby1) = freqz(nCheby1, dCheby1, 2048);
(hCheby2,wCheby2) = freqz(nCheby2, dCheby2, 2048);
(hElip,wElip) = freqz(nElip, dElip, 2048);
%To see phase response, type (for example)
%fvtool(nCheby2, dCheby2)
%To change sample rate click on
% "analysis -> sample frequency"
% group delay is in samples, divide by Fss for seconds
(hgButter,wgButter)...
= grpdelay(nButter, dButter, 2048, Fss);
(hgElip,wgElip) ...
= grpdelay(nElip, dElip, 2048, Fss);
(hgCheby1,wgCheby1) ...
= grpdelay(nCheby1, dCheby1, 2048, Fss);
(hgCheby2,wgCheby2) ...
= grpdelay(nCheby2, dCheby2, 2048, Fss);
```

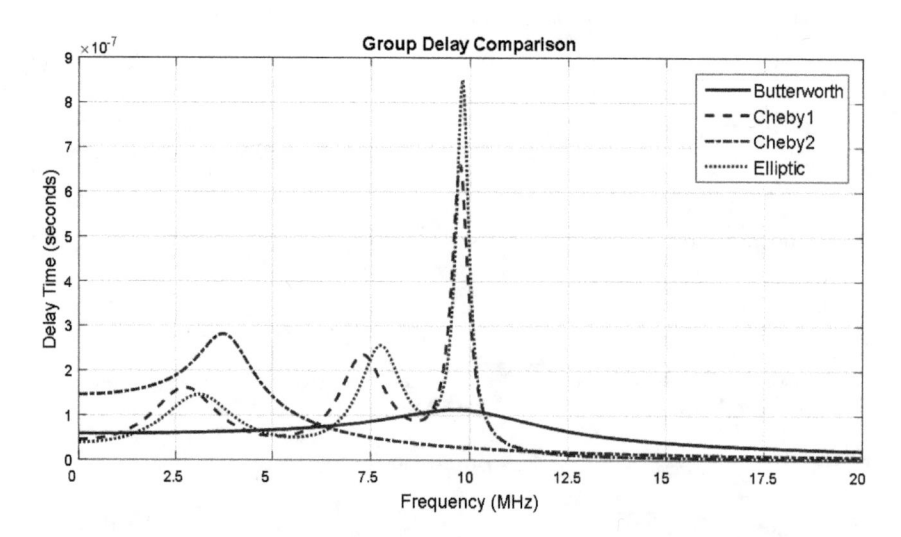

Figure 9.26 Digital filter group delay, sixth order, F_{sample} = 80 MHz and cutoff = 10 MHz.

CHAPTER 9. DIGITAL SIGNAL PROCESSING

Figure 9.26 further illustrates the cost of a sharp transition band for the four filters discussed. Note how the Butterworth has the smoothest filter frequency response and also the least amount of maximum time delay variation (i.e., group delay) at the cutoff frequency. The elliptic filter has the steepest transition band but also the most time delay variation (i.e., group delay) at the cutoff frequency. The Chebyshev 1 filter group delay is similar to the elliptic filter; however, notice how unique the Chebyshev 2 filter group delay is. Group delay should be taken into account, especially when filtering digitally modulated signals such as QAM. Generally, the symbol rate should be not more than 10% of maximum group delay. For example, the Butterworth filter has about 1e-7 seconds maximum group delay. Therefore, the maximum symbol rate should be around $F_{\max} = 0.1 * 10^7 = 1.0 MHz$.

9.2.10 Example 10: Arbitrary Digital Filters

Digital filters with arbitrary frequency response require more complicated calculations than have been presented here so far. MATLAB FilterDesigner is an excellent tool for this task. The FilterDesigner workspace is accessed by typing filterDesigner at the MATLAB prompt (versions of MATLAB prior to 2021a called this fdatool). Figure 9.27 shows a calculation of the halfband filter from the previous example. To dive deeper into digital filter design, [2] is highly recommended.

9.2.11 Example 11: Hilbert Transform

The Hilbert transform is a highpass FIR filter with the ability to affect a +90° phase shift (phase advance) for positive frequencies and a -90° phase shift for negative frequencies. Equation (9.21) is the Hilbert transform frequency response[4] and Figure 9.28 shows the frequency, phase, and impulse response as designed in the MATLAB code next.

To demonstrate the $\pm(\pi/2)$ phase shift property, the Figure 9.28 center plot shows the phase response of a noncausal[5] version of the Hilbert transform. Consider that a pure delay must have a phase shift proportional to frequency (review the beginning of this chapter if necessary), the plotted phase response of the causal Hilbert (i.e., the design we can actually implement) will be dominated by a delay response ramp. The only way to see the $\pm(\pi/2)$ phase shift property is to counteract the delay, making the filter noncausal. The MATLAB code shows how this is done.

---

[4] Note that $\pm j$ is the ±90 complex phase shift operator.
[5] A noncausal FIR filter has coefficients at time indexes less than zero.

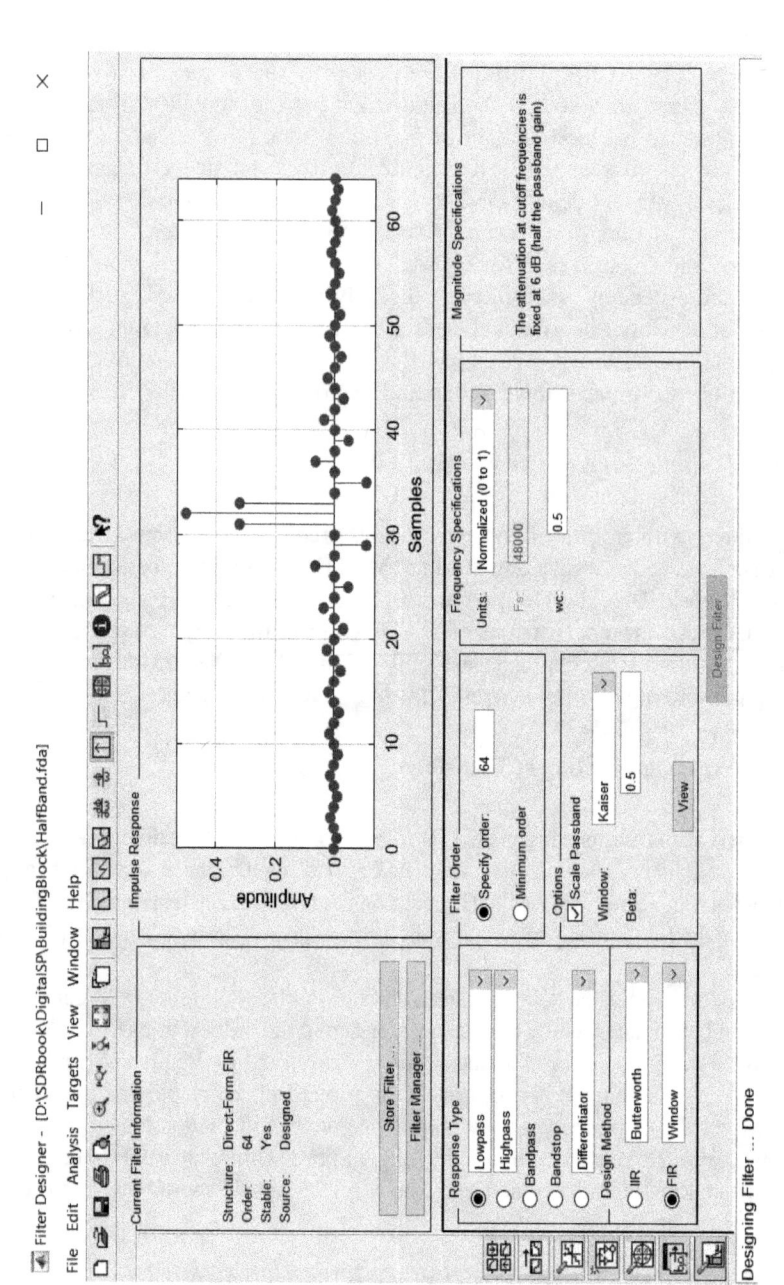

Figure 9.27 Halfband filter using MATLAB filter designer tool.

CHAPTER 9. DIGITAL SIGNAL PROCESSING

$$H(\omega) = \begin{cases} -j & 0 < \omega < \dfrac{\pi}{T_{sample}} \\ +j & -\dfrac{\pi}{T_{sample}} < \omega < 0 \end{cases} \quad (9.21)$$

```
% Design and plot a Hilbert Transformer
%
N = 48; % Must be even number
F = [0.05 0.95]; % Frequency Vector
A = [1 1]; % Amplitude Vector
W = 1; % Weight Vector
L = 4096; % Sampling rate
Nu = firpm(N, F, A, W, 'hilbert');
% Setup a compensating delay to make filter noncausal
% Not for a practical design, for phase plotting only
dly = (zeros(1,N/2) 1);
dlyresp = conj(fftshift(freqz(dly,1,L,'whole')));
resp = dlyresp.*fftshift(freqz(Nu,1,L,'whole'));
Step = 256;
figure(1)
subplot(3,1,1)
pl1 = plot(20*log10(abs(resp)));
set(gca,'XLim',[0 L]);
set(gca,'XTick',(0:Step:L));
set(gca,'XTickLabel',(0:Step:L)-(L/2));
yh1 = ylabel('Freq Resp (dB)');
set(yh1,'FontSize',12);
grid
subplot(3,1,2)
pl2=plot(unwrap(atan2(real(resp),imag(resp)))-(pi/2));
set(gca,'XLim',[0 L]);
set(gca,'XTick',(0:Step:L));
set(gca,'XTickLabel',(0:Step:L)-(L/2));
set(gca,'YLim',[-pi +pi]);
set(gca,'YTick',-pi:(pi/4):+pi);
```

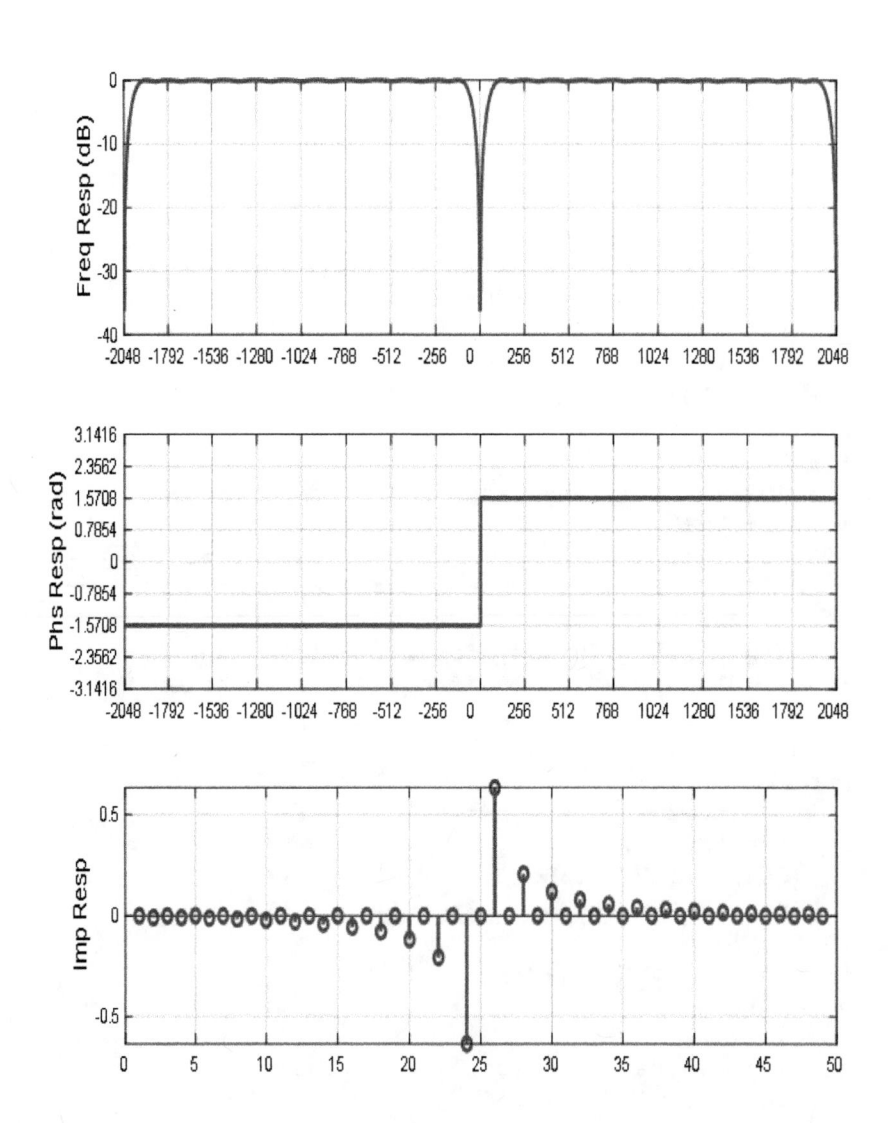

Figure 9.28 Hilbert transform responses.

CHAPTER 9. DIGITAL SIGNAL PROCESSING 367

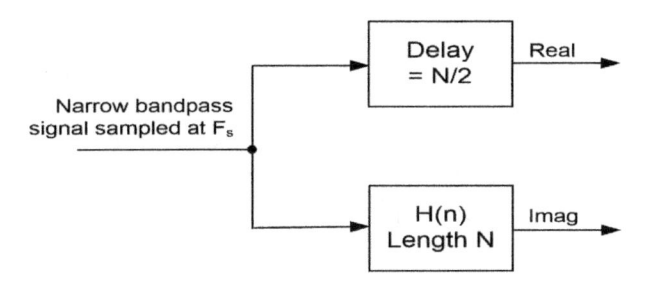

Figure 9.29 A Hilbert transform practical circuit.

```
yh2 = ylabel('Phs Resp (rad)');
set(yh2,'FontSize',12);
grid
subplot(3,1,3)
sh = stem(Nu);
yh3 = ylabel('Imp Resp');
set(yh3,'FontSize',12);
grid
```

As shown in Figure 9.29, the Hilbert transform can be used to convert a real bandpass signal into a complex analytic signal.[6]

9.2.12 Example 12: Allpass Filter

Allpass filters have flat frequency response; no effect on the signal amplitude. Allpass filters are useful for frequency subband coding, such as in MPEG (Motion Picture Experts Group) audio compression; see [5]. However, here we will see how the allpass phase response can be useful as a group delay equalizer.

We start by reviewing group delay. Figure 9.30 is a conceptual definition of group delay. For an analog filter, group delay is measured in seconds. For a digital filter, group delay is often measured in samples (divide samples by sampling rate

---

[6] Complex analytic means complex and differentiable. A complex analytic signal example is: $e^{j\left(\frac{\omega n}{\omega_{sample}}\right)}$ which is infinitely differentiable. Band-limited complex signals are also analytic.

to get seconds). The term group delay refers to the delay imposed on a group of frequencies, such as a band-limited amplitude modulated signal.

The subject of our experiment is shown in Figure 9.31. The IIR Chebyshev 2 bandpass filter we choose has a flat passband and a guaranteed attenuation stopband, in this case -50 dB. The frequency response looks ideal for a narrowband signal centered at about 0.69 normalized frequency (normalized frequency can be multiplied by half the sampling rate to get actual frequency). However, the two vertical lines at the -3 dB points indicate group delay peaks on either side of the passband. There is also group delay variation inside the passband. This can cause severe distortion to a random data bauded signal, especially if the symbol rate is fast enough so that the spectrum takes up most of the passband (recall Section 2.7.1).

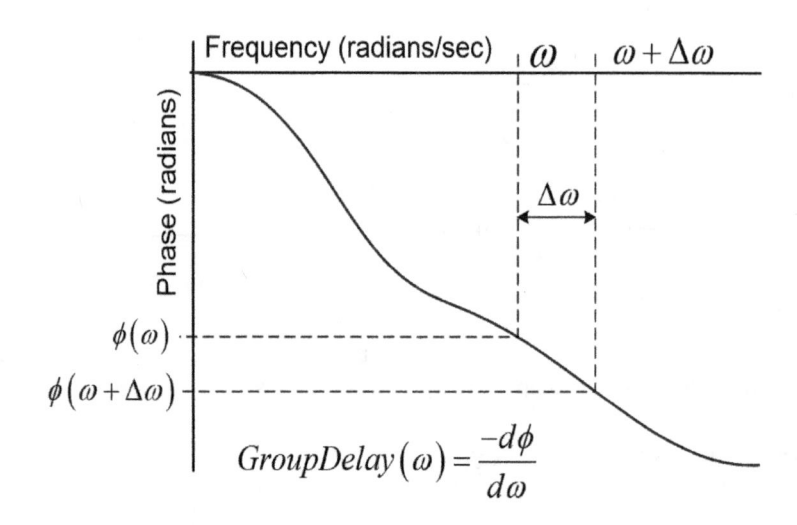

Figure 9.30 Group delay definition.

$$GroupDelay(\omega) = \frac{-d\phi}{d\omega}$$

How can we flatten the passband group delay in the top plot of Figure 9.31? The answer is to cascade an allpass filter with the bandpass filter. This allpass filter is known as a group delay equalizer. In Figure 9.32 we have added a second trace to the lower plot showing equalized flat group delay between the pass band -3 dB points. Notice, in equalizing, the passband group delay is also increased. This is typical and simply adds to the overall receiver delay.

Designing a group delay equalizer allpass filter is not a trivial problem. Fortunately, MATLAB has again come to our rescue. The code below outlines

CHAPTER 9. DIGITAL SIGNAL PROCESSING

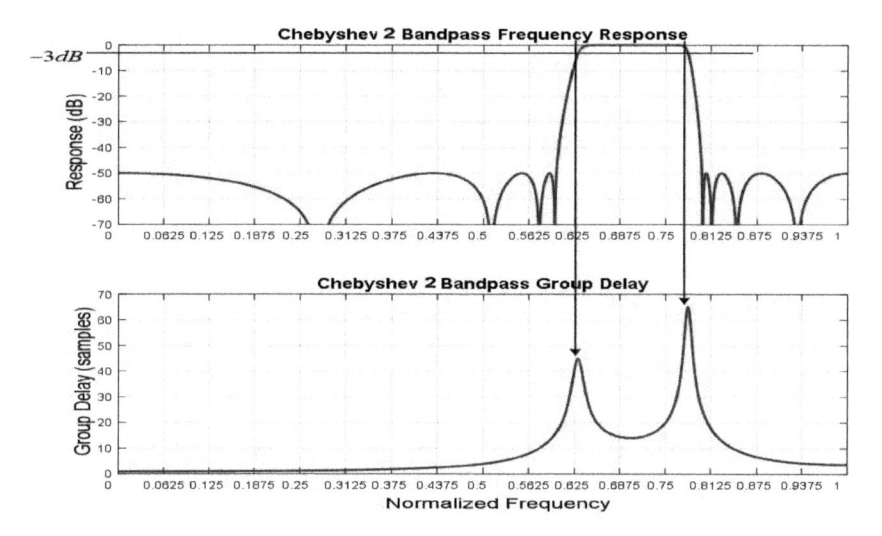

Figure 9.31 ChebyShev 2 bandpass filter response.

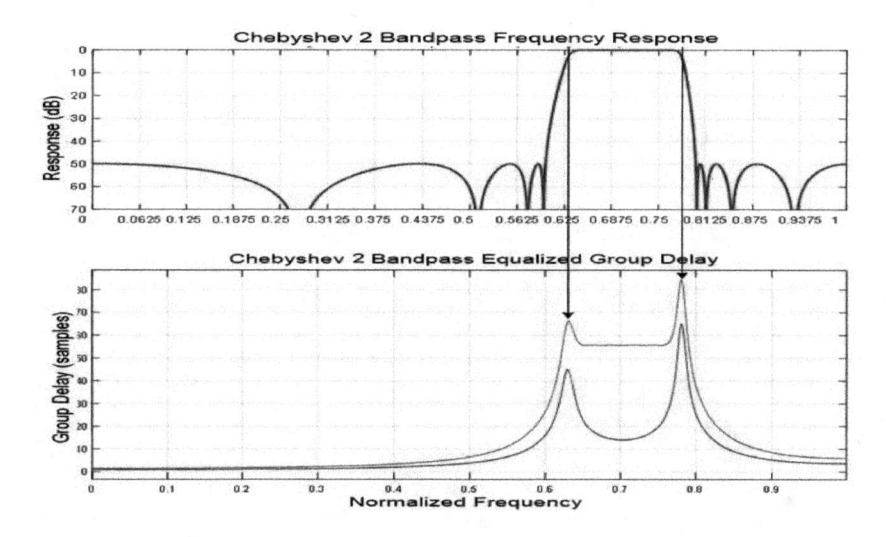

Figure 9.32 ChebyShev 2 bandpass filter, passband group delay equalized.

the entire procedure. Notice in the code that we carefully choose the region of the passband that we want to group delay equalize and we ignore the remaining part of the passband. This limits the constraints and results in better equalization; attempting to group delay equalize the entire sampling band is generally not effective.

```
% Design a bandpass filter and group delay
% equalize passband frequencies
SBripple = 50; % Chebyshev 2 guaranteed attenuation
Order = 8; % Number of poles, zeros
Band = (0.6 0.8); % Normalized passband, Nyquist = 1;

% Calculate zeros, poles, gain result into
% Return as SOS (Second Order Sections)
% for better stability
% SOS have very good fixed point stability also.
% Chebyshev II has a nice flat passband
(z,p,k) = cheby2(Order,SBripple,Band);
(SOSbpf,gain) = zp2sos(z,p,k);
%SOSbpf is a MATLAB system object
%Biquad is the same as second order section:
%2 poles, 2 zeros
%Biquad is short for two quadratic equations:
%neum and denominator
biquadBPF = ...
dsp.BiquadFilter('Structure', 'Direct form I', ...
'SOSMatrix', SOSbpf,'ScaleValues', gain);
scale(biquadBPF,'linf','scalevalueconstraint',...
'none','maxscalevalue',2)
% focus the GDE on passband we need
f = 0.65:0.001:0.755; % Nyquist BW goes to 1, Fs = 2
g = grpdelay(SOSbpf,f,2); % group delay of passband
%f = frequencies corresponding to a
StartingGD = max(g)-g+5;
%a = group delay profile required (in samples)
edges = (f(1) f(end)); %edges= band-edge frequencies
(num,den) =iirgrpdelay(8, f, edges, StartingGD); %AP
% Second order sections and biquad are the same DSP
```

CHAPTER 9. DIGITAL SIGNAL PROCESSING 371

```
% System objects can be used in Simulink directly.
(SOSgde,gg) = tf2sos(num,den);
% Convert to MATLAB system object,
% like C++ object but for MATLAB components
biquadGDE = dsp.BiquadFilter('Structure',...
'Direct form I', 'SOSMatrix', SOSgde,'ScaleValues', gg);
scale(biquadGDE,'linf','scalevalueconstraint',..
'none','maxscalevalue',2)
%Final group delay equalized bandpass filter:
biquadCAS...
= cascade(biquadBPF,biquadGDE);
```

Figure 9.33 shows the unequalized (left) and equalized (right) pole-zero plots for our Chebyshev 2 bandpass. Notice how the added equalizer poles and zeros (right) are opposite each other around the unit circle. This arrangement is meant to result in no effect on the amplitude through the equalizer. However, as shown in Figure 9.32, the phase transfer characteristic (i.e., group delay) can be very finely adjusted.

Summing up, we made an IIR filter act like an FIR filter over a limited passband. Figure 9.34 represents the hardware for the entire equalizer filter. MATLAB fvtool (filter view tool) reports that this IIR SOS design costs 59 multipliers and 48 adders. However, what would an FIR filter with the same shape cost? An FIR equivalent will have flat group delay without any extra effort. Figure 9.35 shows an attempt to match both filter amplitudes. As shown in the figure, the FIR design requires more than twice as many parts. At least in this case, the IIR combination of zeros and poles is more effective and efficient than the all-zero FIR design.

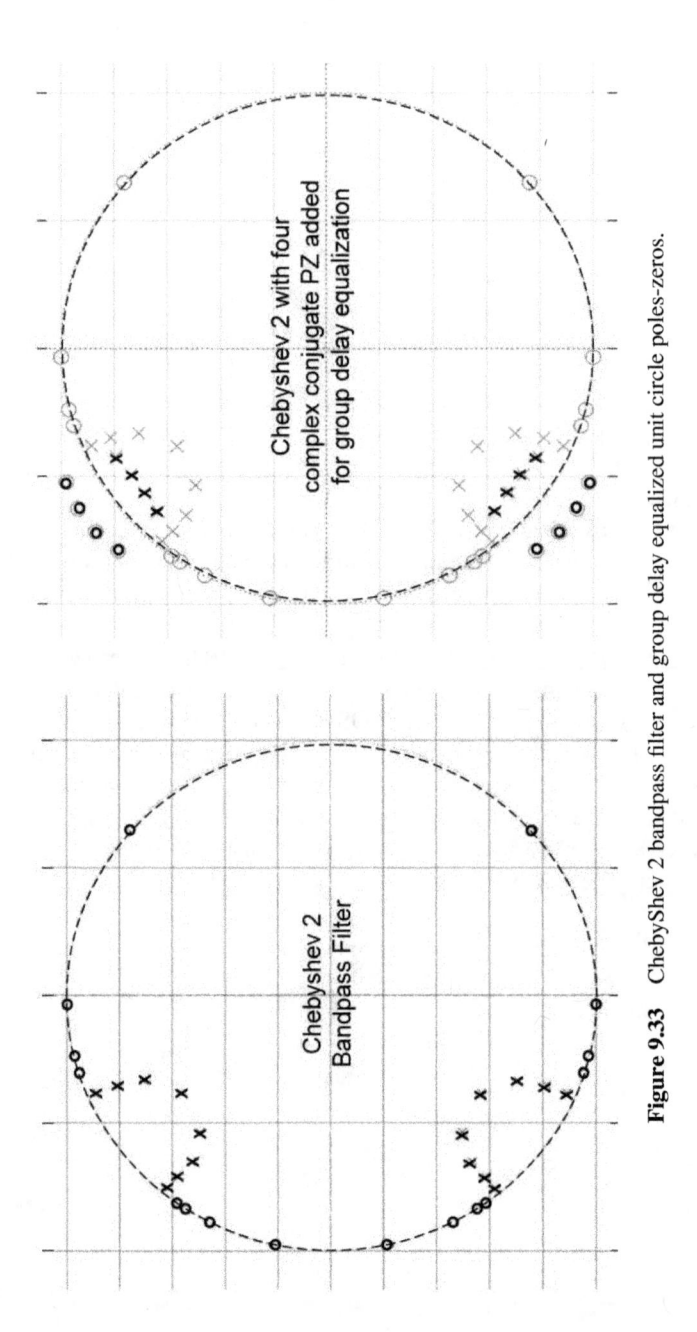

Figure 9.33 ChebyShev 2 bandpass filter and group delay equalized unit circle poles-zeros.

CHAPTER 9. DIGITAL SIGNAL PROCESSING

$$BQ(z) = \frac{a_0 + z^{-1}a_1 + z^{-2}a_2}{1 + z^{-1}b_1 + z^{-2}b_2}$$

FBQ = Chebyshev 2 Bandpass Filter BiQuad
EBQ = Group Delay Equalizer BiQuad

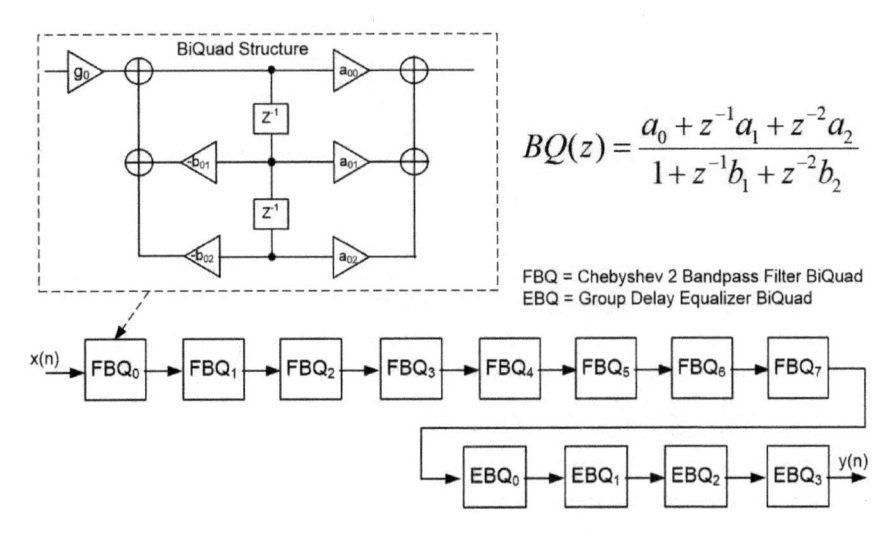

Figure 9.34 ChebyShev 2 bandpass filter and group delay equalizer schematic.

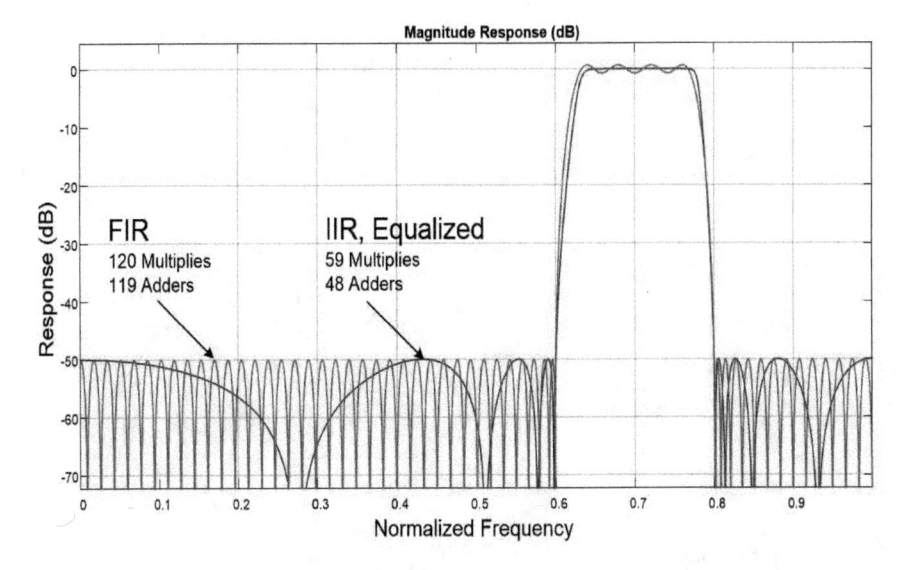

FIR
120 Multiplies
119 Adders

IIR, Equalized
59 Multiplies
48 Adders

Figure 9.35 ChebyShev 2 equalized IIR and FIR equiripple bandpass comparison.

9.3 DIGITAL SIGNAL PROCESSING CIRCUITS

This section describes DSP circuits that are more complicated than just filters.

9.3.1 Example 1: Spectral Inversion of Sampled Signal

As described in Section 7.3.1.1, receiver analog frequency downconversions can sometimes result in a reversed sampled spectrum at the ADC output. This situation may need to be corrected. As shown in Figure 9.36, the ADC output signal can be multiplied by an $F_s/2$ tone to shift the sampling images by $F_s/2$ and reverse the spectrum. To visualize this $F_s/2$ tone, refer to either trace of Figure 9.38, an $F_s/4$ tone. By deleting all the zero samples (every other sample), we halve that tone frequency to $F_s/2$. This leaves the progression $[1, -1, 1, -1, 1, -1 \ldots]$. Thus, ADC output spectrum inversion is achieved by alternating the polarity of every other sample.

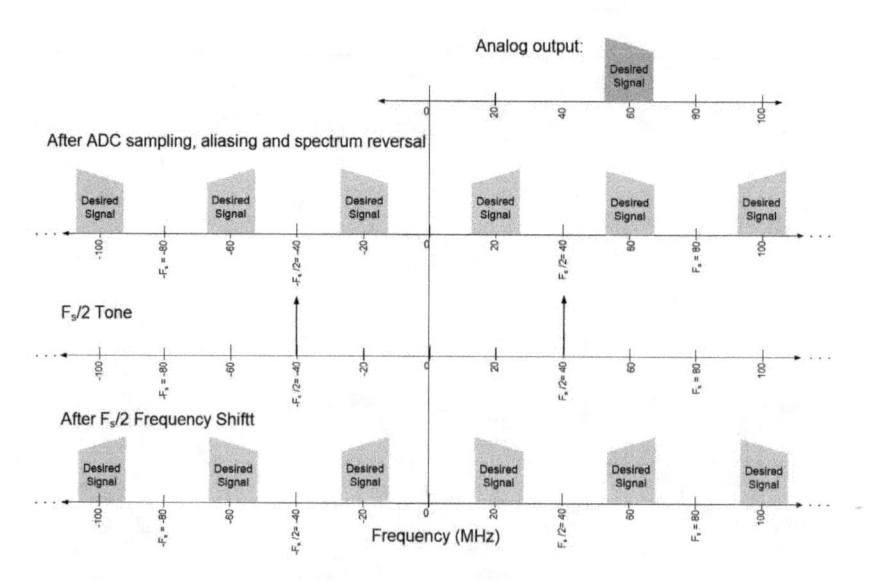

Figure 9.36 Spectrum reversal by frequency shift of $F_s/2$, all signals real.

9.3.2 Example 2: NZIF to Complex Baseband

This section describes a common design technique for converting the NZIF ADC output to filtered and subsampled complex baseband (ZIF ADC outputs are already complex, but just need filtering).

9.3.2.1 Frequency Folding Around $\frac{F_s}{2}$

Our example modem from Section 7.3.1 uses the $F_s = (4/3)F_{IF}$, nonzero IF sampling scheme shown in the frequency domain in Figure 7.29.

Let's discuss an example where the ADC input is centered at $3F_s/4 = 60MHz$ and the sampling frequency is $F_s = 80MHz$. This example is also shown in the first two traces of Figure 9.36. After ADC sampling, the band-limited 20 MHz digital IF signal is an aliased version of the 60 MHz analog IF. In the sampled domain, any signals in the range $[-F_s/2, F_s/2]$ are always repeated at multiples of the sampling rate F_s.

There a few aspects of the NZIF sampling technique to be careful of. First, along with the desired signal, noise in the first Nyquist zone (40 MHz to 80 MHz) is folded back into the zero Nyquist zone (0 to 40 MHz). Second, for sampled signal integrity, the ADC analog input usable bandwidth must extend past the highest sampled frequency at $3F_s/4 = 60MHz$. This was shown in Figure 8.7. Finally, the sampling aperture jitter, t_a, in N1 must be accounted for. At IF passband frequencies, SNR degradation due to increase in sampling clock jitter and/or increasing input frequency is approximated by Equation (9.22). Aperture jitter is due to external ADC sample clock jitter and internal ADC sample and hold uncertainty [6].

$$SNR_{jitter} = -20Log_{10}\left(2\pi t_a f_{sampled}\right) \quad (9.22)$$

Figure 9.37 outlines the frequency-domain operation of an NZIF complex baseband downconverter circuit, similar to the nonzero IF example of Section 7.3.1. All four traces of Figure 9.37 are in the digital domain, after the ADC. The first trace of Figure 9.37 is the same as the last trace of Figure 9.36 and shows the sampled signal at the ADC digital output (after possible spectral inversion). There is indeed an aliased real[7] signal centered at $F_s/4$. Note also that because we are on the output side of the ADC, the complete spectrum centered on zero frequency is replicated at all multiples of the sampling frequency.

---

7 We know this signal is real because it is centered at zero frequency and has symmetric plus and minus frequency images.

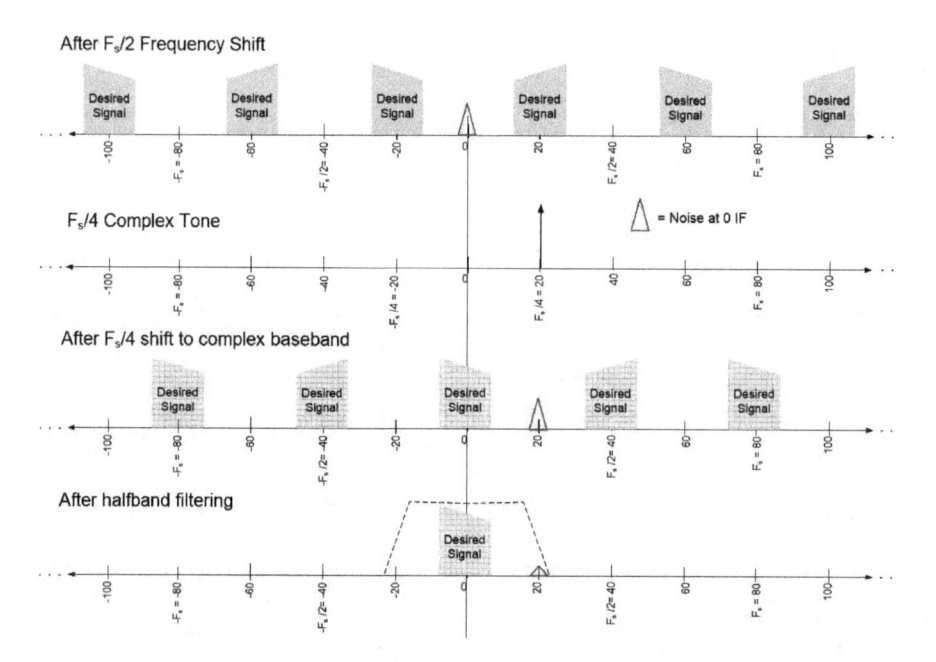

Figure 9.37 Downconversion from $F_S/4$ to complex baseband.

The complex $F_S/4$ LO is in the second trace.[8] The first trace multiplied by the second trace produces the third trace, where the complex LO has shifted the $F_S/4$ signal down to zero frequency. A close look shows that in the third trace the zero centered signal in not symmetric around zero. Thus, it must have I and Q complex signal components. The last trace shows the halfband filtering of this this complex signal. Notice that the noise at zero intermediate frequency is shifted and attenuated. This is an advantage of the NZIF technique discussed here. Many times, the halfband filter output is subsampled by 2 to limit the sample rate into follow on circuits.

9.3.2.2 Complex LO

The $F_S/4$ complex LO in the second trace of Figure 9.37 has very good implementation efficiency because each sample advances the phase exactly 90°:

---

[8] We know this signal is complex because it has only a one-sided frequency image.

CHAPTER 9. DIGITAL SIGNAL PROCESSING

$$u_i(k) = \cos\left(\left(\frac{\pi}{2}\right)k\right), \quad u_q(k) = \sin\left(\left(\frac{\pi}{2}\right)k\right)$$
$$k = 0, 1, 2, 3, 4 \ldots \quad F_{LO} = \frac{F_s}{4} \tag{9.23}$$

$$\left(\frac{F_{LO}\text{cycles}}{\text{second}}\right)\left(\frac{\text{second}}{F_s\text{samples}}\right)\left(\frac{2\pi\text{radians}}{\text{cycle}}\right) = \left(\frac{\pi}{2}\right)\frac{\text{radians}}{\text{sample}} \tag{9.24}$$

Figure 9.38 shows the $F_s/4$ complex LO in the time domain as a set of four $F_s/4$ LO samples (dots) repeating every sample time ($T_s = 1/F_s$). Thus, the complex LO simply becomes a set of two, length four lookup tables that are cycled through repeatedly.

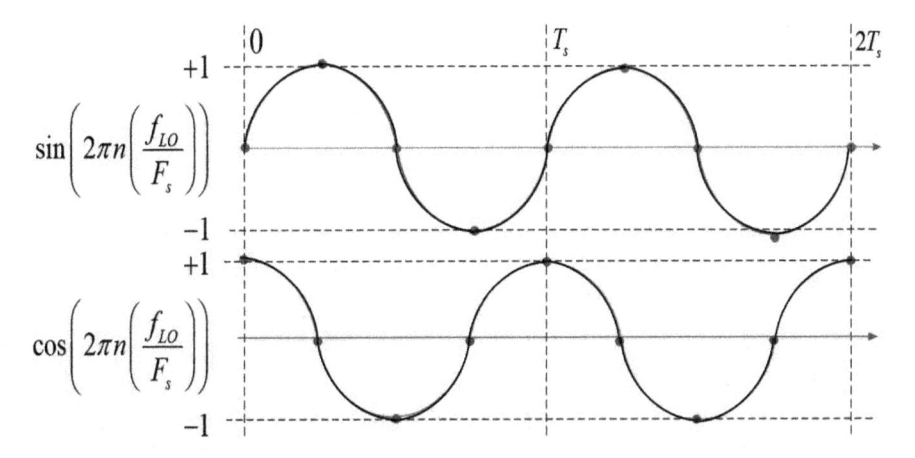

Figure 9.38 Complex local oscillator for $F_s/4$ downconversion.

9.3.2.3 Halfband Lowpass Filter

Recall that Figure 9.36, trace two, is the sampled spectrum after downconversion to $F_s/4$. As discussed, this downconversion is brought about by aliasing. Figure 9.37, third trace, shows how the complex LO is used to further downconvert from $F_s/4$ to 0 Hz.

For further digital processing, we use a halfband lowpass filter to remove other spectral components, such as noise, outside of the frequency range of $\pm F_s/4$. Whatever remains after halfband filtering will again be replicated at multiples of the sampling frequency.

A subsample by 2 on the halfband output will normally cause aliasing of the frequency ranges $[-F_s/2, -F_s/4]$ and $[F_s/4, F_s/2]$ into the ranges $[-F_s/4, 0]$ and $[0, F_s/4]$, respectively. However, a well-designed halfband LPF makes this subsample practically alias-free because it will have very little signal in the frequency ranges $[-F_s/2, -F_s/4]$ and $[F_s/4, F_s/2]$.

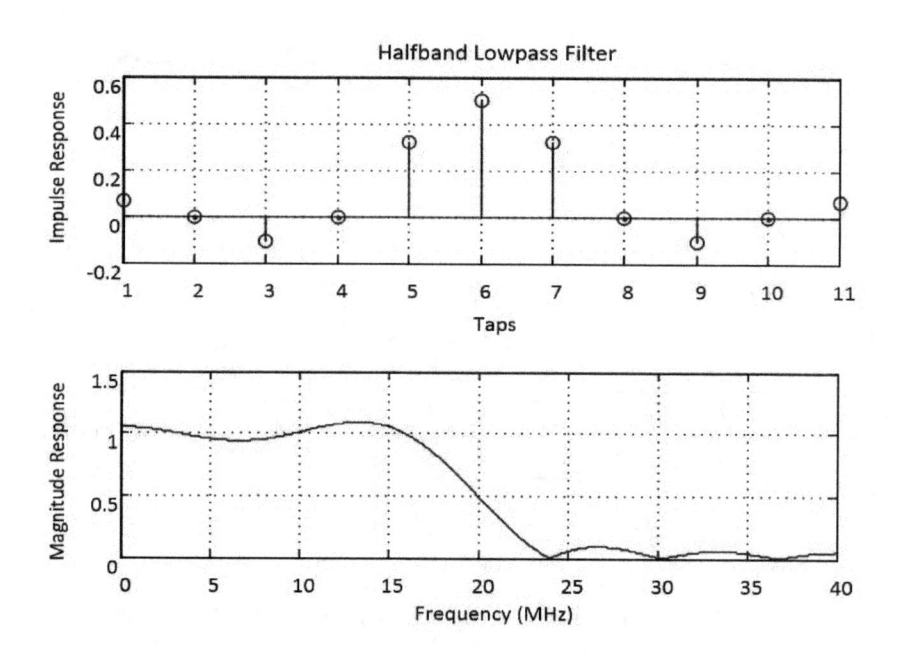

Figure 9.39 Halfband LPF, time and frequency magnitude response.

The halfband LPF in Figure 9.39 has some special properties shown below. There is more information about this important DSP filter in Section 9.2.6

1. The number of taps is odd and even taps are all zero, except for the center.
2. Taps directly adjacent to the center are half the value of the center.
3. At exactly half the sampled bandwidth, the frequency response is 0.5 of the frequency response at 0 Hz; see the center plot of Figure 9.39.

CHAPTER 9. DIGITAL SIGNAL PROCESSING

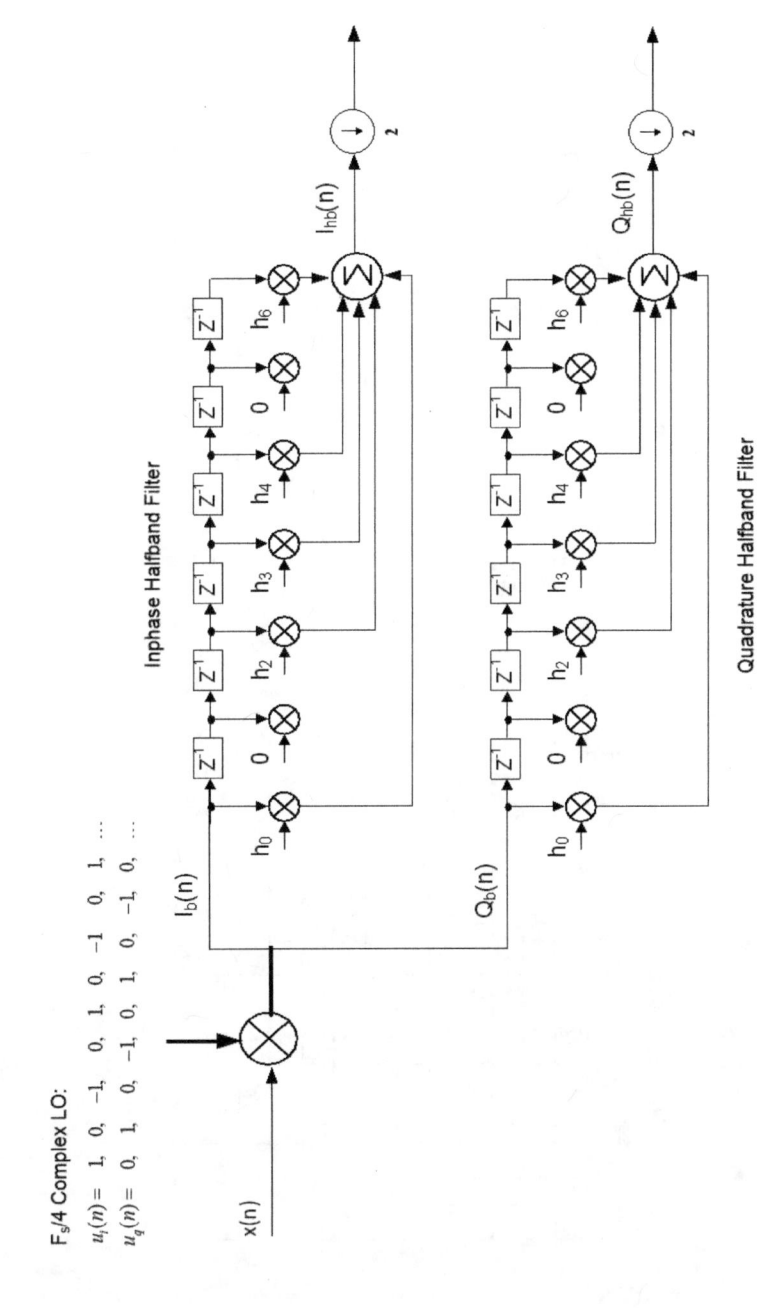

Figure 9.40 Complete downconvert, lowpass filter, and subsample circuit.

Table 9.2 Complex Downconvert, In-Phase Details

Input	LO		Mixer Output		In-Phase Halfband Tap Contents						
$x(n)$	$I(n)$	$Q(n)$	$I_b(n)$	$Q_b(n)$	h_0	$h_1 = 0$	h_2	h_3	h_4	$h_5 = 0$	h_6
$x(0)$	1	0	$x(0)$	0	$x(0)$	0	0	0	0	0	0
$x(1)$	0	1	0	$x(1)$	0	$x(0)$	0	0	0	0	0
$x(2)$	-1	0	$-x(2)$	0	$-x(2)$	0	$x(0)$	0	0	0	0
$x(3)$	0	-1	0	$-x(3)$	0	$-x(2)$	0	$x(0)$	0	0	0
$x(4)$	1	0	$x(4)$	0	$x(4)$	0	$-x(2)$	0	$x(0)$	0	0
$x(5)$	0	1	0	$x(5)$	0	$x(4)$	0	$-x(2)$	0	$x(0)$	0
$x(6)$	-1	0	$-x(6)$	0	$-x(6)$	0	$x(4)$	0	$-x(2)$	0	$x(0)$
$x(7)$	0	-1	0	$-x(7)$	0	$-x(6)$	0	$x(4)$	0	$-x(2)$	0
$x(8)$	1	0	$x(8)$	0	$x(8)$	0	$-x(6)$	0	$x(4)$	0	$-x(2)$
$x(9)$	0	1	0	$x(9)$	0	$x(8)$	0	$-x(6)$	0	$x(4)$	0
$x(10)$	-1	0	$-x(10)$	0	$-x(10)$	0	$x(8)$	0	$-x(6)$	0	$x(4)$
$x(11)$	0	-1	0	$-x(11)$	0	$-x(10)$	0	$x(8)$	0	$-x(6)$	0

CHAPTER 9. DIGITAL SIGNAL PROCESSING

Table 9.3 Complex Downconvert, Quadrature Details

Input	LO		Mixer Output		Quadrature Halfband Tap Contents						
$x(n)$	$I(n)$	$Q(n)$	$I_b(n)$	$Q_b(n)$	h_0	$h_1 = 0$	h_2	h_3	h_4	$h_5 = 0$	h_6
$x(0)$	1	0	$x(0)$	0	0	0	0	0	0	0	0
$x(1)$	0	1	0	$x(1)$	$x(1)$	0	0	0	0	0	0
$x(2)$	-1	0	$-x(2)$	0	0	$x(1)$	0	0	0	0	0
$x(3)$	0	-1	0	$-x(3)$	$-x(3)$	0	$x(1)$	0	0	0	0
$x(4)$	1	0	$x(4)$	0	0	$-x(3)$	0	$x(1)$	0	0	0
$x(5)$	0	1	0	$x(5)$	$x(5)$	0	$-x(3)$	0	$x(1)$	0	0
$x(6)$	-1	0	$-x(6)$	0	0	$x(5)$	0	$-x(3)$	0	$x(1)$	0
$x(7)$	0	-1	0	$-x(7)$	$-x(7)$	0	$x(5)$	0	$-x(3)$	0	$x(1)$
$x(8)$	1	0	$x(8)$	0	0	$-x(7)$	0	$x(5)$	0	$-x(3)$	0
$x(9)$	0	1	0	$x(9)$	$x(9)$	0	$-x(7)$	0	$x(5)$	0	$-x(3)$
$x(10)$	-1	0	$-x(10)$	0	0	$x(9)$	0	$-x(7)$	0	$x(5)$	0
$x(11)$	0	-1	0	$-x(11)$	$-x(11)$	0	$x(9)$	0	$-x(7)$	0	$x(5)$

Table 9.4 In-Phase and Quadrature Outputs after Subsampling by 2

Sample	In-Phase (Table 9.2)	Quadrature (Table 9.3)
0	$x(0)h_0$	0
1	0	$x(1)h_0$
2	$-x(2)h_0 + x(0)h_2$	0
3	$x(0)h_3$	$-x(3)h_0 + x(1)h_2$
4	$x(4)h_0 - x(2)h_2 + x(0)h_4$	$x(1)h_3$
5	$-x(2)h_3$	$x(5)h_0 - x(3)h_2 + x(1)h_4$
6	$-x(6)h_0 + x(4)h_2 - x(2)h_4 + x(0)h_6$	$-x(3)h_3$
7	$x(4)h_3$	$-x(7)h_0 + x(5)h_2 - x(3)h_4 + x(1)h_6$
8	$x(8)h_0 - x(6)h_2 + x(4)h_4 - x(2)h_6$	$x(5)h_3$
9	$-x(6)h_3$	$x(9)h_0 - x(7)h_2 + x(5)h_4 - x(3)h_6$
10	$-x(10)h_0 + x(8)h_2 - x(6)h_4 + x(4)h_6$	$-x(7)h_3$
11	$x(8)h_3$	$-x(11)h_0 + x(9)h_2 - x(7)h_4 + x(5)h_6$

CHAPTER 9. DIGITAL SIGNAL PROCESSING   383

9.3.2.4   Completing the Circuit

Figure 9.40 is a time-domain implementation of an NZIF complex baseband downconverter. We will show how the special properties of the $F_s/4$ complex LO, halfband LPF, and subsample by 2 can work together to obtain significant DSP simplification.

Start with the in-phase signal. Each row of Table 9.2 is a discrete time step at the input sample rate F_s. The first column shows the first 12 input samples, and the second and third columns show the complex $F_s/4$ LO; see Figure 9.38.

The complex mixer output, columns four and five, is simply the multiplication of the input samples by the complex LO; column four for the in-phase product and column five for the quadrature product. For Table 9.2, in-phase mixer output, $I_b(n)$, is input to the in-phase halfband LPF and $Q_b(n)$ is not used in this table. Note the correspondence between $I_b(n)$ and h_0 in columns four and six, respectively.

Now, because we have a subsample by 2 on the output of Figure 9.40, we only retain every other halfband filter output. Table 9.4[9] shows the halfband filter outputs at each input sample step. Due to the halfband output subsample by two, only the gray shaded outputs are used; the unshaded tap values are ignored. Thus, the in-phase subsampled output is just a multiplication of even-numbered input samples by the center tap h_3 along with a sign change for alternate samples. The explicit sample by sample depiction of this process is shown so that the interested reader can convince themselves that this simplification is valid; see the two-tone test of this circuit in Figure 9.43.

Table 9.3 is an explicit sample by sample illustration of quadrature signal processing, using the same simplifications. The right side shows that an FIR filter must be implemented using only even-numbered halfband LPF taps. Thus the quadrature output, shown in the second column of Table 9.4, is slightly more complicated than the in-phase output.

9.3.2.5   Final Filtering

The last trace of Figure 9.37 is repeated in the first trace of Figure 9.41 (for clarity, the spectral replicates centered on multiples of the sampling frequency are omitted). The filter in the second trace makes possible a final subsampling operation, with the condition that the noise aliased into the passband (Nyquist zone zero, N0) must be at an acceptably low level. For example, in Figure 9.41, $F_s/2$ in the second to last line becomes F_s in the last line. Any noise in the range $[F_s/2, F_s]$ in the second to

---

9   The corresponding halfband filter output of Table 9.2 is shown in the first column of Table 9.4.

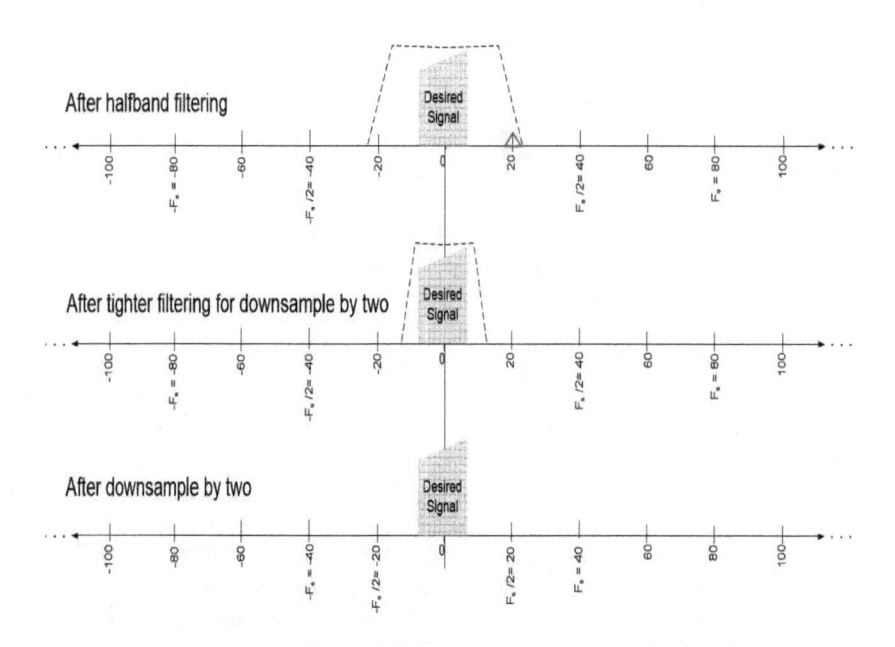

Figure 9.41 Final lowpass filtering of complex baseband signal.

last trace will alias and add to the noise already present in the range $[0, F_s/2]$ in the last trace. The final filter typical bandwidth, second trace in Figure 9.41, is not the same as a matched filter because it has no direct relation to the symbol rate (SR), more on that later. This is also not a half band filter. A popular digital filter circuit for this application is the cascaded integrator comb (CIC) filter, described in detail in Section 9.2.4.

Figure 9.42 is a Simulink simulation of the final design of the $F_s/4$ (ADC sampling rate divided by 4) to complex baseband circuit. Simulink has two uses for this block diagram. First, it can execute a time-driven simulation and second, Simulink can convert the block diagram to VHDL code (or other code, like C) that can flow into the FPGA manufacturer's tools to produce a functioning FPGA circuit for an actual radio modem.

Notice the fixed point attributes (FPA) on every signal, for example, sfix22_En18. Final FPGA implementation will have these same signals with the same FPA. Fixed point signals are represented in a range: too large and they clip, too small and they underflow. Both cases cause distortion. Simulink makes this easy to

CHAPTER 9. DIGITAL SIGNAL PROCESSING

check because signal magnitude is exactly the same in the simulation as in the final hardware. Simulink signal timing also matches in both simulation and hardware. These two properties greatly facilitate design and implementation.

The circuit of Figure 9.42 is often tested by inputting a set of two (or more) tones and analyzing the output. Figure 9.43, upper trace, is the real input with tones offset from the $F_s/4 = 200Hz$ input center frequency. The lower trace is the complex output, centered at 0 Hz but showing the 45 and 25 MHz offset. In the complex output spectrum we are checking several aspects:

1. Input and output tone amplitude the same, i.e., gain = 1.
2. Little or no image frequency, output tones only on one side of 0 Hz.
3. Low DC offset, no bias introduced by circuit.
4. SFDR spurs must be at a specified level below the desired output tones.

9.3.3 Example 3: High Resolution Sinewave Generation

In the early 1970s, when digital signal processing hardware was very limited, the authors of [7] invented a sine wave generator that cut the lookup table length considerably for the same angle resolution. This is illustrated in Figure 9.44. Equation (9.25) shows the basic trigonometry reason why this works.

Let's take a 16 bit example. Resolution is 360/65536 = 0.00546°. For a direct solution, table length is 65536 entries. This might be a problem for a very small low power processor. However, this solution requires four small tables of only 256 entries each.

$$\begin{aligned}\cos(\phi) &= \cos(\phi_{coarse} + \phi_{fine}) \\ &= \cos(\phi_{coarse})\cos(\phi_{fine}) - \sin(\phi_{coarse})\sin(\phi_{fine}) \\ \sin(\phi) &= \sin(\phi_{coarse} + \phi_{fine}) \\ &= \sin(\phi_{coarse})\cos(\phi_{fine}) + \cos(\phi_{coarse})\sin(\phi_{fine})\end{aligned} \quad (9.25)$$

9.3.4 Example 4: Arctangent Approximation

For many years, engineers have been searching for an accurate approximation to the arctangent. The arctangent is the angle represented by a given complex number. The real and imaginary parts are analogous to the adjacent and opposite sides of a triangle. We will look at two approaches, an equation-based technique that uses a set of nonlinear equations to approximate the arctangent and the CORDIC[10] algorithm, a binary search technique.

---
10 CORDIC is an acronym for coordinate rotation digital computer.

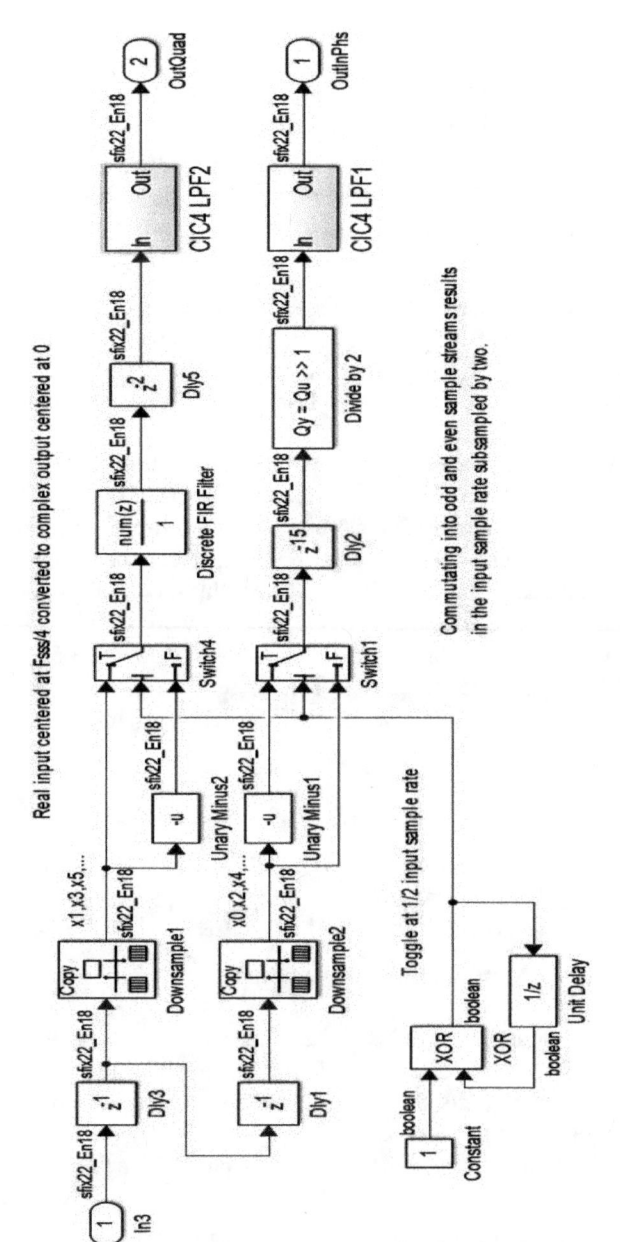

Figure 9.42 Simulink simulation of downconvert and filter circuit, note FPAs.

CHAPTER 9. DIGITAL SIGNAL PROCESSING

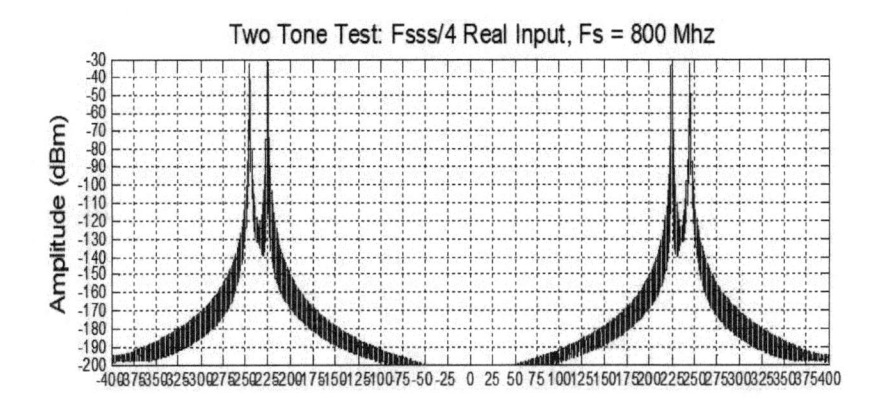

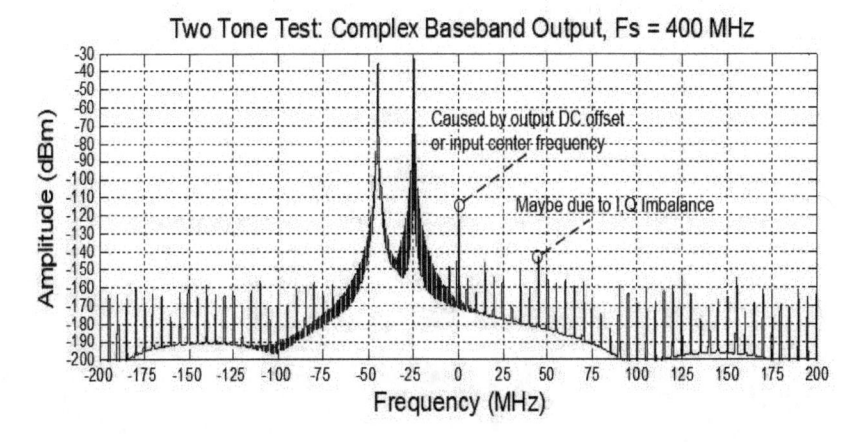

Figure 9.43 Two-tone test of downconvert and filter circuit.

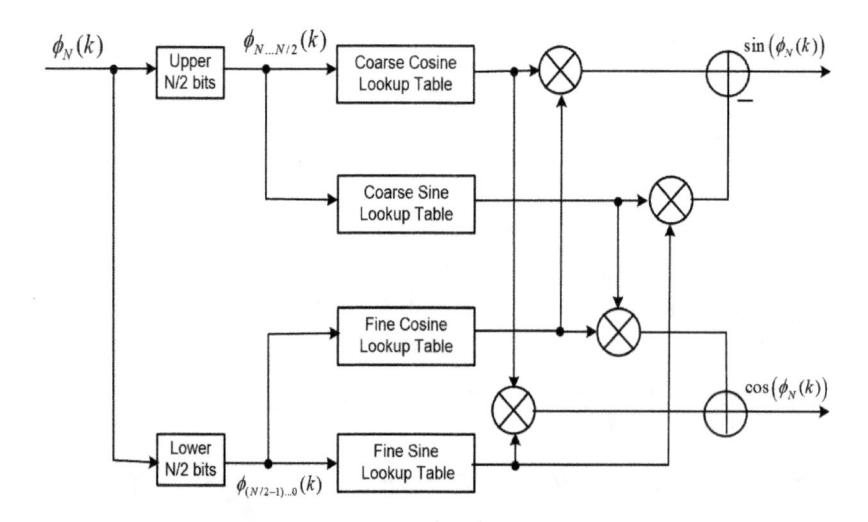

Figure 9.44 A high resolution sinewave generator.

Figure 9.45 shows that both methods are capable of fairly accurate arctangent calculation. The biggest difference is the delay. The equation-based approach takes just a clock cycle or two, whereas the CORDIC approach takes eleven clock cycles in this example. The other difference is component count. The equation-based approach takes five multiplies and two divides, whereas the CORDIC approach takes zero multiplies. The CORDIC, however, has a greater number of low-level parts like logic, shifters, and switches.

9.3.4.1 Arctangent Equation Analysis

The arctangent approximation equations used here are from [2]. Figure 9.46 shows a Simulink implementation of Equation (9.26).

CHAPTER 9. DIGITAL SIGNAL PROCESSING

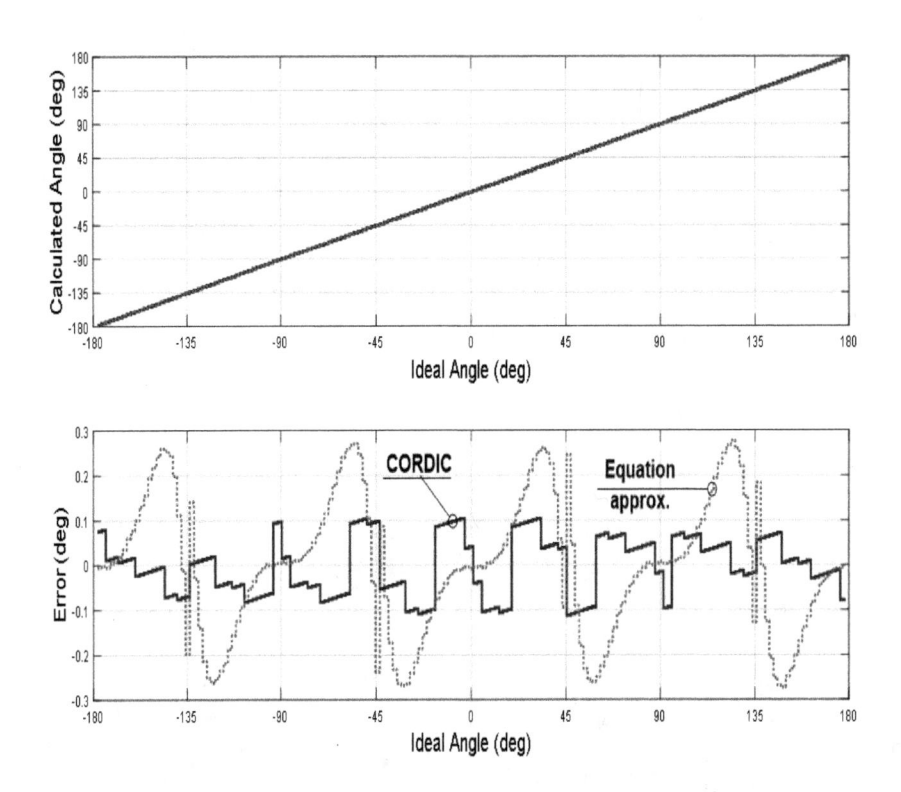

Figure 9.45 Comparison of CORDIC and equation approximation arctangent.

$$\begin{array}{ll} -\dfrac{3\pi}{4} \geq \theta > -\pi & \operatorname{sgn}(Q)\pi + \dfrac{IQ}{I^2 + 0.28125Q^2} \\[6pt] -\dfrac{\pi}{2} \geq \theta > -\dfrac{3\pi}{4} & -\dfrac{\pi}{2} - \dfrac{IQ}{Q^2 + 0.28125I^2} \\[6pt] -\dfrac{\pi}{4} \geq \theta > -\dfrac{\pi}{2} & -\dfrac{\pi}{2} - \dfrac{IQ}{Q^2 + 0.28125I^2} \\[6pt] 0 \geq \theta > -\dfrac{\pi}{4} & \dfrac{IQ}{I^2 + 0.28125Q^2} \\[6pt] 0 < \theta \leq \dfrac{\pi}{4} & \dfrac{IQ}{I^2 + 0.28125Q^2} \\[6pt] \dfrac{\pi}{4} < \theta \leq \dfrac{\pi}{2} & \dfrac{\pi}{2} - \dfrac{IQ}{Q^2 + 0.28125I^2} \\[6pt] \dfrac{\pi}{2} < \theta \leq \dfrac{3\pi}{4} & \dfrac{\pi}{2} - \dfrac{IQ}{Q^2 + 0.28125I^2} \\[6pt] \dfrac{3\pi}{4} < \theta \leq \pi & \operatorname{sgn}(Q)\pi + \dfrac{IQ}{I^2 + 0.28125Q^2} \end{array}$$

(9.26)

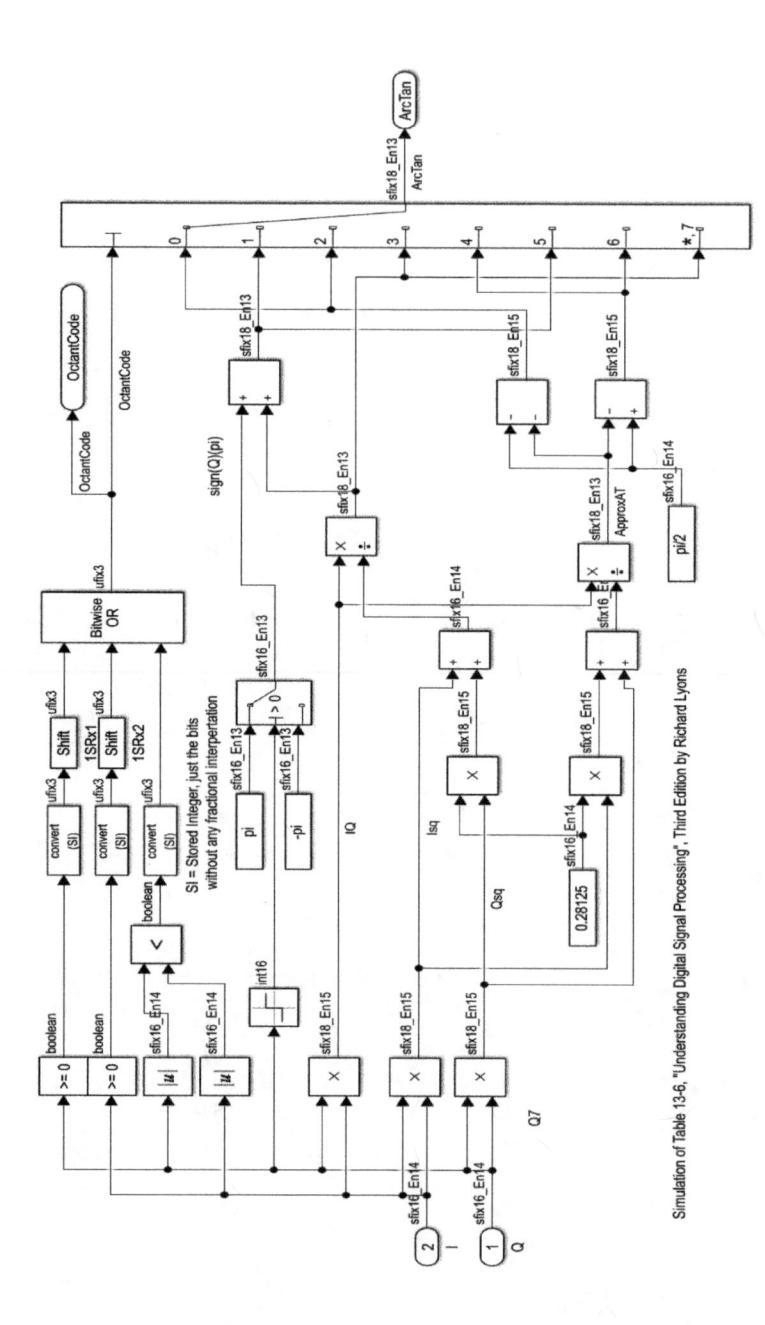

Figure 9.46 Arctangent equation approximation; Simulink model based on the description in [2].

CHAPTER 9. DIGITAL SIGNAL PROCESSING 391

9.3.4.2 Arctangent CORDIC Analysis

The top level construction of a CORDIC algorithm is shown in Figure 9.47. A series of 10 stages each rotate plus or minus (i.e., CCW or CW) by a smaller and smaller angle to converge on the final arctanget approximation. Angular rotations of InX, InY by θ are shown below in Equation (9.27). Each stage θ, see Table 9.6, is represented by $2^{-K} = tan\theta$. The setup stage zero is in Figure 9.48 and the other 10 stages are in Figure 9.49.

$$OutX = InX \cos(\theta) - InY \sin(\theta)$$
$$\frac{OutX}{\cos(\theta)} = InX - InY\frac{\sin(\theta)}{\cos(\theta)}$$
$$OutX = \cos(\theta)(InX - InY \tan(\theta))$$

$$OutY = InY \cos(\theta) + InX \sin(\theta)$$
$$\frac{OutY}{\cos(\theta)} = InY + InX\frac{\sin(\theta)}{\cos(\theta)}$$
$$OutY = \cos(\theta)(InY + InX \tan(\theta))$$

(9.27)

Because for the CORDIC arctangent, we are only interested in the series of angles applied, we can give up the vector magnitude accuracy and simplify Equation (9.27) to end up with with Equation (9.28).

$$OutX = (InX - InY \tan(\theta))$$
$$OutY = (InY + InX \tan(\theta))$$

(9.28)

Each stage has only one angle it can rotate by, positive or negative. These are shown in the third column of Table 9.6. These angles get smaller as the stage numbers get higher, but why the irregular fractional values? It turns out there is a series of angles where $tan(\theta) = \pm 2^{-K}$. Now we can simplify the stage rotation to Equation (9.29). Here K is the stage number and $A = +1$ if $Ain>0$ and $A = -1$ if $Ain<0$.

$$OutX = (InX - InY(A * 2^{-K}))$$
$$OutY = (InY + InX(A * 2^{-K}))$$

(9.29)

Table 9.6 is an example of estimating the arctangent of $0.8242+j0.5664$. Note that each stage can only apply its assigned angle. The sign of this angle depends on the polarity of InY. As shown in Figure 9.50, alternating between quadrants 0 and 3 is the source of the convergence to the final output of $1+j0$. To get the arctangent, the applied angles are simply summed up; the sum of the series of angles that rotated

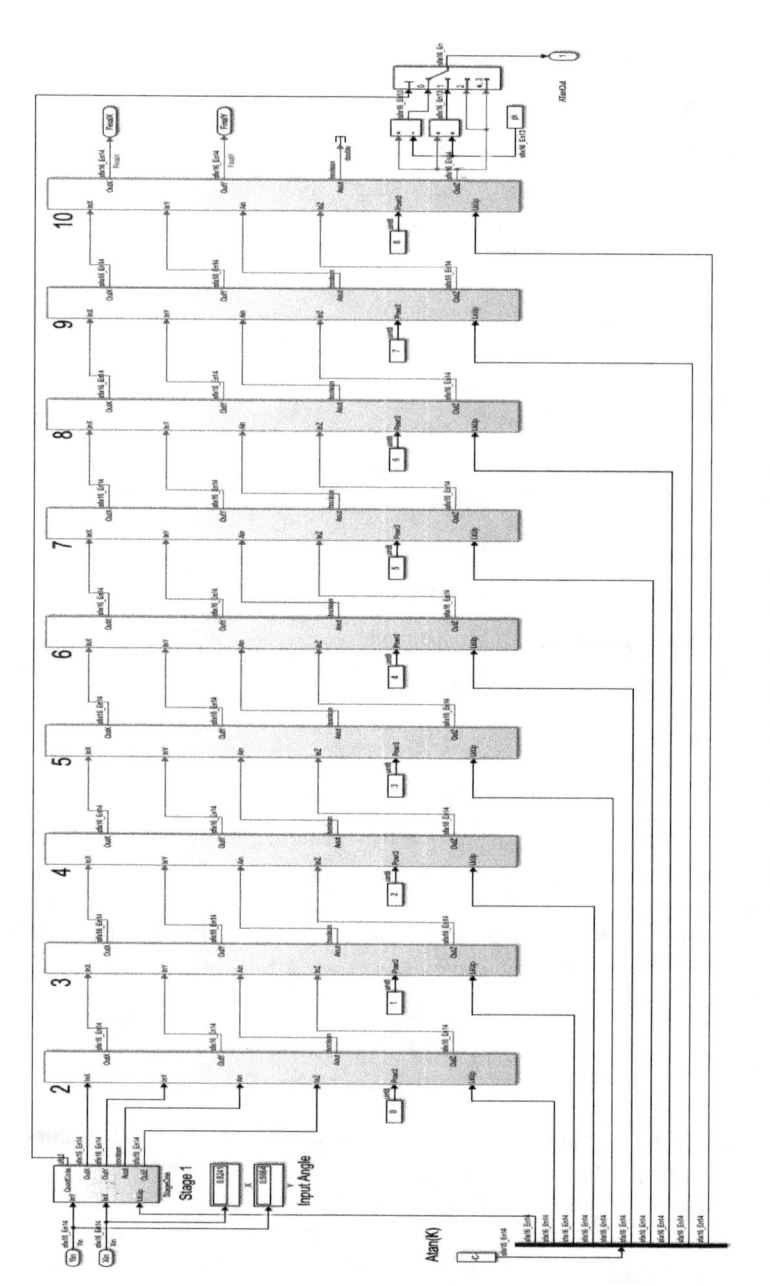

Figure 9.47 CORDIC top level construction.

CHAPTER 9. DIGITAL SIGNAL PROCESSING 393

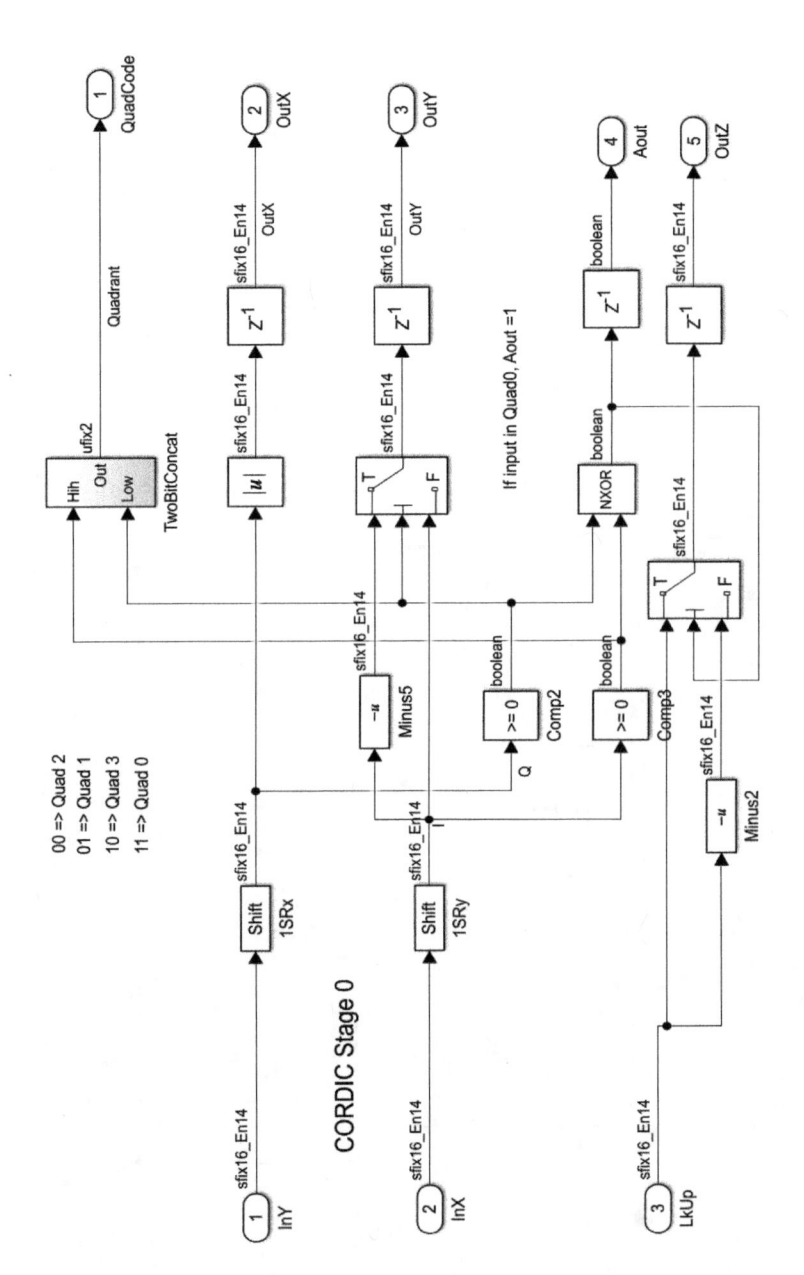

Figure 9.48 CORDIC stage 0 details.

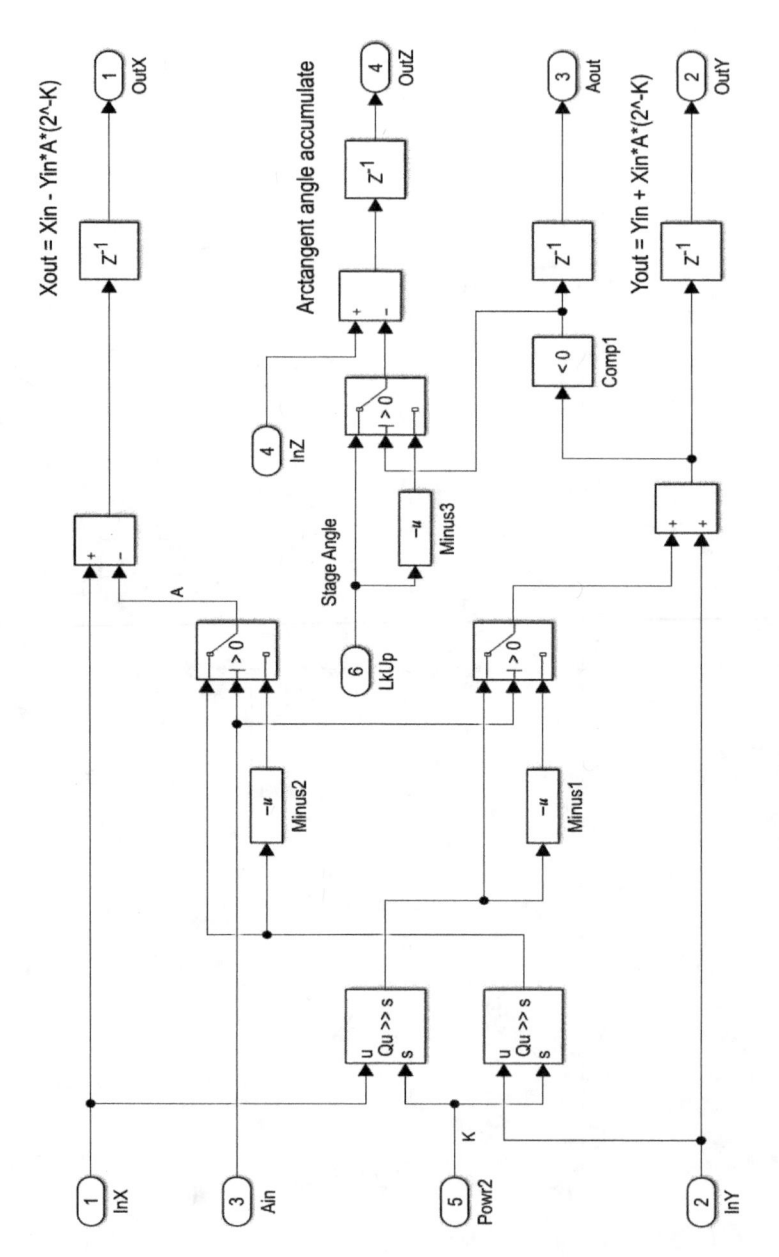

Figure 9.49 CORDIC stage 1 to 10 details.

CHAPTER 9. DIGITAL SIGNAL PROCESSING

Table 9.5 CORDIC Stage Angles

Stage K	θ	$\tan\theta = 2^{-K}$
0		
1	45°	1.0
2	26.565°	0.5
3	14.036°	0.25
4	7.125°	0.125
5	3.576°	0.0625
6	1.7899°	0.0325
7	0.8952°	0.015625
8	0.4476°	0.007813
9	0.1119°	0.00195312
10	0.0559°	0.00097656

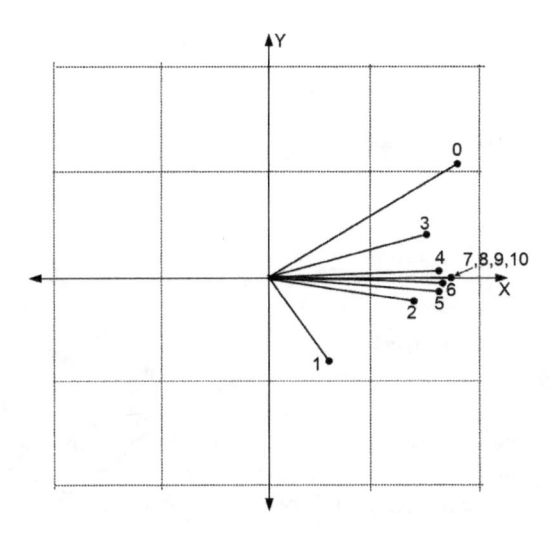

Figure 9.50 Series of CORDIC stage 1 to 10 output angles, converging to 1+j0.

the input to the horizontal axis ends up as the arctangent. See the center right side of Figure 9.49. The CORDIC is accurate between -90° and +90°. To obtain the full -180° to +180°, the components on the lower right side of Figure 9.47 will add or subtract 180° for quadrants 3 and 1, respectively.

Note that the choice of 10 stages was somewhat arbitrary. It is easy to trade off complexity for precision with CORDIC. Fewer stages can be used to reduce parts count; however, the precision will suffer because the last stage angle θ will be larger. Adding more stages increases precision at the cost of additional parts.

Table 9.6 CORDIC Example, Input = X + jY = 0.8242 + j0.5664, Arctangent = 34.5°

Stage K	InX	InY	sign(InY)	Ain	Change	OutZ
0	0.8242	0.5664				
1	0.2832	-0.412	-	0	45°	45°
2	0.6953	-0.128	-	0	-26.559°	18.44°
3	0.7597	0.2188	+	1	14.0299°	32.47°
4	0.8144	0.0288	+	1	7.130°	39.60°
5	0.818	-0.0728	-	0	-3.579°	36.02°
6	0.8226	-0.0218	-	0	-1.790°	34.23°
7	0.8233	0.0039	+	1	0.890°	35.12°
8	0.8234	-0.0089	-	0	-0.439°	34.68°
9	0.8235	-0.0025	-	0	-0.2299°	34.45°
10	0.8235	0.00067	+	1	0.1099°	34.56°

9.4 QUANTIZING NOISE

Unlike analog filters constructed of discrete electronic components and integrated circuits, digital filters do not have ground current noise, semiconductor noise, or other sources. Instead digital filters have three primary noise sources:

1. Quantization noise due to sampling;
2. Multiplier roundoff noise;
3. Overflow.

CHAPTER 9. DIGITAL SIGNAL PROCESSING

9.4.1 Quantization Noise Due to Sampling

Sampling quantization noise starts at the ADC. Analog signals on the ADC input can take on an infinite number of voltage levels. Digital signals on the ADC output are confined to 2^B different levels where B is the number of ADC quantizing bits. If the ADC quantizes the input by rounding to the nearest quantization step, an error is generated between the actual signal value and the chosen step, as shown in Figure 9.51.

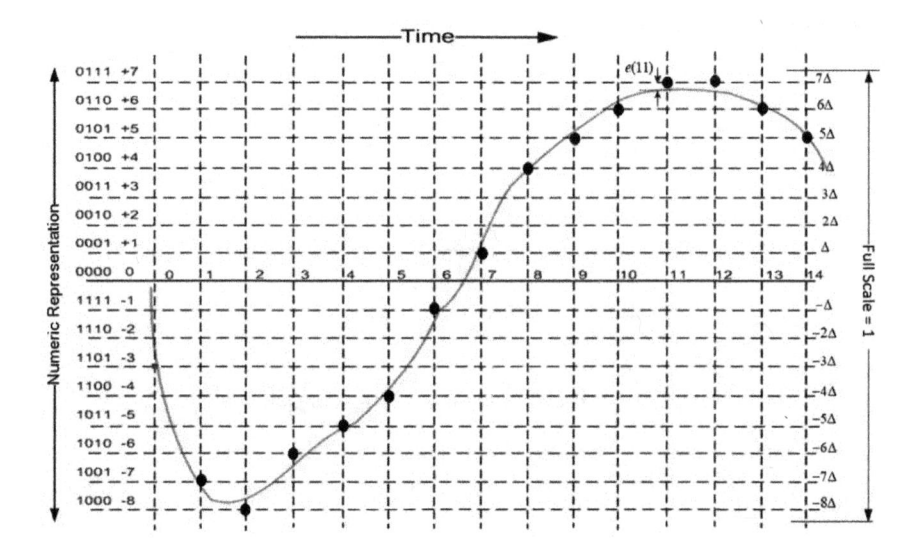

Figure 9.51 Example ADC quantizing grid, $B = 4$.

For Δ equal to the quantizing step voltage (see Figure 9.51), where we have $\Delta = 2^{-B}$, assuming full-scale voltage equals 1.0. For sample index n, the quantizing error is thus $-\Delta/2 \leq e(n) < +\Delta/2$. Samples of this error are independent, uniformly distributed with zero mean, as shown in Figure 9.52. Assuming there is no ADC input clipping, the average power or error variance is given by Equation (9.30), for $\bar{e} = 0$. Note that as the number of quantization steps B increases, this average error power decreases. Another important consideration is the frequency spectrum of the quantizing noise. This is usually considered flat over the sampling bandwidth, with a height of $\left(\dfrac{\Delta^2}{12}\right) \bigg/ \left(\dfrac{F_{sample}}{2}\right)$.

The validity of quantization noise variance in Equation (9.30) is highly dependent on the sampled analog signal. Obviously, a constant, zero frequency input signal will have a constant quantization error. Also, for example, a sampled sine wave frequency that happens to be an integer factor of the sample rate may not have a uniformly distributed error density. For this variance to be valid, we assume a signal that randomly visits most of the quantization levels. The mean value of noise in Equation (9.30) is $\bar{e}$, generally zero. A very thorough description of quantizing error is in [8].

$$E[\sigma^2] = \int_{-\Delta/2}^{\Delta/2} (e-\bar{e})^2 f(e)de = \frac{\Delta^2}{12} = \text{mean squared noise power} \quad (9.30)$$

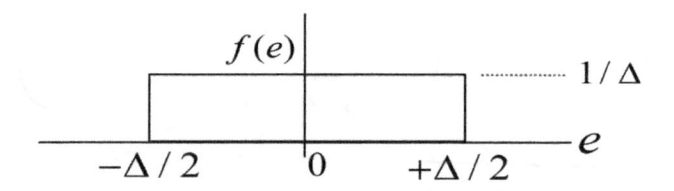

Figure 9.52 Uniform quantizing error probability density.

9.4.2 Multiplier Roundoff Noise

DSP multipliers generate noise because their outputs have to be truncated or rounded, causing a loss in precision equivalent to noise. For example, using Simulink notation, we multiply two numbers with FPA (fixed point attribute) of $sfix22\_en18$. The full precision product will be $sfix44\_en36$, a 44-bit signed number with 36 fractional bits; double the fractional bits are needed to represent the ideal precision. If that product must be truncated back to $sfix22\_en18$, then 18 bits of fraction precision have been lost. That is equivalent to introducing a new quantization error, with associated noise, at the multiplier output.

An effective way to minimize this error is shown in Figure 9.53. Here the three multiplier outputs are added together at full precision plus two guard bits to prevent overflow. The full precision adder output may be truncated later if needed. Full precision adders are easy to implement in FPGAs and sometimes come built into DSP multiply-accumulate hardware. Another tip is to use coefficients that can

CHAPTER 9. DIGITAL SIGNAL PROCESSING   399

be represented simply, for example, 0.125 is just a right shift by three. Filter design software (such as MATLAB "fdatool" or equivalent) may calculate coefficients that are close to these simple coefficients. Change the coefficients and rerun the filter simulation to see if the filter response is still useful.

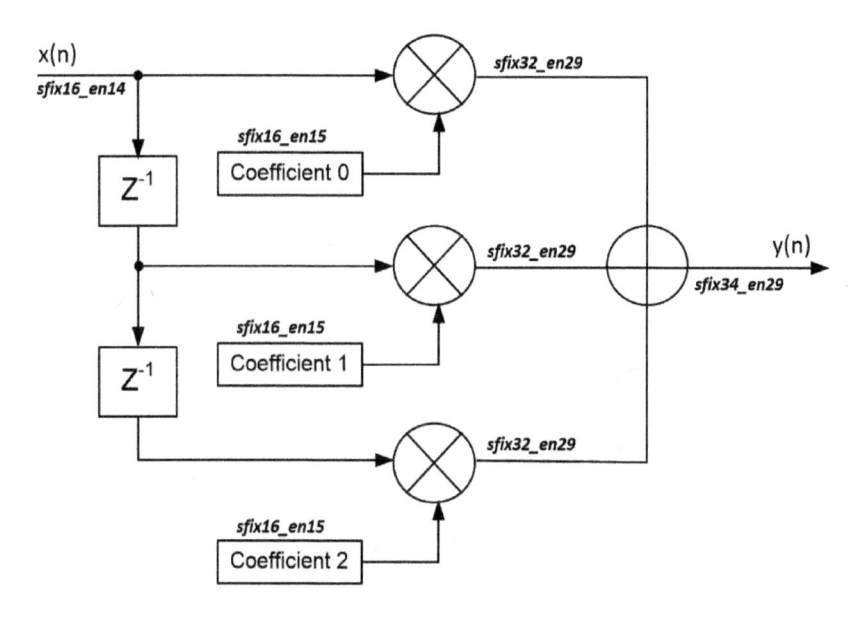

Figure 9.53 Simple FIR filter showing multiplier fixed point outputs.

9.4.3 Overflow

Overflow at the ADC input is called clipping, or saturation, and is minimized by careful AGC design. Overflow can also occur at intermediate stages of filtering. The adder output in Figure 9.53 can be prone to overflow if the output FPA is returned to $sfix32\_en29$. A common location for overflow is at adder or multiplier outputs that attempt to return the FPA to that of the inputs.

For two's complement, arithmetic overflow results in wraparound. Examine Figure 9.51 carefully and notice how, on the left side, a signal exceeding +7 (0111) wraps around to -8 (1000). This can be avoided by telling the FPGA synthesis tools to add extra circuits to detect wraparound and convert it into saturation.

Finally, completely eliminating these problems may not be necessary. For example, a high peak to average signal, such as voice, may be allowed to occasionally saturate a few samples with no noticeable effect on quality of reproduction. Perhaps the best example of this is a Gaussian amplitude distribution. Here the tails of the bell curve go to infinity so they are impossible to represent in a fixed point system. In practice, a clipping level can be specified in terms of multiples of the standard deviation, for example 3σ.

9.5 QUESTIONS FOR DISCUSSION

1. Figure 9.54 shows the zeros in an M point averaging circuit. What is the value of M?

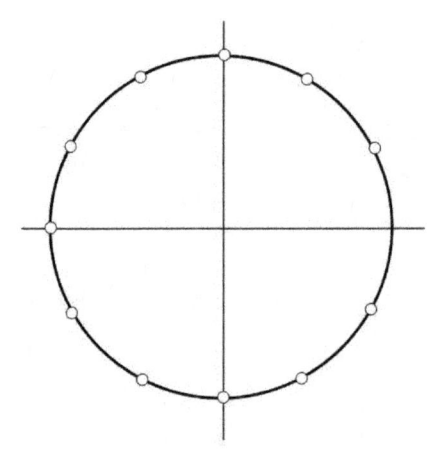

Figure 9.54 Digital averaging zero plot.

2. The halfband lowpass filter in Section 9.2.6 can easily be converted to a halfband highpass filter, shown below in Figure 9.55. Outline a procedure for this.

3. Consider the downsampling procedure in Figure 9.24. Let's say the trace A signal centered around $F_s^*/2$ is missing. In this case, it appears that the halfband lowpass

CHAPTER 9. DIGITAL SIGNAL PROCESSING

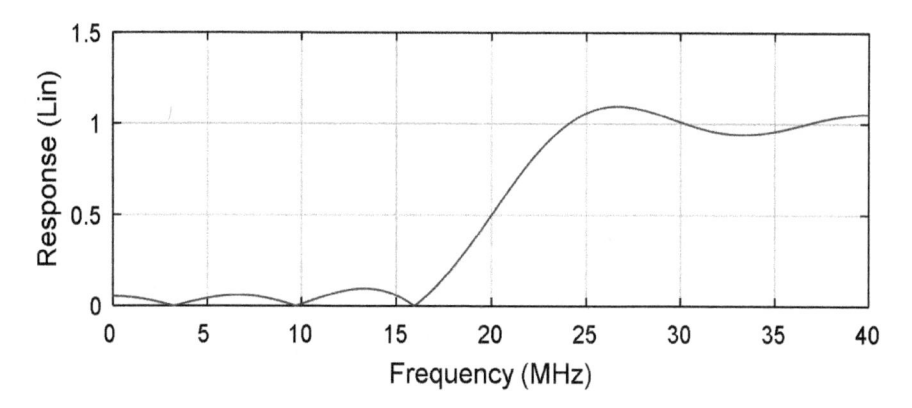

Figure 9.55 Halfband highpass filter.

filter is not needed. What might be a reason for using it anyway?

4. Is the following filter stable: $H(z) = \frac{1-z^{-2}}{1-1.43z^{-1}+1.02z^{-2}}$?

5. Compare the effect on a bauded communications signal of frequency-dependent filter group delay (e.g., Figure 9.26) and channel-dependent intersymbol interference.

6. Consider the convolution $x_{conv}(k) = \sum_{n=-\infty}^{\infty} p(k-n)c(n)$. $p(n)$ is a four tap FIR filter with constant coefficients (i.e., linear time-invariant) and $c(n)$ is a received sampled signal. Explain, in words, how the convolution equation works and what it is used for. Consider the correlation $x_{corr}(k) = \sum_{n=-\infty}^{\infty} p(k+n)c(n)$. $p(n)$ is a sampled reference signal and $c(n)$ is a received sampled signal. The correlation peaks when the received signal is most similar to the reference signal. Explain, in words, how the correlation equation works and what it is used for.

7. When do the convolution and correlation equations have the same result?

REFERENCES

[1] L. B. Jackson. *Digital Filters and Signal Processing*. Kluwer Academic Publishers, 1989.

[2] R. G. Lyons. *Understanding Digital Signal Processing*. Prentice Hall, 2011.

[3] E. Hogenauer. "An Economical Class of Digital Filters for Decimation and Interpolation". *IEEE Transactions on Acoustics, Speech and Signal Processing*, 29, April 1981.

[4] A. V. Oppenheim and R. W. Schafer. *Discrete-Time Signal Processing*. Prentice Hall, 1989.

[5] P. P. Vaidyanathan. *Multirate Systems and Filter Banks*. Prentice Hall, 1993.

[6] J Miller and T. Pate. "Driving High-Speed Analog-to-Digital Converters". *Texas Instruments Application Note SLAA416A*, 2010.

[7] J. Tierney and C. M. Rader. "A Digital Frequency Synthesizer". *IEEE Transactions on Audio and Electroacoustics*, 19(1), March 1971.

[8] W. T. Padgett and D. V. Anderson. *Fixed-Point Signal Processing*. Morgan & Claypool, 2009.

Chapter 10

Symbol and Carrier Tracking

10.1 SYMBOL TRACKING

Transmit symbols are generally modulated and transmitted symbol by symbol. At the receiver symbol tracking primarily refers to tracking symbol phase, that is, finding the symbol boundaries in time.[1] The block diagram in Figure 10.1 shows the parts of a typical modem responsible for symbol tracking.

This chapter starts by mentioning a couple of different ways to select the required samples from the incoming oversampled ADC samples. We call that symbol boundary adjustment. Most of the chapter is concerned with measuring the time error in the selected samples. We call that timing error detection.

10.1.1 Symbol Boundary Adjustment

The first block in the closed-loop timing diagram of Figure 10.1 adjusts each new set of $M = 2$ or 4 samples to conform to symbol boundaries.[2] Figure 10.1 shows a sample interpolator performing this function. The use of the sample interpolator technique in an APSK modem is discussed at length in Chapter 12. The sample stacker, which is little more than a serial-to-parallel converter, is also described in Chapter 12.

An alternative symbol boundary adjustment is shown in Figure 10.2 along with the required adjustment table in Table 10.1. The circuit of Figure 10.2 can be

---

[1] Symbol time duration typically varies slowly, if at all, because symbol frequency (i.e., rate) error for a stationary system is usually very low. Doppler shift between transmitter and receiver can complicate this, however; see Section 14.9.
[2] Recall that the ADC outputs a sample stream and knows nothing about symbols.

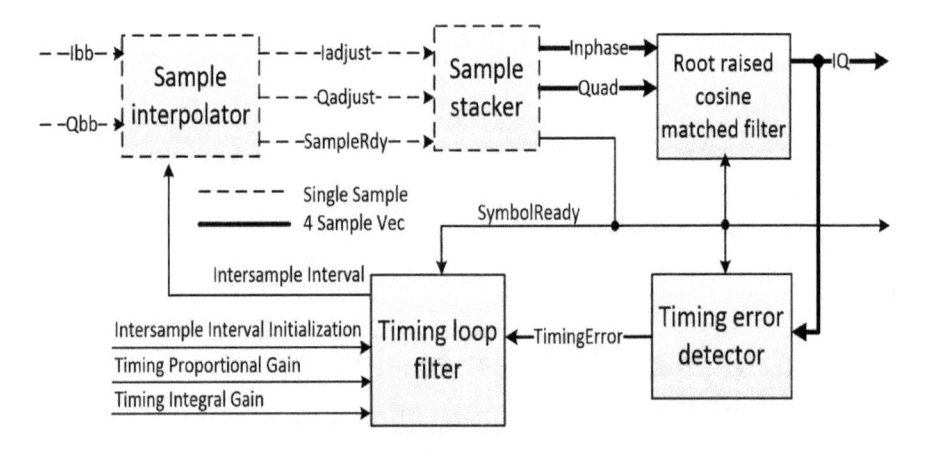

Figure 10.1 The big picture of closed-loop symbol tracking.

used to slide through the sampled ADC data forward or backward. When the symbol timing loop is converged, the output $y(nT_{samp})$ will be adjusted in time to match the incoming symbol boundaries.

This design uses a set of L delay filters. L is the resolution of the adjustment. For example, if $L = 4$, each ADC sample can be divided into four parts with a resolution of $1/4$; see Figure 10.2. If $L = 8$, we have eight filters and a resolution of $1/8$. For a more detailed version of this design; see [1].

The bank of four lowpass FIR filters is not difficult to design. Say the filter is required to cover N symbols, at M samples per symbol and resolution $L = 4$. Start with $K = LNM$ taps; call that filter $g(k)$ for $k = 0, 1, 2, 3...K$. The first filter, Q_0, taps are simply $Q_0(k) = g(0, L, 2L, 3L...K)$. The second filter, Q_1, taps are simply $Q_1(k) = g(1, L + 1, 2L + 1, 3L + 1...K)$, and so on for a total of L filters. When converged, the circuit of Figure 10.2 will deliver samples lined up with symbols to the Figure 10.1 sample stacker for formation into multisample symbols.

10.1.2 Time-Domain Based Timing Error Detector

Symbol timing error is defined as the difference between the assumed start of symbol and the actual start of the symbol in fractions of a sample. Normally, timing error detectors (TED) output a symbol error in the range $\begin{bmatrix} -0.5 & +0.5 \end{bmatrix}$.[3] There

---

[3] The symbol timing is only responsible for lining up symbols, not messages. Thus, the ±0.5 symbol range is all that is needed.

CHAPTER 10. SYMBOL AND CARRIER TRACKING

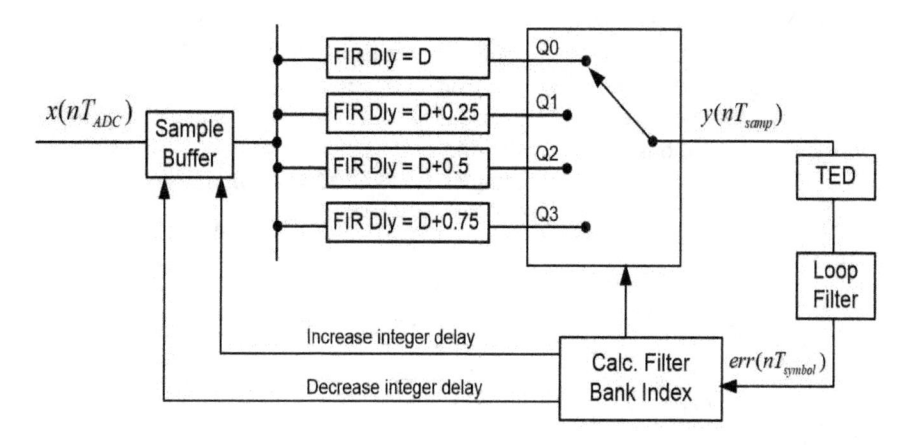

Figure 10.2 An alternative symbol boundary adjustment circuit, $L = 4$.

Table 10.1 Alternative Symbol Boundary Adjustment Table, $L = 4$

Delay	Integer	Fraction
0	0	0
0.25	0	0.25
0.5	0	0.5
0.75	0	0.75
1.0	1	0
1.25	1	0.25
1.5	1	0.5
1.75	1	0.75
2.0	2	0
2.25	2	0.25
...	...	...

are many different ways to measure the symbol timing error, however, they divide roughly into two groups. The first group operates in the time-domain and includes Gardner, modified Gardner, and transition tracking. Most of these will operate on $M = 2$ or 4 receiver samples per symbol (see Section 14.6). The second group operates in the frequency-domain by deriving, from the received symbols, a tone at the symbol rate. These can be very useful because they are constellation agnostic, whereas the time-domain methods tend to break down for complicated constellations. Frequency-domain methods require $M = 4$ samples/symbol ($M = 2$ will not support a tone at F_{symbol}).

Table 10.2 A Timing Error Detector (TED) Summary

TED	Domain	Closed/Open-Loop	M (Minimum)
Gardner	Time	Closed	2
Modified Gardner	Time	Closed	2
Transition	Time	Closed	4
Bandedge (Godard)	Freq	Closed	4
Tone resonator	Freq	Open	4

The time-domain TEDs are usually part of a closed-loop symbol tracking system, as shown in Figure 10.3. The ADC sampled input, on the left, has no alignment to symbol boundaries. The output, on the right, is an integer number of samples per symbol (probably 2 or 4). After symbol tracking is converged, the output is aligned with the symbol boundaries. In summary, the light shaded parts are the symbol error estimate and tracking. The dark shaded part is the symbol error resampler.

In this section, we focus on several approaches to timing error detection. We should note that all of these methods tend to work better when the received signal has excess bandwidth, i.e., signal power at and beyond $(\pm F_{symbol}/2)$[4]. See Figure 3.15 for an illustration of excess bandwidth.

10.1.2.1 Gardner and Modified Gardner TED

Gardner symbol timing was first published in the mid-1980s and has been used extensively since then. Gardner TED provides reasonable performance for very little circuitry. The circuit diagram, including the loop filter, is shown in Figure 10.5.

---

[4] However, for a raised cosine signal, there is no signal power at and beyond $\pm F_{symbol}$.

CHAPTER 10. SYMBOL AND CARRIER TRACKING

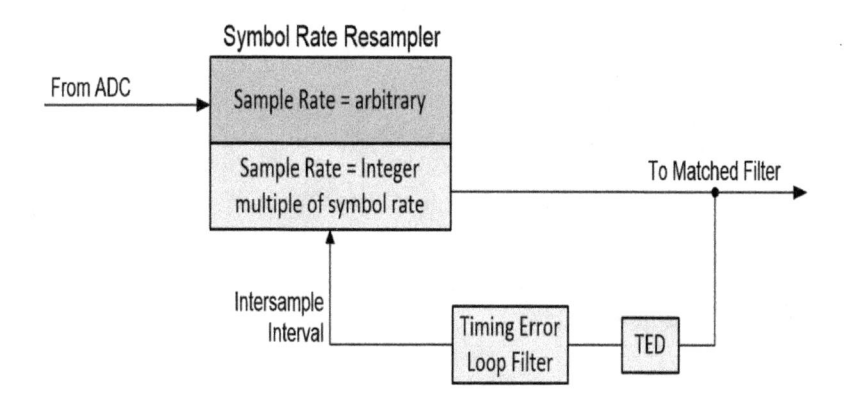

Figure 10.3 Primary blocks required for closed-loop symbol tracking.

These correspond to the blocks Timing Error Detector and Timing Loop Filter in Figure 12.1.

Figure 10.4 is called Gardner timing error detector; see [2]. Figure 10.4 shows a four sample per symbol baseband waveform at the matched filter output (either I or Q). The symbol counter is k. The TED equations are shown below the late timing example (top), on-time (middle), and early timing example (bottom). Note that the zero timing error really means the first samples of each symbol are equal and the energy from each symbol is completely confined to the groups of four samples: $y(4k)$, $y(4k + 1)$, $y(4k + 2)$, and $y(4k + 3)$. The polarity of the center sample is used to prevent the symbol data pattern polarity changes from affecting the timing error. No doubt the reader has already figured out that certain data patterns can cause inaccurate timing errors. That is why the Gardner TED output is often averaged. For this simple example, which could be BPSK or QPSK, the timing error is filtered, updated at the symbol rate, and fed back to the symbol rate resampler to close the tracking loop; see Figure 10.5.

A commonly used indicator of TED performance is the S-curve. Figure 10.7 is an S-curve plot of Gardner and modified Gardner timing error versus TED input. A Simulink simulation is used to construct the S-curve by sweeping the timing error from minus one half sample to plus one half sample in steps of one sixty-fourth sample. Because the symbol values are random, many sweeps are overlayed to show the average performance.

In Figure 10.7, notice that for 16APSK (see Figure 12.23) the original Gardner TED is struggling, whereas the modified Gardner TED seems to be performing well.

Figure 10.7, top plot, shows that the unmodified Gardner in Figure 10.5 may still be a good choice for a simpler signal, such as QPSK. The gain (timing error vertical excursion) in the two circuits is about the same. Note that the primary manifestation of jitter is uncertainty in the S-curve zero crossing.

The modified Gardner circuit is shown in Figure 10.6 for easy comparison with Figure 10.5. The complete Simulink design is in Figure 12.11. The modified Gardner TED equations and more description are in Section 12.4.6.

10.1.2.2 Transition Tracking

Transition tracking is a simple technique used for closed-loop timing recovery; see [3]. If most of the carrier frequency error has been tracked out, transition tracking works well for BPSK and QPSK. Transition tracking may work for higher-order QAM, although modified Gardner may be a better choice.

Referring to Figures 10.8 and 10.9, $x(t)$ is the receiver's symbol boundary estimation. The symbol polarity is sampled in the center of $x(t)$. If there is no polarity transition, $x(t)$ goes to zero.

Generally, the symbols received over the RF channel have a random symbol phase but a stable symbol rate, so we assume $x(t)$ frequency is correct but its phase won't be lined up correctly with the incoming symbols. A phase correction signal must be fed back to the circuit that generates $x(t)$. Phase correction is generated by comparing $x(t)$ with $y_1(t)$, the matched filter output delayed by one half symbol. For example, for $M = 4$ samples per symbol, $y_1(t)$ is delayed by two samples. As shown in the bottom trace of Figure 10.9, if $x(t)$ phase is correct, the symbol center will correspond to the center of $x(t)$ and the average polarity will be zero. If $x(t)$ phase is incorrect, the average polarity of the half symbol delayed symbol within the $x(t)$ window will be positive for an early symbol; and negative for a late symbol; see Figure 10.10. The word average implies a lowpass filter, seen in Figure 10.8. The lowpass filter output is fed back to the circuit generating $x(t)$ to adjust its phase and try to drive the average $u(t)$ to zero.

10.1.3 Frequency-Domain TED

Time-domain TEDs tend to use less hardware and work well for many applications. Frequency-domain TEDs can work well, however, they use more hardware, for example, resonator filters.

CHAPTER 10. SYMBOL AND CARRIER TRACKING

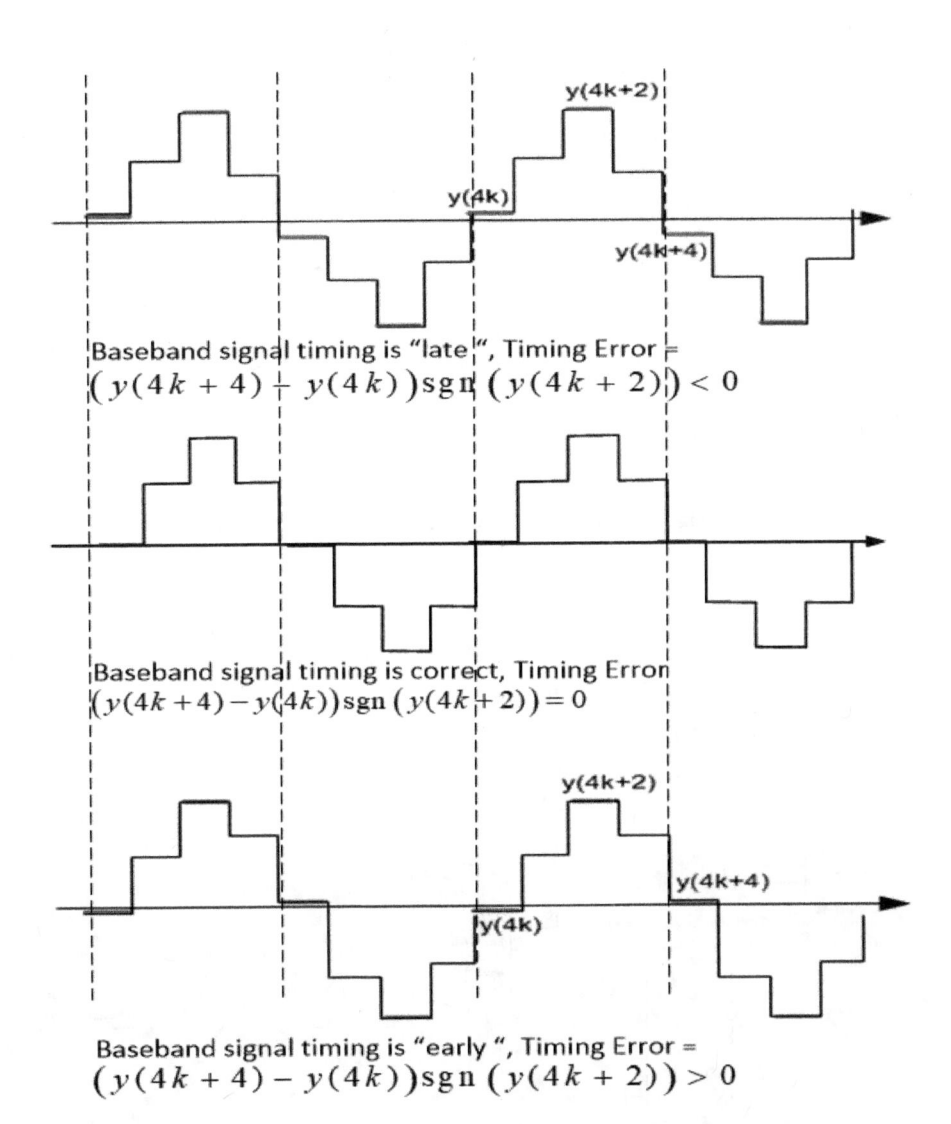

Baseband signal timing is "late", Timing Error =
$(y(4k+4) - y(4k))\text{sgn}(y(4k+2)) < 0$

Baseband signal timing is correct, Timing Error =
$(y(4k+4) - y(4k))\text{sgn}(y(4k+2)) = 0$

Baseband signal timing is "early", Timing Error =
$(y(4k+4) - y(4k))\text{sgn}(y(4k+2)) > 0$

Figure 10.4 Basic operation of Gardner symbol tracking.

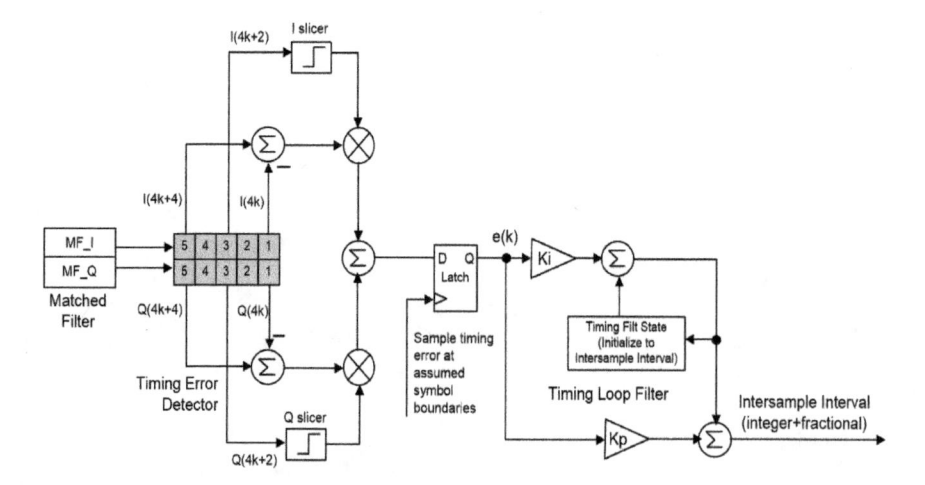

Figure 10.5 Circuit for Gardner TED and loop filter.

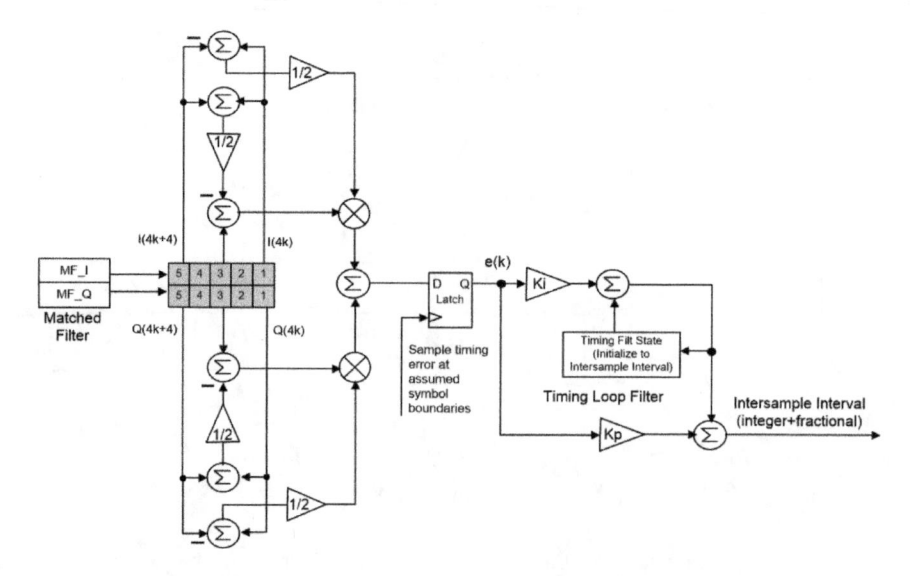

Figure 10.6 Circuit for modified Gardner TED and loop filter.

CHAPTER 10. SYMBOL AND CARRIER TRACKING 411

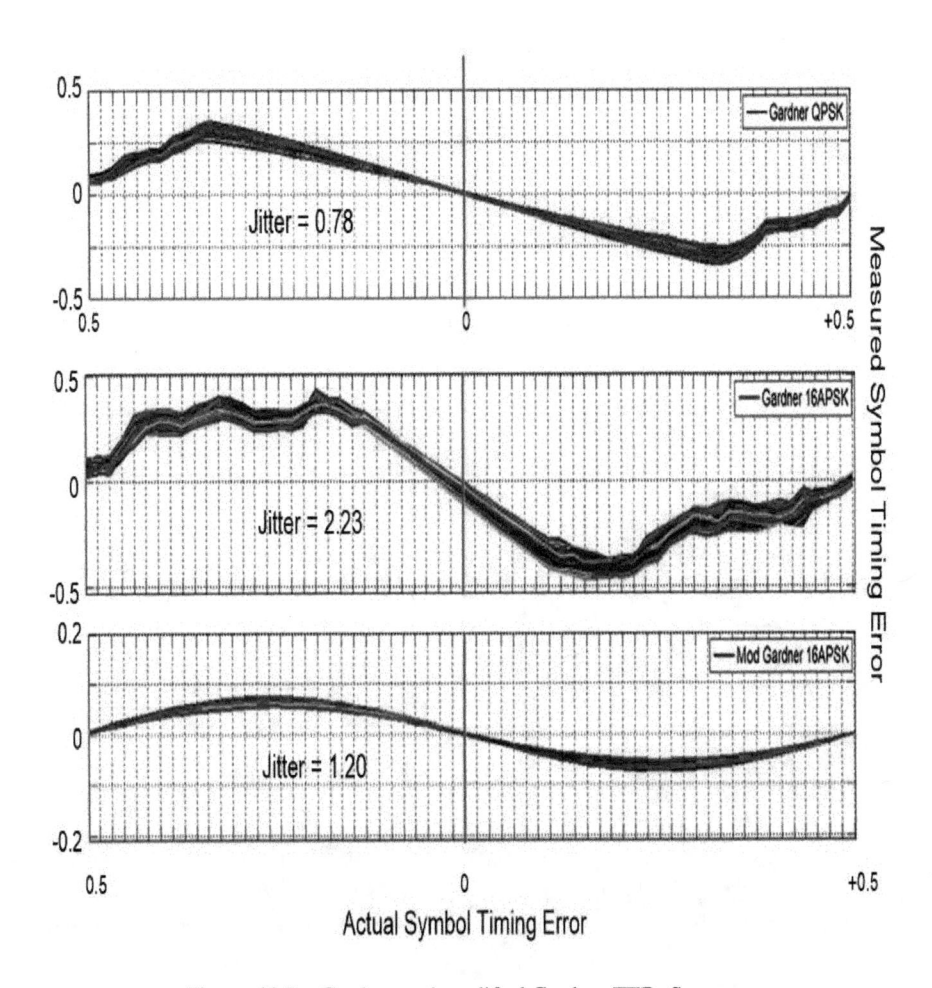

Figure 10.7 Gardner and modified Gardner TED, S-curves.

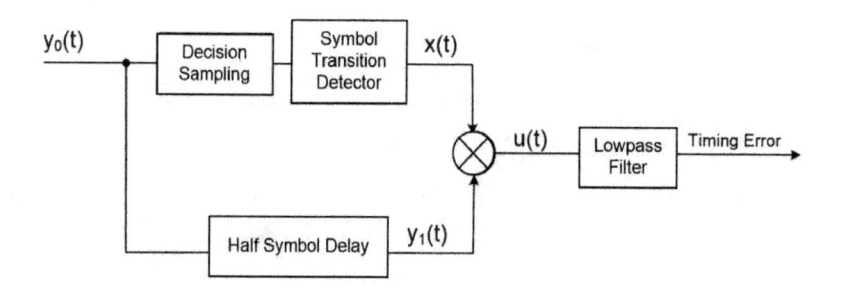

Figure 10.8 A transition tracking TED circuit.

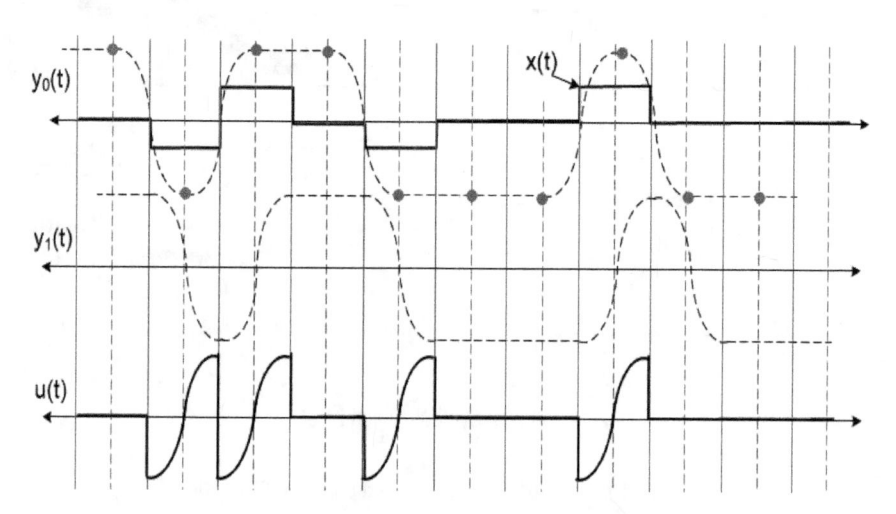

Figure 10.9 Transition tracking TED: on time.

CHAPTER 10. SYMBOL AND CARRIER TRACKING

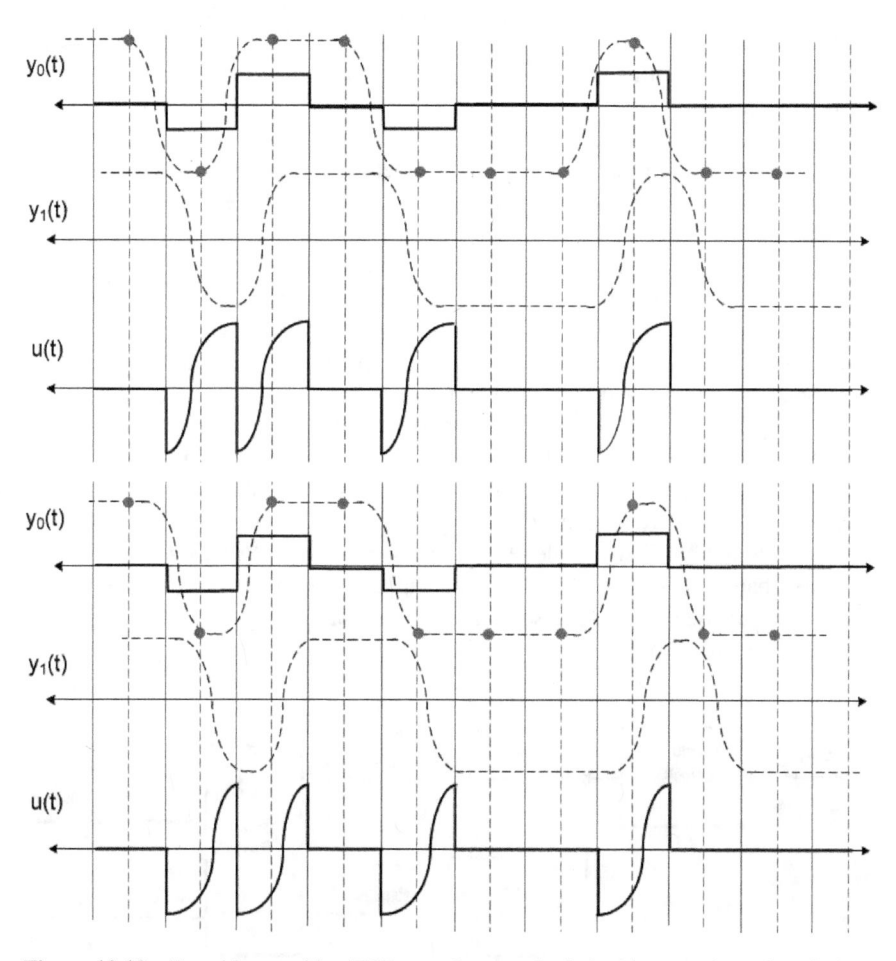

Figure 10.10 Transition tracking TED: top shows early timing, bottom shows late timing.

10.1.3.1 Single Timing Tone

All the TEDs discussed so far are used in closed-loop symbol tracking, as shown in Figure 10.3. Now we discuss an open-loop symbol tracking approach. This was one of the first approaches to modem symbol tracking and was used on early modems such as the 1963 vintage Bell 212A QPSK modem, one of the first commercially available modems.

In this system, data modulated modem signals (also called bauded or symbol based signals) are passed through a nonlinearity, typically a squaring circuit, to obtain a timing tone. Widely used QAM signals are raised cosine filtered with excess bandwidth extending between $-F_{symbol}$ and $-F_{symbol}/2$ as well as between $F_{symbol}/2$ and F_{symbol}; see Figure 3.15. This excess bandwidth must be greater than zero for this technique to work. Another requirement is that the symbol data should be randomized. For example, long strings of constant symbols may cause the symbol timing to fail. These simple requirements allow symbol timing to work without adding bandwidth for pilot signals or synchronization patterns.

Figure 10.11 shows the placement of the timing tone symbol tracking in the receiver digital front end. The contents of the box labeled timing tone tracking are shown in Figure 10.12. The F_{symbol} BPF output will be strongest when the received symbols are random. The zero crossing detector measures the time between presumed symbol first samples.[5] The intersymbol time PLL is acting like a very narrow bandpass filter to stabilize the estimate. The symbol rate resampling uses this symbol time estimate, also called the intersample interval, to interpolate the next four symbol samples.

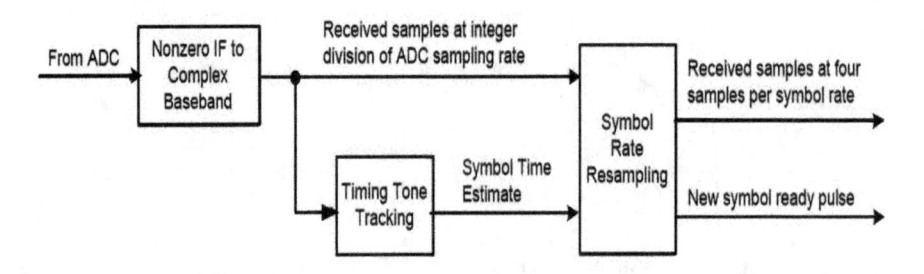

Figure 10.11 Open-loop timing tone tracking placement.

---

5 The first sample of each symbol likely corresponds to a baseband symbol zero crossing.

CHAPTER 10. SYMBOL AND CARRIER TRACKING

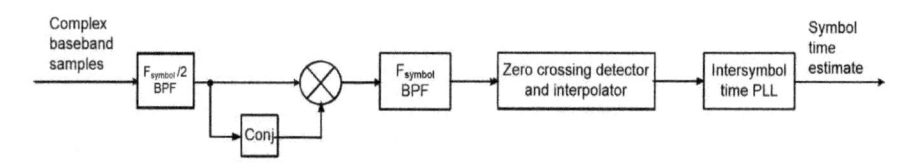

Figure 10.12 Open-loop timing tone tracking.

Figure 10.12 shows the timing tone tracking circuit in more detail. $F_{symbol}/2$ bandpass filtering followed by complex squaring followed by F_{symbol} narrow bandpass filtering is intended to isolate the timing tone with as little noise as possible. The intersymbol time PLL is acting as a very high Q bandpass filter. Performance depends on excess raised cosine bandwidth, randomized data, and $> 4F_{symbol}$ input sampling rate. While similar to bandedge timing, this system does not have the same inherent immunity to carrier frequency offset.

The timing tone typically has envelope variations called pattern noise caused by passing, for example, a QAM signal through a narrow bandpass filter, as in Figure 10.12, These are basically an undesired FM - AM conversion due to the varying spectral content about the resonator filter center frequency. The circuit of Figure 10.12 minimizes this problem by focusing on timing tone zero crossings.

10.1.3.2 Bandedge Timing

This is also called Godard timing recovery from D. N. Godard's original paper; see [4]. Godard's insight was that bandedge tones at $\pm F_{symbol}/2$ can be combined to make a timing tone useable for symbol tracking. As we will see, this technique has built-in immunity to carrier frequency offset.

Figures 10.13 and 10.14 show the entire process of generating Godard's timing tone. The lettered boxes in Figure 10.14 are matched to spectral plots in Figure 10.13. A brief description is below.

A, raised cosine (RC) QAM signal after matched filtering. Notice a slight carrier frequency offset. This will not affect timing error measurement.
B$_{hih}$, RC signal shifted up in frequency to F_{symbol}. The tall box represents a narrow BPF bandwidth. Note the frequency offset appears inside the filter bandwidth.
C$_{hih}$, after narrow BPF.
B$_{low}$, RC signal shifted down in frequency to $-F_{symbol}$.
C$_{low}$, after narrow BPF.
D, Complex product of **C$_{hih}$** and $conj(\mathbf{C_{low}})$. Output is the timing tone at F_{symbol}.

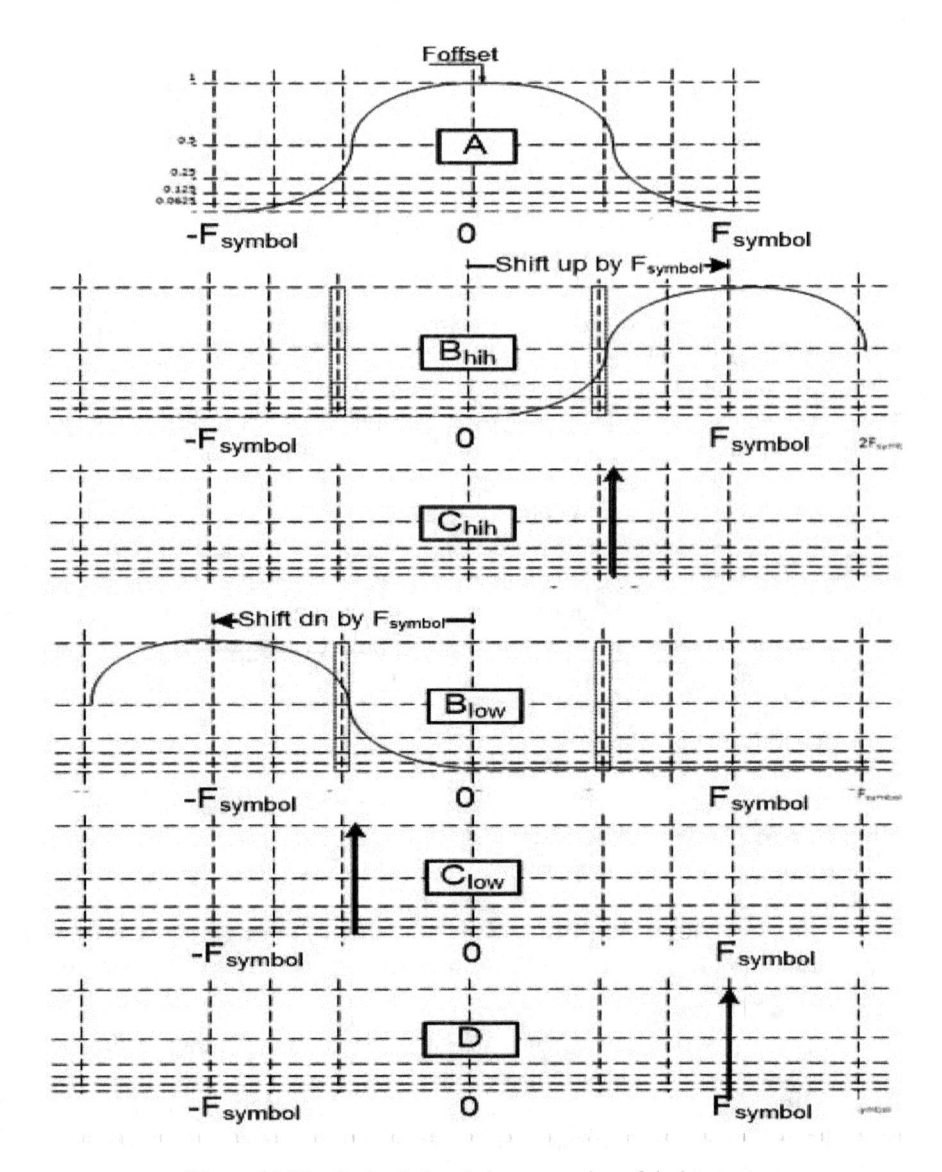

416   *Software Defined Radio: Theory and Practice*

Figure 10.13 Godard's bandedge generation of timing tone.

CHAPTER 10. SYMBOL AND CARRIER TRACKING   417

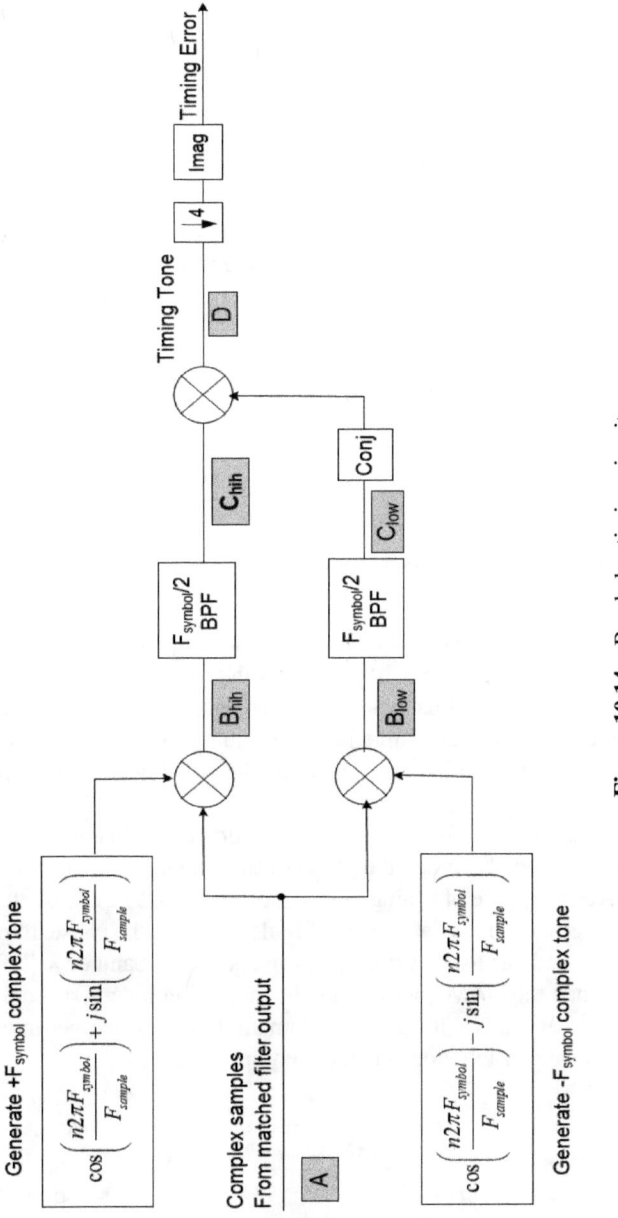

Figure 10.14 Bandedge timing circuit.

In Figure 10.13, the Nyquist frequency limits are at the edges of the figure, $-2F_{symbol}$ and $+2F_{symbol}$. Thus the sampling rate is $4F_{symbol}$ or $M = 4$ samples per symbol. At two samples per symbol the symbol rate timing tone will be at the Nyquist fold over frequency and thus not supported. Four samples per symbol is required for this TED.

Note in Figure 10.13 the input spectrum (A) has a carrier frequency offset F_{offset}. After the narrow bandpass filtering, the two bandedge tones, C_{low} and C_{hih}, also have this frequency error. However, Equation (10.1) shows how the timing tone, D, has no frequency offset.[6] This is because the timing tone is the lower sideband (i.e., difference frequency) of the product of the two bandedge tones. This is shown in Equation (10.1).

$$D(t) = C_{hih}C_{low}^* = e^{j\left(\left(2\pi \frac{F_{symbol}}{2} + 2\pi F_{offset}\right)n\right)} \text{conj}\left(e^{j\left(\left(-2\pi \frac{F_{symbol}}{2} + 2\pi F_{offset}\right)n\right)}\right)$$

$$= e^{j\left(\left(\frac{\pi}{8} + 2\pi F_{offset}\right)n\right)} e^{j\left(\left(\frac{\pi}{8} - 2\pi F_{offset}\right)n\right)} = e^{j\left(\frac{\pi}{4}n\right)}$$

(10.1)

Item D in Figure 10.14, Figure 10.13, and Equation (10.1) is the complex symbol rate timing tone. Figure 10.15 shows the four state, 90° per sample nature of this tone. Subsampling this by 4 leaves just sample 0 at one sample per symbol. The angle around 0° of this sample indicates the timing error. To avoid calculating an arctangent, we use the imaginary part as the small angle approximation to the arctangent.

Figure 10.16 shows the important connection between the phase of the timing tone angle discussed above and the phase of the actual data symbols. Symbol time tracking drives the phase of timing tone sample zero to 0°. Because the timing tone is derived directly from the data signal, this also lines up the phase of the actual data symbols. This leads us to the primary advantage of the band edge TED; no matter how complicated the QAM constellation is, the timing tone is still only four states, 90° apart. Compare this with the time-domain TEDs, which get more difficult to make work as the constellation gets more complicated.

---

6 Equation (10.1) is continuous time for simplicity. The accurate discrete time expression for the high bandedge output is $C_{hih}(n) = e^{j\left(\frac{n(\pi F_{symbol} + 2\pi F_{offset})}{F_{sample}}\right)}$. The carrier offset immunity is the same.

CHAPTER 10. SYMBOL AND CARRIER TRACKING

The bandedge TED and the remaining symbol time tracking components in Figure 10.3 converge such that every four new samples represent a new data symbol. The bandedge TED locks in the symbol boundaries and then the equalizer phase rotates and subsamples to the optimum symbol value at one sample per symbol; the complex constellation points from which symbol soft decisions can be derived.

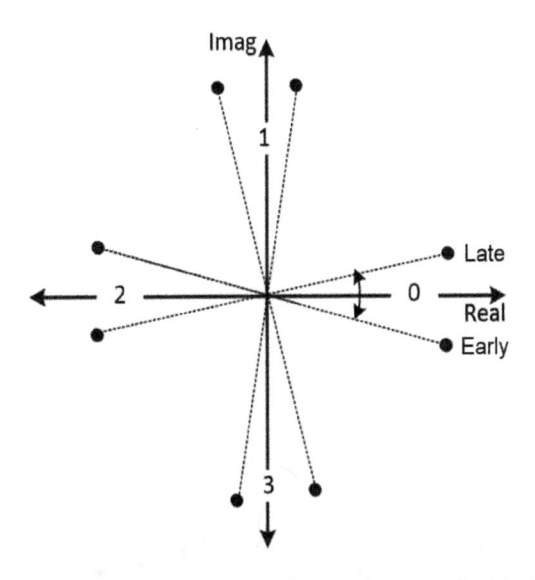

Figure 10.15 Four phase nature of four sample per symbol timing tone.

Figure 10.17 is a simulation diagram that implements the $\pm\dfrac{F_{symbol}}{2}$ frequency conversions in Figure 10.13. The blocks in Figure 10.17 all execute at the symbol rate. However, because of the four sample vector processing, the sampling rate is actually $4F_{symbol}$.

The two complex local oscillators at $-F_{symbol}$ and $+F_{symbol}$ are both implemented by pairs of fixed four sample vectors. To see why these implement a complex sine wave at the symbol rate, consider that each vector contains a complete sampled sine cycle, [0, +1, 0, −1] or cosine cycle, [+1, 0, −1, 0]. Because all the four sample cycles are identical every symbol time, they can be constants. Figure 9.38 and surrounding text illustrate this in more detail.

The circuit in Figure 10.17 is outputting four sample vectors to the $F_{symbol}/2$ narrow bandpass filters, also called resonator filters. These will need to be a little

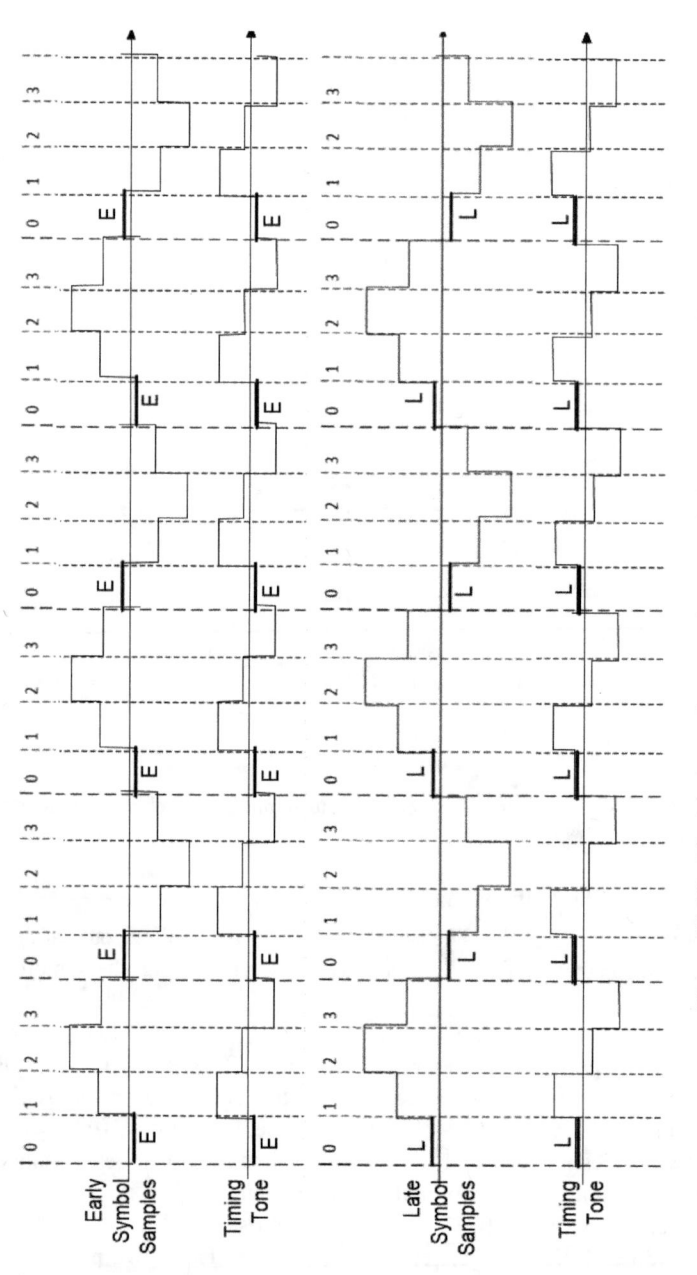

Figure 10.16 Timing errors, four sample per symbol timing tone (I or Q).

CHAPTER 10. SYMBOL AND CARRIER TRACKING

more complicated than a single sample filter, such as in Figure 9.17. Modern FPGA tools can increase the resonator execution speed by parallelizing the resonator circuit.

The signals in Figure 10.17 all have a fixed-point attribute of sfix16en15. This is a 16-bit signed number with 15 fractional bits; see Section 14.2. The magnitude of this number must be less than one or greater than minus one. Simulink can directly translate a circuit like this to hardware description language (HDL) code for implementation in hardware.

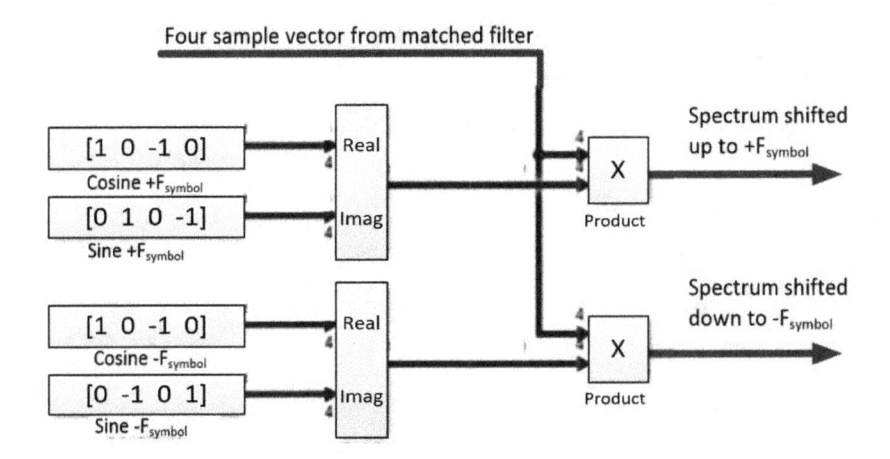

Figure 10.17 Raised cosine baseband signal, frequency shift circuit.

10.1.4 Timing Closed-Loop Dynamics

Figure 10.18 can be used to study the closed-loop response to symbol timing phase changes. Let's look around this tracking loop.

Input $\theta(k)$ is the received intersymbol time interval; the measured number of integer and fraction samples between symbol boundaries. Feedback $\phi(k)$ is the local estimate of the intersymbol time interval. The difference between these is $e(k)$, the symbol timing error; see Figures 10.5 and 10.6. The TED in Figure 10.18 is a generalization of the various timing error detectors described previously. Gain and delay around the loop are modeled by K_d and z^{-D}, respectively. Some of the delay is due to implementation. For example, the subtractor output may need to be registered, especially in an FPGA implementation. The register is a one sample storage circuit that will require a one sample delay.

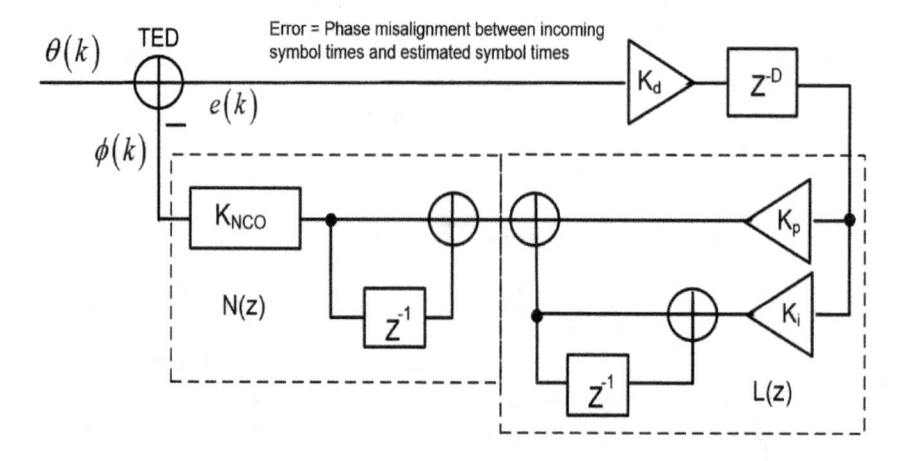

Figure 10.18 Symbol time tracking, closed-loop second order model.

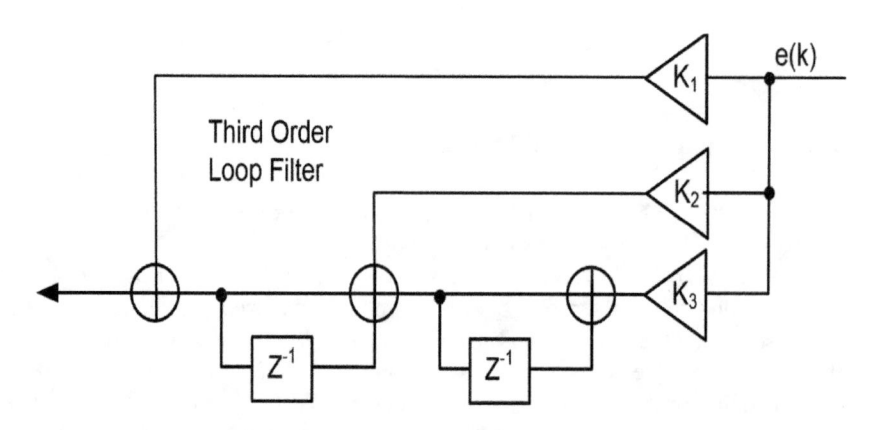

Figure 10.19 Symbol time tracking, alternative third order loop L(z).

CHAPTER 10. SYMBOL AND CARRIER TRACKING

Loop dynamics is controlled primarily by parameters K_p and K_i. Finally, the numerically controlled oscillator (NCO) by itself is a first-order ramp generator, i.e., a single unstable pole on the unit circle. Equation (10.2) shows the transfer functions of the loop filter, $L(z)$, and the NCO, $N(z)$. Equation (10.3) is the complete closed-loop transfer function of the circuit in Figure 10.18.

The closed-loop response has some dependence on delay D. Generally, FPGA operations (e.g., add, multiply, shift) must be complete in one clock cycle. For high-speed operation, this can be accomplished by breaking up operations into multiple clock cycles. For example, say a multiplier output must be transferred to an adder input. One clock cycle is used for the multiply and another clock cycle is used for the add.[7]

$$L(z) = \frac{(K_P + K_I) - z^{-1}K_P}{1 - z^{-1}}$$

$$N(z) = \frac{K_{NCO}}{1 - z^{-1}}$$

(10.2)

$$\frac{\phi(z)}{\theta(z)} = \frac{z^{-D}K_d L(z) N(z)}{1 + z^{-D} K_d L(z) N(z)} = \frac{z^{-D} K_d \left(\frac{K_p + K_i - z^{-1} K_p}{1 - z^{-1}} \right) \left(\frac{K_{NCO}}{1 - z^{-1}} \right)}{\left(\frac{1 - z^{-1}}{1 - z^{-1}} \right)^2 + \left(\frac{K_{NCO} K_d z^{-D} \left(K_p + K_i - z^{-1} K_p \right)}{(1 - z^{-1})^2} \right)}$$

$$= \frac{K_{NCO} K_d z^{-D} (K_p + K_i) - K_{NCO} K_d K_p z^{-(D+1)}}{1 - 2z^{-1} + z^{-2} + K_{NCO} K_d (K_p + K_i) z^{-D} - K_{NCO} K_d K_p z^{-(D+1)}}$$

(10.3)

10.1.4.1 Timing Closed-Loop Parameters

We start by assuming the values of D, K_{NCO}, and K_d in Figure 10.18. Reasonable values are $D = 1$, $K_{NCO} = 1$ and $K_d = 1$. The next constant we need is B_n/F_{sample}, the noise bandwidth in hertz as a fraction of the sampling frequency. A typical value might be $B_n/F_{sample} = 0.1$. Finally, we need the damping factor, a common starting value is $\varsigma = 1/\sqrt{2}$. This is referred to as critically damped. As the damping

---

[7] We say the multiplier and adder outputs are clock registered. This allows the composite multiply and add two clock cycles to complete. This of course increases the closed-loop delay D.

factor increases the settling time increases and the closed-loop frequency response becomes tighter. The opposite is also true. The damping factor thus affects a trade-off between response time and noise bandwidth. Equation (10.4) is a summary of damping factors. Slightly underdamped is a common way to get fast response with just a little ringing (oscillation that dies out over time).

Second-order loop filters will drive a constant or slowly changing carrier frequency error to zero. An alternative third-order loop filter, shown in Figure 10.19, will drive an accelerating frequency error to zero. Where receiver movement dynamics can be complicated, such as on a ship or a road navigation system (i.e., car GPS) third-order loop tracking may be needed. However, a modem that sits on a shelf may only need second-order tracking.

$$\varsigma = \begin{cases} < 1 & underdamped \\ = 1 & critically\ damped \\ > 1 & overdamped \end{cases} \quad (10.4)$$

The code below calculates and simulates the loop model of Figure 10.18. The simulation results are in Figure 10.20. Notice how increasing D from 1 to 4 increases the closed-loop bandwidth and makes the step response peak higher. The higher delay could be compensated for by raising the damping factor. Note that the MATLAB code below has references to equations for calculating K_i and K_p. Reference [5] has additional helpful PLL analysis.

```
% Runs the digital PLL simulation and plots results
LL=256;
NFFT = 1024;
Fs = 4096; % 4 samples/symbol at Fsym = 1024 Hz
Ts = 1/Fs;
Fsym = 1024;
SampPerSym = Fs/Fsym;
%Knco = 1=> 1 radian/sample, sample phase advances by 1
BnTs = 0.05; % Approx 32.6Hz Loop bandwidth
dmp = 1/sqrt(2); %damping factor, critically damped.
Kn = 1;
Kd = 1;

% Next three equations implement equation C.58 from [5].
```

CHAPTER 10. SYMBOL AND CARRIER TRACKING 425

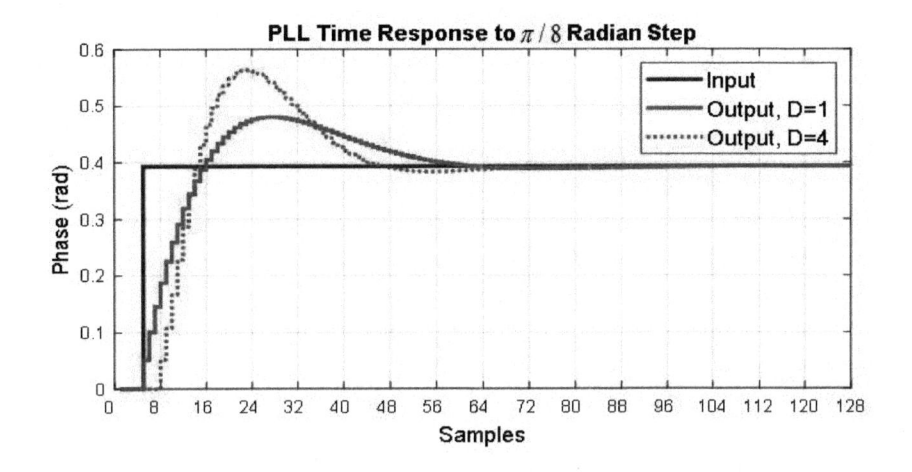

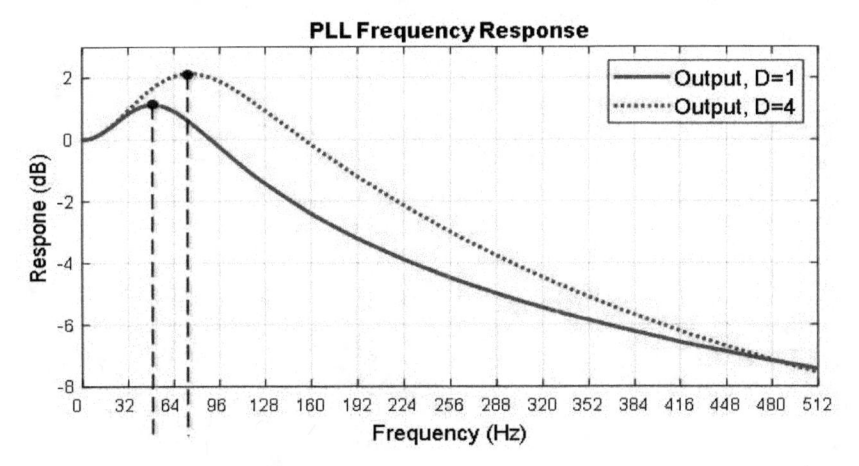

Figure 10.20 Symbol tracking closed-loop response, $\varsigma = 0.707$.

$$dn = 1+2*dmp*(BnTs/(dmp+(1/(4*dmp))))+(BnTs/(dmp+(1/(4*dmp))))^2$$
$$Kp = 4*(dmp*BnTs/(dmp+(1/(4*dmp))))/(dn*Kn*Kd)$$
$$Ki = 4*(BnTs/(dmp+(1/(4*dmp))))^2/(dn*Kn*Kd)$$

% A simpler approach to calculating Ki and Kp,
% requires starting with higher Wn = 58Hz; see [6]

$$Wn = 2*pi*58$$
$$Kp = 2*dmp*Wn/(Fs*Kd*Kn)$$
$$Ki = (Wn^2)/(Kd*Kn*Fs^2)$$

```
% Generate a step input for testing:
Input=[zeros(1,SampPerSym) (pi/8)*ones(1,LL-SampPerSym)];
Output = zeros(4,LL);
Fresp = zeros(4,NFFT);
Fscale = zeros(4,NFFT);
zb = zeros(1,D-3);
for D=1:4
PLLnum = [zeros(1,D-1) Kd*Kn*(Kp+Ki) -Kd*Kp*Kn ];
switch(D)
case1
PLLden = [1 (Kd*Kn*(Kp+Ki)-2) (1-Kd*Kp*Kn) ];
case2
PLLden = [1 -2 (1+(Kd*Kn*(Kp+Ki))) -Kd*Kp*Kn ];
case3
PLLden = [1 -2 1 ((Kd*Kn*(Kp+Ki))) -Kd*Kp*Kn ];
otherwise
PLLden = [1 -2 1 zb ((Kd*Kn*(Kp+Ki))) -Kd*Kp*Kn];
end
Output(D,:) = filter(PLLnum, PLLden, Input);
(Fresp(D,:),Fscale(D,:)) = freqz(PLLnum,PLLden,NFFT);
end
figure(1)
MM=8;
subplot(2,1,1);
hh=stairs(Input,'b');
hold on
hh=stairs(Output(1,:),'r');
```

```
hh=stairs(Output(4,:),'r:');
hold off
set(gca,'XLim',[0 LL/2]);
set(gca,'XTick', 0:MM:LL);
set(gca,'XTickLabel', (0:MM:LL));
legend('Input','Output, D=1','Output, D=4');
title(sprintf('PLL Time Response to pi/8 Radian Step'));
subplot(2,1,2);
hh=plot(Fscale(1,:),10*log10(abs(Fresp(1,:))),'r');
hold on
hh=plot(Fscale(4,:),10*log10(abs(Fresp(4,:))),'r:');
hold off
set(gca,'XLim',[0 pi/4]);
set(gca,'XTick',[0: pi/64: pi/4]);
set(gca,'XTickLabel',[0: Fs/128: Fs/8]);
set(gca,'YLim',[-8 3]);
set(gca,'YTick',[-8:2:2]);
title(sprintf('PLL Frequency Response'));
legend('Output, D=1','Output, D=4');
```

10.1.5 Timing Resilience to Carrier Frequency Offset

After the process of downconverting a received signal to complex baseband, some nonzero carrier frequency offset will generally remain. Sometimes the receiver needs to converge the symbol time tracking prior to removing all carrier frequency offset. Here we investigate how three important symbol tracking algorithms: Gardner, modified Gardner, and Godard bandedge timing perform in the presence of carrier frequency offset.

Figure 10.21 shows the results of a carrier frequency offset test. The modulation is 16APSK, symbol rate = 12.5 Mbps, no added noise. The three plots on the left have zero and the three plots on the right have 100 kHz carrier frequency offset. Figure 10.22 displays six plots with 28.5 dB $E_s N_0$ (22.5 dB $E_b N_0$) signal to noise. In addition, the plots on the right have 100 kHz carrier frequency offset for a realistic test. Jitter is the average S-curve zero crossing location distance from center zero. Since there are only 64 error steps between -0.5 and +0.5, linear interpolation is used to increase accuracy. Ninety-six trials are averaged for each graph. The modified

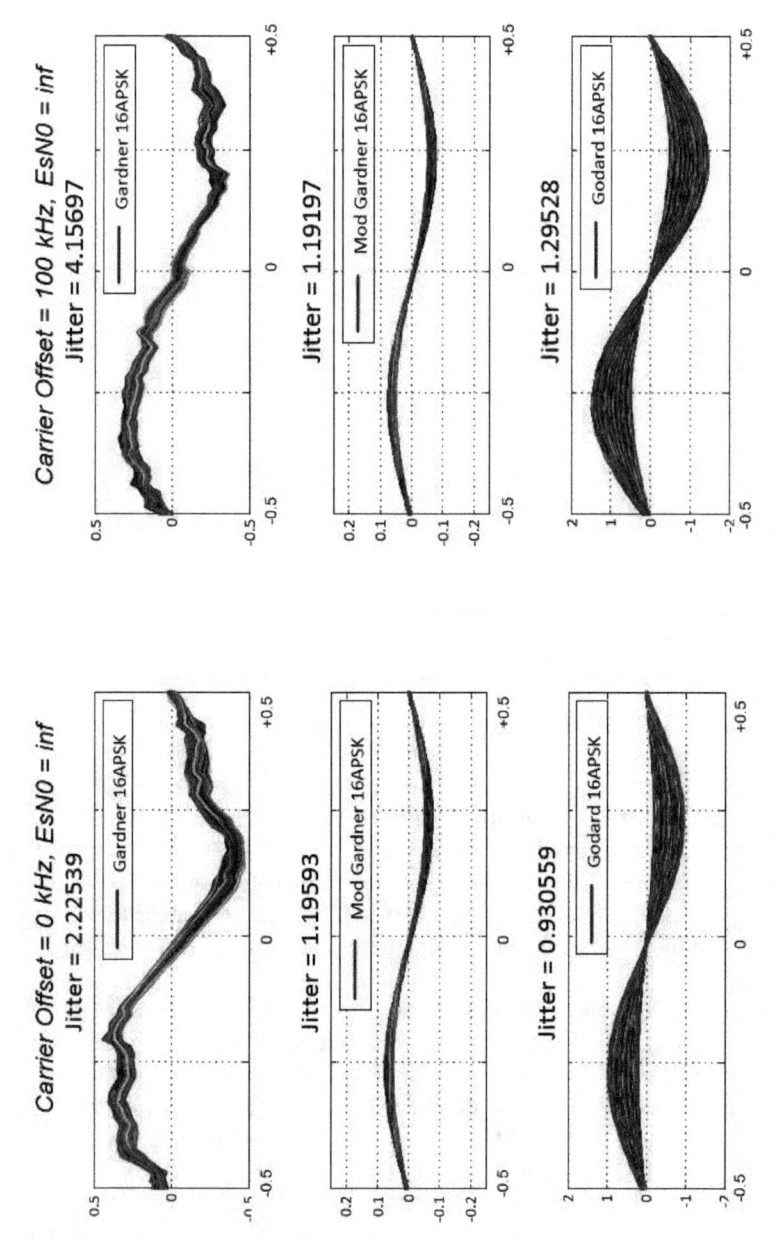

Figure 10.21 Symbol time tracking jitter, carrier frequency offset with no noise.

CHAPTER 10. SYMBOL AND CARRIER TRACKING

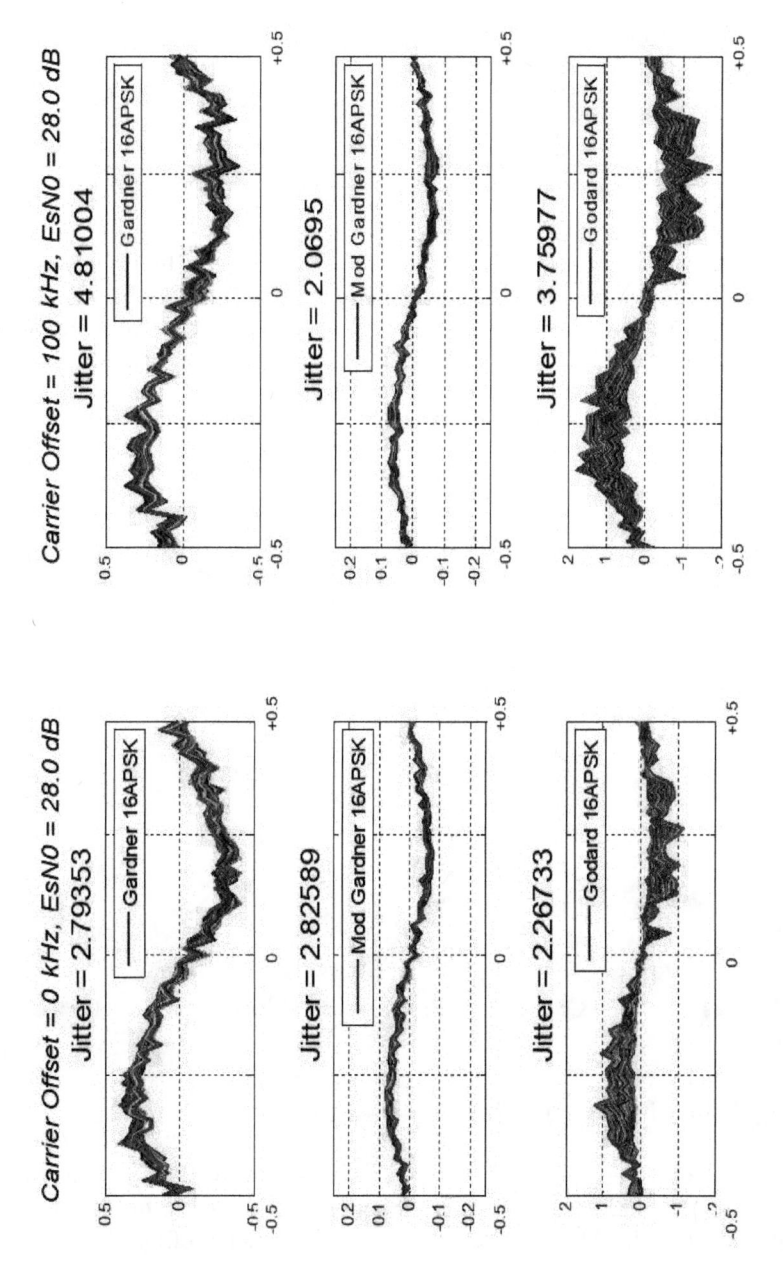

Figure 10.22 Symbol time tracking jitter, carrier frequency offset, 22.5 dB $E_b N_0$.

Gardner TED seems to have the best overall performance. This TED also has the advantage of requiring less hardware than the bandedge TED.

10.2 CARRIER TRACKING

While symbol tracking is concerned primarily with phase, carrier tracking is focused on completely removing the carrier frequency and phase offsets. In Figure 10.23, the ideal phase dots represent where each received symbol + sign must be for optimum soft decisions. However, as seen in the figure, a phase error will cause a static angular offset. Because phase is the integral of frequency, a carrier frequency error will cause a continuously rotating received constellation point (i.e., received phase).

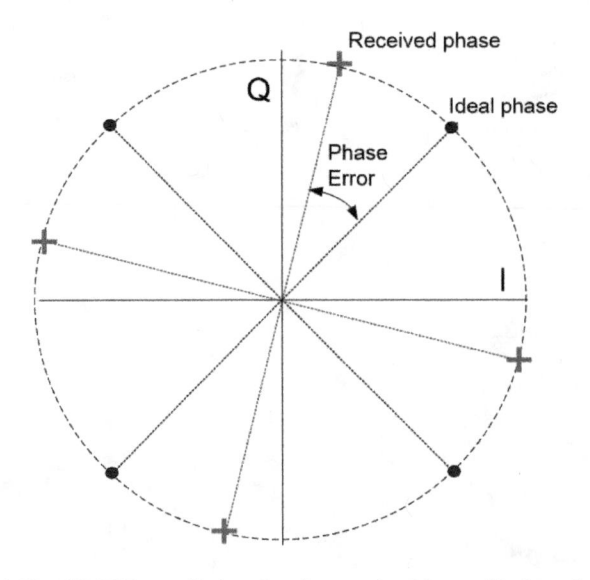

Figure 10.23 QPSK constellation showing received versus ideal symbol points.

10.2.1 Coarse Frequency Offset

Processing signals received over a channel with a large Doppler offset can benefit greatly from an initial estimate and correction of the carrier frequency offset. Frequency offset should be measured up front at the output of the IF to complex baseband circuits. By canceling most of the high Doppler carrier shift, the reliability of packet detection may be greatly increased.

We cover two ways this is accomplished. First, the same preamble sequence occurring at different times can be cross-correlated to detect frequency offset. This is how IEEE 802.11a works and is covered in detail in Section 13.6.2. Second, continuous blocks of received complex baseband samples can be converted to the frequency-domain. The signal of interest is detected and the FFT bin(s) it occupies is used to measure the carrier frequency offset. This is discussed in Section 14.8.

10.2.2 Symbol Rate Decision Directed Carrier Tracking

Decision directed carrier tracking generates carrier corrections at the symbol rate. The phase error shown in Figure 10.23 becomes the input to a phase lock loop (PLL). The PLL output drives a numerically controlled oscillator (NCO) that directly corrects the received signal frequency and phase. Carefully chosen loop dynamic parameters as well as the use of complex numbers throughout ensures very tight control over the received constellation points.

Decision directed means that a hard decision must be made on each received symbol to generate the closest ideal phase point in Figure 10.23. This takes more hardware; however, decision-directed carrier tracking is versatile and can handle many different constellation types. For more details, please refer to Section 12.5.1.

10.2.3 Costas Loop

The Costas loop does not require symbol decisions to track carrier phase[8]. Costas loops are a hardware-efficient way to track BPSK and QPSK carriers. Costas loops are used for carrier tracking in many Global Positioning System (GPS) receivers; see [8].

Figure 10.24 is an example of a simple receiver based on a Costas loop. This would be useful for a short BPSK coherently modulated message with a robust preamble. At the start of the preamble the Costas loop locks onto the IF carrier. Then the preamble detector finds the first sample of the first message symbol. For a fixed symbol rate, known a priori, the symbol counter provides a latch pulse when the next set of symbol samples have gone by. For a short message there will be no need for closed-loop symbol tracking. Many pager devices, popular in the 1990s, worked like this. Also, the ADSB short message discussed in Chapter 12 uses this approach for noncoherent PPM modulation.

---

8 The Costas loop was invented by John P. Costas in 1956 while working for General Electric; see [7].

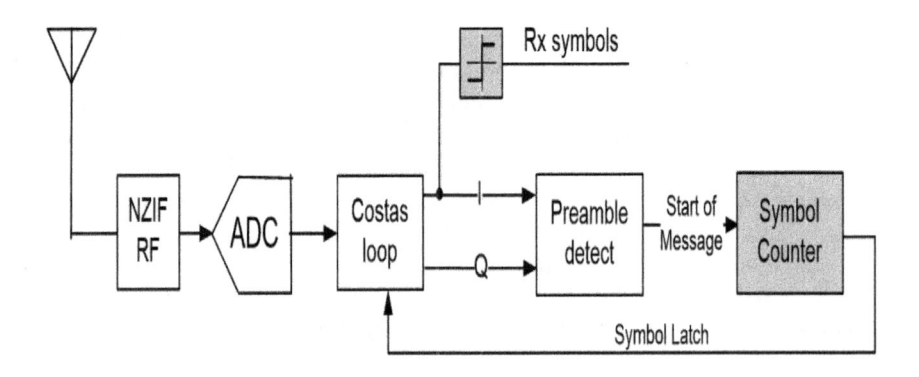

Figure 10.24 Simple, low power receiver based on a Costas loop.

10.2.3.1 BPSK Costas Loop

The received BPSK constellation is in Figure 10.25. Ideal received constellation points are circled dots on the horizontal axis. The actual received symbols, **X**, have a phase error of ϕ. The Costas loop phase error discriminator simply multiplies baseband I and Q. Equation (10.5) shows the phase error discriminator (multiplier) output, assuming a received magnitude of A. As described in [8], this discriminator is commonly used in GPS receivers.

$$v_{imag}(nT) = i(nT)\,q(nT) = A\cos(\phi)\,A\sin(\phi) = \frac{A^2}{2}\sin(2\phi) \qquad (10.5)$$

In Figure 10.26, the BPSK Costas loop, the unshaded blocks run at the input IF sampling rate $1/T_s$ and the shaded blocks run at the symbol rate $1/T$. An important design constraint is that the ADC sample time, ideally T_s, should be an integer divisor of the symbol time T. The pair of multipliers and lowpass filters translate the IF to complex baseband (this is called carrier wipe-off in GPS parlance). For a correctly timed symbol latch, the integrate and dump filter samples the received constellation points. The phase error discriminator output drives a loop filter to set the acquisition and tracking dynamics (loop filters are covered in Section 10.1.4).

CHAPTER 10. SYMBOL AND CARRIER TRACKING

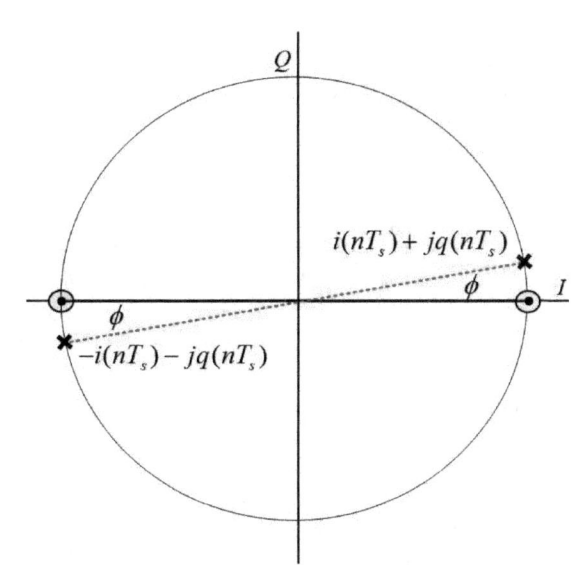

Figure 10.25 Typical BPSK received constellation.

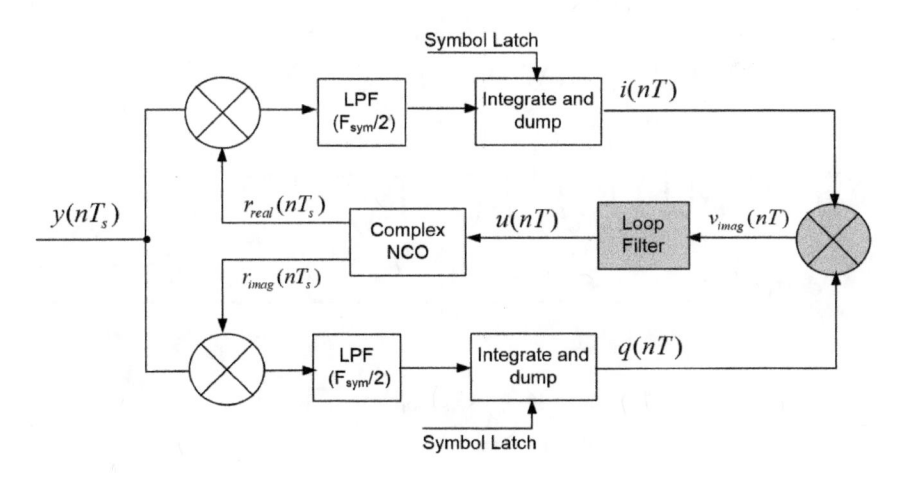

Figure 10.26 A Costas loop for BPSK.

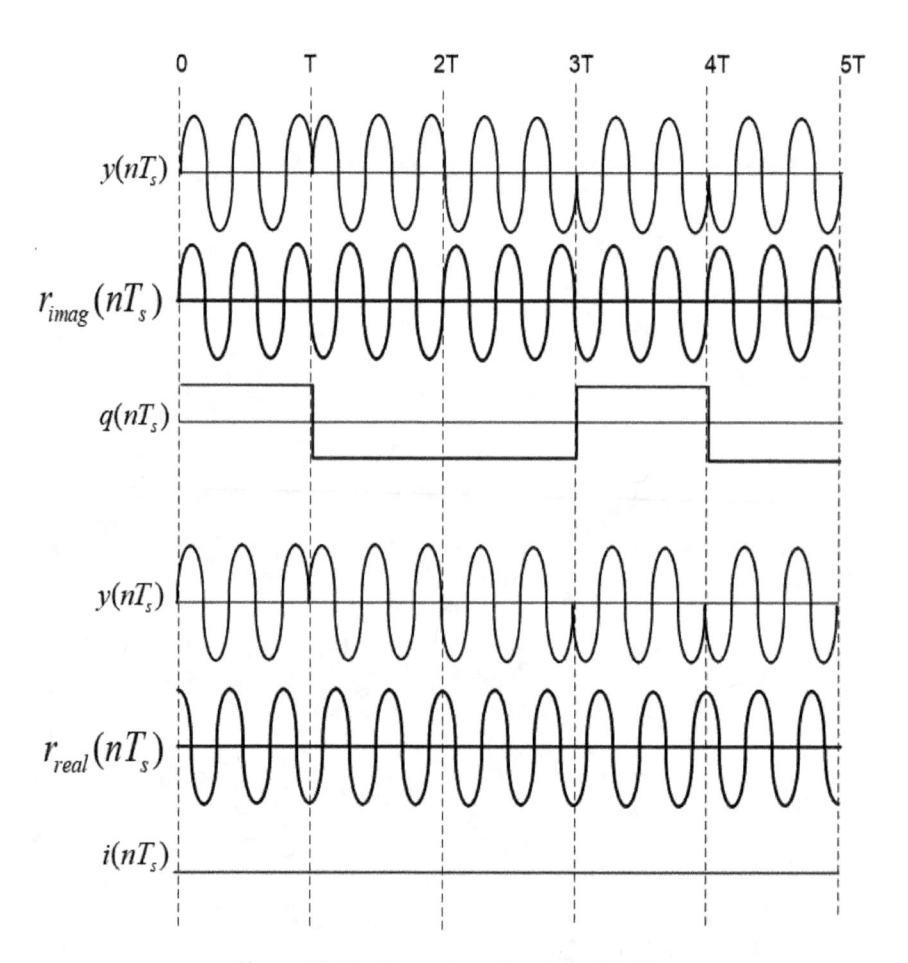

Figure 10.27 Costas loop signals for BPSK.

CHAPTER 10. SYMBOL AND CARRIER TRACKING

The loop filter is at least second order, so that the phase error can be driven to zero. Feedback $u(nT)$ controls the NCO frequency output for carrier removal. When the Costas loop is locked, the signal $i(nT)$ and/or $q(nT)$ will provide a soft decision for received symbols.

In the example in Figure 10.27, the imaginary NCO output is in-phase with the input and thus the lower rail filtered product has the demodulated bits. The real NCO output is 90° out of phase and thus the upper rail filtered product is zero. The two multipliers each have sum and difference frequencies. The lowpass filter removes the sum frequency and leaves the difference frequency. Note that, due to phase ambiguity between the NCO and the input $y(nT_s)$, the symbol soft decisions could be on the in-phase or quadrature outputs. The preamble detector in Figure 10.24 could be used to initialize the loop correctly.

10.2.3.2 QPSK Costas Loop

The received QPSK constellation is in Figure 10.28. A QPSK Costas loop is shown in Figure 10.29. Ideal received constellation points are circled dots. The actual received symbols, **X**, have a phase error of ϕ. I and Q slicers produce ideal constellation points $I(nT) + jQ(nT)$. The imaginary part, $v_{imag}(nT)$, of the complex product of the ideal and actual received symbol is proportional to the phase angle ϕ.

The Costas loop phase error discriminator calculates the imaginary part, $v_{imag}(nT)$, of the complex product of the ideal and actual received symbol; see Equation (10.6). This is proportional to the phase angle ϕ. We could calculate the real and imaginary parts of $v(nT)$ and take the inverse tangent, but that would cost a lot for a minor increase in performance.

$$\begin{aligned} v_{imag}(nT) &= imag\left[(I(nT) - jQ(nT))(i(nT) + jq(nT))\right] \\ &= (I(nT)q(nT) - Q(nT)i(nT)) \end{aligned} \quad (10.6)$$

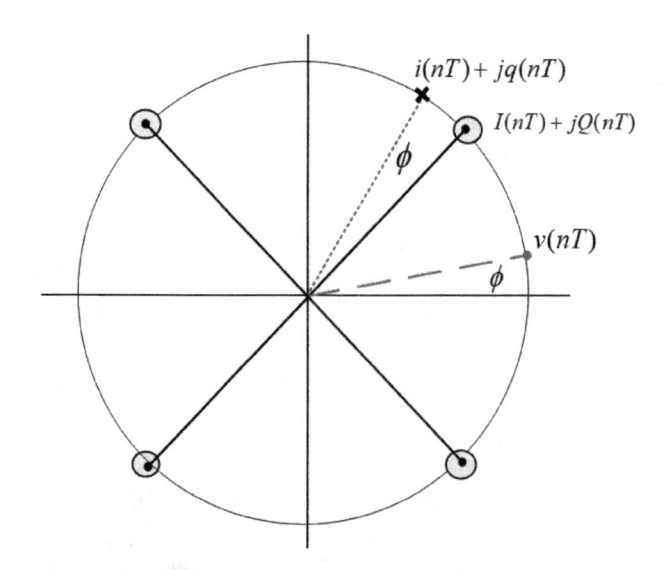

Figure 10.28 Typical QPSK received constellation.

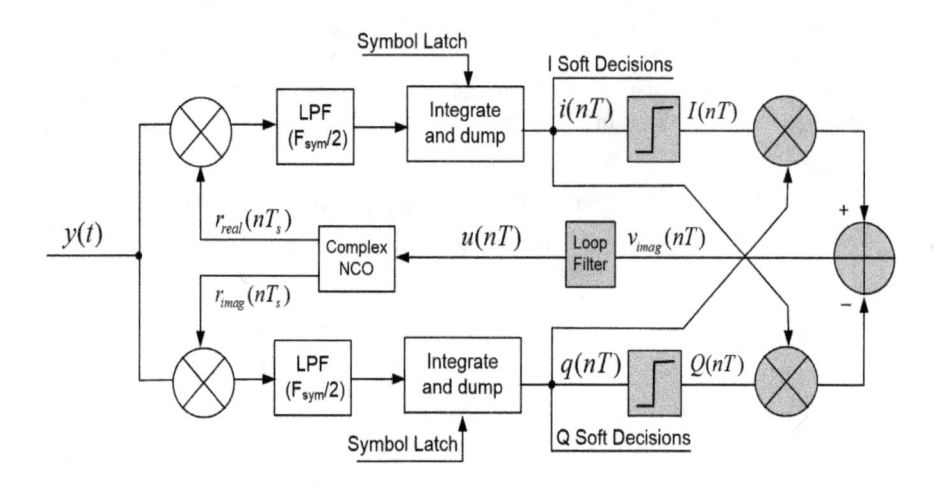

Figure 10.29 A Costas loop for QPSK.

10.3 QUESTIONS FOR DISCUSSION

1. Show how Equation (10.3) changes as the loop delay D increases. Write a MATLAB program to show how the poles move as D gets larger.

2. Closed-loop band-edge and open-loop timing tone symbol tracking use very narrow bandpass filters that may require a large fixed-point precision. True or false?

3. For decision-directed closed-loop carrier tracking, longer feedback decision delay implies:
a. Smaller loop bandwidth;
b. Smaller Doppler tracking range;
c. Longer loop transient response;
d. All of the above.

4. What is the best timing error detector for the following signals:
a. QAM signal with 1,024 states and unknown Doppler frequency offset.
b. BPSK/QPSK signal from an orbiting space craft to a ground station. Note: Velocity is always known so most Doppler can be removed prior to symbol tracking.
c. FSK signal from a pager to a cell site base station.
d. QAM signal with 1,024 states between two fixed mountain-top sites. Symbol rate is less than 15 kHz.

REFERENCES

[1] M. Rice and F. Harris. "Polyphase Filterbanks for Symbol Timing Synchronization in Sampled Data Receivers". *IEEE MILCOM Proceedings*, 2002.

[2] F. M. Gardner. "A BPSK/QPSK Timing-Error Detector for Sampled Receivers". *IEEE Transactions on Communications*, 34(5), 1986.

[3] J. W. Bergmans. *Digital Baseband Transmission and Recording*. Kluwer Academic Publishers, 1996.

[4] D. N. Godard. "Passband Timing Recovery in an All-Digital Modem Receiver". *IEEE Transactions on Communications*, 26(5), 1978.

[5] M. Rice. *Digital Communications: A Discrete-Time Approach*. Second Edition, 2023.

[6] J. Wilson. "Parameter Derivation of Type-2 Discrete-Time Phase-Locked Loops Containing Feedback Delays". *IEEE Transactions on Circuits and Systems II: Express Briefs*, 56(12), 2009.

[7] J. P. Costas. "Synchronous Communications". *Proceedings of the IRE*, 44(12), 1956.

[8] E. D. Kaplan. *Understanding GPS, Principles and Applications*. Artech House, 1996.

Chapter 11

ADSB Digital Signal Processing

11.1 INTRODUCTION

ADSB is a new air traffic control paradigm that lets aircraft identify themselves continuously and provide position and velocity. In this chapter, we study the transmission parameters and structure of the ADSB messages. We also study the design of a functioning ADSB receiver based on the Analog Devices, Inc. ADALM-Pluto SDR.[1] Here is where the ADSB acronym comes from:

Automatic Aircraft ID, heading, altitude, position, and velocity information.
Dependent The transmission of position and velocity depends on proper operation of on-board GPS equipment.
Surveillance Surveillance refers to the flight information transmitted.
Broadcast The information is broadcast to any nearby airplanes or ground stations.

Each ADSB-equipped aircraft sends a regular series of ADSB packets (called squitter). There are two types of ADSB packets, extended squitter (112 bits) and short squitter (56 bits). The extended squitter has the aircraft ID plus a 56-bit combined type code and message while the short squitter only has the aircraft ID. Most aircraft are required to broadcast extended squitter packets on a 1,090 MHz carrier. The repetition time varies from less than a second to several seconds, depending on the message contents. ADSB-equipped aircraft can transmit multiple types of ADSB messages for a complete set of information. This simulation considers extended squitter only.

---

[1] In this chapter we present a receive only design. Under no circumstances is the design information presented here to be used for transmitting ADSB messages.

440 Software Defined Radio: Theory and Practice

11.2 ADALM-PLUTO SDR HARDWARE

With a fixed 1,090 MHz carrier and simple noncoherent pulse position modulation (PPM), the ADSB signal can be processed by many different SDRs. The ANALOG DEVICES ADALM-Pluto SDR has the advantage of reasonable cost and excellent documentation (see www.analog.com). The AD9363 internal radio integrated circuit is widely used in SDR and other radio products. To understand the flexibility and usefulness of the AD9363, consult the AD9363 Design File Package. This can be requested on www.analog.com. Figure 11.1 shows the equipment setup we assume for this chapter.

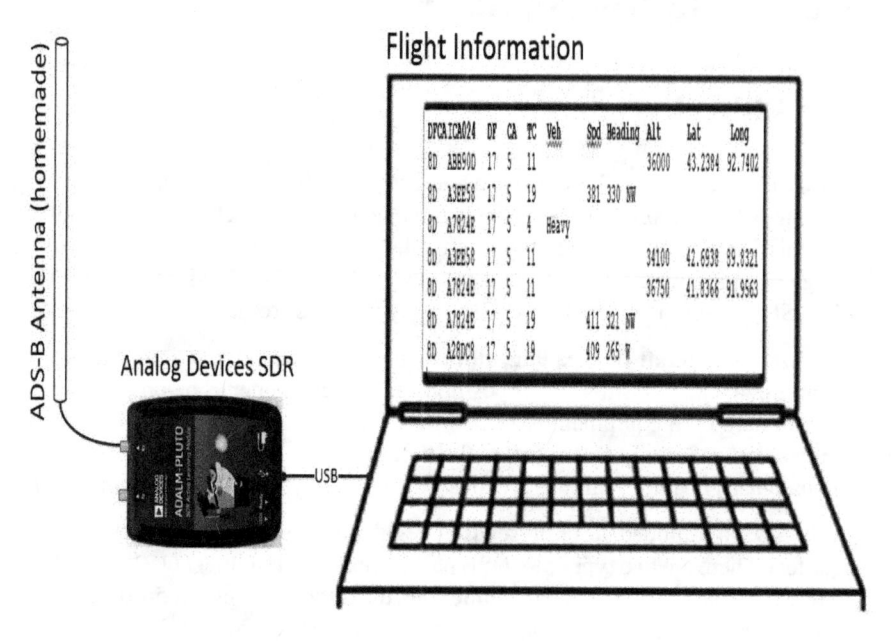

Figure 11.1 Typical SDR setup for receiving and decoding ADSB.

11.2.1 AD9363 Transceiver Interface

The first step in transferring samples from the AD9363 receiver to a Simulink model is to install Simulink Communications Toolbox Support Package for Analog Devices ADALM-Pluto Radio Find this by clicking on "Add-ons" at the top of MATLAB

CHAPTER 11. ADSB DIGITAL SIGNAL PROCESSING 441

workspace window. The next step is to connect the SDR to a spare USB port. Then type A = findPlutoRadio() in the MATLAB workspace. Structure A will be returned with the radio serial number and radio ID number. This verifies that MATLAB can find your ADALM-Pluto.

11.2.1.1 MATLAB Approach

We start with an example of MATLAB code that interfaces to the ADALM-Pluto receiver. The code below gets a new sample from the SDR, checks for sample valid and then calls follow on DSP processing. This code is not used in the Simulink simulation studied in the rest of the chapter. It is only presented for readers who want to control the ADALM-Pluto SDR without using Simulink.

```
rx = sdrrx('Pluto'); %Create SDR receiver System object
% methods(rx) Returns a list of rx object functions
% Change a few object property settings:
% Set receiver center frequency
rx.CenterFrequency = 1.2e+9;
rx.GainSource = 'AGC Fast Attack'; % Set the AGC mode
rx.info %Queries Pluto and returns hardware settings
disp(rx); %Shows object property settings
% DeviceName: 'Pluto'
% RadioID: 'usb:0' What USB on PC?
% CenterFrequency: 1.2000e+09
% (Output frame rate =
% (BasebandSampleRate/SamplesPerFrame)
% BasebandSampleRate: 1000000
% SamplesPerFrame: 20000
% EnableBurstMode: false
% Asynchronous interface between SDR and MATLAB
for k = 1:MaxFrames
[data,valid,overflow] = rx();
% If not valid, sample not ready, have to try again
if(valid)
% Missing samples can't be corrected,
% all we can do is warn user
if(overflow)
```

```
disp('OverFlow Warning');
end
% Call the Receiver DSP implementation
% Hope it completes fast enough to prevent overflow
ReceiverDSP(data);
```

11.2.1.2 Simulink Approach

The "ADALM-Pluto Receiver" in Figure 11.2 and its associated GUI in Figure 11.3 illustrate the Simulink approach to interfacing to the ADALM-Pluto SDR receiver. The "ADALM-Pluto Receiver" block on the left transfers samples through the USB to the Simulink simulation. This block will have available on its data output a 51,840 sample vector at the real world rate of 231.4815 Hz (see the comment in Figure 11.2 about actual sample rate). There are two modes of operation, discussed below, burst and continuous mode.

In *burst mode*, 16,777,216/51,840 = 323 blocks of 51,840 samples each are transferred before a new burst is needed. These 323 blocks will be contiguous, just like a radio in real time. Thus, the overflow flag will be zero during a burst. The radio will continue to receive bursts and output sets of 323 blocks. However, between bursts the overflow flag will probably go high, indicating missed samples. Burst mode is useful for complicated simulations that are not expected to keep up with the SDR in continuous sample mode but still want a block of guaranteed contiguous samples.

In *continuous mode* (i.e., burst mode not enabled in Figure 11.3) the radio simply outputs a continuous series of 51,840 sample blocks. The Simulink model is responsible for keeping up with these. If there are no missed samples and the overflow flag stays low, the radio to Simulink interface has achieved a form of stream processing. In this case, continuous mode may be the best choice. The fairly simple ADSB simulation discussed here will probably work in either mode.

Note in Figure 11.3 that the AGC is set to AGC Fast Attack. This is appropriate for these short ADSP squitters. The AGC starts at a high gain with no received signal present. When the signal starts, the AGC quickly ramps down to the correct level for the AD9363's ADC processing. We would like the AGC to be settled by the end of the preamble; see Figure 11.7. The setting AGC Slow Attack is for longer transmissions that can afford to wait for a slower, but more stable, AGC adjustment.

At the simulation start, when Simulink starts to update all the block states from input to output, the PHY Layer Simulink block looks for a 51,840 sample vector. In burst mode, 323 of these vectors will be already sampled and available for transfer,

CHAPTER 11. ADSB DIGITAL SIGNAL PROCESSING 443

so the simulation can take its time. In continuous mode, only one 51,840 sample vector will be available and the simulation needs to finish processing and input the next sample vector to avoid overflow.

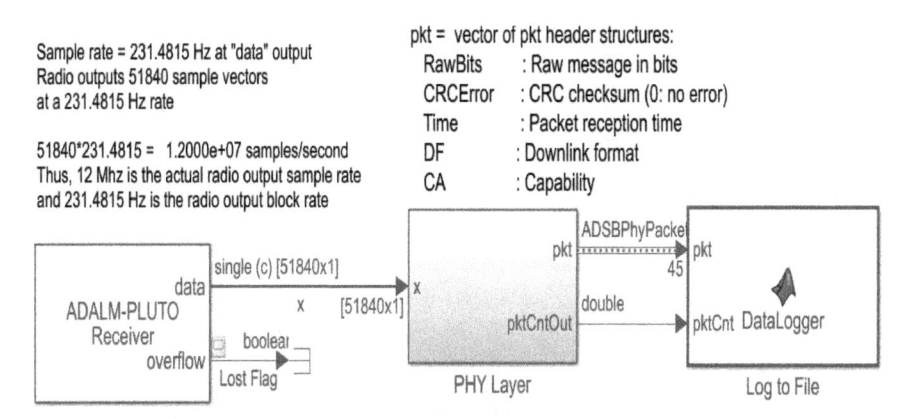

```
Sample rate = 231.4815 Hz at "data" output
Radio outputs 51840 sample vectors
at a 231.4815 Hz rate

51840*231.4815 = 1.2000e+07 samples/second
Thus, 12 Mhz is the actual radio output sample rate
and 231.4815 Hz is the radio output block rate
```

pkt = vector of pkt header structures:
RawBits   : Raw message in bits
CRCError : CRC checksum (0: no error)
Time   : Packet reception time
DF   : Downlink format
CA   : Capability

ADLAM-Pluto radio interface is a combination of Simulink and actual radio hardware
The "data" output represents a hardware interface not driven by Simulink
(AD9363 sample clock is a multiple of the "data" interface sample clock).
Sample vectors can be missed if not read fast enough. Overflow flag indicates this.

Figure 11.2 ADSB receiver, top level.

Below is the MATLAB code to set up and run the simulation in Figure 11.2. Note that the RxStpTim calculation is based on one burst with the largest number of burst samples that Pluto can provide. As discussed, this burst is guaranteed to have no missing samples. This simulation may run in continuous mode on a fast enough PC.

```
% This script sets up and simulates an ADPS decoder
% sdrsetup modifies system PATH and LD_LIBRARY_PATH
% variables in the current MATLAB session to enable
% running the support package sdrsetup
% Setup 'adsbParam' struct to control ADSB decoding
% AdsbConfig is a copyrighted MathWorks function
radioSampleRate = 2.4e6;
adsbParam = AdsbConfig(radioSampleRate);
```

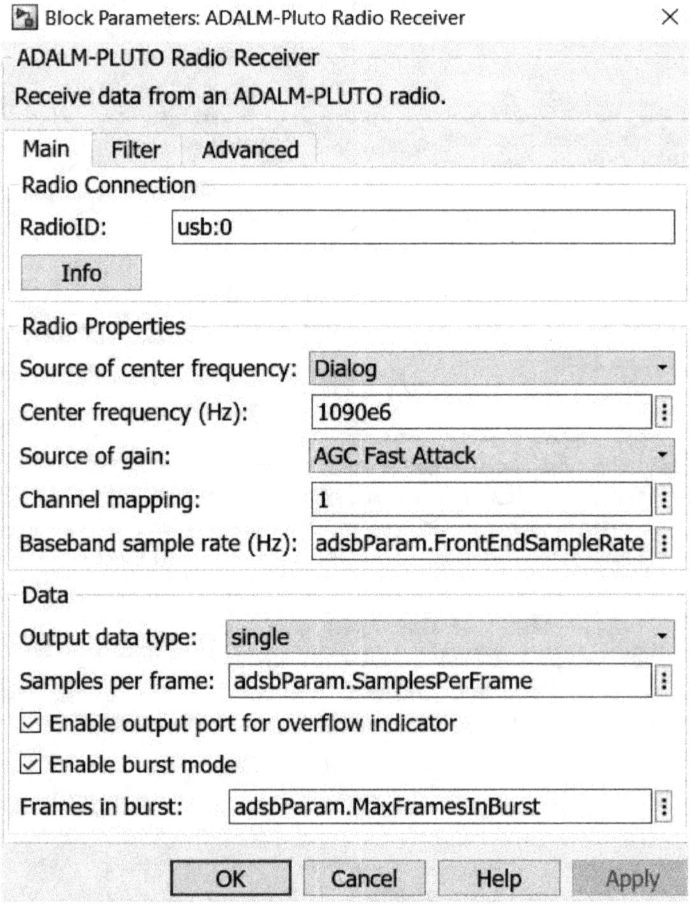

Figure 11.3 ADALM-Pluto Simulink setup GUI.

CHAPTER 11. ADSB DIGITAL SIGNAL PROCESSING

```
load dataAdsbBusObjects
% Calculate the simulation stop time based on
% maximum samples/burst and samples/frame:
% Max samples/burst =16777216 (Pluto SDR specification)
% (51840 samples/frame) * (N frames/burst)
%< (16777216 samples/burst)
adsbParam.MaxFramesInBurst=floor(16777216/51840); %322;
% stop timeneeded to process max samples in a burst
RxStpTim =...
((adsbParam.MaxFramesInBurst...
*adsbParam.SamplesPerFrame)/12e6);
% Run ADSB for burst length (about 1.3 seconds)
% Log results in ADSBlogger.txt file
set_param('ADSBdemoV1','Solver',...
'VariableStepDiscrete',...
'StopTime',num2str(RxStpTim));
sim('ADSBdemoV1', 'SimulationMode', 'normal');
```

11.2.2 AD9363 Receive Antenna

A properly constructed ADSB receive antenna is essential. We describe here a coaxial colinear (COCO) array of halfwave dipoles. The antenna will be resonant at 1,090 MHz, the ADSB carrier frequency.[2] A commercial version of this antenna can be purchased. However, some readers may want to construct their own.

Start with several feet of 50 Ohm coax cable[3] (if you work in a lab, check the recycle bin). Construct an even number of the sections in Figure 11.4; more sections result in better performance. The ADSB signal wavelength is (299,792,000 meters/second) / (1,090,000,000 cycle/second) = 27.5 cm. From this a half-wavelength is 13.75 cm.

For length X we need the coax velocity factor; in this case, we used RG-59A/U PE (Belden 8241), velocity factor = 0.66. For best performance, check the velocity factor carefully.

---

2 "Is COCO Your Cup of Tea," by L.B. Cebik, available at http://on5au.be/content/a10/vhf/coco.pdf provides interesting technical insights into how this antenna works.

3 RG-58 is an example of 50 Ohm and RG-59 is an example of 75 Ohm. Either cable can be used.

For our antenna, X = 0.66*13.75 = 9.075cm. Trimmed bare copper length Y is about a half inch. A long feed line to the ADALM-Pluto receiver input should be trimmed the same way.

As shown in Figure 11.5, connect the half-wavelength sections by pushing the exposed center conductor into the next section's braided shield area (be careful; it is easy to push the center conductor into your finger). Leave the far end of the cable open. These connections should be wrapped with electrician's tape or covered with heat shrink tubing. An even number of sections, say, 6 or 8, should work well. Finally, for added durability, the entire assembly can be secured inside a length (several feet) of half-inch diameter plastic pipe.

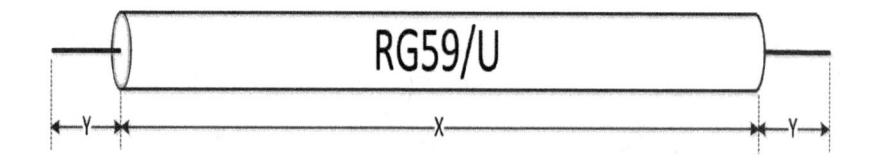

Figure 11.4 ADSB receive antenna section.

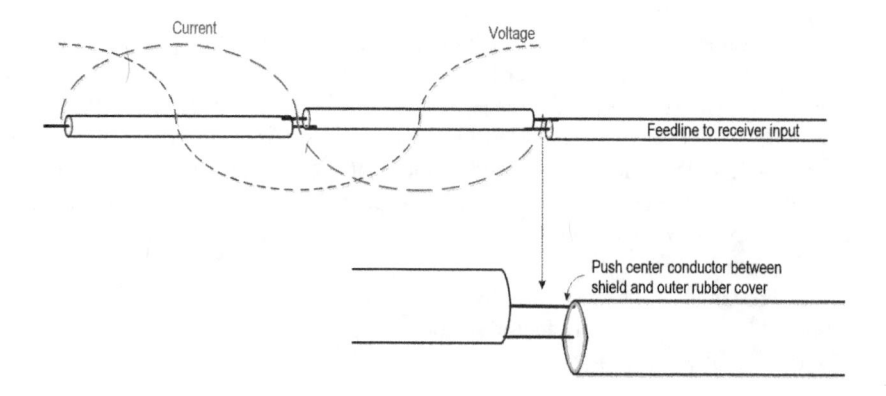

Figure 11.5 ADSB receive antenna section coupling.

11.3 ADSB DIGITAL SIGNAL PROCESSING

The ADSB design presented here is a vastly simplified version of the MathWorks ADSB receiver described in [1], [2], [3], and [4]. Readers interested in the full-featured design may want to start with the design presented here as an easier-to-understand preliminary learning step[4].

Most of the DSP is in the PHY Layer block in the center of Figure 11.2. The contents of that block are shown in Figure 11.6. While the simulation is running in burst mode, the PHY Layer block inputs 323 frames of 51,840 samples each (see MATLAB code above).

The first block on the left side of Figure 11.6 calculates the power in each received complex sample, discarding the phase. The buffer block increases the 51,840 sample frame size to 53,280 samples by taking 1,440 samples from the end of the current output and repeating them in the beginning of the next output. This is done to ensure that an ADSB packet that straddles two 51,840 sample buffers is not missed. Figure 11.7 shows a typical output of the 51,480 sample buffer at the top and the ADSB packet samples in that buffer at the bottom. The job of the ADSB receiver is to find messages and extract the information that they contain.

11.3.1 ADSB Preamble Detect

Figure 11.8 shows that each bit beyond the preamble is composed of a chip that can be at the bit start half (for a 1) or the bit end half (for a 0). The radio sample rate is 12 MHz and the bit rate is 1 MHz, so the number of samples per bit is 12 and the samples per chip is 6. The preamble chips look like the data, however, they are not meant to be decoded the same way.

The PHY Packet Search block in Figure 11.6 has two inputs, xFilt for preamble searching and xBuff for validating the preamble and also for extracting packet data beyond the preamble. XFilt is driven by the preamble sync pattern matched filter with coefficients shown in Figure 11.6. Note that, due to the filter convolution, the matched filter coefficients are reversed from the preamble pattern in Figure 11.8. This FIR filter is being used as a pattern detector. An example of a preamble correlation hit is shown in Figure 11.9, at about sample number 450. This hit is not reliable enough to declare a preamble detect. The position of this ADSB packet must be checked and refined by sync pattern validator.

---

[4] Some of the detailed ADSB technical specifications are also found in "The 1090 MHz Riddle, A Guide to Decoding Mode S and ADSB," from an article by J. Sun published in 2021 by TU Deft Open Publishing.

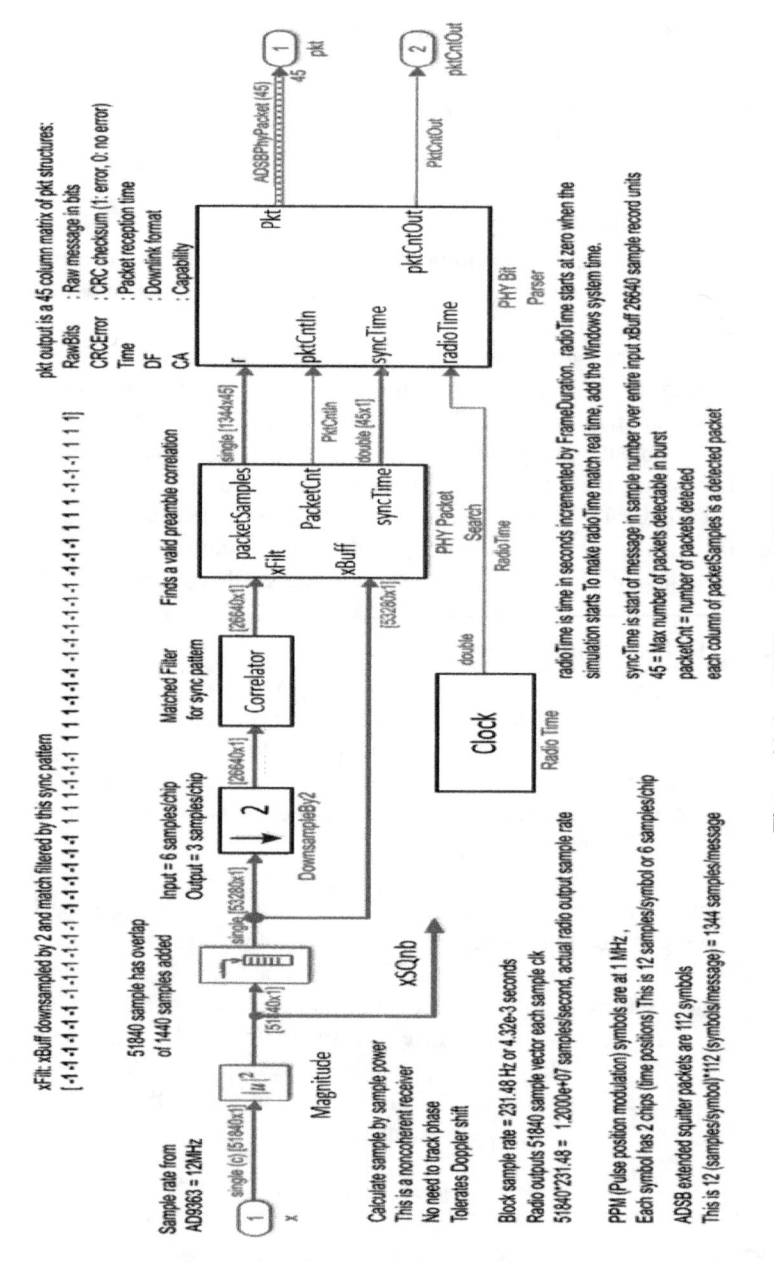

Figure 11.6 ADSB receiver, PHY layer.

CHAPTER 11. ADSB DIGITAL SIGNAL PROCESSING 449

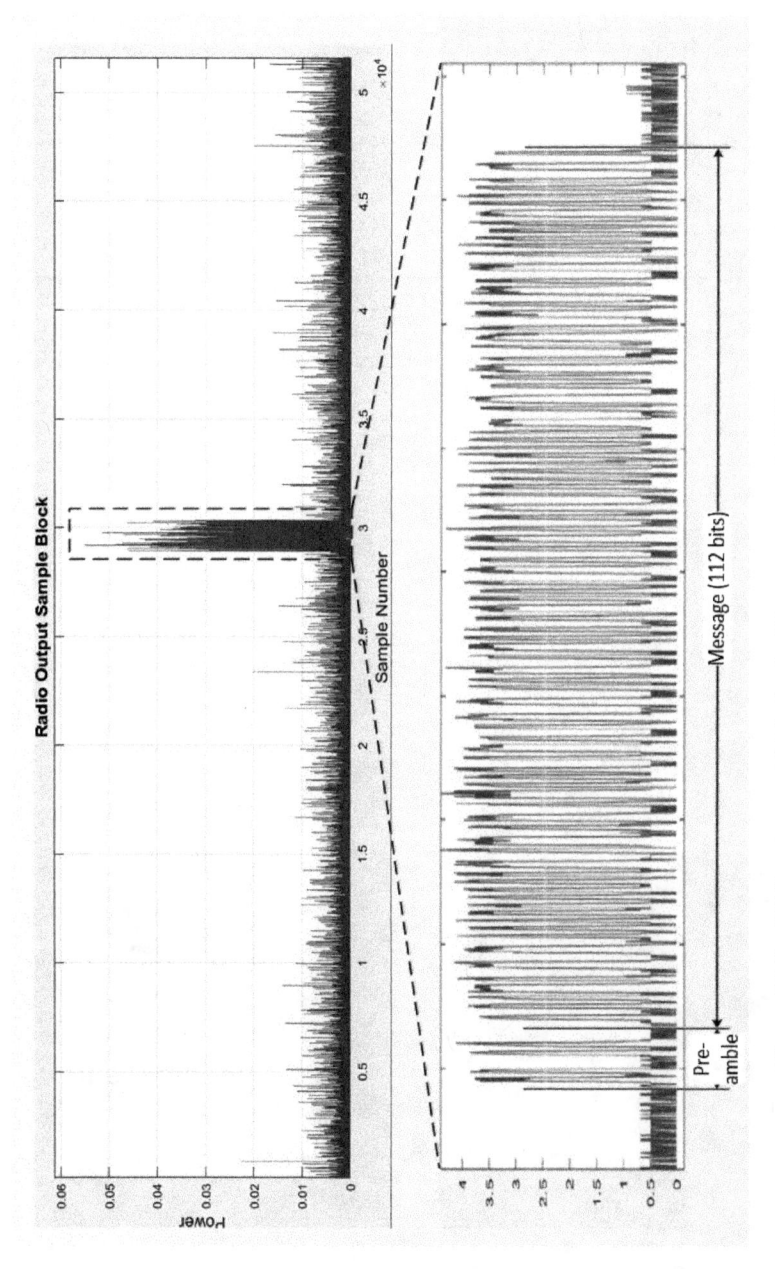

Figure 11.7 PHY packet search block, xbuff input vector, and ADSB message extraction.

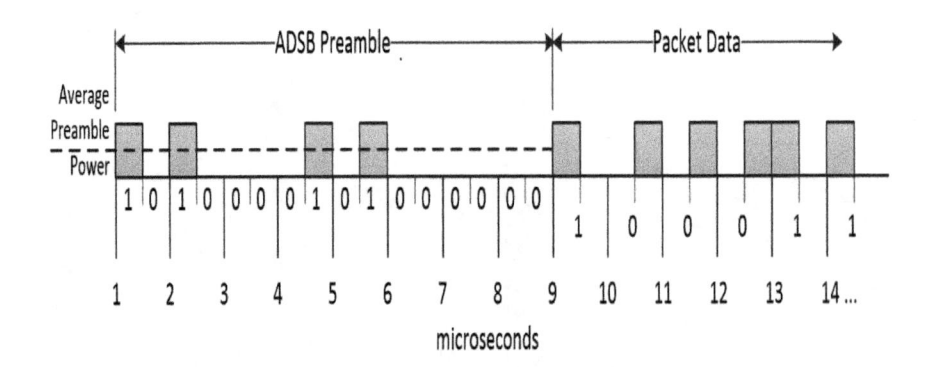

Figure 11.8 ADSB message preamble.

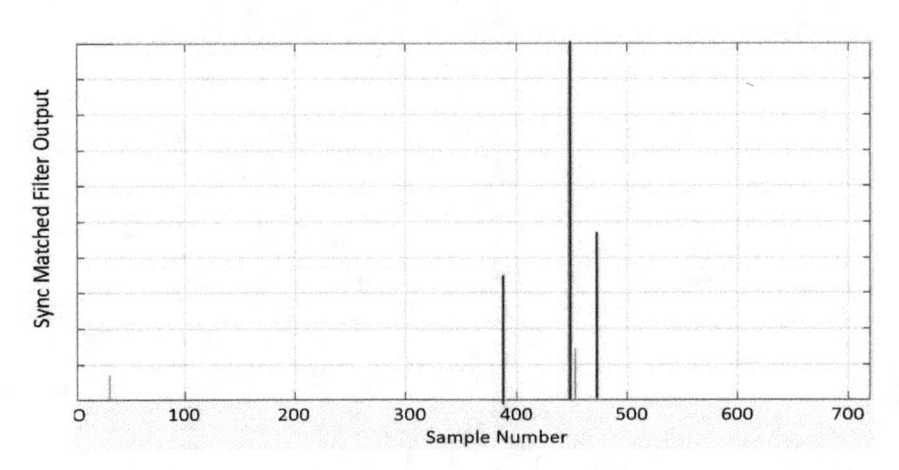

Figure 11.9 ADSB sync matched filter output.

CHAPTER 11. ADSB DIGITAL SIGNAL PROCESSING

Figure 11.10 is the sync validator circuit. The signal xSQnb is a single sample unbuffered stream shown in Figures 11.6 and 11.10. Suffix nb means not buffered and xSQnb provides a sample-by-sample input to Figure 11.10 for this explanation. In Figure 11.6, xSQnb implies a parallel to serial converter that is not shown. xSQnb is for explanatory purposes only. Likewise, signal xFiltnb on Figure 11.10 is a single sample version of xFilt in Figure 11.6.

Consider the 16 sum blocks down the center of Figure 11.10. Notice that each set of six blocks is summing a different set of time indices that taken together span the 96 preamble samples (6 samples per chip times 16 chips; see Figure 11.8). When 96 samples from a complete preamble happen to occupy the shift register circuit on the left of the figure, then the series of block outputs, top to bottom, are the sums of the series of six sample chips in Figure 11.8. Now consider that the Low Avg, High Avg, and Average blocks in the lower right. When a valid preamble is shifted in, the low average is for the zero chips and the high average is for the nonzero chips in Figure 11.8. This average is shown in Figure 11.8. The comparators on the far right of Figure 11.10 validate the preamble by detecting when $c(1)$, $c(3)$, $c(8)$, $c(10)$ are above average power and the others are below average power. Note that for this to work, n must be the current sample (at the preamble end) and $n - 95$ must be 96 samples back in time (at the preamble beginning).

11.3.2 ADSB Packet Samples Tabulation

The preamble validation circuit is able to accurately establish both chip and packet boundaries. The PHY Packet Search block in Figure 11.6 is able to search the xBuff input sample vector for ADSB messages. For every vector of 53,280 input samples, three outputs are produced from the PHY Packet Search block:

packetSamples For each valid preamble detect pulse, 1,344 real message samples (power only, no phase) are extracted from the received sample block. Note that 1,344 = (12 samples/bit)(112 bits/message). Each new ADSB message has its packet samples arranged in a column of the packetSamples matrix; see Table 11.1
packetCnt is the number of valid packet columns in the packetSample matrix, max = 45.
SyncTime is a vector of up to 45 XBuff starting indexes, one for each valid column in packetSamples. The number of valid columns is usually less than 45.

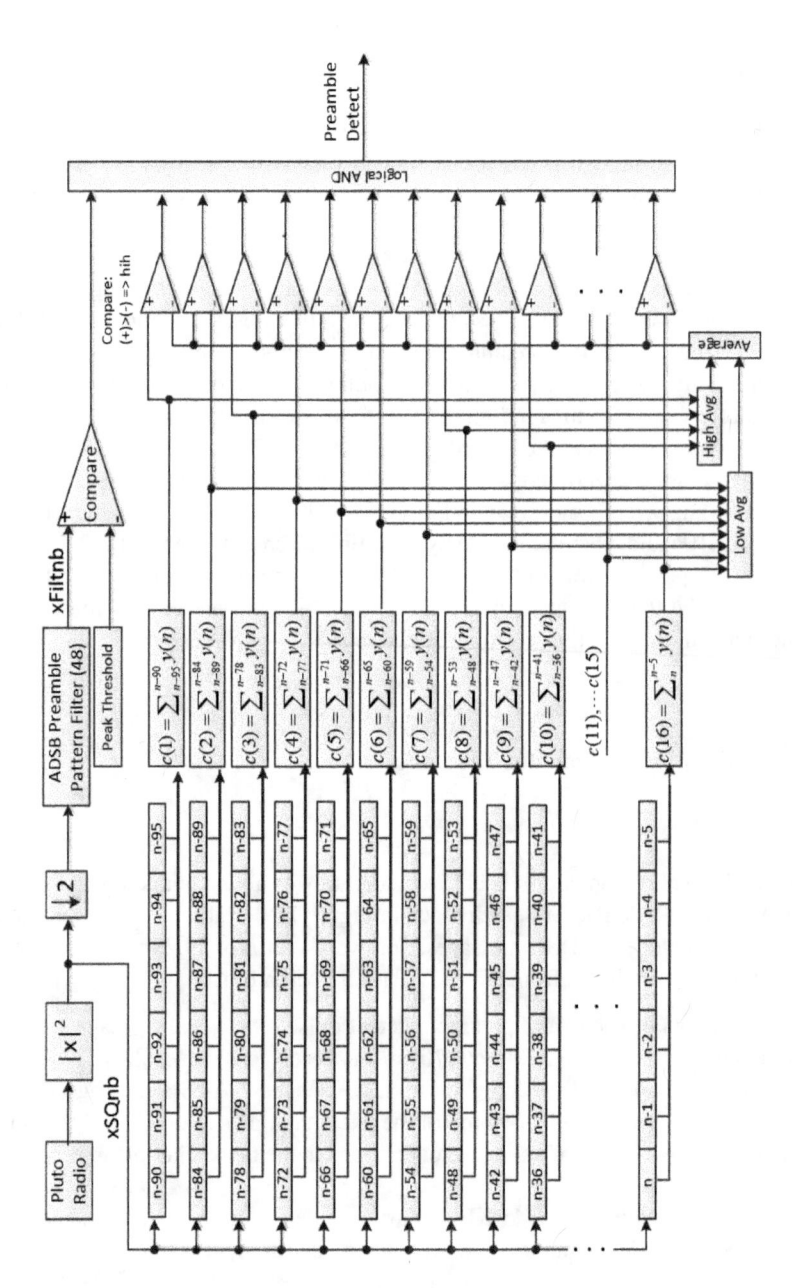

Figure 11.10 ADSB sync validator circuit.

CHAPTER 11. ADSB DIGITAL SIGNAL PROCESSING

Table 11.1 PHY Packet Search, packetSamples Output

Packet 1	Packet 2	$\cdots$	Packet 45
$x(1,1)$	$x(2,1)$	$\cdots$	$x(45,1)$
$x(1,2)$	$x(2,2)$	$\cdots$	$x(45,2)$
$x(1,3)$	$x(2,3)$	$\cdots$	$x(45,3)$
$\cdots$	$\cdots$	$\cdots$	$\cdots$
$x(1,1344)$	$x(2,1344)$	$\cdots$	$x(45,1344)$

11.3.3 bitParser Block

Further processing is needed to convert the 12 samples/bit columns of packetSamples into 112 single bit values, 1 or 0. As shown in Figure 11.8 and discussed previously, for each bit either the first chip has six power samples > 0 and the second chip has six zero samples or vice versa. This is called pulse position modulation or PPM (see Section 3.3.6). The matrix multiply approach in Equation (11.1) is used to convert PPM position into an antipodal soft decision. The first matrix on the left is simply the received packet samples where each row contains 12 bits (as aligned by preamble detection, similar to the columns of Table 11.1). The column vector of +1s followed by -1s in the center converts the 12 samples to a single ±1 soft decision for each of the 112 message bits. In this system, these soft decisions are simply sliced around zero to produce 1 (for > 0) or 0 (for < 0) hard decisions. Each set of 112 bits are organized into different fields of a pkt(n) structure, where $n = 1$ to 45. The final output of the PHY Layer circuit in Figure 11.6, ADSBPhyPacket(45), is an array of these structures.

11.4 ADSB PHY LAYER OUTPUT DETAILS

Now that we have organized up to 45 separate 112-bit PHY packets into ADSBPhyPacket(45), we can start to look at bit fields within each packet. This section has details on how the information in Table 11.2 was extracted from the 112 bits in each packet. Note that pkt(n) is a structure and can have multiple bits assigned to each field. For example, pkt(n).Header.RawBits has a length of 112.

$$\begin{bmatrix} x_{1,1} & x_{1,2} & x_{1,3} & x_{1,4} & x_{1,5} & x_{1,6} & x_{1,7} & x_{1,8} & x_{1,9} & x_{1,10} & x_{1,11} & x_{1,12} \\ x_{1,13} & x_{1,14} & x_{1,15} & x_{1,16} & x_{1,17} & x_{1,18} & x_{1,19} & x_{1,20} & x_{1,21} & x_{1,22} & x_{1,23} & x_{1,24} \\ x_{1,25} & x_{1,26} & x_{1,27} & x_{1,28} & x_{1,29} & x_{1,30} & x_{1,31} & x_{1,32} & x_{1,33} & x_{1,34} & x_{1,35} & x_{1,36} \\ \cdots & \cdots & \cdots & \cdots & \cdots & \cdots & \cdots & \cdots & \cdots & \cdots & \cdots & \cdots \\ x_{1,1333} & x_{1,1334} & x_{1,1335} & x_{1,1336} & x_{1,1337} & x_{1,1338} & x_{1,1339} & x_{1,1340} & x_{1,1341} & x_{1,1342} & x_{1,1343} & x_{1,1344} \end{bmatrix} \begin{bmatrix} 1 \\ 1 \\ 1 \\ 1 \\ 1 \\ 1 \\ -1 \\ -1 \\ -1 \\ -1 \\ -1 \\ -1 \end{bmatrix} = \begin{bmatrix} s_{1,1} \\ s_{1,2} \\ s_{1,3} \\ \cdots \\ s_{1,112} \end{bmatrix}$$

$$\begin{bmatrix} y_{1,1} \\ y_{1,2} \\ y_{1,3} \\ \cdots \\ y_{1,112} \end{bmatrix} = \text{sgn} \begin{bmatrix} s_{1,1} \\ s_{1,2} \\ s_{1,3} \\ \cdots \\ s_{1,112} \end{bmatrix} \tag{11.1}$$

CHAPTER 11. ADSB DIGITAL SIGNAL PROCESSING 455

Table 11.2 ADSB PHY Packet bitParser Output

Struct	Field	Subfield	Contents
pkt(n)	.Header	.RawBits	112 message bits
pkt(n)	.Header	.CRCError	True/false
pkt(n)	.Header	.Time	RadioTime pkt detected
pkt(n)	.Header	.DF	Downlink format
pkt(n)	.Header	.CA	Capability
pkt(n)	.ICA024		Aircraft unique ID
pkt(n)	.TC		Type Code: msg struct
pkt(n)	.AirborneVelocity	.IntentChange	0: None, 1: intent chg.
pkt(n)	.AirborneVelocity	.IFRCapability	Instrument flight rules
pkt(n)	.AirborneVelocity	.VelocityUncertainty	Velocity error estimate
pkt(n)	.AirborneVelocity	.Speed	Knots
pkt(n)	.AirborneVelocity	.Heading	Degrees, rel. to north
pkt(n)	.AirborneVelocity	.HeadingSymbol	N, NE, ... W, NW
pkt(n)	.AirborneVelocity	.VerticalRateSource	0: GNSS, 1: barometric
pkt(n)	.AirborneVelocity	.VerticalRate	Feet/minute
pkt(n)	.AirborneVelocity	.TurnIndicator	
pkt(n)	.AirborneVelocity	.GHD	Geometric height dist.
pkt(n)	.Identification	.VehicleCatagory	Light, medium, heavy
pkt(n)	.Identification	.FlightID	Airline and flight no.
pkt(n)	.AirbornePosition	.Status	Alert status
pkt(n)	.AirbornePosition	.DiversityAntenna	1: single, 0: dual
pkt(n)	.AirbornePosition	.Altitude	Feet
pkt(n)	.AirbornePosition	.UTCSynchronized	Universal coord. time
pkt(n)	.AirbornePosition	.CPRFormat	Compact position report
pkt(n)	.AirbornePosition	.Longitude	Deg, min, sec
pkt(n)	.AirbornePosition	.Latitude	Deg, min, sec

Let's examine the 112-bit ADSB pkt(n).Header.RawBits structure outlined; see Table 11.3. The first three nonpreamble rows, as well as the CRC, are on every ADSB squitter. The shaded rows of Table 11.3 vary with the message. The left column shows the bit position in the 112 packet bits.

11.4.1 Aircraft ID, Type Code = 1,2,3,4

The shaded rows in Table 11.3 show the 56-bit ADSB message. These 56 bits include the type code and message bits. The type code is there to show how to decode the message bits; for this example, the type code indicates an Aircraft ID message; see Table 11.4. The aircraft category subtype has to do with the size and type of the plane (e.g., heavy is the kind of multiaisle plane you might fly coast to coast on).

Each 6-bit character of the aircraft call sign must be expressed as an index into to a $2^6 = 64$ element lookup table to reveal the text of the airline and flight number. For small x = unused and _ = separation, the lookup table is shown below. For example, AAL586 means American Airlines flight number 586 and is coded as: 1,1,12,53,56,54,0,0.

```
%Look up Table for Aircraft Call Sign:
%xABCDEFGHIJKLMNOPQRSTUVWXYZ
%xxxxx_xxxxxxxxxxxxxxxx0123456789xxxxxx
```

11.4.2 Aircraft Velocity, Type Code = 19

Whereas the previous type code 1,2,3,4 was simply information read directly from the msg (except for the lookup table), the velocity type code 19 requires a little more interpretation. Table 11.5 parses out the message contents.

Consider velocity subtype 1, subsonic ground speed reported as cartesian coordinates. Instead of the usual (x, y) from algebra, we use V_{EW}, V_{NS} as coordinates of the aircraft velocity vector.[5] The MATLAB code, next, illustrates the conversion of the velocity subtype 1 message bits into the two final velocity results: ground speed in knots and dir = compass point symbol.

---

5 This is a two-dimensional velocity vector, as if the aircraft is traveling in a straight line along flat ground.

CHAPTER 11. ADSB DIGITAL SIGNAL PROCESSING

Table 11.3 ADSB Top Level Bit Fields

Bit Field	Field Name	Description
N/A	Eight preamble bits	Discarded after detection
1:5	DF	Downlink format: 17 for transponder, 18 for nontransponder
6:8	CA	Capability (IFR)
9:32	ICA024	Aircraft unique ID, used to look up model number
33:37	TC	Type code: indicates ADSB message structure
38:88	Message	TC indicates how to extract position, velocity, lat, and long
89:112	CRC	Cyclic redundancy check, used for bit error detection

Table 11.4 ADSB Aircraft ID Message Contents

Bits	Msg	Field Name	Description
33:37	1:5	Type code = 1,2,3,4	Indicates aircraft ID msg
38:40	6:8	Aircraft category subtype	0: No data
			1: Light
			2: Medium
			3: Heavy
			4: High vortex
			5: Very heavy
			6: High speed
			7: Rotocraft
41:46	9:14	0 Character	
47:52	15:20	1 Character	
53:58	21:26	2 Character	
59:64	27:32	3 Character	Aircraft call sign
65:70	33:38	4 Character	
71:76	39:44	5 Character	
77:82	45:50	6 Character	
83:88	51:56	7 Character	

CHAPTER 11. ADSB DIGITAL SIGNAL PROCESSING 459

Table 11.5 ADSB Aircraft Air Velocity Message Contents

Bits	Msg	Field Name	Description
33:37	1:5	Type code = 19	Indicates aircraft velocity msg
38:40	6:8	Velocity subtype	0: Reserved
			1: Ground, Cartesian, subsonic
			2: Ground, Cartesian, supersonic
			3: Air, polar, subsonic
			4: Air, polar, supersonic
			5: Reserved
			6: Reserved
			7: Reserved
41	9	Intent	0: No change, 1: change
42	10	IFR capability	Instrument flight rules
43:45	11:13	Velocity uncertainty	0: Unknown
			1: < 10 m/s
			2: < 3 m/s
			3: < 1 m/s
			4: < 0.3 m/s
46	14	EW sign	0: Flying E to W, 1: flying W to E
47:56	15:24	EW value	East-west velocity, knots
57	25	NS sign	0: Flying N to S, 1: flying S to N
58:67	26:35	NS value	North-south velocity, knots
68	36	Vert. rate source	0: GNSS, 1: barometric
69	37	Vert. rate sign	0: Ascending, 1: descending
70:78	38:46	Vert. rate value	Vertical rate (ft/min) = value*64
79:80	47:48	Turn indicator	
81	49	GHD sign	
82:88	50:56	GHD	Geometric height distribution

```
% For each Cartesian direction, convert
% message bits to unsigned integer
% Then convert to signed and apply
% sign from message bit
value = [512 256 128 64 32 16 8 4 2 1]*adsbBits(15:24);
VelEW = (1-2*adsbBits(14))*(value-1);
value = [512 256 128 64 32 16 8 4 2 1]*adsbBits(26:35);
VelNS = (1-2*adsbBits(25))*(value-1);
% Compute vector length (speed in Knots)
% and angle (heading in radians)
(heading,speed) = cart2pol(VelEW,VelNS);
heading = heading*180/pi; % Convert to degrees
if heading<0
heading = 360+heading; % Convert to unsigned
end
% Convert heading in degrees to compass points
% Note that ADSB has put North at 0 degrees
% Dir is chosen from
% [N, NE, E, SE, S, SW, W, NW]
% by comparing heading to angle sectors.
```

11.4.3 Aircraft Position, Type Code = 9:18

The ADSB position message is shown below in Table 11.6. Much of the decoding is straightforward, for example, altitude is obtained by first converting bit concatenation [9:15 17:20] into an unsigned integer, call it N. Then the altitude in feet is simply $N*$(altitude step multiplier)-1,000. However, the latitude and longitude are somewhat complicated because they require an understanding of CPR (Compact Position Reporting, also known as Complicated Position Reporting).

11.4.3.1 CPR

CPR is how latitude and longitude are derived from ADSB aircraft position messages. In this example, calculation of unambiguous aircraft position requires two, close in time and the second following the first, messages from the same aircraft[6] (the same aircraft means the same ICA024 number, not call sign). From each message, we store

---

6 There is a version of CPR, not covered here, that only needs one message.

CHAPTER 11. ADSB DIGITAL SIGNAL PROCESSING

Table 11.6 ADSB Aircraft Position Message Contents

Bits	Msg	Field Name	Description
33:37	1:5	Type code = 9:18	Indicates an aircraft position message
38:39	6:7	Surveillance status	0: No emergency
		1: Permanent alert	
		2: Temporary alert	
		3: SPI	
40	8	Single antenna flag	0: Dual antenna, 1: Single antenna
41:47	9:15	Altitude bits 10:4	
48	16	Altitude step multiplier	0: 100 feet, 1: 25 feet
49:52	17:20	Altitude bits 3:0	
53	21	Time	0: Not UTC synchronized, 1: UTC synch.
54	22	CPR format (F bit)	0: Even, 1: Odd
55:71	23:39	CPR encoded latitude	Lat = (CPR encoded latitude)/131072
72:88	40:56	CPR encoded longitude	Lng = (CPR encoded longitude)/131072

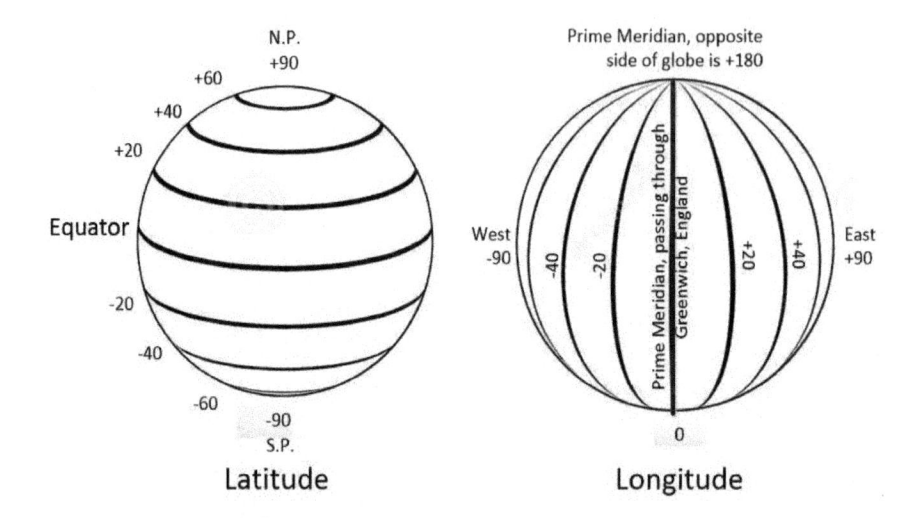

Figure 11.11 Review of latitude and longitude.

the information in the three shaded rows of Table 11.6. One additional requirement is that the F bit in each message must be different.

Start by calculating variables Lat and Lng in the last two rows of Table 11.6. This requires first converting the 17-bit binary field into an unsigned integer. Because the maximum unsigned integer will be $2^{17} - 1 = 131,071$, Lat and Lng will end up being in the range [0 131,071) and both Lat and Lng will be ≥ 0 and < 1. Table 11.7 is an example for two different pairs of aircraft position messages, one in each row. These were received from the same aircraft. Either pair below will be useful.

Table 11.7 ADSB Compact Position Reporting, Initial Information

Situation 1			Situation 2		
Msg	F bit	Lat,Lng Designations	Msg	F bit	Lat,Lng Designations
1	0	LatEv, LngEv	1	1	LatOd, LngOd
2	1	LatOd, LngOd	2	0	LatEv, LngEv

Figure 11.11 hints at how latitude and longitude can be used to pinpoint a location on the globe. We start by using the data from Table 11.7 to calculate latitude from CPR data. In this system, latitude is divided into zones and each zone is divided into 2^{17} bins. Equation (11.2) starts by calculating the latitude zone

CHAPTER 11. ADSB DIGITAL SIGNAL PROCESSING

number, z, as well as the number of degrees covered by each even or odd zone. Then Equation (11.3) provides the final latitude in degrees.

Longitude calculation is slightly more complicated because we need a way to calculate a reduction in the number of longitude zones for latitudes near the Earth's poles; see Figure 11.11. The number of longitude zones (called NL) corresponding to a specific latitude is calculated in Equation (11.4). To make the calculation of NL easier, MathWorks has implemented Equation (11.4) as a lookup table function $NL = helperAdsbNL(lat)$. As we did for latitude, Equation (11.5) first calculates the longitude zone number, m, and then calculates constants equal to the number of degrees covered by each even or odd zone. Equation (11.6) provides the final longitude calculation.

$$\begin{aligned} z &= \lfloor 59 LatEv - 60 LatOd + 0.5 \rfloor \\ DegLatOdd &= 360/59 = 6.1 \\ DegLatEven &= 360/60 = 6.0 \end{aligned} \qquad (11.2)$$

$$\begin{aligned} LatitudeOdd &= DegLatOdd \, (\mathrm{mod}(z, 59) + LatOd) \\ LatitudeEven &= DegLatEven \, (\mathrm{mod}(z, 60) + LatEv) \\ Latitude &= \begin{cases} LatitudeOdd & if (F \to 0 \text{ to } 1) \\ LatitudeEven & if (F \to 1 \text{ to } 0) \end{cases} \\ Latitude &= Latitude - 360 \quad if \, (Latitude \geq 270) \end{aligned} \qquad (11.3)$$

$$NL(\text{latitude}) = \begin{cases} 59 & \text{if (latitude} = 0) \\ \left\lfloor 2\pi \left(\arccos \left(1 - \frac{1 - \cos(\frac{\pi}{30})}{\cos^2(\frac{\pi}{180} |\text{latitude}|)} \right) \right)^{-1} \right\rfloor & \text{if (|latitude|} < 87) \\ 1 & \text{if (|latitude|} > 87) \\ 2 & \text{otherwise} \end{cases}$$

$$(11.4)$$

$$m = \begin{cases} \lfloor LngEv(NL(LatitudeOdd)-1) - LngOd(NL(LatitudeOdd)) + 0.5 \rfloor & if(F = 0 \to 1) \\ \lfloor LngEv(NL(LatitudeEven)-1) - LngOd(NL(LatitudeEven)) + 0.5 \rfloor & if(F = 1 \to 0) \end{cases}$$

(11.5)

$$DegLonOdd = \frac{360}{\max(NL(LatitudeOdd)-1, 1)}$$

$$DegLonEven = \frac{360}{\max(NL(LatitudeEven), 1)}$$

$$Longitude = \begin{cases} DegLongEven(m + Lng\_od) & if(F = 0 \to 1) \\ DegLongOdd(m + Lng\_ev) & if(F = 1 \to 0) \end{cases}$$

(11.6)

$$Longitude = Longitude - 360 \quad if(Longitude \geq 180)$$

CHAPTER 11. ADSB DIGITAL SIGNAL PROCESSING 465

Table 11.8 ADSB Messages Logged to a File

ICAO24	Call Sign	CA	TC	Vehicle	Speed	Direction	Altitude	Latitude	Longitude
ACC3BD	ASA1022	5	4	Heavy					
A9F0CE		5	11				43000		
A5FF8A		7	19		393	104,E			
AD63A1		5	11				37000		
A56634		5	19		282	266,W			
AD63A1		5	19		492	122,SE			
A37A02		5	11				39025		
A521C8	N43LJ	5	4	Medium					
AD63A1	AAL586	5	4	Heavy					
A4D99A		5	11				35000		
A56634		5	11				15625	42.9388	91.7866
A5FF8A		7	11				39000		
A4D99A		5	19		520	86,E			
A5FF8A		7	19		393	104,E			
A9F0CE		5	11				43000	42.0846	89.0075
A9F0CE		5	19		402	271,W			
A5FF8A	N486GS	7	4	Light					
A842F8		5	11				42975		
ACEF18	SWA1386	5	4	Heavy					
ABFE12		5	19		360	274,W			
A56634		5	19		282	266,W			
AD63A1		5	11				37000	41.983	89.57

11.4.4 ADSB Message Logging

The Log to File block in Figure 11.2 stores data from each ADSB message in lines of a text file. An example of messages received over several seconds of operation is shown in Table 11.8. The shaded entries are from the same aircraft, American Airlines Fight 586. From the cluster of ADSB message received we can learn that this flight is traveling at altitude 37,000 feet, 492 knots, SE at 122°. At the time the messages were received, it was over central Iowa (ADSB messages are meant to be monitored only by nearby planes or by the closest airport). Indeed, we can look up AAL 586 on a web site such as flightaware.com and check our information. As a further development, the information in Table 11.8 could be transferred to mapping software. However, we will stop here because the ADSB example in this chapter is meant to be simple and straightforward to facilitate learning the basics of ADSB.

REFERENCES

[1] A. Cosma and D. Pu. "Four Quick Steps to Production, Part 1 - Mode S Detection and Decoding Using Matlab and Simulink". *Analog Dialogue*, 49(09), September 2015.

[2] A. Cosma and D. Pu. "Four Quick Steps to Production, Part 2 - Using Model-Based Design for Software Defined Radio". *Analog Dialogue*, 49(10), October 2015.

[3] A. Cosma and D. Pu. "Four Quick Steps to Production, Part 3 - Mode S Signals Decoding Algorithm Validation Using Hardware in the Loop". *Analog Dialogue*, 49(11), November 2015.

[4] A. Cosma and D. Pu. "Four Quick Steps to Production, Part 4 - Rapid Prototyping Using the Zynq SDR Kit and Simulink Code Generation Workflow". *Analog Dialogue*, 49(12), December 2015.

Chapter 12

APSK Digital Signal Processing

12.1 APSK OVERVIEW

Digital video broadcasting, second generation (DVB-S2), was adopted in 2009 as a standard for digital television. DVB-S2 defines multiple modulation formats and error control coding to standardize adaptive coding and modulation (ACM) on satellite communications channels; see Section 2.7.3.1. Details of DVB-S2 are shown in [1].

DVB-S2 includes well-known waveform constellations such as QPSK and 8PSK. A newer waveform, 16APSK (16 amplitude phase shift keying), has a round constellation that is better suited for satellite communications.[1] The modem that we design here features processing for QPSK, 8PSK, and 16APSK.

12.2 MODEM OVERVIEW

Here we describe the block diagram view of a typical SDR digital modem. This modem can be separated into a front end for finding the symbol boundaries and a back end for estimating the transmitted value of each symbol. Let's go into more detail.

---

1 For example, given the same error performance the 16APSK constellation achieves about 1.5 dB lower peak to average power (PAPR) than the original square 16QAM (my testing).

12.2.1 Modem Front End

A simplified APSK modem view is Figure 12.1. In this discussion, M = the number of samples assigned to one received symbol, four in this design (in some modem designs, M can be 2, there is a detailed discussion of this choice in Section 14.6). In Figure 12.1, the modem is separated into front and back ends. The modem front end has several jobs:

IF to Complex Baseband: Translate the intermediate frequency (IF) input centered at $F_{sss}/4 = 200 MHz$ to complex baseband, filter, and subsample by 2. The subsample by 2 from $F_{sss} = 800\ MHz$ to $F_{ss} = 400\ MHz$ simply reduces the processing load.

Sample Interpolator: Resample the IF to Complex Baseband output to generate symbol vectors of M samples each. SymbolReady = 1 indicates a valid symbol. The symbol rate can be anywhere from 50 MHz down to almost zero.

Preamble Detect: Detect the packet preamble and estimate the interpolator initialization (needed to line up symbol boundaries) as well as the first sample of the first symbol (used to enable the sample stacker).

Sample Stacker: Using the current estimate of symbol boundaries, assemble (stack) the new samples into vectors of M samples to represent each new symbol.

Symbol Ready: When a new vector of M symbol samples is collected, generate a SymbolReady = 1 pulse to enable symbol processing circuits.

Matched Filter: Match filter the received symbols to optimize the signal to noise ratio. DVB-S2 specifies a root raised cosine filter.

Symbol Timing Error: Refine the estimate of symbol boundaries in the timing error generator and feedback adjustments of intersample interval to effect symbol tracking.

12.2.2 Modem Back End

The modem back end inputs are complex baseband M sample vectors, I and Q, and the SymbolReady pulse. When the symbol timing loop converges correctly, each Sample Stacker output vector will carry the M samples of only one symbol. Although the back-end blocks are running at F_s sample rate, they are not enabled until a SymbolReady pulse occurs. Fixing the sample rate and only enabling the circuits when needed for a new symbol is how this modem can work at many different symbol rates.

The modem back end has several primary jobs:

CHAPTER 12. APSK DIGITAL SIGNAL PROCESSING

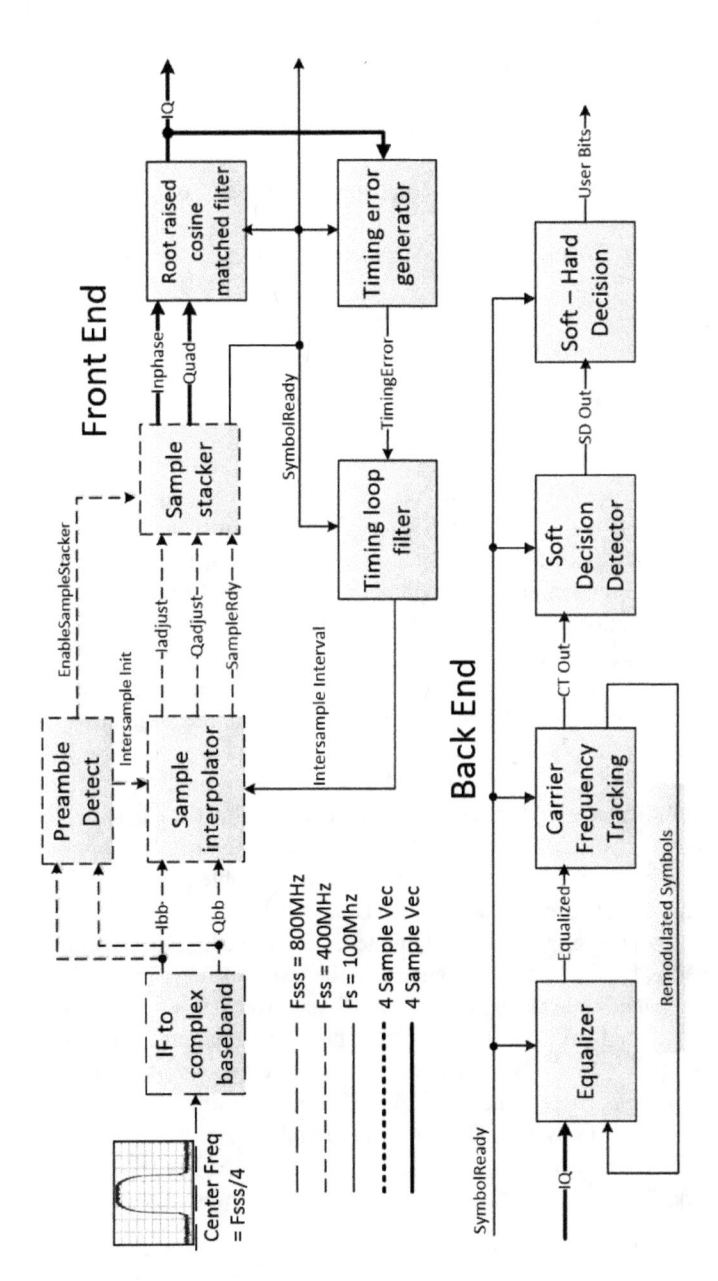

Figure 12.1 Digital modem front and back end processing.

Equalizer: Adaptively equalize the received signal to mitigate the effects of channel multipath distortion.
Carrier Frequency Tracking: Adaptively remove any residual carrier frequency error.
Soft Decision Detector: Generate soft decision estimates of the symbol constellation points.
Forward Error Correction: Decode forward error correction to generate final symbol hard decisions.

12.3 MODEM TRANSMITTER SIGNAL

The simulated complex baseband signal used for testing the receiver is shown in Figure 12.2. The preamble, starting on the left, is a BPSK signal on the real axis. After the preamble end, the 16APSK packet data test signal starts.

The 16APSK constellation for this signal is shown in Figure 12.23. Note that each constellation point corresponds to a specific 4-bit pattern. Notice also that this constellation can be stable (like in the figure) and the data bit patterns the receiver assigns to each detected symbol can be wrong when compared with those assigned by the transmitter. This is because symmetric constellations used here have a 90° ambiguity. To get around that problem, the simulation uses a gear shift preamble. The preamble starts at one symbol state in quadrant 0 and then moves to two symbol states, quadrant 0 and 2 and then finally to the desired data modulation indicated by ModType; see Figure 12.2. This trick gets the receiver bit hard decisions correct. However, if the start of message (SOM) is important, the preamble will need an additional correlator for a preamble pattern lineup. For BER testing in this simulation, we use a bit correlator to measure the delay and line up the transmitter-receiver symbols without necessarily needing the exact pattern start.

The MATLAB function that generates the test packet in Figure 14.21 is shown below. Note that complex sample vector RefSourceSignal is stored in the MATLAB workspace and read out continuously during the receiver simulation by a Read From Workspace block.[2] Since we know the exact characteristics of the transmit signal we need to receive, there is no need to use simulation time regenerating it on the fly. The data vector TxSyms is simply the unmodulated baseband bits to be used for BER testing. Note that createApskWaveform(...) is not a MATLAB built-in function. Although the Simulink APSK receiver will use any symbol rate, the transmitter in

---

2 There is an option at the simulation top level to add noise, carrier frequency offset, and other impairments.

CHAPTER 12. APSK DIGITAL SIGNAL PROCESSING

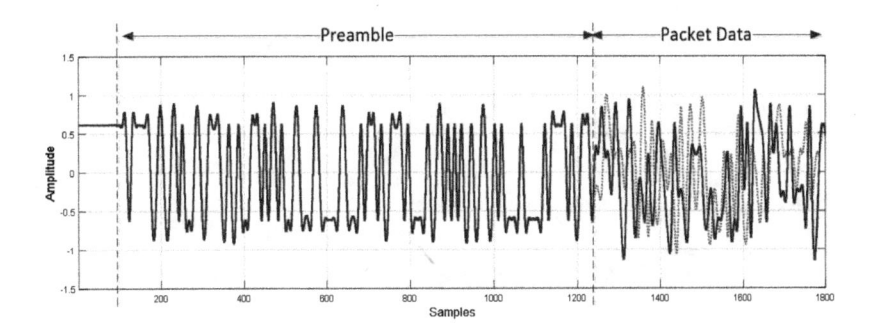

Figure 12.2 APSK packet used for testing the receiver.

the following code needs an integer number of samples per symbol. This is because the receiver has a continuously adapting symbol tracking loop while the transmitter does not.

```
% Generate APSK Packet for transmit
NumberSymbols = 4070;
NumberPreambleSyms =140;
TxSampsPerSym = 8;
ModType = 8; % 4=QPSK, 8=8PSK, 16=16APSK
% Calculate one packet of APSK use
% Sq. Root Raised Cosine filtering
% because receiver will also have
% Sq. Root Raised Cosine filtering
% Note that the sample rate is not specified,
% however the number TxSampsPerSym implies it
% Sample rate and symbol rate depend
% on rate samples are read
(TxSyms, RefSourceSignal ) ...
= createApskWaveform(ModType,TxSampsPerSym,
NumberSymbols, NumberPreambleSyms);
```

12.4 MODEM RECEIVER FRONT END BLOCKS

Here we provide a careful description of all the modem front end blocks; see Figure 12.1.

12.4.1 IF to Complex Baseband

The block labeled IF to Complex Baseband in Figure 12.1 was described in detail in Section 9.3.4. The same circuit is used here.

12.4.2 Sample Interpolator

We start with a detailed description of the sample interpolator, shown in Figure 12.4. The purpose of this circuit is to output samples with a time spacing equal to intersample.[3] The sample interpolator input sample rate (400 MHz in this design) is an integer multiple of the ADC sampling rate. The output sample rate is a multiple (4 in this design) of the symbol rate. The ADC and symbol rate are generally not an integer ratio. Intersample is the current estimate of the sample spacing required between each output sample.[4] Intersample has an FPA (fixed point attribute) of sfix48_40 (see Section 14.2) so it can affect a very fine timing adjustment.

The sample interpolator design starts at the top left of Figure 12.4 by assembling in-phase and quadrature vectors of the latest four-input samples. Note that this four input sample spread is needed by the cubic interpolator adjust (CFA) and has nothing to do with four samples/symbol on the sample stacker output.

The in-phase and quadrature CFA run continuously at the 400 MHz input sample rate. However, the outputs, Iadjust and Qadjust, only change when a NextSample pulse occurs. This will be four times each symbol for $M = 4$. The detailed description of the sample interpolator will be presented below. An alternative to this interpolator approach is the polyphase filterbank described in [2].

The circuit in Figure 12.4 inputs intersample on the lower left and controls the generation of NextSample output on the right. We will use the example in Figure 12.3 to explain this operation. Let's say, for example, that intersample = 4.3 and thus the sample interpolator circuit needs to construct one output sample for every 4.3 input samples. In Figure 12.3 FTA (fractional time accumulation) is the leftover fractional time that must be accounted for after the integer time in PresetInt has been

---

3   The units of intersample are integer and fractional samples.
4   Time between each sample is $F_{sym}/4$ in this design. However, timing recovery feedback can make this vary slightly.

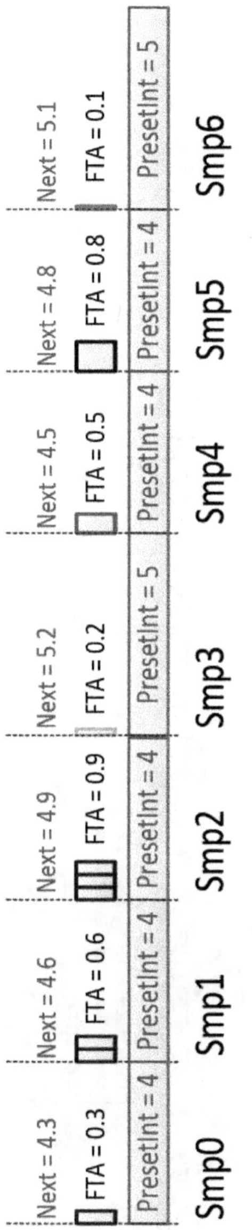

CHAPTER 12. APSK DIGITAL SIGNAL PROCESSING

Figure 12.3 Sample interpolator signal flow.

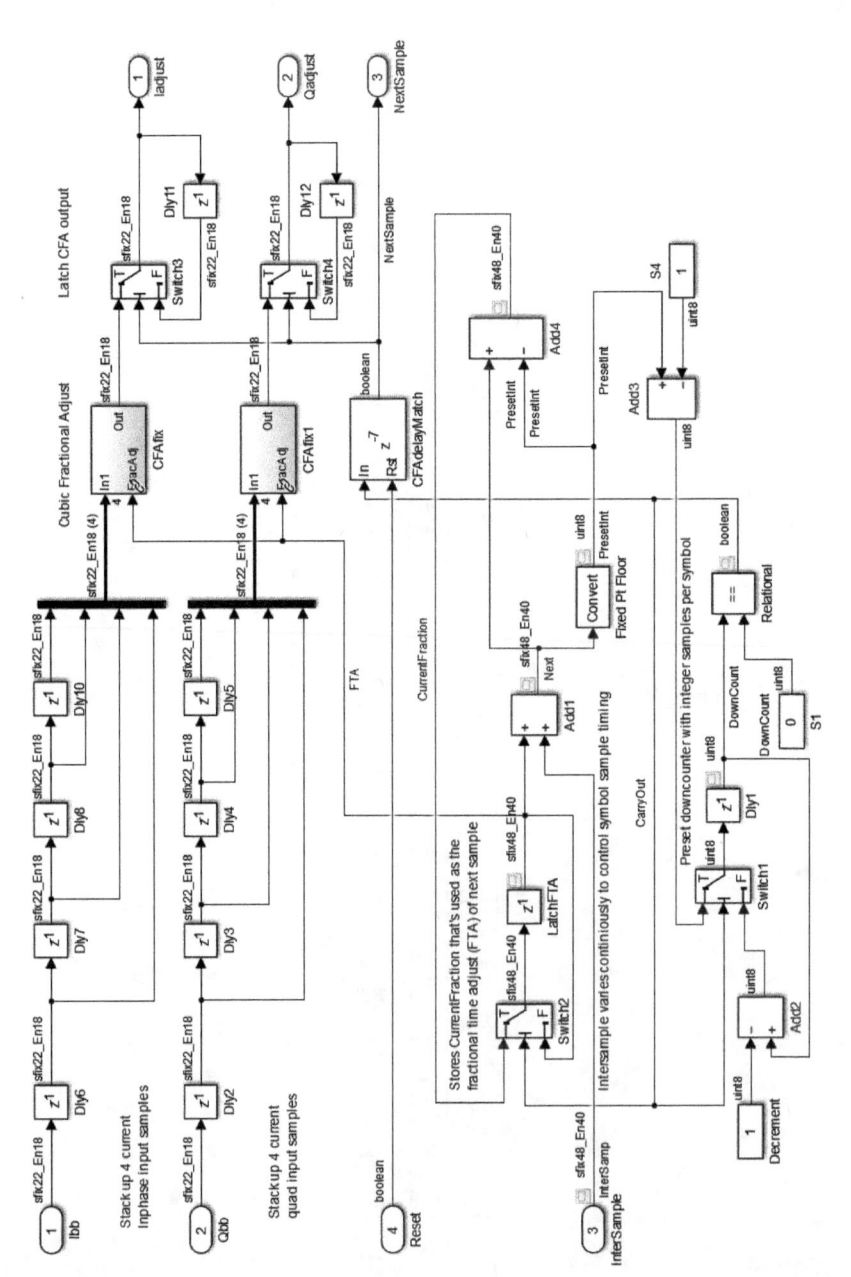

Figure 12.4 Sample interpolator design.

counted out. The CFA (cubic farrow adjust) slides the sample time position forward by FTA to bring about the fraction advance. Note how for Smp0 in Figure 12.3, a 4.3 input sample advance to the next output sample consists of an integer count of 4 and an interpolator slide forward of 0.3 input samples.

When FTA accumulates to greater than 1, it must have 1 subtracted, so it gets back to a fraction in the range, (0 + 1) (CFA only works with fractions). The 1 that was subtracted gets accounted for by adding an integer 1 to PresetInt for that sample. In Smp2, FTA starts at 0.9 and must increase by 0.3; 0.9+0.3 = 1.2. Thus, in Smp3, FTA starts at 1.2. Now we must reduce Smp3 FTA by subtracting 1 from Smp3 FTA and adding a 1 to PresetInt. The final Smp3 FTA equals 0.2 and PresetInt goes to 5.

In the actual circuit of Figure 12.4, at the lower left, the timing (LatchFTA) starts at 0, so Next = 4.3. Next is immediately split into PresetInt = 4 and CurrentFraction = 0.3 (similar to Figure 12.3). PresetInt loads the downcounter (very bottom of the figure) to 4 − 1 = 3. The downcounter counts 3, 2, 1, 0 for a total of 4 integer samples and then CarryOut pulses high and adjusts the CFAs to interpolate 0.3 samples forward for a total of 4.3 input samples. CarryOut is delayed to match the delay through the CFAs and then output as NextSample along with the complex sample that has a calculated time spacing of 4.3 samples.[5]

As described, upon reaching 0, the downcounter produces CarryOut = 0. CarryOut has two functions: store the CurrentFraction into the LatchFTA memory element and input a pulse into CFAdelayMatch. The CFAs need seven clocks to calculate the FTA and thus NextSample is CarryOut delayed by 7 clocks. NextSample also latches the computed complex sample after it has propagated through the CFAs.

Note that although we did output a fractional sample, we really only accounted for 4 samples on the input. The FTA = 0.3 samples interpolated forward to get to the end of Smp0 were not counted so we keep track of them by storing 0.3 CurrentFraction in the LatchFTA and adding it to the Smp1 Intersample input to get 4.6. LatchFTA is now 4 Smp1 integer samples plus 0.3 fractional sample left over from Smp0 plus 0.3 fractional part of Smp1. Variable Next, the output of Add1, is now the 0.3 previous fraction plus the new InterSample of 4.3 to equal 4.6. To get to the true end of Smp1, we must interpolate forward FTA = 0.6 samples. CurrentFraction now becomes 0.6. This continues with Smp2, FTA = 0.9. For Smp3, FTA = 1.2. Now FTA has an integer part of 1. We can now account for this integer by making PresetInt = 5 to skip forward an extra sample and reduce FTA to 0.2. This step is also necessary because the CFA inputs must be in the range (0 + 1).

---

[5] An output sample 4.3 input samples forward of zero has a certain numeric value. However, it can still be output at 4 integer sample clocks. The time value of a sample and the time it is sent forward can be different.

Smp4 adds another Intersample = 4.3 to make Next = 0.2 + 4.3 = 4.5 so we count out 4 integer samples again, interpolate forward 0.5 samples, and store FTA = 0.5. Smp5 continues this process with FTA = 0.8 and finally Smp6 has FTA exceeding 1 so we increase the down count to 5 and store FTA = 1.1 - 1 = 0.1.

This circuit has great utility for symbol timing adjustment because, depending on the FPA of Intersample, very tiny fractional parts can be accounted for.

12.4.3 Cubic Farrow Interpolator, Fixed Point (CFAfixed)

The circuit used for fractional sample adjustment is shown in Figure 12.5. Note that the four samples input at the top left are adjacent in time to the output fractional sample to be calculated. They have nothing to do with $M = 2$ or 4. A thorough and well written description is in [3].

12.4.4 Sample Stacker

As mentioned, the sample interpolator is not concerned about the choice of $M = 2$ or 4 samples per symbol. That is the job of the sample stacker, described here.

For $M = 4$, used here, the sample stacker circuit shown in Figure 12.6, shifts in-phase and quadrature samples from the sample interpolator into serial to parallel four sample shift registers. These samples are only shifted in when the NextSample input pulses high. As discussed for the sample interpolator, input samples are only valid when the sample interpolator calculates the next one and makes NextSample = 1.

In this design, $M = 4$ so the upcounter at the top left of Figure 12.6 keeps track of when four samples have been collected and produces SymbolLatch to latch vectors of four samples into InphsVector and QuadVector on the right side of Figure 12.6. SymbolLatch goes high when UpCount has incremented to 4. At that point SymbolLatch feeds back flipping StartSwitch and subtracting 4 from UpCount. This enables the count loop to start up again. For synchronization, UpCount only increases when NextSampleReady goes high.

The last operation is to count down sample SymbolLatch by four. This produces a SymbolReady pulse output. The vectors of four samples are also latched and downsampled by 4 to account for their total time value. Thus, as the current four sample symbol is output, the next four samples are being shifted in. An overarching assumption is that, if symbol tracking is working properly, the four sample vectors represent just one symbol and not the last part of a symbol followed by the first

CHAPTER 12. APSK DIGITAL SIGNAL PROCESSING 477

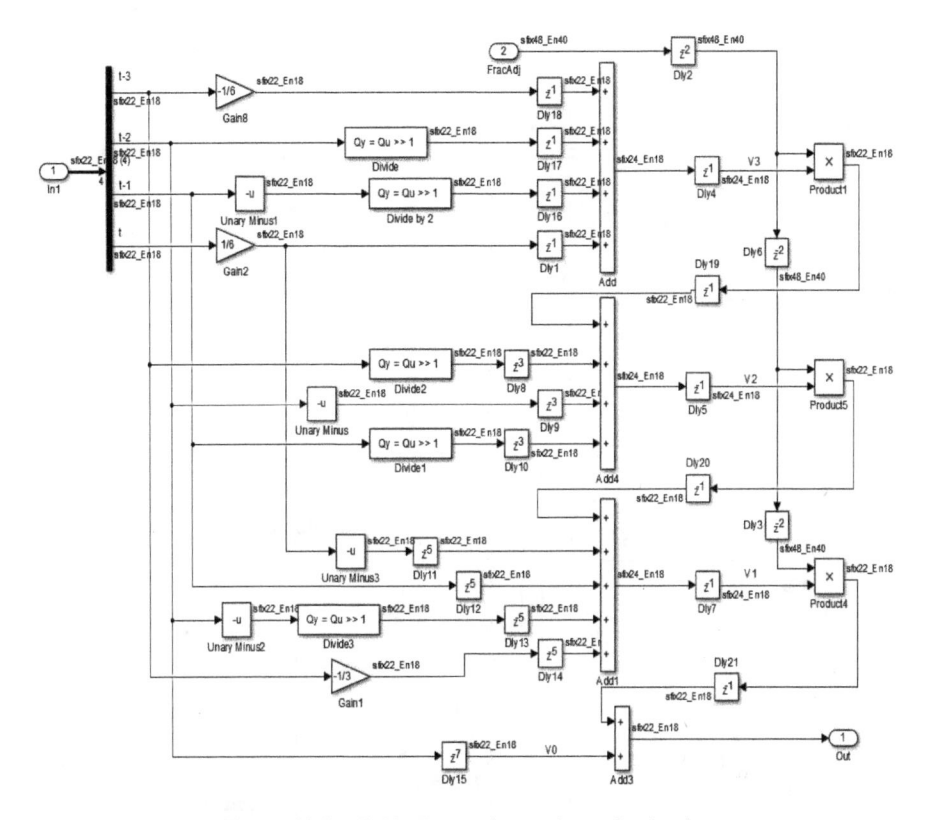

Figure 12.5 Cubic Farrow interpolator, fixed point.

part of the next symbol. Note that, throughout this chapter, the index k is used to increment to the next symbol when SymbolReady goes high.

12.4.5 Root Raised Cosine Matched Filter

The received matched filter affects the optimum trade-off between noise rejection and receive signal bandlimiting. This is proved in [4]. Thus, the matched filter impulse response matches the transmit pulse shape. For example, consider the sequence of square received symbols represented by 4 samples. A matched filter for this sequence is simply [1/4, 1/4, 1/4, 1/4]. As a sliding average, the matched filter lines up the four coefficients with the next four samples, multiplies each, and adds

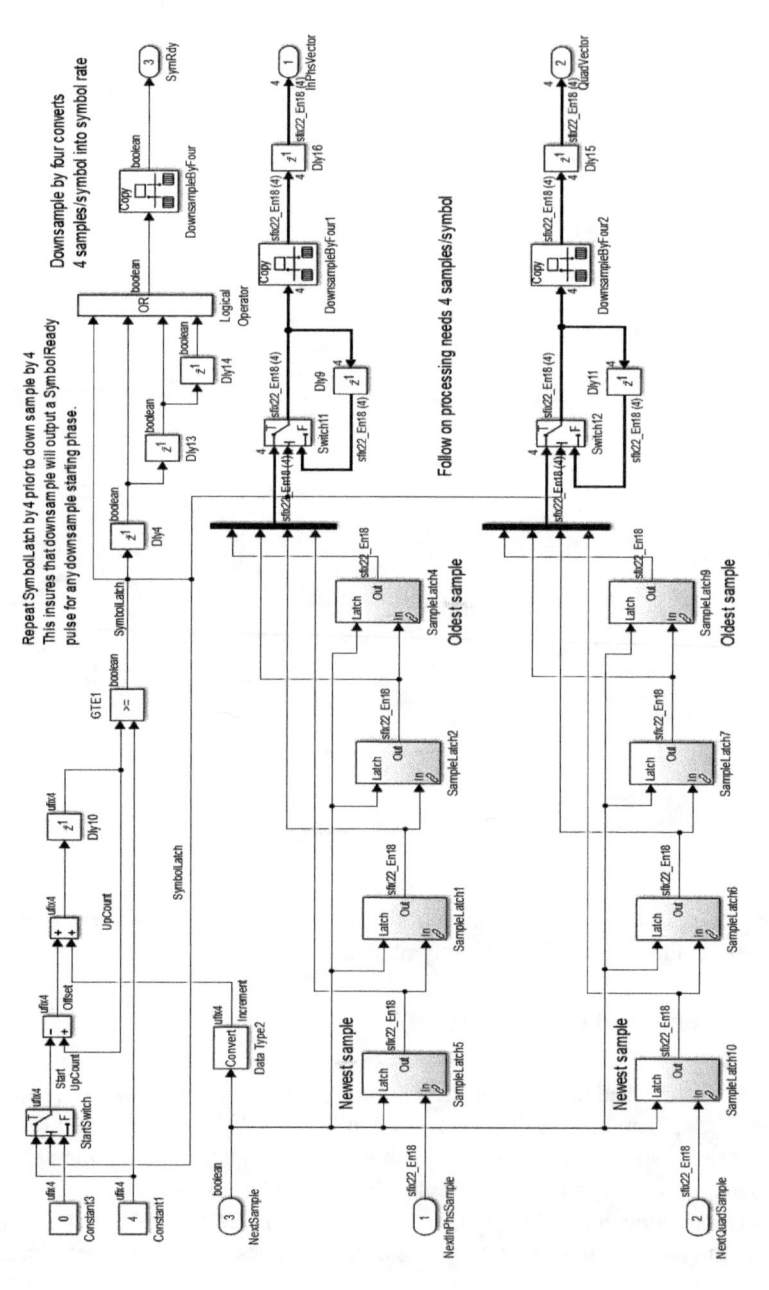

478   *Software Defined Radio: Theory and Practice*

Figure 12.6   Sample stacker design.

CHAPTER 12. APSK DIGITAL SIGNAL PROCESSING 479

the four products. Every four sample clocks, the matched filter output is detected to determine the symbol value. Note that matched filters are for linear modulation, AM or QAM. Nonlinear modulations, such as FM, use a more complicated predetection filter.

This modem design utilizes a root raised cosine (RRC) matched filter. Figure 12.7 shows the RRC filter circuit. Note the Enable block on the left. Enable is connected to SymbolReady from the sample stacker so that the RRC only runs when the SymbolReady input goes high. When enabled, the RRC matched filter inputs the four sample in-phase and quad vectors from the sample stacker and outputs four sample in-phase and quad vectors, RRC filtered.

A useful Simulink feature is the muxes (vertical black bars) in the center that allow easy conversion of four separate signals into four element signal vectors. Finally, the real and imaginary parts of the complex signal are combined into one output on the right side of Figure 12.7.

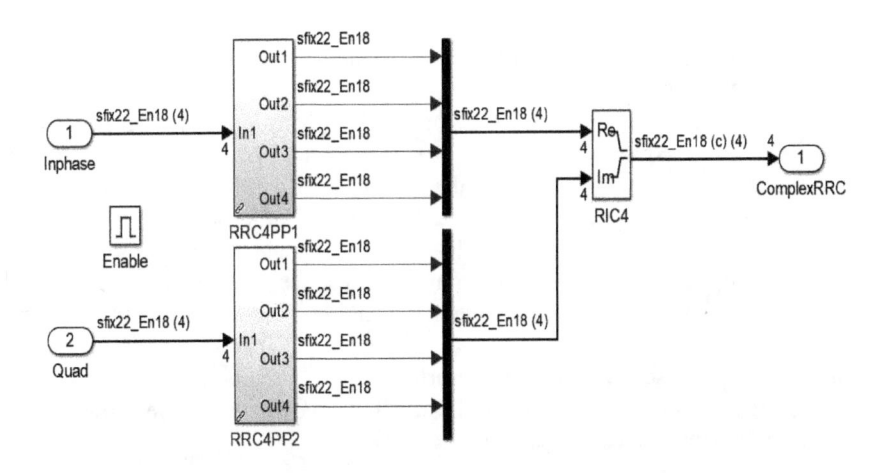

Figure 12.7 Root raised cosine matched filter top level circuit.

We stated above that the matched filter inputs four samples and outputs four samples. To show how this works, we present a simple 13-coefficient implementation example. Figure 12.8 shows the time domain-coefficients that must be convolved with the series of four-symbol sample vectors. For $M = 4$, there are four separate filter circuits, represented in four separate tables next. These filter circuits run in parallel, taking in four inputs and generating four outputs each time they are enabled. This

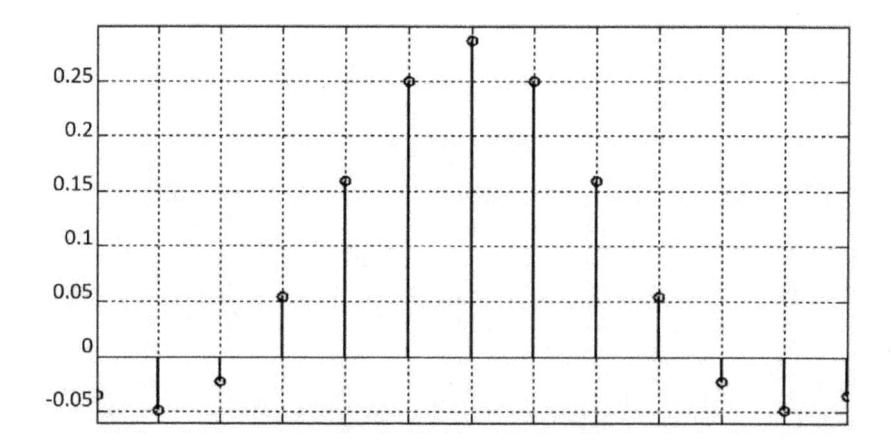

Figure 12.8 Root raised cosine matched filter coefficients.

technique is sometimes called FIR filter parallelization. Some FPGA tools will implement this automatically.

Table 12.1 represents a fixed point in time where four input sample vectors have been collected in RRC memory. InVec0 was the first and InVec3 is current. The four element vectors came in on the left side of Figure 12.7, either in-phase or quad. Table 12.1 illustrates the calculation of the first output sample. This is a finite impulse response (FIR) filter, a simple weighted average. Coefficients h_0 through h_{12} are multiplied by samples $x(0)$ through $x(12)$, respectively. The 16 products are summed to form Out1, the first sample in the RRC matched filter output vector.

The next three Tables 12.2, 12.3, and 12.4, illustrate the simultaneous calculation of the second, third, and fourth samples in the output vector. Note that the input samples have remained fixed and only the coefficients have shifted for the four sample output calculations.

Table 12.1 Calculation of the First Output Sample

InVec 3	InVec 2	InVec 1	InVec 0
$x(12)h_{12}$	$x(8)h_8$	$x(4)h_4$	$x(0)h_0$
$x(13)(0)$	$x(9)h_9$	$x(5)h_5$	$x(1)h_1$
$x(14)(0)$	$x(10)h_{10}$	$x(6)h_6$	$x(2)h_2$
$x(15)(0)$	$x(11)h_{11}$	$x(7)h_7$	$x(3)h_3$

CHAPTER 12. APSK DIGITAL SIGNAL PROCESSING

Table 12.2 Calculation of the Second Output Sample

InVec 3	InVec 2	InVec 1	InVec 0
$x(12)h_{11}$	$x(8)h_7$	$x(4)h_3$	$x(0)(0)$
$x(13)h_{12}$	$x(9)h_8$	$x(5)h_4$	$x(1)h_0$
$x(14)(0)$	$x(10)h_9$	$x(6)h_5$	$x(2)h_1$
$x(15)(0)$	$x(11)h_{10}$	$x(7)h_6$	$x(3)h_2$

Table 12.3 Calculation of the Third Output Sample

InVec 3	InVec 2	InVec 1	InVec 0
$x(12)h_{10}$	$x(8)h_6$	$x(4)h_2$	$x(0)(0)$
$x(13)h_{11}$	$x(9)h_7$	$x(5)h_3$	$x(1)(0)$
$x(14)h_{12}$	$x(10)h_8$	$x(6)h_4$	$x(2)h_0$
$x(15)(0)$	$x(11)h_9$	$x(7)h_5$	$x(3)h_1$

Table 12.4 Calculation of the Fourth Output Sample

InVec 3	InVec 2	InVec 1	InVec 0
$x(12)h_9$	$x(8)h_5$	$x(4)h_1$	$x(0)(0)$
$x(13)h_{10}$	$x(9)h_6$	$x(5)h_2$	$x(1)(0)$
$x(14)h_{11}$	$x(10)h_7$	$x(6)h_3$	$x(2)(0)$
$x(15)h_{12}$	$x(11)h_8$	$x(7)h_4$	$x(3)h_0$

When the next SymbolReady pulse occurs, shift InVec 1 to InVec 0, InVec 2 to Invec 1, InVec 3 to InVec 2, and input four new samples into InVec 3. Then the four-output sample calculation starts over in the same way as described. High computational speed can be achieved as all four tables can be calculated at the same time in FPGA hardware.

12.4.5.1 Alternative Matched Filter Design Ideas

As discussed above, M can be 2 or 4. Table 12.1, 12.2, 12.3, and 12.4 are for $M = 4$. If $M = 2$, we only need calculate Table 12.1 and Table 12.3. There is even a possibility, for short messages where there is no symbol tracking because the preamble lines up the symbol boundaries and there is no equalizer, that only Table 12.1 needs to be calculated.

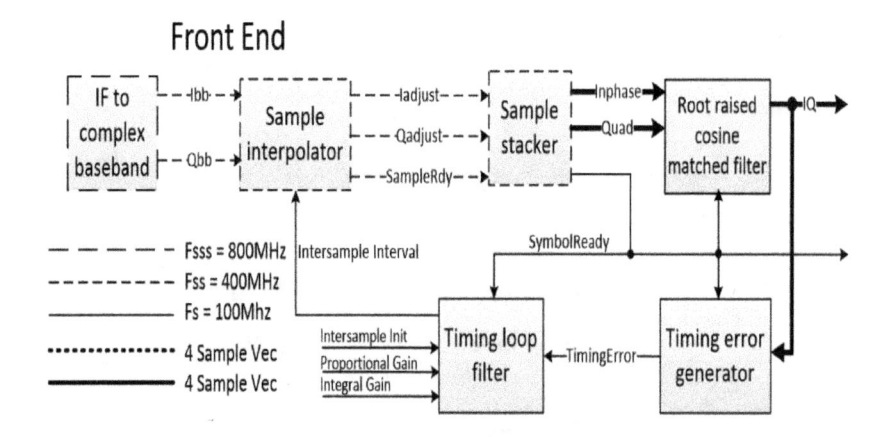

Figure 12.9 APSK front end symbol timing recovery components.

12.4.6 Timing Error Detector

Figure 12.9 shows all the symbol tracking modem front end blocks. These comprise a second-order phase locked loop; see Section 10.1.4. The TED measures the error between ideal and actual symbol boundaries. This error is feedback through the timing loop filter (TLF) and updates the intersample interval for interpolated samples. Some readers may notice that we are correcting the timing error prior to removing the residual carrier frequency offset. The modified Garner TED that we are using is a little better at handling carrier frequency offset; see Section 10.1.5 and also [5].

Figure 12.10 shows four samples for each baseband symbol. The peak valued symbol sample is just prior to each grid line. Just prior to that sample is a one sample/symbol TED output, shown as a pulse that is nonzero every four samples.

The modified Gardner TED does a better job of handling multilevel QAM, such as 16APSK, than the original Gardner TED [6]. Figure 12.10 provides a clue. Every grid line covers four samples or one symbol. The baseband symbols are the dark lines and the TED outputs are the lighter lines.

Symbols aligned to grid lines 32, 36, and 40 cannot provide much timing information because the corresponding baseband waveform is almost flat. Therefore, the TED outputs corresponding to these three are zero. The modified Gardner TED attempts to assign the largest weighting to the most reliable timing information. This effectively filters out unreliable timing information. In Equation (12.1), k is the

CHAPTER 12. APSK DIGITAL SIGNAL PROCESSING

symbol index. Because $M = 4$, $i(4k)$ is the first symbol sample, $i(4k+2)$ is the first symbol center sample and $i(4k+4)$ is the first sample of the next symbol. Notice that the symbol aligned to grid line 56 is clearly late (see Figure 10.16).

$$\begin{aligned}T_{error}(k) &= T_{Ierror}(k) + T_{Qerror}(k) \\ &= \left(i(4k+2) - \left(\frac{i(4k+4)+i(4k)}{2}\right)\right)\left(\frac{i(4k+4)-i(4k)}{2}\right) \\ &+ \left(q(4k+2) - \left(\frac{q(4k+4)+q(4k)}{2}\right)\right)\left(\frac{q(4k+4)-q(4k)}{2}\right)\end{aligned} \quad (12.1)$$

12.4.7 Timing Loop Filter

The timing loop filter (TLF), along with the delay around the loop, controls the dynamics of the symbol timing second-order response. In control theory, this TLF is called a proportional integral (PI) second-order control filter [7]. The transfer function corresponding to Figure 12.12 can be calculated as:

$$L(z) = \frac{K_P + K_i - z^{-1}K_P}{4(1-z^{-1})} \quad (12.2)$$

K_P = Proportional feedback simply feeds back the scaled timing error. As the actual symbol timing error goes to zero the timing error feedback goes to zero. This contradiction means that proportional feedback can reduce symbol timing error but cannot drive it to zero.

K_I = Integral feedback can drive the timing error to zero. Integrator memory stores a scaled timing error per symbol estimate that results in a ramp output that counteracts the received signal's timing error ramp (if the receive symbol time error estimate is incorrect, the timing error will be a ramp because the error will accumulate every symbol). InterSample output, discussed in Section 12.4.2, is the estimate of the number of ADC input samples between matched filter input samples. Note that the integrator memory delay block must be initialized on reset to RxSpan, the number of samples/symbol (in this case, RxSpan = 4 * Intersample).

More detailed design information can be found in Section 10.1.4.1.

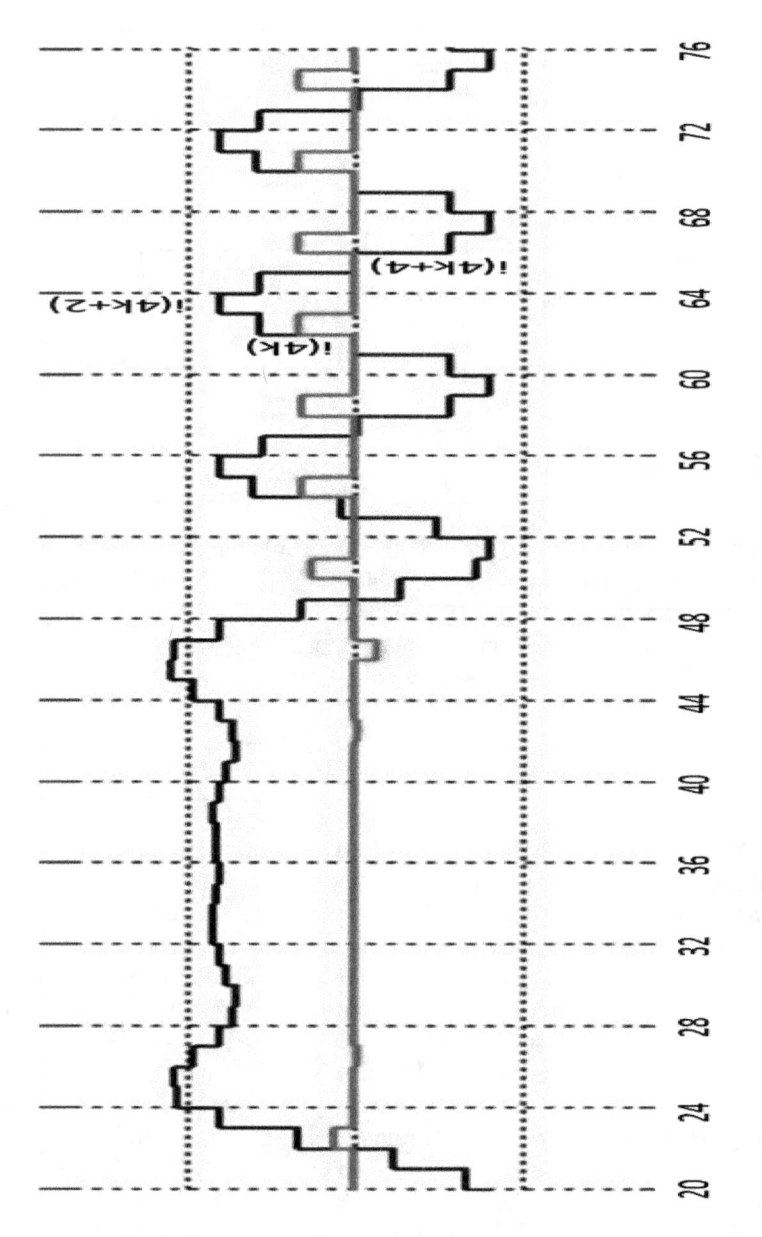

Figure 12.10 Received symbol waveform and modified Gardner TED output.

CHAPTER 12. APSK DIGITAL SIGNAL PROCESSING

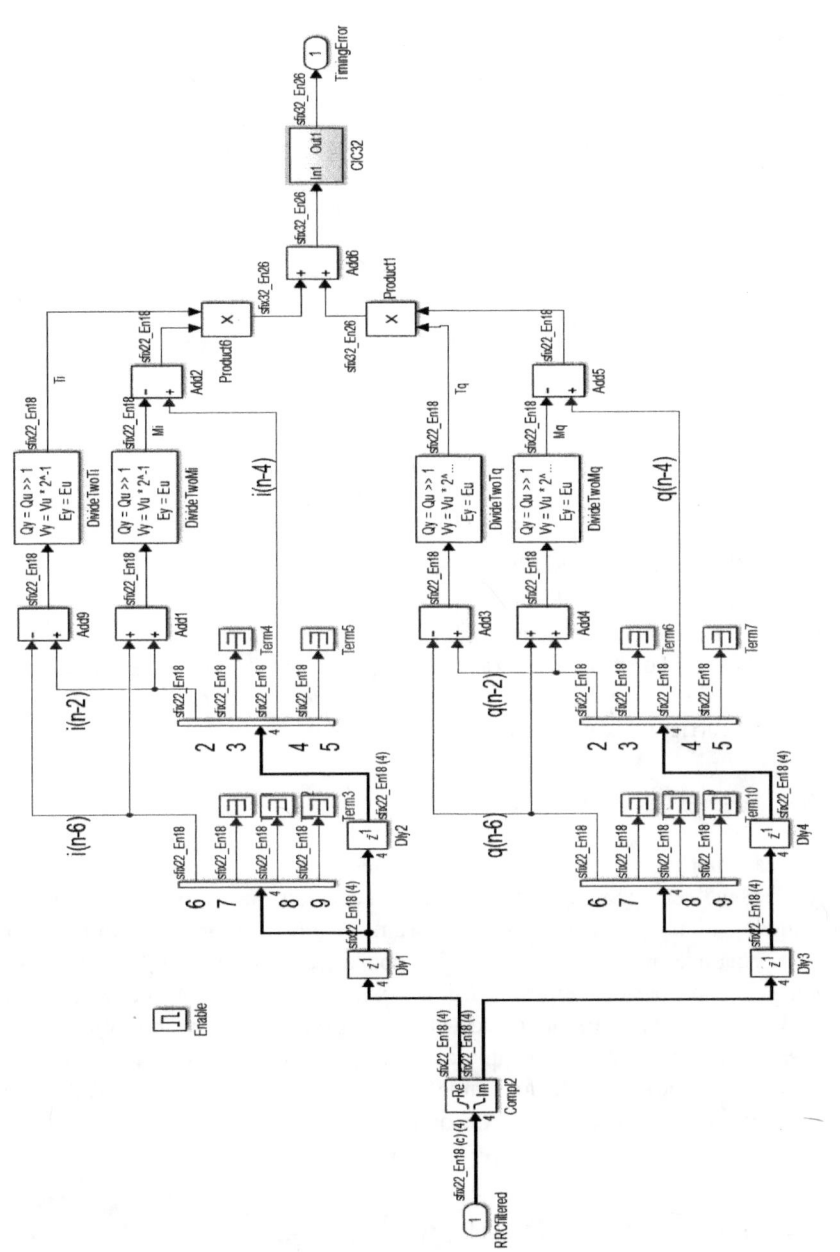

Figure 12.11 Modified Gardner TED circuit, Simulink model.

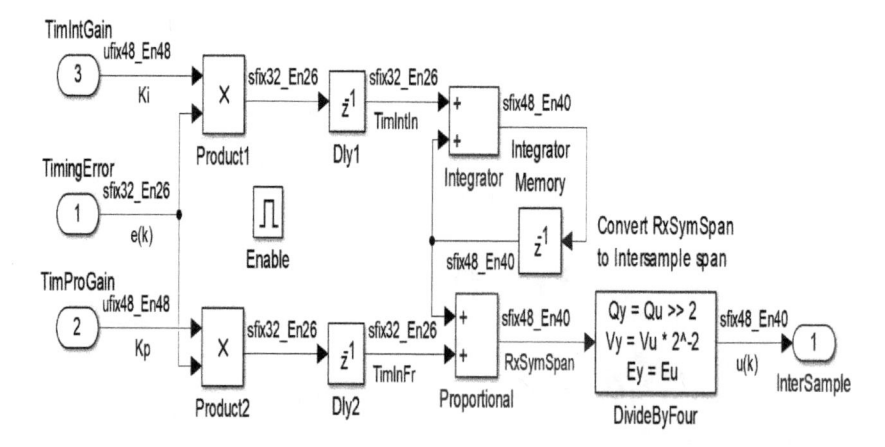

Figure 12.12 Second-order timing loop filter, Simulink model.

12.5 MODEM BACK-END BLOCKS

The equalizer and carrier frequency tracking shown in Figure 12.13 work together to clean up the signal distortion and track out frequency offset that causes constellation spin. The decision detector blocks estimate the value of each received symbol. We start with a carrier tracking discussion.

12.5.1 Carrier Tracking

First recall that the modem back end processes a new $M = 2$ or 4 sample vector only when enabled by SymbolReady. When not enabled, these blocks simply go into hold mode with no signal changes. The M sample/symbol complex vector IQ on the left side equalizer input of Figure 12.13 is converted to one sample/symbol on the equalizer output (besides removing ISI, the equalizer functions as a downsampler; see Section 14.6). Thus, carrier tracking processes one sample/symbol constellation points. Figure 12.14 shows a simple example of a QPSK constellation. The four signal points, indicated by the four large dots on the circle, are at the correct QPSK angles and can be represented by Equation (12.3).

$$u(k) = e^{j\phi(k)} \text{ for } \phi(k) \text{ from the set: } \left[\frac{\pi}{4} \quad \frac{3\pi}{4} \quad -\frac{3\pi}{4} \quad -\frac{\pi}{4} \right] \quad (12.3)$$

CHAPTER 12. APSK DIGITAL SIGNAL PROCESSING

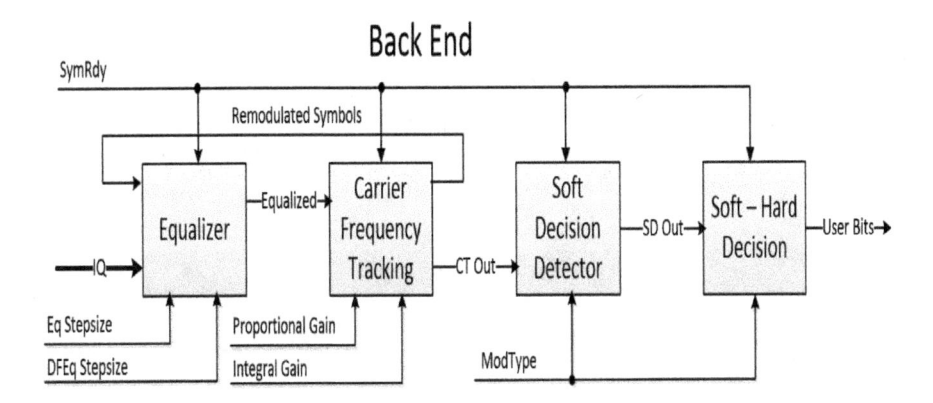

Figure 12.13 APSK equalizer, carrier tracking and decision components.

The four carrier tracking input crosses, $u'(k)$ are at incorrect QPSK angles and can be represented by Equation (12.4).

$$u'(k) = e^{j(\phi(k) + \theta_{err} + \omega_{err} k T_{sym})} \qquad (12.4)$$

Readers unfamiliar with exponential notation for complex signals should review Section 14.3. Note that $u'(k)$ has two extra exponential terms compared with $u(k)$. These are the static phase error, θ_{err}, and the time-varying phase error per symbol, $\omega_{err} k T_{sym}$. To drive these errors to zero, we must first measure the symbol-by-symbol phase error between the ideal circles and the crosses, for example, see Figure 12.14.

Figure 12.15 is a more detailed look at carrier tracking. When converged, the NCO output provides a close estimate of the static and time-varying inverse phase errors. Therefore, at the M1 (multiplier one) output, these errors are counteracted and we have the received constellation point with only a small residual error, $e^{j(\phi(k) + \tilde{\theta}_{err} + \tilde{\omega}_{err} k T_{sym})}$

At the M2 (multiplier two) output, we only want the errors, not the constellation points. To remove the constellation point, the slicer estimates the likely point and the baseband remodulator generates the complex representation. If the remodulator output is correct, then the M2 output is given by Equation (12.5).

$$e^{j(\tilde{\theta}_{err} + \tilde{\omega}_{err} k T_{sym})} = \cos\left(\tilde{\theta}_{err} + \tilde{\omega}_{err} k T_{sym}\right) + j \sin\left(\tilde{\theta}_{err} + \tilde{\omega}_{err} k T_{sym}\right) \qquad (12.5)$$

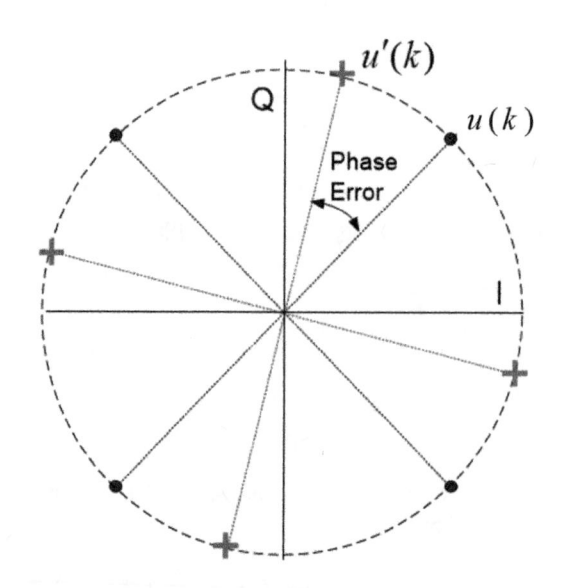

Figure 12.14 Example QPSK received constellation.

$$u'(k) = e^{j(\phi(k)+\theta_{err}+\omega_{err}kT_{sym})} \qquad e^{j(\phi(k)+\tilde{\theta}_{err}+\tilde{\omega}_{err}kT_{sym})}$$

$$e^{-j(\hat{\theta}_{err}+\hat{\omega}_{err}kT_{sym})} \qquad \sin(\tilde{\theta}_{err}+\tilde{\omega}_{err}kT_{sym})$$

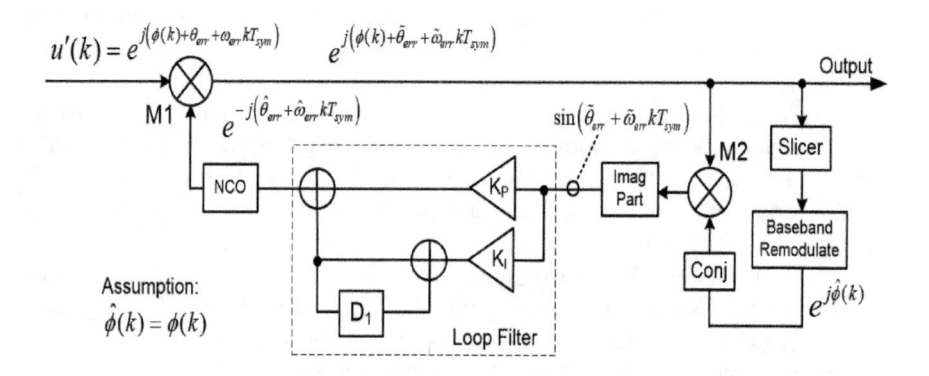

Assumption:
$$\hat{\phi}(k) = \phi(k)$$

$$e^{j\hat{\phi}(k)}$$

Figure 12.15 Carrier tracking details.

CHAPTER 12. APSK DIGITAL SIGNAL PROCESSING

We retain only the imaginary part so that the sine function forms an S curve (see, for example, Figure 10.7). The S curve error discriminator works best when the feedback errors are small and the loop is in tracking (as opposed to acquisition) mode. Additional discussion of why the imaginary part only is retained is in Section 13.6.6.5.

When converged, if the phase of the input error goes in one direction, then the phase of the NCO output is driven in the opposite, counteracting, direction. Note that the NCO output, $e^{-j(\hat{\theta}_{err} + \hat{\omega}_{err}kT_{sym})}$ is the conjugate of the phase and frequency error estimate. We would like this to be correct, regardless of magnitude. When the loop is converged, the M1 output residual error, $\tilde{\theta}_{err} + \tilde{\omega}_{err}kT_{sym}$, is due to wideband channel noise attached to each constellation point, i.e., a corrupted estimate.

This second-order carrier tracking circuit can correct for a nonzero constellation rotation rate; that is, $\omega_{err}kT_{sym} \neq 0$. The loop filter in Figure 12.15 has a one symbol delay element D_1 that can ramp up and store the value needed to produce a continuous counteracting complex phasor on the NCO output.

12.5.1.1 Carrier Tracking Loop Phase Model

If the loop filter parameters are selected correctly, we have a stable tracking loop. An important step in carrier tracking design is to model the loop dynamic response. Assuming that the loop is locked (tracking mode), we can linearly model carrier tracking by considering only the phase. Equation (12.6) is the transfer function from input phase $\alpha(k)$ to feedback phase $\beta(k)$ for the model of Figure 12.16.

$$\beta(z) = (\alpha(z) - \beta(z))\left(z^{-1}K_d L(z) N(z)\right)$$

$$\frac{\beta(z)}{\alpha(z)} = \frac{z^{-1}K_d L(z) N(z)}{1 + z^{-1}K_d L(z) N(z)} \quad (12.6)$$

Loop filter transfer function $L(z)$ was already presented in Equation (12.2). The NCO transfer function is a simple ramp generator, $N(z) = \frac{K_n}{1 - z^{-1}}$. Plugging these into Equation (12.6) we end up with Equation (12.7).

$$\frac{\alpha(z)}{\beta(z)} = \frac{K_n K_d z^{-1}(K_p + K_i) - K_n K_d K_p z^{-2}}{1 + z^{-1}(K_n K_d (K_p + K_i) - 2) - z^{-2}(1 - K_n K_d K_p)} \quad (12.7)$$

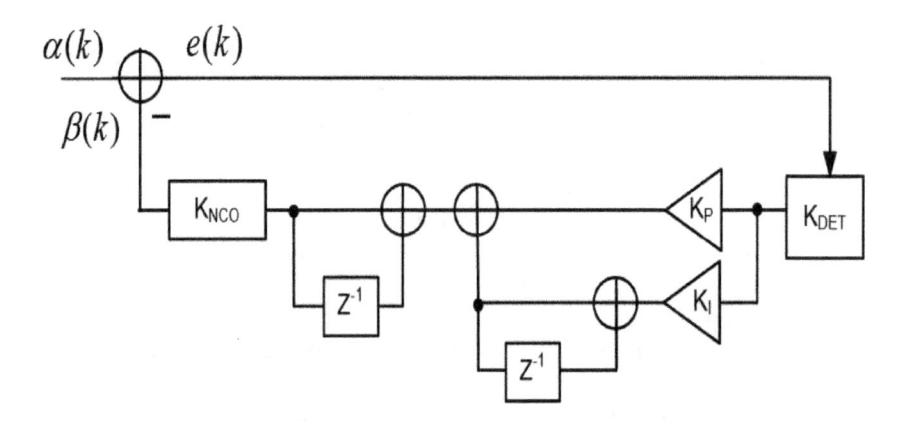

490   *Software Defined Radio: Theory and Practice*

Figure 12.16 Carrier tracking loop phase error model.

MATLAB can facilitate loop analysis. The MATLAB damp() command displays the natural frequency, damping factor zeta, and poles for the second-order system. In MATLAB, set: $Kp = 0.4$, $Ki = 0.03$, and $Kn = Kd = 1$. Then execute the script shown below:

```
PLLnum = [Kd*Kn*(Kp+Ki) -Kd*Kp*Kn ];
PLLden = [1 (Kd*Kn*(Kp+Ki)-2) (1-Kd*Kp*Kn) ];
% Ts is the sampling time in seconds
PLLsys = tf(PLLnum,PLLden,Ts);
(W, zeta, p) = damp(PLLsys);
```

Table 12.5 Carrier Tracking Loop Dynamics Using MATLAB damp( ) Function

Pole	Natural Frequency (MHz)	Damping Factor
$0.965 + j0.169$	1.39	0.12
$0.965 - j0.169$	1.39	0.12

Additional guidance on estimating loop parameters Kp and Ki can be found in Section 10.1.4.1 as well as various references, such as [8] and [9]. For various selections of loop parameters, a simulation can be constructed based on Figure 12.16. The simulated response of $\beta(k)$ to a step input on $\alpha(k)$ is shown in Figure 12.17.

CHAPTER 12. APSK DIGITAL SIGNAL PROCESSING 491

The horizontal axis is a symbol count. This can be converted to time by dividing by the symbol rate.

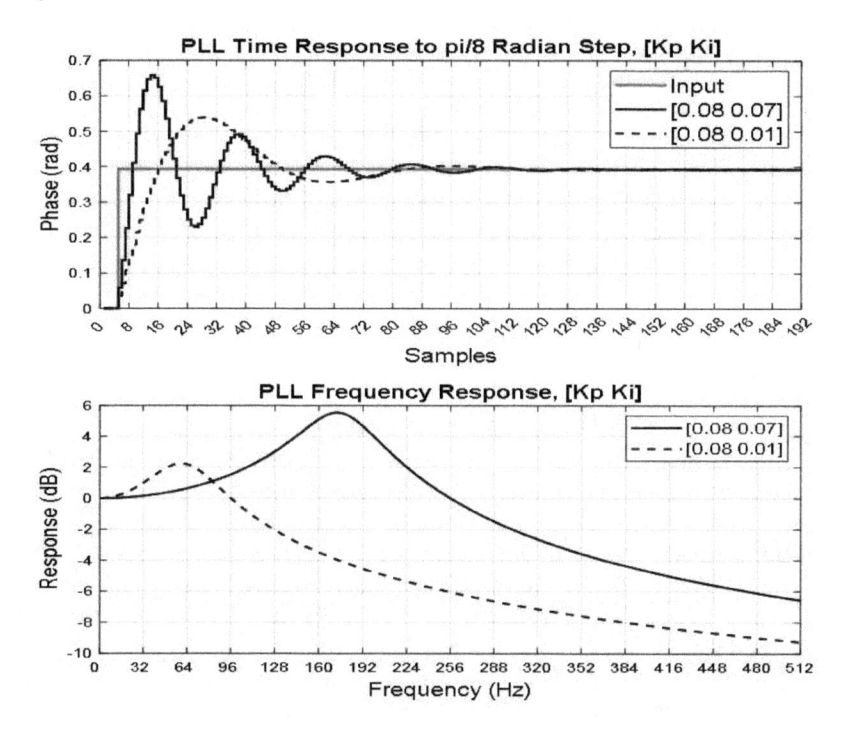

Figure 12.17 Carrier tracking loop step and frequency response.

12.5.1.2 Carrier Tracking Implementation

Figure 12.18 is a complete Simulink carrier tracking diagram. This figure is meant to be compared with Figure 12.20; the multiply blocks have the same labels. Note the NCO complex output going to M1 and conjugate to M3. Multiplier M2 provides the carrier phase error, the phase difference between the carrier tracking output and the remod feedback. SW1 selects one of four different modulation types (32APSK is not implemented here). SW2 zeros the phase error for constellation points not on the outer ring (16APSK only). Figure 12.19 is the inside of the block labeled Loop Filter and NCO. Notice the similarity with the time tracking loop filter in Figure 12.12.

The DataActive input, not shown in Figure 12.20, goes high when the preamble has completed. After the preamble, the remod constellation is switched from QPSK to the full constellation; see Section 12.3. The correct full constellation is preloaded in the lookup table labeled EntireConst4Data before the simulation is run.

12.5.2 Equalizer

First let's examine Figure 12.20 to see how the equalizer works with the carrier tracking. Similar to Figure 12.14, $u'(k)$ is a series of constellation points, with phase and frequency error. If the carrier tracking is converged, then multiplier M1 removes most of this error.

M2 in Figure 12.20 has the same function as M2 in Figure 12.15. The M2 output has the constellation point errors, not the constellation points. As in Figure 12.15, the slicer estimates the likely point and the baseband remodulator generates the complex representation. If the remodulator output is correct and the modulation is BPSK or QPSK, then the M2 output is simply given by Equation (12.5). If SW1 is down for 16APSK modulation, then the constellation point error will be zero unless the current constellation point is on the 16APSK outer ring; see Figure 12.23. 16APSK carrier tracking only uses the outer 12 constellation points.

The noise-free constellation point estimate from the baseband remodulator is $r(k) = e^{j\hat{\phi}(k)}$. In a departure from Figure 12.15, we introduce a third multiplier, M3. The conjugate of the NCO ouput multiplies $r(k)$ in M3 and reintroduces the rotation that was removed by M1. Thus, M1 input is a rotating noisy constellation and the M3 output is the same thing, only without noise. Thus, we have rotating and noisy $u'(k)$ and rotating and noise-free estimate $\hat{u}'(k)$. If carrier tracking is converged, then these two are synchronized and lined up ($\hat{u}'(k) \approx u'(k)$). Subtracting them results in S2 output $e(k)$, the noise and distortion on u'(k). If the equalizer output is very noisy then slicer mistakes may occur and ($\hat{u}'(k) \neq u'(k)$) may cause discontinuities in $e(k)$. This method of keeping the carrier tracking and equalizer independent is from [10]. Figure 12.20 uses 16APSK constellation pictures to show the synchronized rotation at the input of M1 and output of M3.

In the left side of Figure 12.20, the equalizer is divided into two major parts. The least mean square (LMS) is on the top half and the decision feedback (DF) is in the lower half (see also Figure 12.21). The LMS is a linear transversal filter with 32 taps (covering 16 symbols at 2 samples/symbol) that are adjusted to minimize error $e(k)$. In the lower half of the figure, the DF inputs the 6 previous symbol decisions (remod outputs), scales, and sums them to recreate the intersymbol interference inflicted on each symbol by the channel precursors. This ISI is canceled from

CHAPTER 12. *APSK DIGITAL SIGNAL PROCESSING* 493

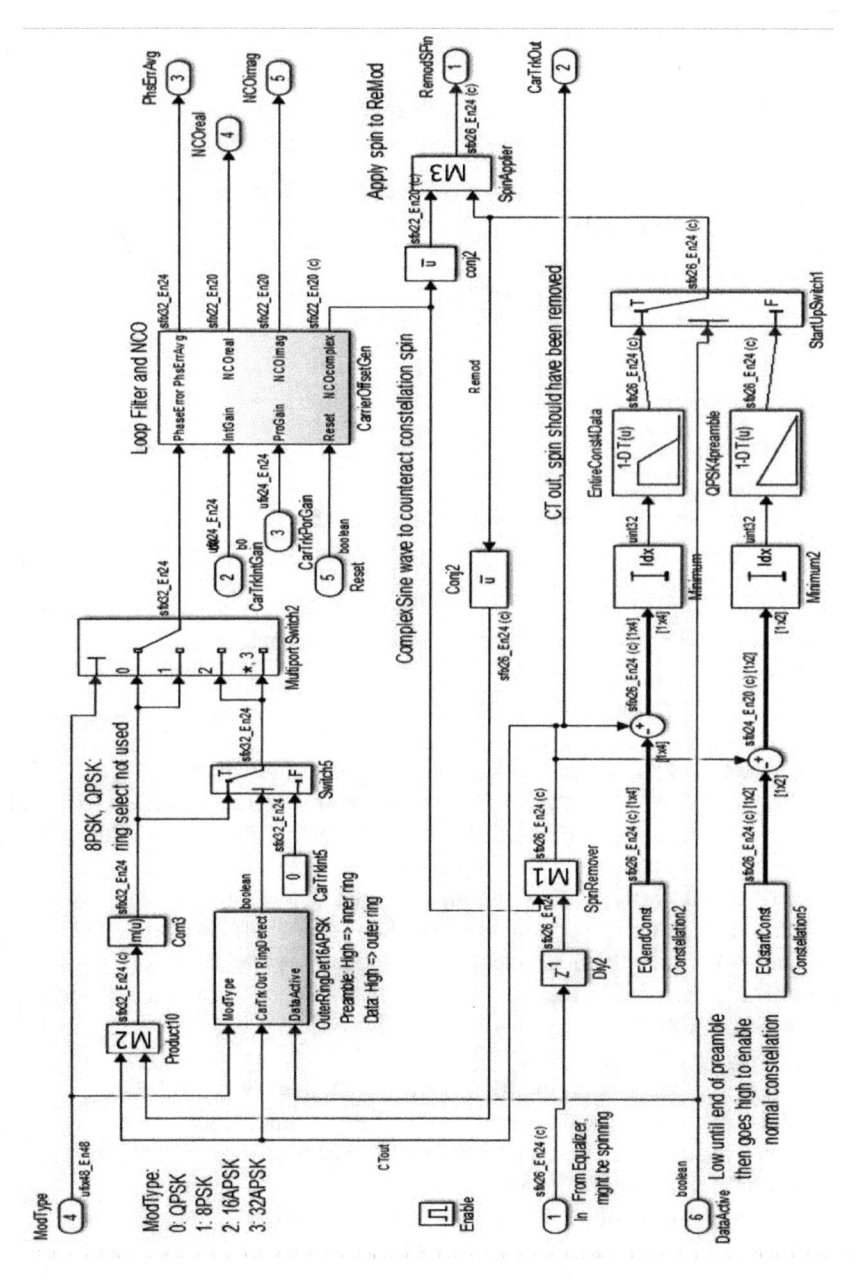

Figure 12.18 Carrier tracking Simulink circuit diagram.

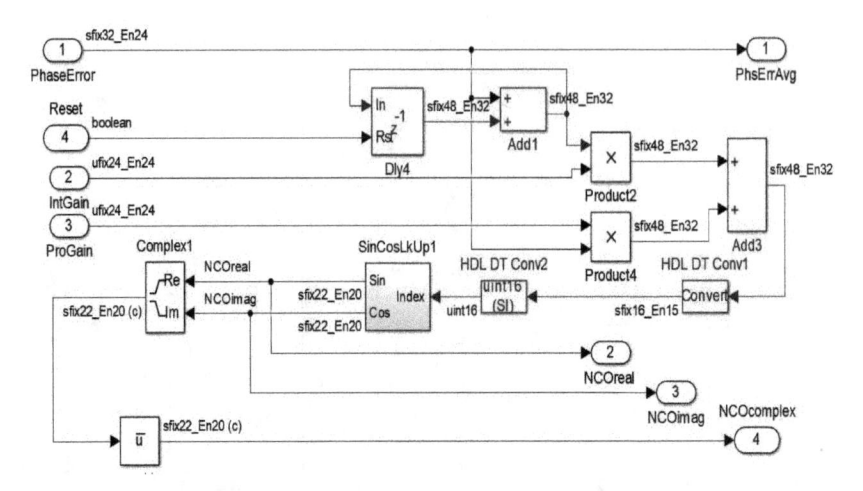

Figure 12.19 Carrier tracking loop filter and NCO, Simulink.

the LMS output in S1 in Figure 12.20. Chapter 5 has a more detailed equalizer mathematical background. Here the focus is on implementation.

12.5.2.1 Equalizer Implementation

Now let's examine the actual equalizer Simulink circuit diagram in Figure 12.21. This is an $M = 2$, two samples/symbol instead of the $M = 4$ circuits we have discussed previously. The equalizer output is still 1 sample/symbol. When SymbolReady goes high and enables the circuit, two samples for the current symbol are input on the left side of the LMS (top half). These are concatenated with the current 32 sample equalizer vector to make 34 samples at the Selector1 block input. The Selector1 block deletes the oldest two samples so that the concatenate and select block acts like a shift register. The current 32 sample selector output vector is sent to the Flip block, which simply flips the vector left to right. The 32-element flipped output is used two ways:

1. The 32 elements are multiplied by the 32 coefficients (filter taps) in the Eq Product block (upper right corner). The 32 Eq Product block outputs are then summed to generate the one sample/symbol LMS equalizer output. This is similar to the multiply-add operation of an FIR filter.

2. The conjugate of the 32 elements are all multiplied by the scaled error in the Err2Tap Correlator (see $e(k)$ in Figure 12.20 and also Figure 5.6). LMS training

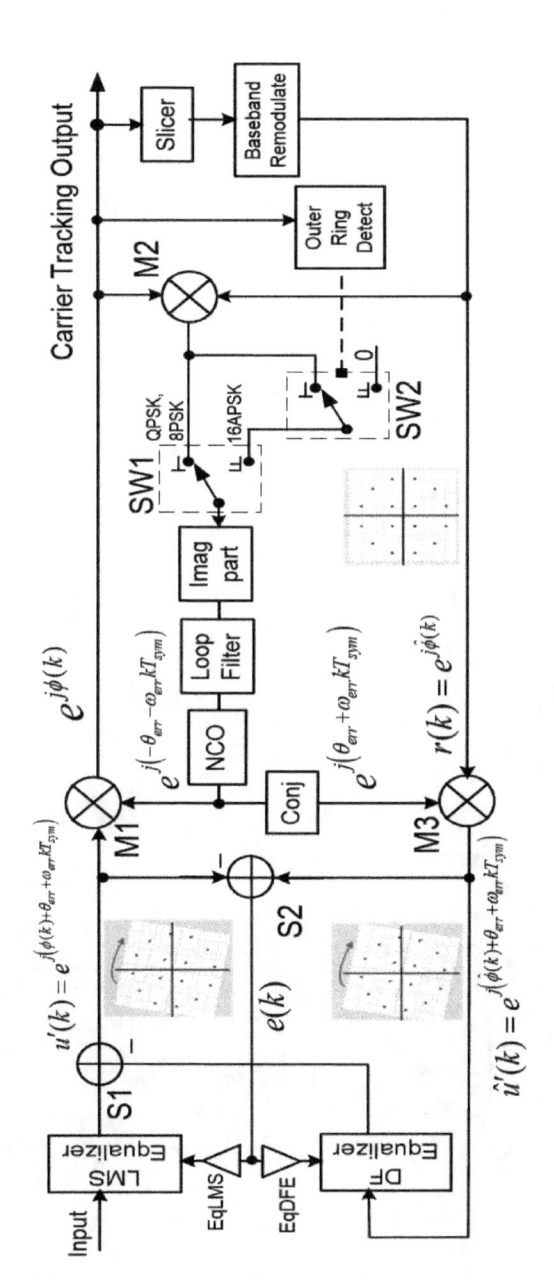

Figure 12.20 Equalizer relation to carrier tracking.

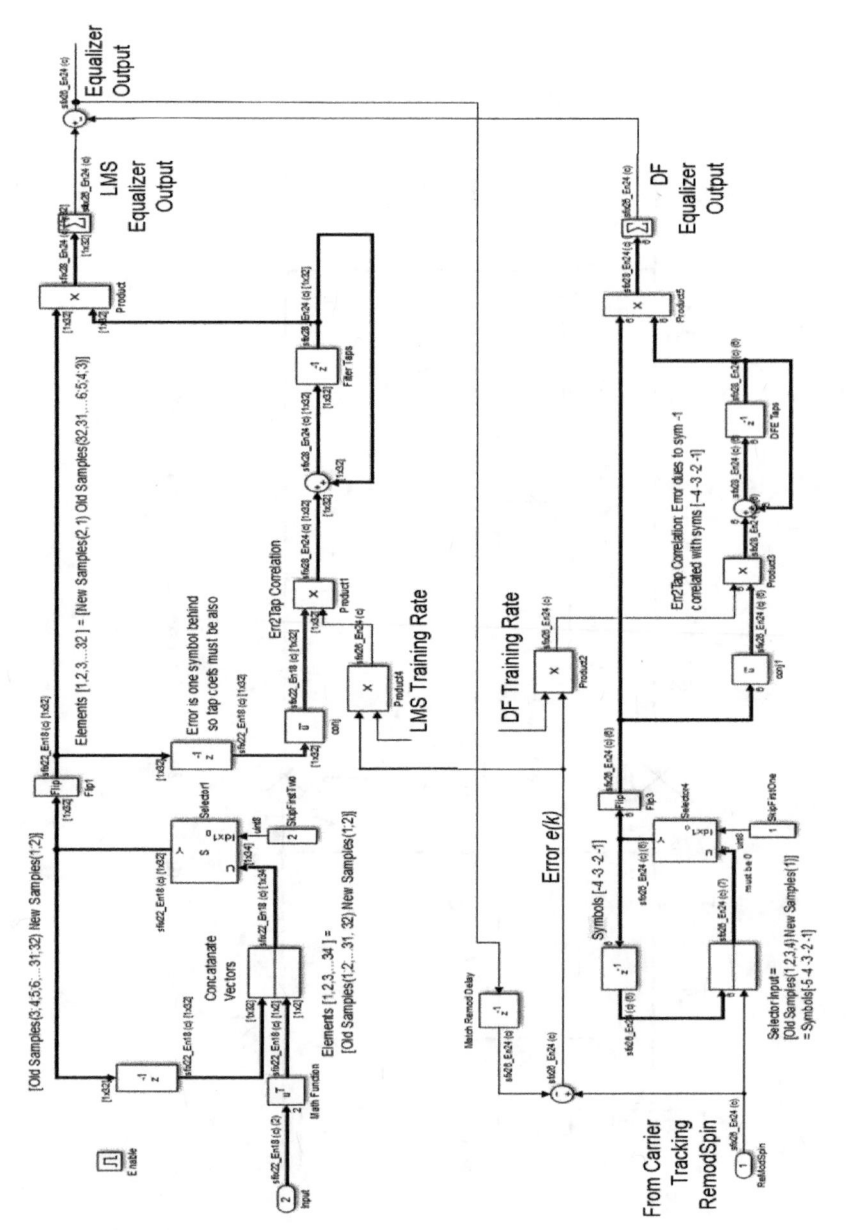

Figure 12.21 LMS and DF equalizer Simulink circuit diagram.

CHAPTER 12. APSK DIGITAL SIGNAL PROCESSING 497

rate, the error scaling, is a constant less than 1. The LMS equalizer coefficients are stored in 32 element memory, Filter Taps, and updated every symbol.

The DF, lower part of Figure 12.21, is very similar to the LMS with one exception: The DF input in the lower left corner consists of already decided symbols[6] from the carrier detector remodulator (M3 output in Figure 12.20). These are input to a shift register loop based on Selector4. The LMS equalizer output is subtracted from the ReModSpin input to generate an error for training the DFE.

Assuming that the remodulated constellation points are correct, they can each be scaled and summed into a model of the multipath ISI that the original symbols from the transmitter encountered on the channel. Product5 block scales and sums 6 symbols so if the multipath ISI is 6 symbols or less, we can make an estimated model of the multipath and subtract it from the LMS equalizer output in the upper right corner of Figure 12.21. A comparison of the upper and lower halves of Figure 12.21 shows that the DF coefficient adjustment to minimize $e(k)$ is very similar to the LMS coefficient adjustment. We should note that with a high noise channel, where many incorrect remodulator symbol decisions are input, the DF will be less effective.

12.5.2.2 Equalizer, Carrier Tracking, and Interconnections

A closer look at the equalizer and carrier tracking interconnections is provided by the Simulink diagram in Figure 12.22. The left block is detailed in Figure 12.21 and the right block is shown in Figure 12.18. The block I/O (inputs and outputs) match the I/O labels on these figures. The unconnected I/O are only for testing.[7] One addition block, StartupTiming, drives DataActive. After a startup delay, this signal switches the carrier tracking from preamble to normal data mode.

Input parameters, such as the equalized step size parameters, EqLMS and EqDFE, are in Simulink constant blocks. They are set in the MATLAB workspace before the simulation is run. Also set up are the carrier tracking integral and proportional second-order phase lock loop parameters CarTrkInt and CarTrkPro as well as the modulation type code, ModType (see Figure 12.18).

Except for SymbolReady, the block interconnecting lines carry points from the constellation indicated by ModType. Notice that the carrier tracking block output, RemodSPin, is the M3 output in Figure 12.18. This feedback connection contains the remodulated spinning constellation to compare with the equalizer input. The

---

6   These symbols are called precursors because they preceded the current symbol.
7   Note that in Simulink, inputs are always on the block left side and outputs are on the right side. Also, the block enable signal is always at the top.

Equalized output is still spinning but the CarTrkOut output on the right block should be stable.

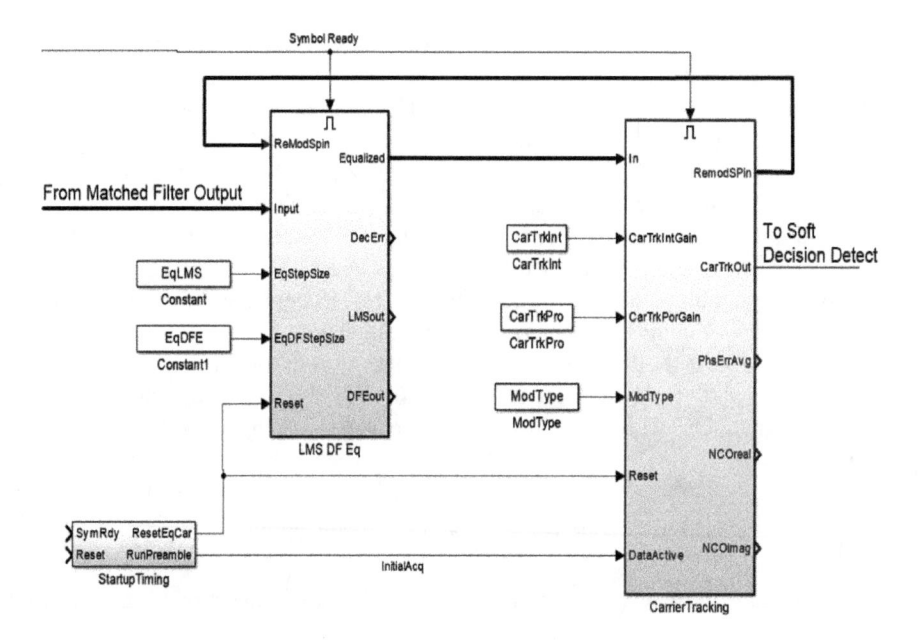

Figure 12.22 Equalizer to carrier tracking Simulink block connections.

12.5.2.3 Symbol Decision Detector

A typical carrier tracking output constellation is shown in Figure 12.23. In a perfect noise-free environment, each complex symbol value will correspond to one of the tiny crosses. The correspondence between symbol states and the bits they carry is planned. Since errors most often happen between adjacent symbols, these only have 1 bit difference. This is called Gray coding and is chosen to reduce bit error rate (BER).

The "clouds" of small dots around each ideal symbol value are the actual received symbol values for an SNR of about 25 dB. Because the clouds of dots are clustered close to each ideal symbol value,[8] the decision detector usually reports the correct closest ideal constellation point for each received symbol.

8 A very careful look at Figure 12.23 reveals a slight phase error in an otherwise correct tracking.

CHAPTER 12. APSK DIGITAL SIGNAL PROCESSING 499

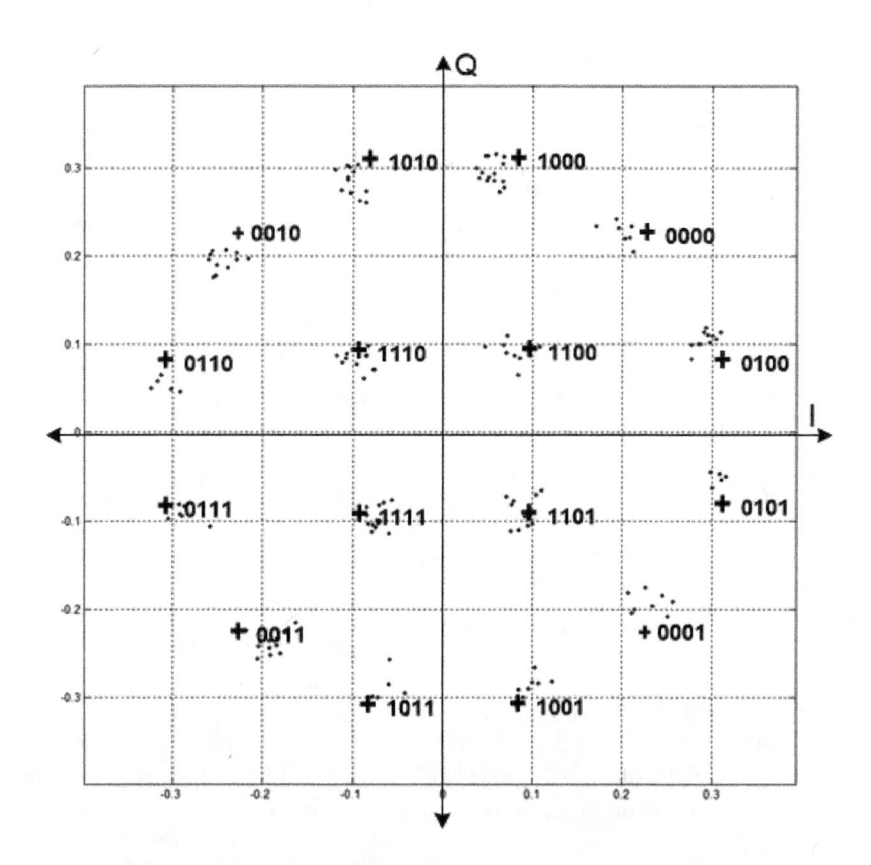

Figure 12.23 16APSK constellation showing symbol values and ideal symbols.

12.5.2.4 16APSK Detection Details

We start with 16APSK because it is the most difficult to detect. Referring to Figure 12.23, for 16APSK hard decision detection, the four bits are labeled $[b_3 b_2 b_1 b_0]$. For example, the inner point in quadrant 0 is $[b_3 b_2 b_1 b_0] = 1100$. There are 16 symbols and they each represent the $4 = log_2(16)$ bits shown.

To detect b_0, the least significant bit, simply slice around the horizontal axis. Received symbols with a positive imaginary part have $b_0 = 0$ and vice versa. To detect b_1, the next to least significant bit, simply slice around the vertical axis. Received symbols with a positive real part have $b_1 = 0$ and vice versa. Close examination of Figure 12.23 should reveal this.

Unfortunately, detecting b_2 and b_3 is not as simple. Start by noting the actual received quadrant and then reflecting the received symbol to quadrant zero by taking the absolute values of the two coordinates. For received symbol $p_{rec} = s_i + j s_q$, calculate $r_i = |s_i|$ and $r_q = |s_q|$. As outlined in Figure 12.24 and Equation (12.8), the preliminary step to deciding b_2 is to decide the following:

If $b_2 = 0$, what is the most likely received constellation point?
If $b_2 = 1$, what is the most likely received constellation point?

$$b_{2zero} = \begin{cases} 1000 & r_q > \sqrt{3} r_i \\ 0000 & r_q \leq \sqrt{3} r_i \end{cases}$$
$$b_{2one} = \begin{cases} 0100 & r_i > R_{mid} \\ 1100 & r_i \leq R_{mid} \end{cases} \quad (12.8)$$

Define s_{2zero} as the complex vector corresponding to the point selected for b_{2zero} and s_{2one} as the complex vector corresponding to the point selected for b_{2one}, both decided above. Points s_{2zero} and s_{2one} are ideal constellation points, not noisy received points.

The hard decision for b_2 is shown in Equation (12.9). This equation is simply deciding which vector, s_{2zero} or s_{2one} that p_{rec} is closest to.

$$b_2 = \begin{cases} 1 & \|p_{rec} - s_{2one}\| < \|p_{rec} - s_{2zero}\| \\ 0 & \|p_{rec} - s_{2zero}\| < \|p_{rec} - s_{2one}\| \end{cases} \quad (12.9)$$

Finally, we must make a hard decision for b_3. As outlined in Figure 12.25, and similar to the hard decision for b_2, the first step is to decide the following:

If $b_3 = 0$, what is the most likely received constellation point?
If $b_3 = 1$, what is the most likely received constellation point?

CHAPTER 12. APSK DIGITAL SIGNAL PROCESSING

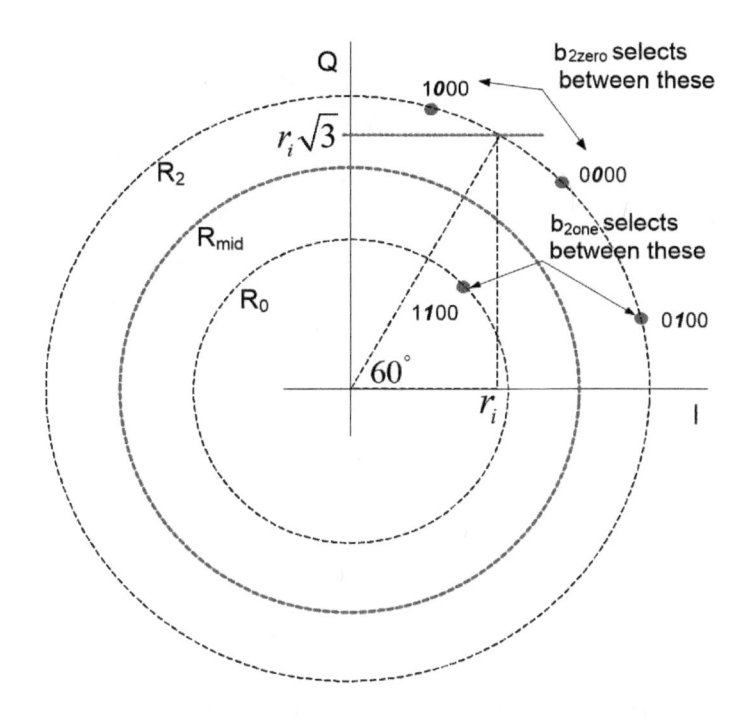

Figure 12.24 16APSK constellation showing detection of bit b_2.

$$b_{3zero} = \begin{cases} 0100 & r_i > \sqrt{3} r_q \\ 0000 & r_i \leq \sqrt{3} r_q \end{cases}$$
$$b_{3one} = \begin{cases} 1000 & r_q > R_{mid} \\ 1100 & r_q \leq R_{mid} \end{cases} \quad (12.10)$$

Define s_{3zero} as the complex vector corresponding to the point b_{3zero} and s_{3one} as the complex vector corresponding to the point b_{3one}, both decided above. The hard decision for b_3 is simply:

$$b_3 = \begin{cases} 1 & \|p_{rec} - s_{3one}\| < \|p_{rec} - s_{3zero}\| \\ 0 & \|p_{rec} - s_{3zero}\| < \|p_{rec} - s_{3one}\| \end{cases} \quad (12.11)$$

The Simulink block diagram for the 16APSK symbol decision detector is shown in Figure 12.26. The input is the series of one sample per symbol constellation points plus noise. When the carrier tracking is converged, these are not spinning.

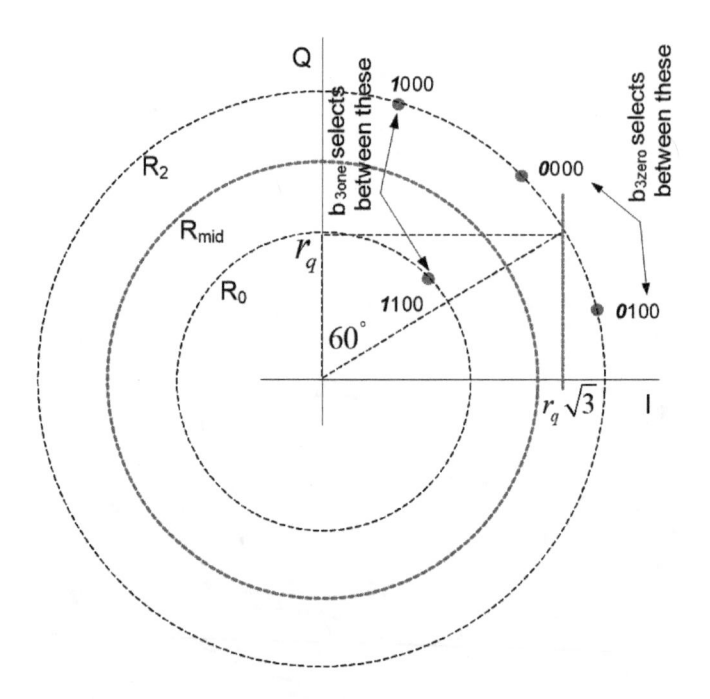

Figure 12.25 16APSK constellation showing detection of bit b_3.

For each received point, the decision detector tries to estimate the corresponding transmit point. For 16APSK, each symbol represents four bits $[b_3 b_2 b_1 b_0]$.

The center of Figure 12.26 shows a stack of 4 Simulink lookup tables. The first table, Sd02, calculates b_{2zero} by implementing Equation (12.8). The contents of this look up table are in Table 12.6. Likewise, the second look up table, Sd12, calculates b_{2one} by implementing Equation (12.10). For a final value of b_2, the distance between the carrier tracking output and both b_{2zero} and b_{2one} are calculated and subtracted. This is an implementation of Equation (12.9). However, the b_2 result is not 1 or 0 but a continuum of values where positive corresponds to 1 and negative corresponds to 0. This is referred to as a soft decision.

To convert $[b_3 b_2 b_1 b_0]$ soft decisions on the right side of Figure 12.26 into hard decisions, simply slice each of the four soft decisions around 0. This means that results > 0 are assigned to a binary 1 and results < 0 are assigned to a binary 0.

The next two look up tables, Sd03 and Sd13, perform the same function for b_3 that the first pair of look up tables did for b_2.

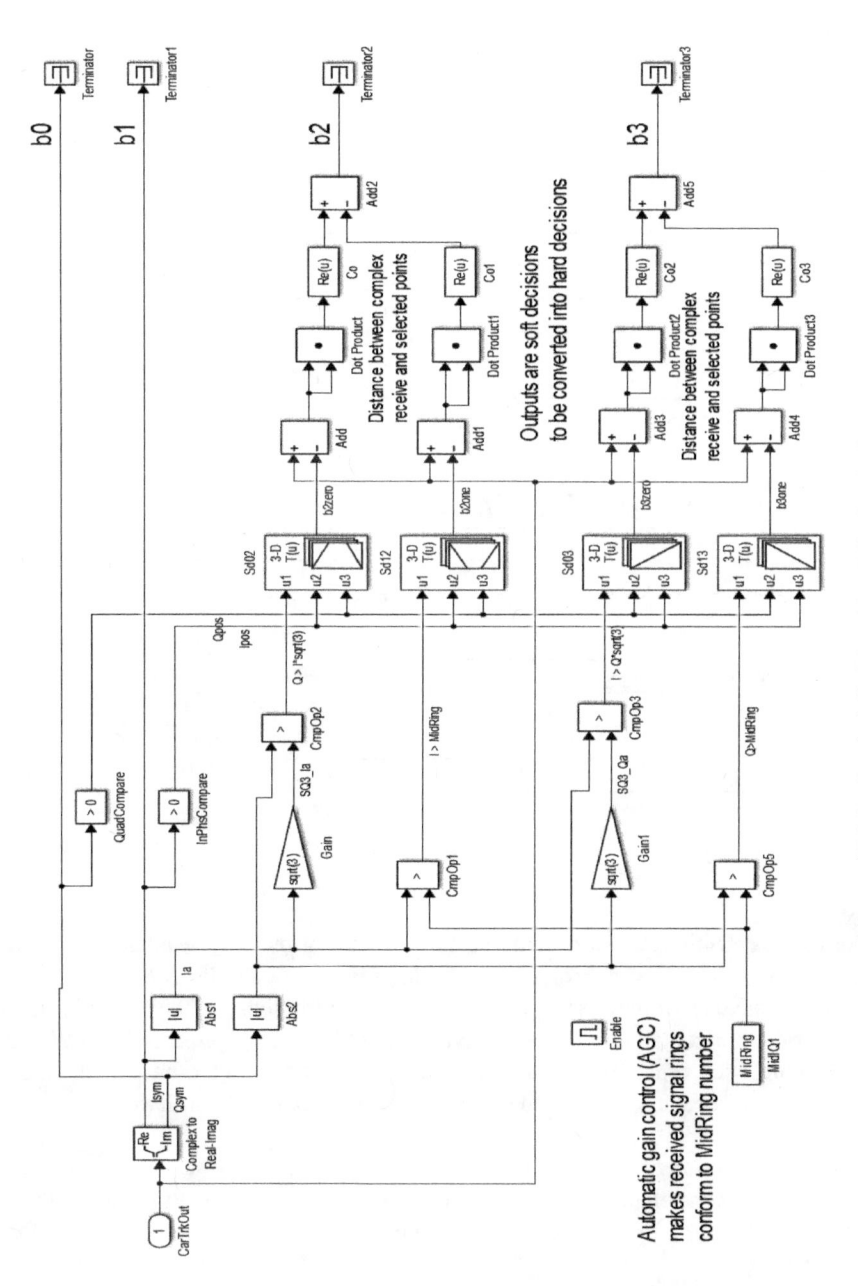

Figure 12.26 Symbol decision Simulink circuit diagram.

Table 12.6 b_{2zero} Look Up Table

$I > 0$	$Q > 0$	$Q > \sqrt{3}$	b_{2zero}	s_{2zero}
1	1	1	1000	$0.083 + j0.31$
1	1	0	0000	$0.23 + j0.31$
0	1	1	1010	$-0.083 + j0.31$
0	1	0	0010	$-0.23 + j0.23$
0	0	1	1011	$-0.083 - j0.31$
0	0	0	0011	$-0.23 - j0.23$
1	0	1	1001	$0.083 - j0.31$
1	0	0	0001	$0.23 - j0.23$

Table 12.7 b_{2one} Look Up Table

$I > 0$	$Q > 0$	$R_i > R_{mid}$	b_{2one}	s_{2one}
1	1	1	0100	$0.31 + j0.083$
1	1	0	1100	$0.094 + j0.094$
0	1	1	0110	$-0.31 + j0.083$
0	1	0	1110	$-0.094 + j0.094$
0	0	1	0111	$-0.31 - j0.083$
0	0	0	1111	$-0.094 - j0.094$
1	0	1	0101	$0.31 - j0.083$
1	0	0	1101	$0.094 - j0.094$

12.5.2.5 8PSK Detection Details

For 8PSK detection, we work with the carrier detector output constellation rotated by 22.5° CW. This is shown on the right side of Figure 12.27. 8PSK b_0 and b_1 soft decisions are now straightforward. To detect b_0, the least significant bit, simply slice around the horizontal axis. Received symbols with a positive imaginary part have $b_0 = 0$ and vice versa. To detect b_1, the next to least significant bit, simply slice around the vertical axis. Received symbols with a positive real part have $b_1 = 0$ and vice versa.

For detection of the 8PSK most significant bit (MSB), we start again with the 22.5° rotated constellation and plot the absolute value of the I and Q. The result is shown in Figure 12.28. Here it is clear the the MSB = 1 for $Q > I$ and MSB = 0 for $I > Q$.

CHAPTER 12. APSK DIGITAL SIGNAL PROCESSING 505

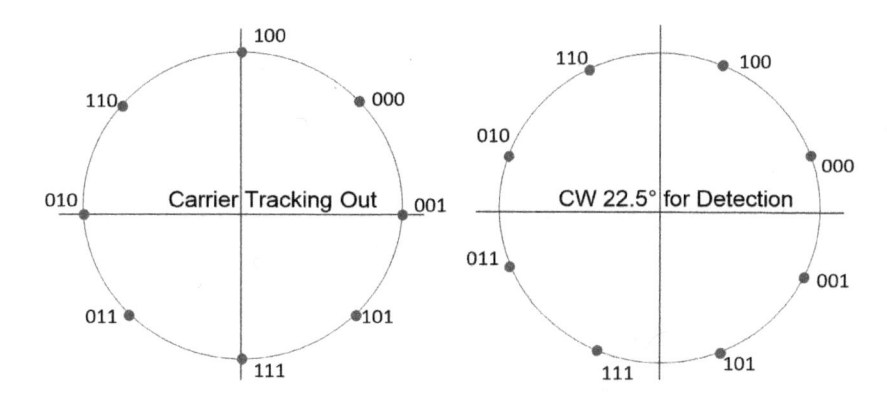

Figure 12.27 8PSK constellation showing 22.5° CW detection rotation.

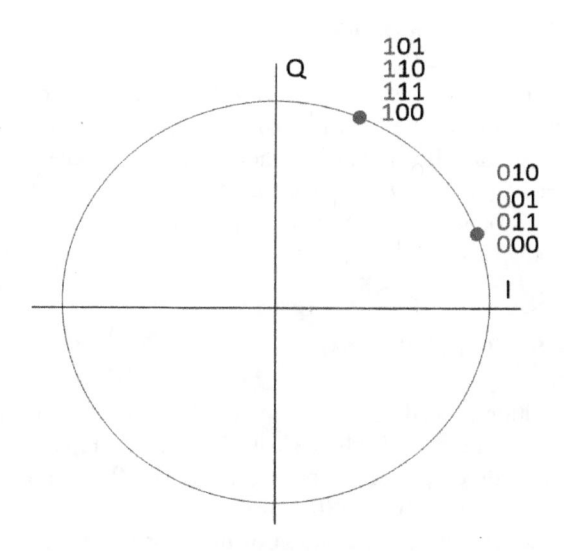

Figure 12.28 8PSK most significant bit detection.

```
                    Q
                    |
      10            |        00
       •            |         •
                    |
                    |
────────────────────┼──────────────── I
                    |
                    |
                    |
      11  •         |         •  01
                    |
```

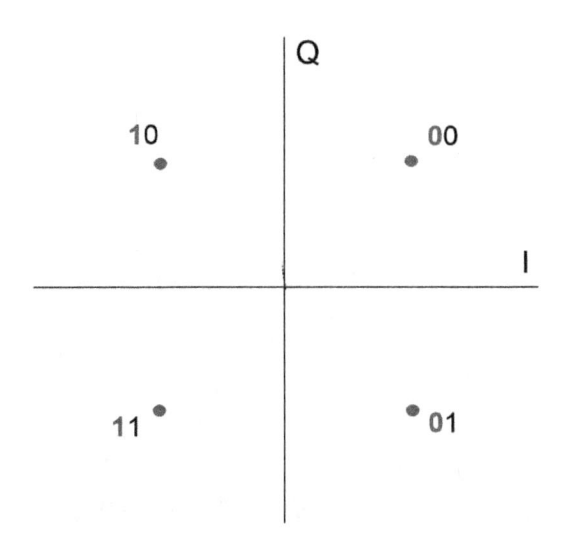

Figure 12.29 QPSK bit detection.

12.5.2.6 QPSK Detection Details

QPSK is easy to detect. As shown in Figure 12.29, the signs of the received I and Q can be used to determine both b_0 and b_1, respectively. To detect b_0, the least significant bit, simply slice around the horizontal axis. Received symbols with a positive imaginary part have $b_0 = 0$ and vice versa. To detect b_1, the most significant bit, simply slice around the vertical axis. Received symbols with a positive real part have $b_1 = 0$ and vice versa.

12.6 APSK SYSTEM TESTING

A good verification procedure is to compare the symbol error rate (SER) of the complete APSK simulation with the calculated SER. Figures 12.30 and 12.31 are examples. Both graphs show performance in AWGN only. The calculated performance is based on Equation (6.13).

Some readers may prefer a graph of bit error rate vs. $E_b N_0$. Note that, for QPSK, BER is approximated by SER/2 because QPSK has 2 bits per symbol. Likewise for 8PSK with 3 bits/symbol. We say approximated because, assuming Gray coding, most symbol errors only result in 1 bit error (e.g., symbol 11 errors to

CHAPTER 12. APSK DIGITAL SIGNAL PROCESSING 507

01 or symbol 001 errors to 101), although this is not always true. For QPSK, $E_b N_0$ (dB) = $E_s N_0$ (dB) - 3 dB and for 8PSK, $E_b N_0$ (dB) = $E_s N_0$ (dB) - 4.77 dB.

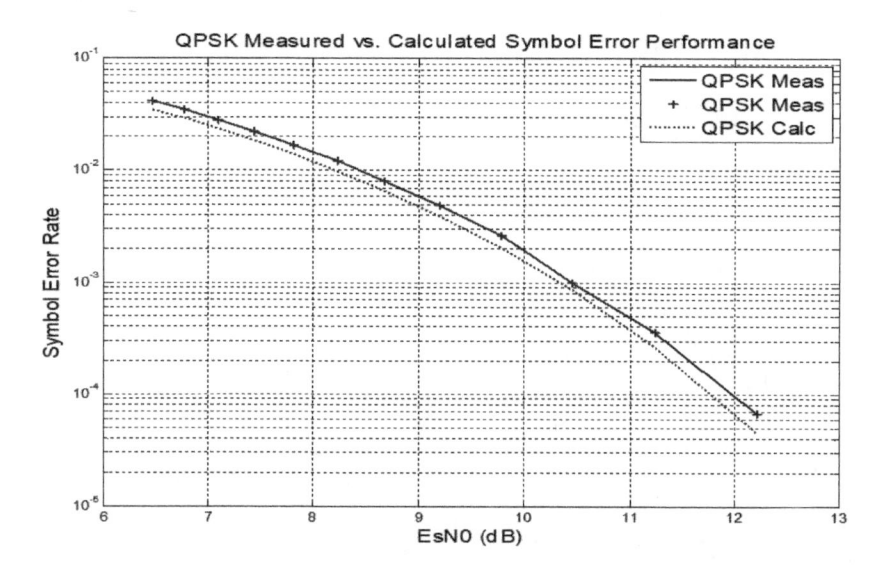

Figure 12.30 QPSK symbol error rate testing.

12.6.1 APSK Hardware Testing

We will start with the basic hardware configuration shown in Figure 13.41. The ADALM-Pluto SDR (as used for ADSB in Chapter 11) will be used for this test. Other SDR transceivers (as opposed to receive-only SDRs, see Table 1.1) can be used in place of the ADLAM-Pluto, for example, the USRP E310. Not all of these will interface to MATLAB/Simulink however.

The SDR antennas should be installed prior to turning on the transmitter. The default frequency is in the unlicensed ISM band so radiating small amounts of power from the transmit antenna is no problem.

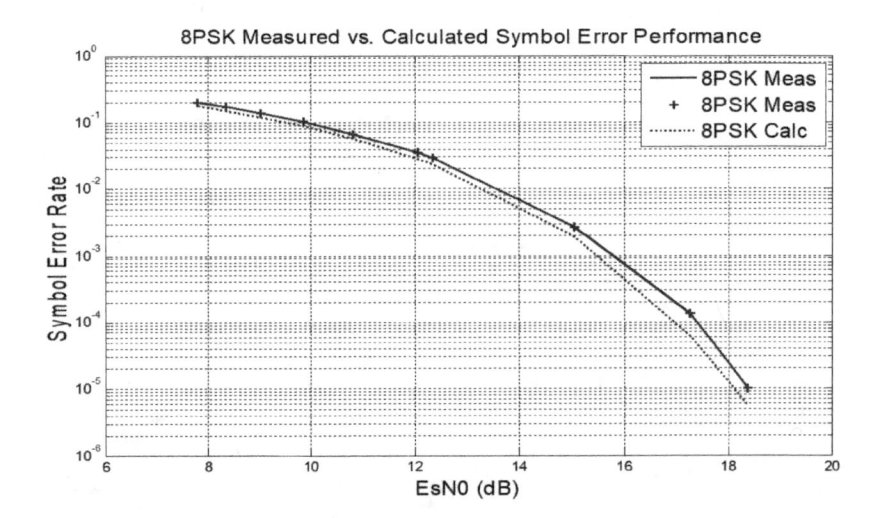

Figure 12.31 8PSK symbol error rate testing.

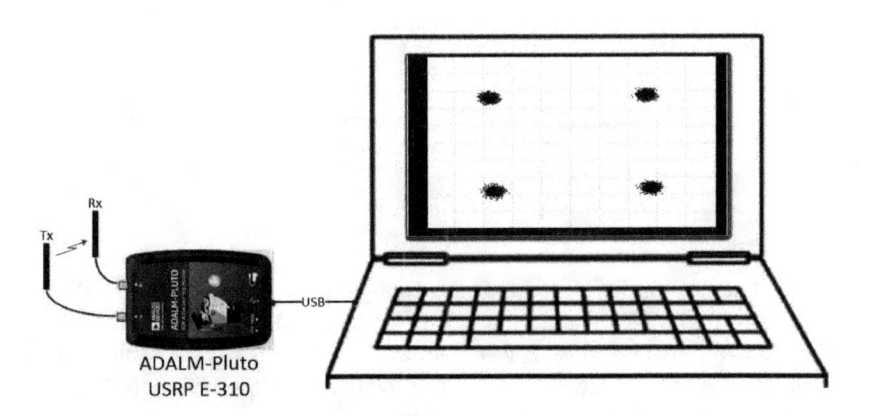

Figure 12.32 SDR basic system test configuration.

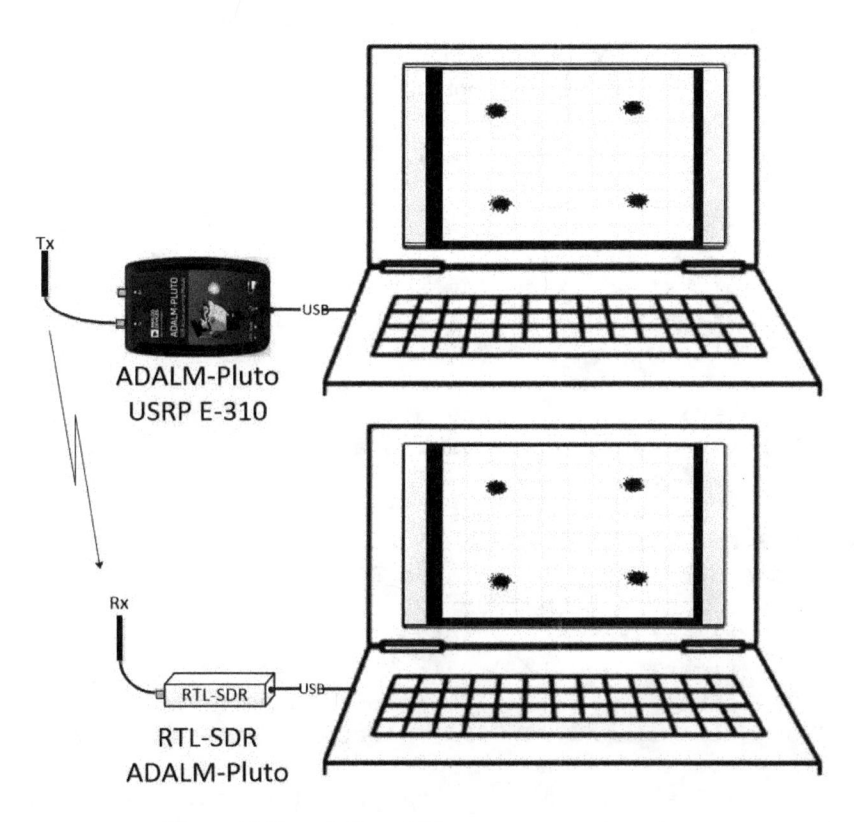

Figure 12.33 Alternate SDR system test configuration.

12.6.1.1 Transmit Setup

The APSK simulation setup file stores a copy of the APSK simulated baseband samples in "APSKbaseband.mat." This file stores two sample vectors, BBreal and BBimag.

The MATLAB code below reads file APSKbaseband.mat, sets up a tx object to control the ADALM-Pluto transmitter and then downloads the samples to the transmitter. Because MATLAB is not a multithreaded program, we have to start the transmitter in repeat mode and then we can focus on the receiver. The code below sets up the ADALM-Pluto transmitter and then returns to MATLAB.

```
%Load ADALM-Pluto transmit buffer with APSK samples
%Transmit repeatedly and end program because we can't
%control tx and rx at the same time.
if(exist('tx','class')) % Stop current transmitters
tx.release
end
Fs50 = 50; %AD9363 baseband sample rate
TxLength =84000; %Tx buffer space needed
TxStart = 1;
TxEnd = TxStart + TxLength-1;
%APSK QPSK transmit test signal, 400MHz sampling rate
load APSKbaseband % contains BBreal BBimag
%Reduce sample rate from 400 to 50 MHz
%This is to accomodate AD9363 sample rate limitations
HB1 = firhalfband(12,0.25); %First downsample filter
HB2 = firhalfband(18,0.25); %Second downsample filter
HB3 = firhalfband(24,0.25); %Second downsample filter
%Fs = 400 to 200 MHz
BBreal200 = downsample(filter(HB1,1,BBreal),2);
BBimag200 = downsample(filter(HB1,1,BBimag),2);
%Fs = 200 to 100 MHz
BBreal100 = downsample(filter(HB2,1,BBreal200),2);
BBimag100 = downsample(filter(HB2,1,BBimag200),2);
%Fs = 100 to 50 MHz
BBreal50 = downsample(filter(HB3,1,BBreal100),2);
BBimag50 = downsample(filter(HB3,1,BBimag100),2);
TxR1 = 0.5*BBreal50(TxStart:TxEnd);
```

CHAPTER 12. APSK DIGITAL SIGNAL PROCESSING 511

```
TxIg = 0.5*BBimag50(TxStart:TxEnd);
TxComplex = complex(TxRl, TxIg);
tx = sdrtx('Pluto',...
'Gain',-20,...
'BasebandSampleRate',Fs50*10^6,...
'ShowAdvancedProperties',true);
%Download transmit signal samples (50MHz) into AD9363
%TxComplex contains an APSK packet
%This is repeated continuously
tx.transmitRepeat(TxComplex);
%To stop transmitter type "tx.release" in workspace
```

12.6.1.2 Receive Setup

The MATLAB code below sets up an rx object to control the ADALM-Pluto receiver and then uploads samples from the receiver. The samples received this way could be further processed by the rest of the APSK processing code. The Simulink version of this was discussed at length in this chapter. Although the SDR transmitter is in repeat mode and cannot be changed on the fly, other system aspects such as antenna placement can be tested. The effect of receiver code changes can be easily tested.

```
% 1 if rx not created yet, 0 if modifying existing rx
NewRxSystemObject = 1;

%Setup ADLAM-Pluto receive object and download
%latest samples. Sample rate is 50MHz because
%Transmit and receive samples rates must match
format compact
Fs50 = 50;
RxLength = 2^19; %Grab 524288 receive samples
% If there is already a receiver running, keep it:
% A new receiver object will revert to default options.
if(NewRxSystemObject) rx = sdrrx('Pluto',...
'SamplesPerFrame',RxLength, ...
'BasebandSampleRate',Fs50*10^6);
end
%Two ways to learn about the rx object:
```

```
%Type "rx" to see radio setups.
%Type methods(rx) to see object functions, examples:
%rx.release stops the receiver
%rx.info returns version information
RxSamples = double(rx());

figure(10)
subplot(2,1,1);
plot(real(RxSamples),'r');
hold on
plot(imag(RxSamples),'b');
hold off
set(gca,'YLim',[-2000 2000]);
grid
ht = title('ADALM-Pluto QPSK Receive Samples');
set(ht,'FontSize',14);
NFFT = (1/2)*4096;
HannWin = hann(NFFT/4);
HannWin = HannWin/sum(HannWin);
pSig160Out...
= pwelch(RxSamples,HannWin,[],NFFT,Fs50,'twosided');
pSig160OutLog = 10*log10(abs(fftshift(pSig160Out)));
scale160 =round(((-Fs50/2):((Fs50/16)):(Fs50/2)));
subplot(2,1,2);
h160 = stairs(pSig160OutLog ,'b');
set(h160,'LineWidth',1.5);
ylabel('PSD (dB)');
set(gca,'XLim',[0 NFFT]);
set(gca,'XTick',0:round((NFFT/16)):NFFT);
set(gca,'XTickLabel',scale160);
xlabel('Frequency (MHz)');
yt = ylabel('Amplitude (dB)');
set(yt,'FontSize',14);
ht = title('ADALM-Pluto QPSK Receive Spectrum');
set(ht,'FontSize',14);
xt = xlabel('Frequency (MHz)');
```

CHAPTER 12. APSK DIGITAL SIGNAL PROCESSING 513

12.6.1.3 Results

Figure 12.34 shows the transmit and receive sample records along with their power spectra. The transmit and receive signal spectra should be the same since there is very little RF propagation loss. There are a couple of parameters in the SDR receive object that can be tweaked. For example, receive gain and AGC. A well organized MATLAB receive program provides flexibility to try different simulated receive parameters, for example, carrier tracking loop gains. Note receiver baseband noise bandwidth of about 40 MHz in the lower spectrum. This is also known as a noise "pedestal" in traditional receiver parlance. This is an indicator of the receive bandwidth limitation.

12.6.1.4 Receiver Bandwidth Modification

Figure 12.36 shows the 10 Msyms/sec, 35% excess bandwidth QPSK receive signal. The AD9363 receive bandwidth is about three times the signal bandwidth. We might want to reduce this to eliminate close-in high-power blocking signals.

Here are the steps to reduce the receiver bandwidth. First type the following in the MATLAB workspace:

```
rx.release %Stops the receiver
rx.designCustomFilter % Type methods(rx)
```

This brings up the Filter Wizard screen. Make the changes shown in Figure 12.35 in the order shown. Now run the receiver code above. Result is in Figure 12.37.

12.6.1.5 Simulink Configuration

Figure 12.38 shows how the Simulink components discussed earlier in the chapter can be coupled with the ADLAM-Pluto SDR to form a Simulink based receiver. If the Simulink simulation has trouble keeping up with the real-time streaming data from the ADLAM-Pluto, the ADLAM-Pluto may have to be run in block mode. This is where a guaranteed contiguous block is produced, but in between blocks there could be missed samples. See Section 11.2.1 for more details.

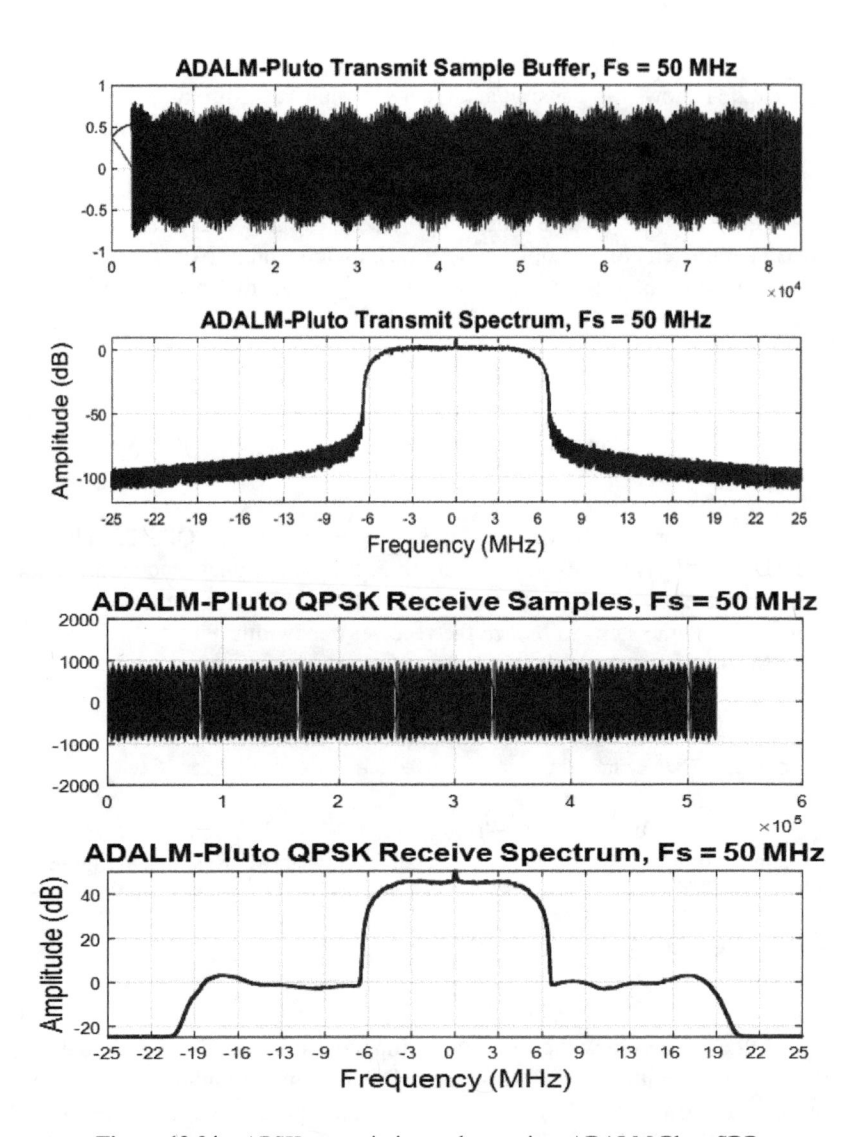

Figure 12.34 APSK transmission and reception, ADALM-Pluto SDR.

CHAPTER 12. APSK DIGITAL SIGNAL PROCESSING 515

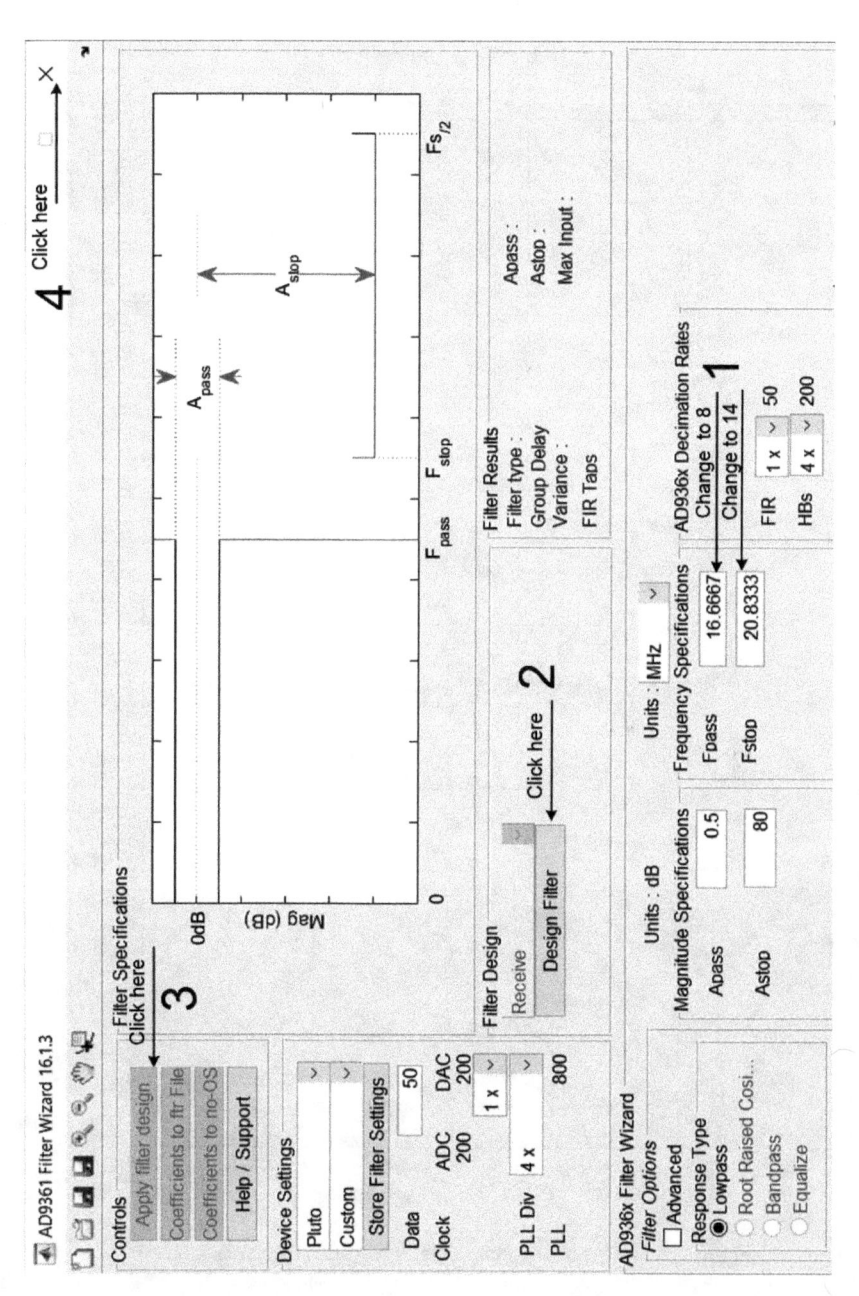

Figure 12.35 MATLAB filter design wizard.

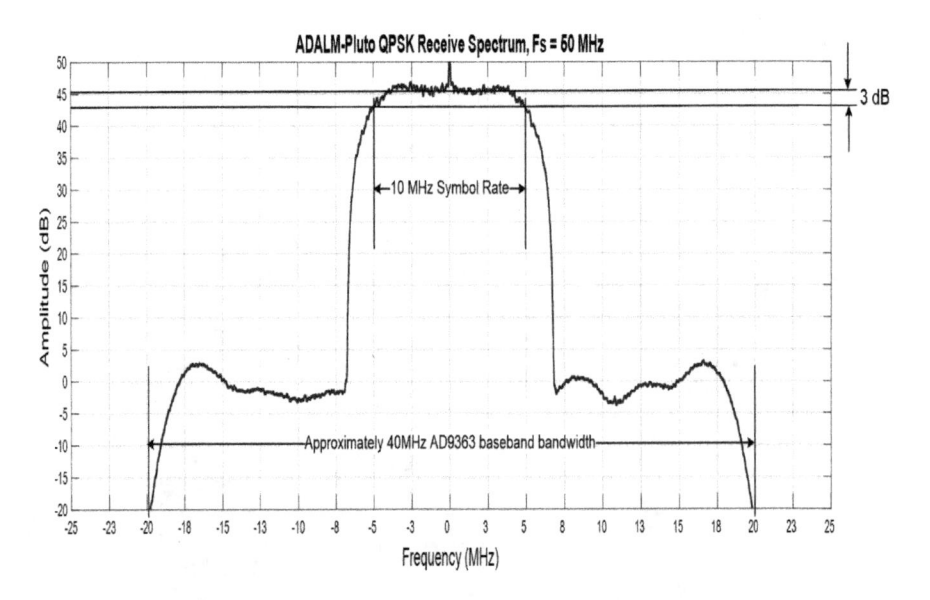

516 *Software Defined Radio: Theory and Practice*

Figure 12.36 ADALM-Pluto default receive bandwidth.

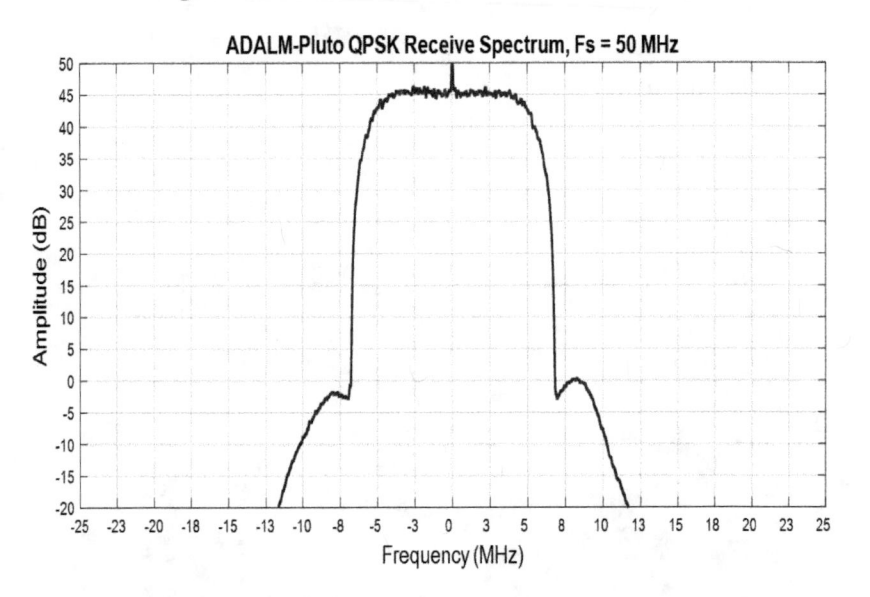

Figure 12.37 ADALM-Pluto default modified receive bandwidth.

CHAPTER 12. APSK DIGITAL SIGNAL PROCESSING

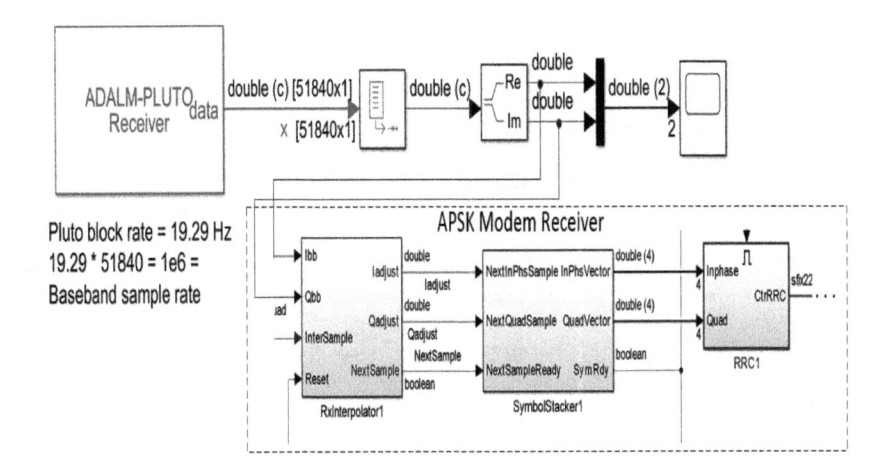

Figure 12.38 APSK receiver using ADALM-Pluto and Simulink.

12.6.1.6 Alternate Test Configuration

The ADALM-Pluto SDR uses a common PLL (phase locked loop) clocking circuit. For a more real-world simulation, we can use the alternate configuration in Figure 12.33. This requires more equipment, however, listed below are the advantages:

1. **Carrier Frequency Tracking** Transmit and receive LOs (local oscillators) are set up separately in the Pluto SDR, therefore, the alternative setup will have no advantage.
2. **Channel Impairments** In the basic setup, Figure 13.41, the receive and transmit antennas are only a few centimeters apart. The alternative setup allows these to be any distance apart. This may be very helpful for measuring the effectiveness of the receive equalizer.
3. **Transmitter Simulation** In the basic setup, we had to generate one transmit frame and repeat it; see MATLAB code above. In the alternative setup the transmit simulation can have a dedicated ADALM-Pluto Transmitter system object block that will allow a series of changing transmit frames.

REFERENCES

[1] European Technical Standards Institure. "Digital Video Broadcasting (DVB-s2)". *ETSI EN 302 307*, April 2009.

[2] F. Harris. "Polyphase Filterbanks for Symbol Timing Synchronization in Sampled Data Receivers". *MILCOM 2002. Proceedings*, October 2002.

[3] Mathworks. "Fractional Delay Filters Using Farrow Structures". *MILCOM 2002. Proceedings*, June 2002.

[4] J. G. Proakis. *Communication System Engineering*. Prentice-Hall, 2002.

[5] J. C. Song. "The Performance of Symbol Timing Algorithm for Multi-level Modulation Scheme". *IEEE, Proceedings of the Vehicular Technology Conference*, 1996.

[6] F. M. Gardner. "A BPSK/QPSK Timing Error Detector for Sampled Receivers". *IEEE Transactions on Communications*, 34(5), 1986.

[7] G. F. Franklin. *Feedback Control of Dynamic Systems*. Prentice Hall, 2002.

[8] V. Vilnrotter. "Performance Analysis of Digital Tracking Loops for Telemetry Ranging Applications". *IEEE Transactions on Circuits and Systems II: Express Briefs*, August 2015.

[9] J. Wilson. "Parameter Derivation of Type-2 Discrete Time Phase Locked Loops Containing Feedback Delays". *IEEE Transactions on Circuits and Systems II: Express Briefs*, 56(12), December 2009.

[10] J. A. C. Bingham. *The Theory and Practice of Modem Design*. John Wiley and Sons, 1988.

Chapter 13

IEEE802.11a Digital Signal Processing

13.1 IEEE802.11A OVERVIEW

IEEE802.11a is a wireless LAN (local area network) OFDM based standard, first released in 1999. This introduction provides a summary of the signal characteristics.

13.1.1 IEEE802.11a OFDM Basics

IEEE802.11a has a maximum data rate of 54 Mbps and is transmitted in a 20 MHz wide channel in the 5 GHz unlicensed ISM (Industrial, Scientific, Medical) band. Table 13.1 shows the complete set of IEEE802.11a modulation and coding options. Both the IEEE802.11a transmitter and receiver are based on fast Fourier transforms of length 64 (see Section 3.2.10). The transmitter uses an IFFT (inverse FFT), the inputs to the IFFT are called symbols, and the outputs are called samples. This will hopefully become more clear as we proceed.

In the transmitter, the coded symbols per IFFT input frame are always 48 but the coded bits per frame in Table 13.1 varies with the modulation. This is due to the nature of QAM where multiple bits are carried by one symbol; see Section 3.2.8. In this chapter, we study the implementation of only the first line of Table 13.1.

Readers unfamiliar with OFDM may benefit from a review of Section 3.2.10.

Knowing the rate of customer input bits is 6 MHz and the output sample rate, which determines the output bandwidth, is 20 MHz allows us to calculate a bandwidth efficiency of $6/20 = 0.3$. Bandwidth efficiency ((bits/second)/(Hz of bandwidth)) can be increased by using the higher data rate modes in Table 13.1. For example, the last line in Table 13.1 has a bandwidth efficiency of $54/20 = 2.7$.

Table 13.1 IEEE802.11a Basic Modes

Mbps	Modulation	Code Rate (R)	Coded Bits	Data Bits
6	BPSK	1/2	48	24
9	BPSK	3/4	48	36
12	QPSK	1/2	96	48
18	QPSK	3/4	96	72
24	16QAM	1/2	192	96
36	16QAM	3/4	192	144
48	64QAM	2/3	288	192
54	64QAM	3/4	288	216

Table 13.2 is a summary of IEEE802.11a signal characteristics for the 6MHz mode only. The first four entries are in units of IFFT input symbols. Note that in the third row, setting the center symbols (28, 29, 30, ... 37) to zero helps to avoid DC offset problems when translating the RF signal to the complex baseband at the receiver.

The next three entries are about prepending output samples used to construct the CP (cyclic prefix). The last six rows of Table 13.2 are concerned with output frame timing. Finally, the last row shows the calculated bandwidth efficiency.

Let's consider the scaled-down and simplified transmitter example in Figure 13.1. This transmitter operates at complex baseband. For every new frame, 12 symbols (bits if BPSK) on the left are shifted into a serial to parallel converter. They are combined with strategically placed zeros to form a frequency-domain input symbol frame to the inverse fast Fourier transform (IFFT). OFDM transmitters use an IFFT to convert a set of frequency-domain symbols at the input to samples of a time-domain signal at the outputs. Each of the 16 inputs is placed at a particular frequency, shown in the frequency placement column (we assume the entire circuit is sampled at F_s). For example, to avoid receiver DC offsets, a zero is placed at 0 Hz, input number nine.

The right side of Figure 13.1 prepends the last four IFFT outputs to the beginning of the parallel to serial output. This is called a cyclic prefix (CP) and is used for equalization, as described in Section 3.2.10.1.

Because the IFFT is a frequency to time-domain converter, the IFFT output must be in the time-domain. This output frame contains the next 16 continuous time-domain samples of an orthogonal multiplex of signals centered at the baseband frequencies shown on the input side of the IFFT. The output spectrum is similar to Figure 3.26.

CHAPTER 13. IEEE802.11A DIGITAL SIGNAL PROCESSING 521

Table 13.2 Physical Layer Operation Characteristics of 6 Mbps IEEE802.11a

Characteristic	Value	Comment
Data symbols/frame	24	Customer data symbols before rate 1/2 coding
Coded symbols/frame	48	Coded data symbols per OFDM frame
Input zero symbols	12	28-37 (avoiding center freq), 1 and 64 (avoiding ends)
Input pilot symbols	4	8,23,43,57 (for carrier tracking and equalization)
IFFT output samples	64	Also the size of the IFFT
Cyclic prefix samples	16	See Figure 3.25
Output frame samples	80	CP + IFFT output samples
Modulation type	BPSK	For this simulation
OFDM output frame time	3.2 usec	Time to unload 64-channel IFFT at 20 MHz clock rate
Cyclic prefix time	0.8 usec	Time to transmit 16 additional CP samples at 20 MHz
Output frame transmit time	4 usec.	Total time per output frame
Input symbol rate	6 Mbps	24/4 Mbps (the inverse of usec is MHz)
Output sample rate	20 MHz	For BPSK, this is also the output bit rate
Bandwidth efficiency	0.3	6/20 (bits/sec)/Hz (BPSK rate 1/2 mode))

In summary, at the transmitter each of the 16 IFFT inputs is placed at a particular complex baseband frequency channel on the output frame. The IFFT outputs are samples of all 16 frequency channels added together (the frequency channels are orthogonal so they do not interfere with each other in the sum). At the receiver a 16-input FFT converts the time-domain frequency multiplex samples back into the frequency-domain channels that the transmitter started with; that is, the transmit data; see Section 3.2.10. Although this simulation is focused on complex baseband signals, a practical OFDM system would upconvert to a radio frequency, as in Figure 7.46.

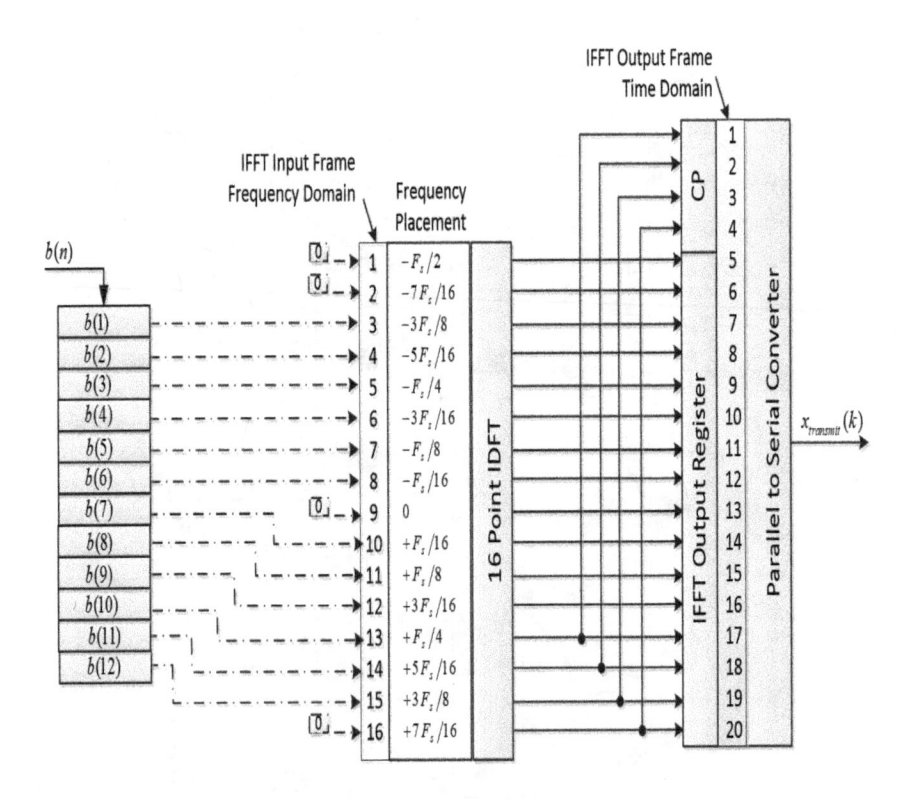

Figure 13.1 OFDM transmitter example.

Getting back to our 6 MHz IEEE802.11a simulation, Table 13.3 shows the layout of customer data, pilots, and zeros at the input to the transmitter 64-element

IFFT (remember, these are frequency-domain IFFT input symbols). Note that a pilot channel can be any symbol pattern known a priori to the receiver, although here they are simply constants. As described above, each table entry is at a particular frequency in the output frame frequency multiplex samples. These output time-domain samples have the desired spectral shape shown in Figure 13.3.

Table 13.3 IFFT Input Configuration for IEEE802.11a

IFFT Inputs	Usage
1	Zero
2-7	Data 1-6
8	Pilot 0
9-21	Data 7-19
22	Pilot 1
23-27	Data 20-24
28-37	Zero
38-42	Data 25-29
43	Pilot 2
44-56	Data 30-42
57	Pilot 3
58-63	Data 43-48
64	Zero

13.1.2 IEEE802.11a Frame Structure

Figure 13.2 shows the preamble structure of the beginning of every IEEE802.11a packet transmission. Ten repetitions of a 16-symbol short training field (STF) start the packet and enable coarse frequency and timing estimations. The symbols used are defined in [1]. The STF is followed by two long training fields (LTF). The first, LTF0, is used for fine frequency and timing estimations. The second, LTF1, is used for frequency-domain equalization. STFs and LTFs have a fixed pattern of input symbols specified by [1]. Finally, the signal frame is attached to the end of the preamble to give the receiver information about the packet.

After the 10 STFs, all data is organized into OFDM output frames. An OFDM output frame is a set of 80 samples, 64 of which are read from the output of an IFFT. The last 16 of the 64 samples are copied and prepended to the frame start. These are called the cyclic prefix (CP) and are covered in detail in Section 3.2.10.1.

We will try to consistently use the word symbols for the frequency-domain IFFT input frame and samples for the time-domain IFFT output frame. The IEEE802.11a Simulink simulation described in this chapter implements all the preamble functions shown (except diversity combining and the signal frame). In this simulation, all post-preamble frames, including the signal frame, are filled with random data for measuring the bit error rate.

13.1.3 IEEE802.11a Transmit Spectrum

The IEEE802.11a transmit spectrum in Figure 13.3 has the on channel (active channel) and associated regulatory mask in the center. The adjacent channel regulatory masks are shown as dotted lines on either side. Transmit channels belonging to individual independent users are always spaced 20 MHz apart. Note that placing zeros at the outer frequencies, see Table 13.3, helps the OFDM signal stay within the regulatory spectral mask. Also note at about 20 dBm down from the flat top, the is a flare out instead of a straight downward plunge in the spectrum sides. This is due to phase discontinuity between frames. Later in this chapter, we will look at a DSP technique to alleviate this.

13.2 IEEE802.11A TRANSMITTER OVERVIEW

The simulated transmitter, shown in Figure 13.4, produces a preamble followed by a 10 data frame IEEE802.11a physical layer packet. No data link and above layers are simulated. Thus, patterns specified by the IEEE for use in the preamble are accurate, but the data frames carry random BPSK symbols.

13.3 IEEE802.11A CHANNEL OVERVIEW

IEEE802.11a signal is often used in an office, school, or home environment that has significant multipath interference. The wireless LAN error rate performance is typically set more by multipath interference than AWGN (additive white Gaussian noise). Thus, our Simulink simulation makes use of both channel simulator blocks; see Figure 13.4.

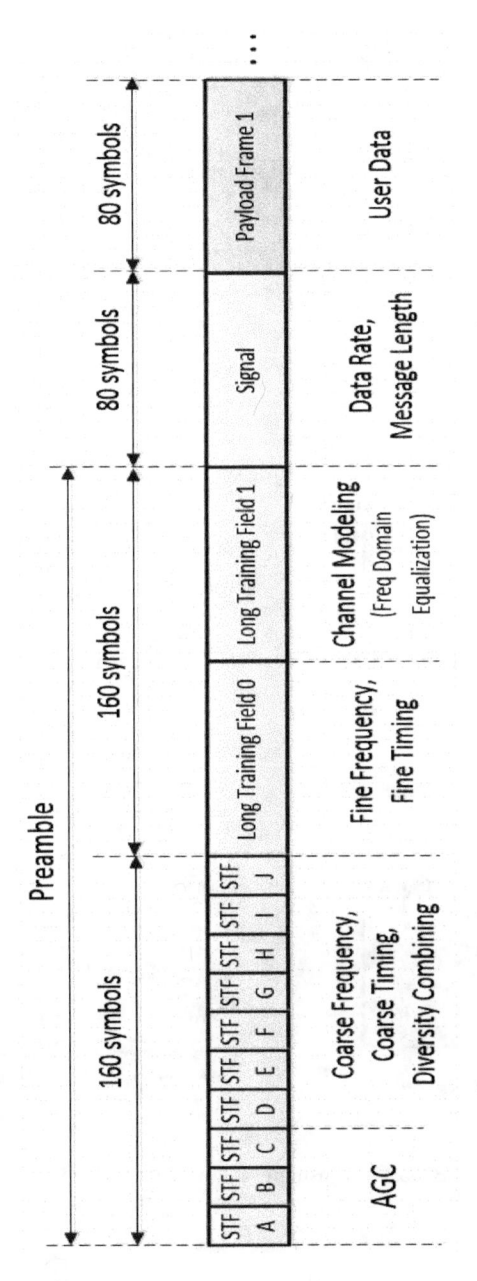

CHAPTER 13. IEEE802.11A DIGITAL SIGNAL PROCESSING

Figure 13.2 IEEE802.11a preamble structure.

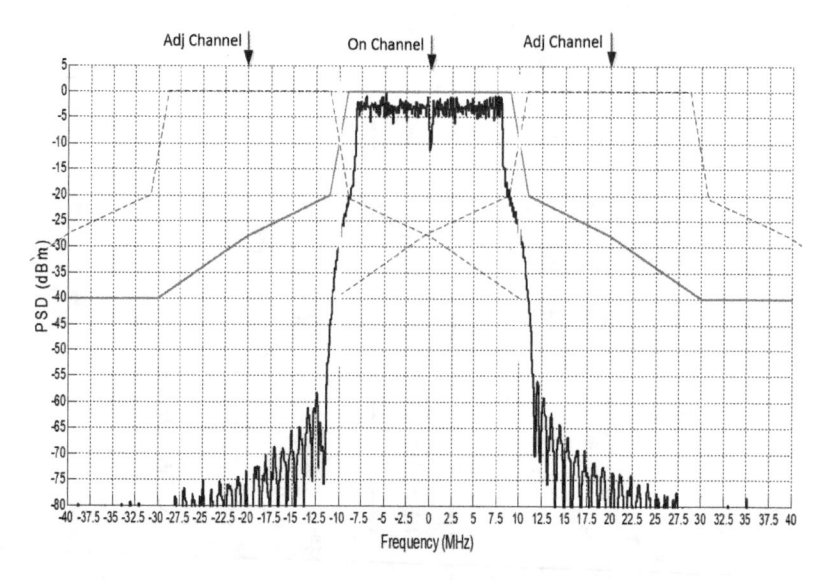

Figure 13.3 IEEE802.11a transmit spectrum.

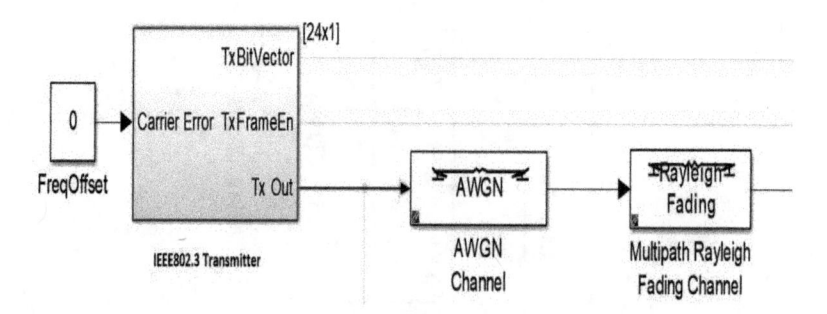

Figure 13.4 IEEE802.11a transmitter followed by channel impairment blocks.

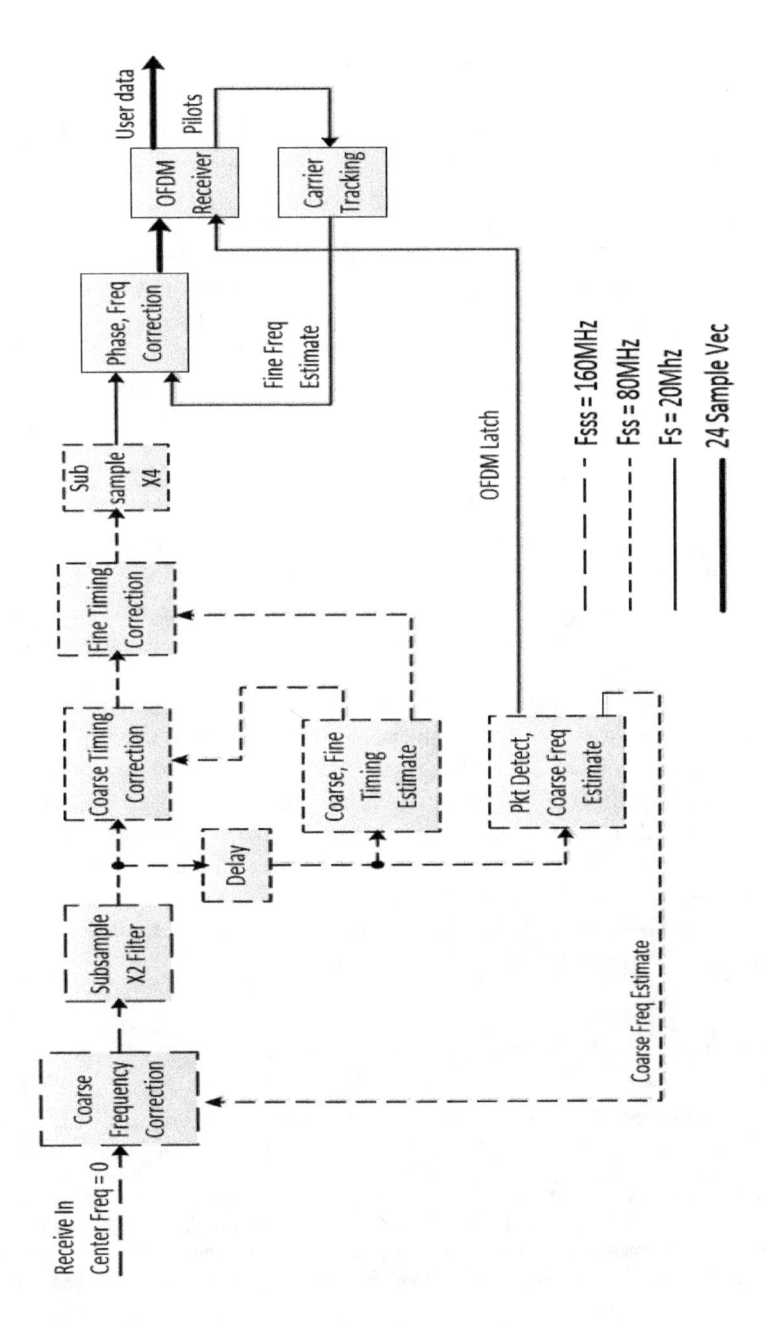

CHAPTER 13. IEEE802.11A DIGITAL SIGNAL PROCESSING

Figure 13.5 IEEE802.11a receiver simplified processing.

13.4 IEEE802.11A RECEIVER OVERVIEW

Figure 13.5 is a simplified IEEE802.11a receive modem block diagram. Let's start at the beginning on the top left of Figure 13.5. The input is a complex baseband IEEE802.11a receive signal, sampled at 160 MHz. This simulated signal starts with the preamble of Figure 13.2 and then runs for as long as the simulation runs. The frequency spectrum is shown in Figure 13.3. For testing, the simulated channel can apply a carrier frequency offset. This is estimated and removed by the receiver in the coarse frequency detect and carrier offset correction blocks, respectively.

The received complex baseband signal with coarse carrier frequency corrected is now filtered and subsampled by 2. This filter removes noise from ±10 MHz to ±80 MHz to improve performance and support subsampling. The X2 Filter output is sampled at 80 MHz.

The next block, Coarse Timing Correction, in Figure 13.5 is used to line up the received OFDM frames, to the nearest integer sample, with the OFDM latch pulse from the Packet Detect block. If the received packet boundary was detected correctly, the OFDM latch pulse goes high for one sample at the beginning of every OFDM frame. The coarse timing correction is followed by fine timing correction which shifts the received signal in time by plus or minus one half sample for more accurate frame timing.

The filter and subsample by 4 applies further lowpass filtering to limit noise and lowers the sample rate to 20 MHz, the sample rate of the transmitter IFFT output. The OFDM receiver block output is a vector of the 24 customer data symbols per frame.[1] In addition, the receiver block extracts the four pilot signals from the received frame and sends them back to the carrier tracking block as an average.[2] Note that a new vector of 24 parallel bits of user data is updated every 4 usec (24/4 = 6 MHz) on the user data output. The input to the phase, freq correction block is a 20 MHz serial received sample stream. Each frame contained in the stream is detected and converted to a 24-symbol parallel user data output.

13.5 IEEE802.11A TRANSMITTER DESIGN DETAILS

Figure 13.7 is the complete transmitter simulation circuit. This entire circuit runs at 160 MHz sample rate. As shown in Figure 13.2, an IEEE802.11a packet consists of

[1] This simulation is BPSK only, so there are 24 bits per frame, or 24 symbols per frame. Note also that when these 24 bits are rate 1/2 coded, the actual number of bits input to the IFFT is 48.
[2] Carrier tracking measures the pilot frequency offset and outputs a counteracting complex sine wave to reduce it to zero. Pilot measurement and feedback comprises a second-order phase lock loop.

CHAPTER 13. IEEE802.11A DIGITAL SIGNAL PROCESSING 529

a preamble followed by multiple 80 sample OFDM frames. After upsampling by 8 (samples/symbol), one frame output consists of 8*80 = 640 samples. Note that 20 MHz is the baseband output symbol rate, see Figure 13.9. So, after upsampling by 8 samples/symbol, the output sampling rate is 8*20 = 160 MHz. The spectrum is still 20 MHz wide, however.

In this simulation, the entire packet is a preamble followed by 10 OFDM frames. The 160 Mz sample clock is shown in the top trace of Figure 13.6. The 160/8 = 20 MHz wide spectrum at TxOut in Figure 13.7, right side, is shown in Figure 13.9. This spectrum has four pilot channels seen popping up (the pilot channel power level is adjusted higher than normal here) across the top and a null in the center. The last trace in Figure 13.6, DataSel, tells the transmitter what kind of frame to generate. The DataSel transition from 3 to 4 indicates a transition from second LTF to Payload frames; see Figure 13.22.

At the very top left of Figure 13.7 is the carrier frequency offset generator, used to test the receiver frequency offset estimation algorithm. Below that is the New Symbol Control. This block outputs a 1 on NewSym every eight sample clocks; see Figure 13.6. The next block down is New Frame Control. Because each frame is 80 samples (64 IFFT output samples preceded by 16 cyclic prefix samples), the New Frame Control block outputs a NewFrame pulse for every 80 NewSym input pulses. SymCount output numbers each symbol, 1 to 80. The NewFrame pulse also enables the OFDM Generator to output OFDMout, a vector of all 80 symbols needed for the next frame.

The OFDM Filter, Figure 13.8, acts as an upsampling interpolator filter for the stream of seven zeros between each NewSym sample; see Section 9.2.7. The OFDM filter also has a very sharp cutoff, which helps to give the transmit signal spectrum the box like shape seen in Figure 13.9. Also shown is the MATLAB code to generate the filter coefficients.

```
function [num] = OFDMfilter()
% Computes IEEE 802a transmit filter, requirements:
% Sharp transition band at +/-10MHz, Fs = 160 MHz
M = 1024; % Number of FFT points
L = 513; % Filter length, odd. Filter order is one less
N = 128; % number of data subcarriers in sub-band
L2 = floor(L/2); n = -L2:L2;
% Sinc function prototype filter
pb = sinc((N).*n./M);
% Sinc truncation window
```

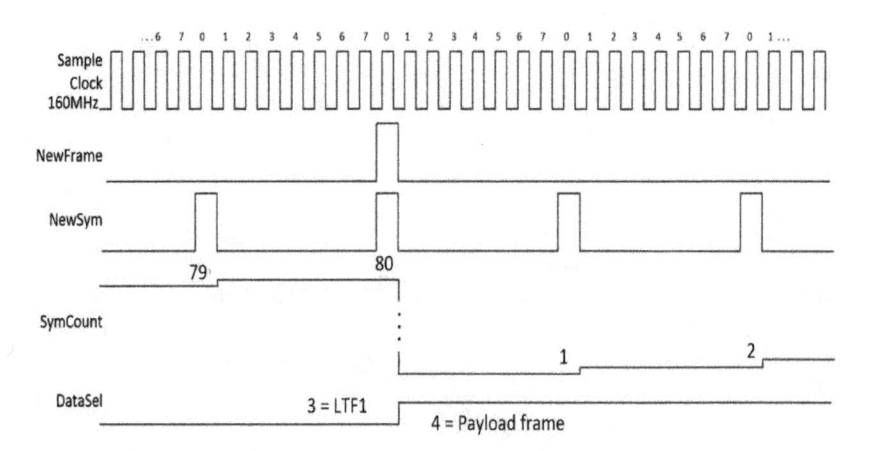

Figure 13.6 IEEE802.11a transmitter timing.

```
w = (0.5*(1+cos(2*pi.*n/(L-1)))).^0.6;
% Normalized lowpass filter numerator coefficients
num = (pb.*w)/sum(pb.*w);
```

13.5.1 OFDM Generator

Digging a little deeper, the OFDM Generator block in Figure 13.10 supplies the next 80 symbol frame when the NewFrame Enable input goes high. Between enables, the block is not active. The kind of frame produced (preamble or data) depends on the DataSelect input (see the table in the lower left of Figure 13.7). Preamble frames are constructed inside the Preamble Construct block. For a data frame, the DataFrameConstruct block output in Figure 13.10 assembles the 48, rate 1/2 encoded, BPSK symbols into the 64-symbol frame shown in Table 13.3. This block has no delay, it simply arranges the selected frame data into an output vector. Details of the DataFrameConstruct block are in Figure 13.11.

13.5.1.1 Preamble Generator

Figure 13.12, Preamble Construct, is based on two five position selector switches. The left switch determines the type of 64-symbol data that makes up each frame and the right switch attaches the cyclic prefix to make a complete 80-symbol frame.

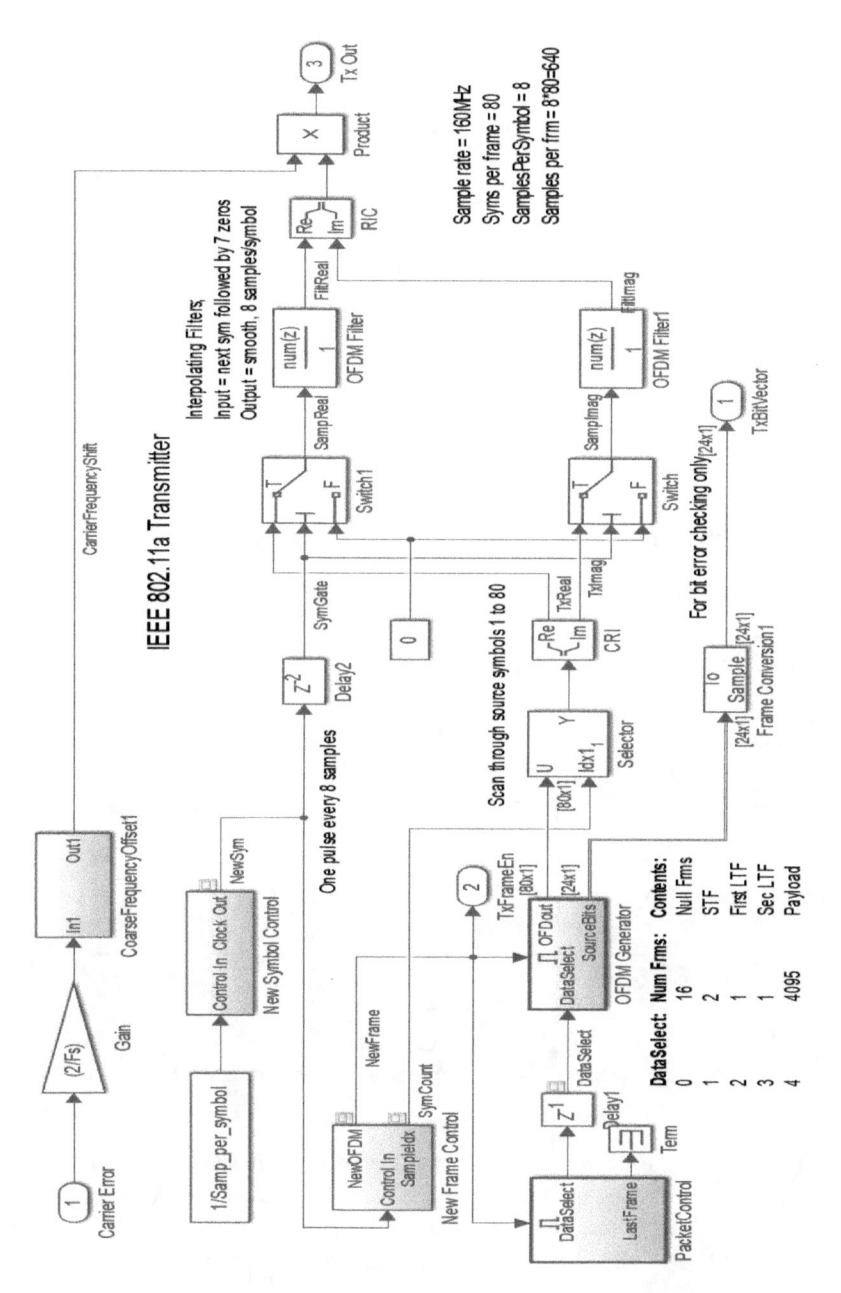

CHAPTER 13. IEEE802.11A DIGITAL SIGNAL PROCESSING 531

Figure 13.7 IEEE802.11a transmitter simulation circuit.

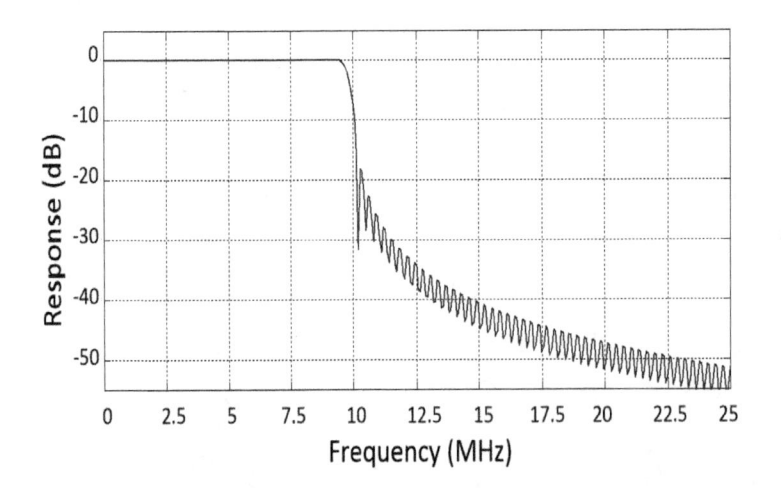

Figure 13.8 IEEE802.11a transmitter output shaping filter response.

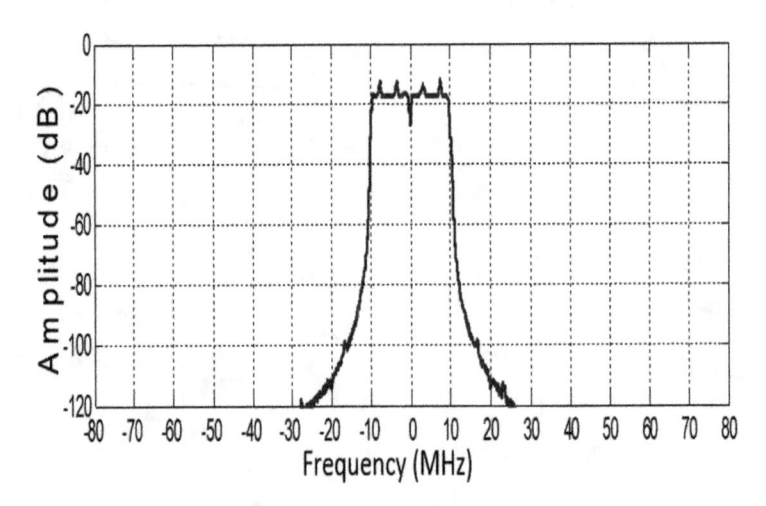

Figure 13.9 IEEE802.11a transmitter output spectrum, $F_s = 160$ MHz.

In Figure 13.13 the unshaded data comes from the left switch and the shaded cyclic prefix data comes from the right switch. Note that STF and LTF (short and long training field) is just a set of 64 complex numbers specified in the 802.11a specification [1].

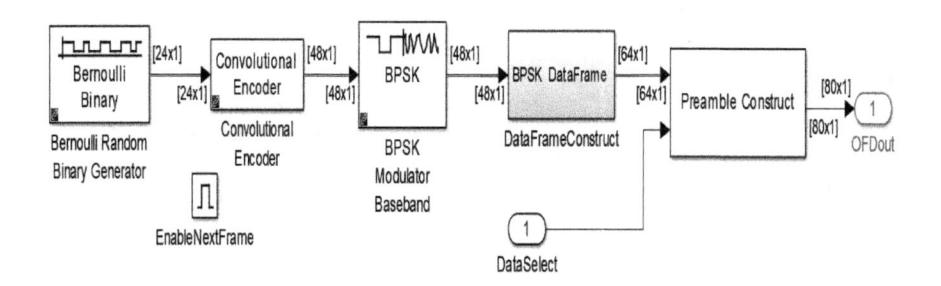

CHAPTER 13. IEEE802.11A DIGITAL SIGNAL PROCESSING 533

Figure 13.10 IEEE802.11a transmitter OFDM generator block.

Referring to Figure 13.13, the very first frame (STF B,C,D,E) has a 16-sample CP prepended as STF A. The same technique is used to construct the next frame (STF F,G,H,I,J). The CP for the first two frames in Figure 13.13 is just another 16-symbol STF. For the third frame, the CP is 32 symbols, in accordance with the specification. The 16-sample CP for the fourth frame is generated by arrangement of LTF symbols; see Section 13.6.3 for more information.

13.5.1.2 CP Transmit Spectrum Control

The CP for payload frames (DataSel = 4) is a little more complicated. Recall from the discussion in Section 3.2.10 that the 16-symbol CP end phase matches the 64-symbol frame start phase (Figure 3.27). So no phase discontinuity results between the end of the CP and the start of the frame to which the CP is attached. However, there is still a phase discontinuity between the end of one 80-sample frame and the start of a new one (i.e., start of the new frame CP). This can cause the OFDM transmit spectrum to broaden somewhat about 20 dB down; see Figure 13.3.

Figure 13.15 shows an attempt to alleviate this phase discontinuity between frames 0 and 1. As expected, the start and end phases, P0, of the 64-sample IFFT output for frame 0 are the same. The IFFT output for frame 1 has start and end phase P1. A CP that starts at phase P0 and ends at P1 would generate no phase discontinuity between frame 0 and 1. We can approximate that situation by a weighted combination of the two 16-symbol segments in Figure 13.15. The weighting, shown in Figure 13.14, ensures that P0 is the phase at the start of the frame 1 CP and P1 is the phase at the end of the frame 1 CP. This special contrived CP of frame 1 should contribute very little phase discontinuity. In Figure 13.15, the weighting function is based on

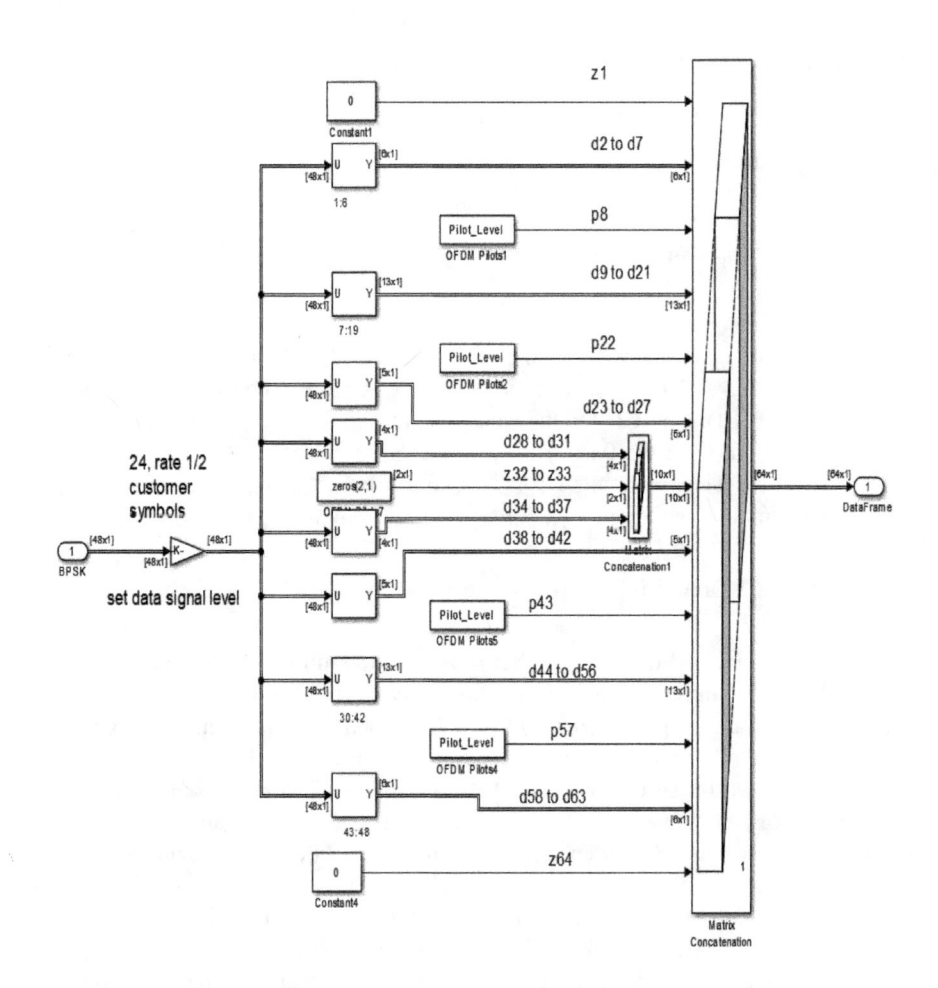

Figure 13.11 IEEE802.11a transmitter data frame constructor.

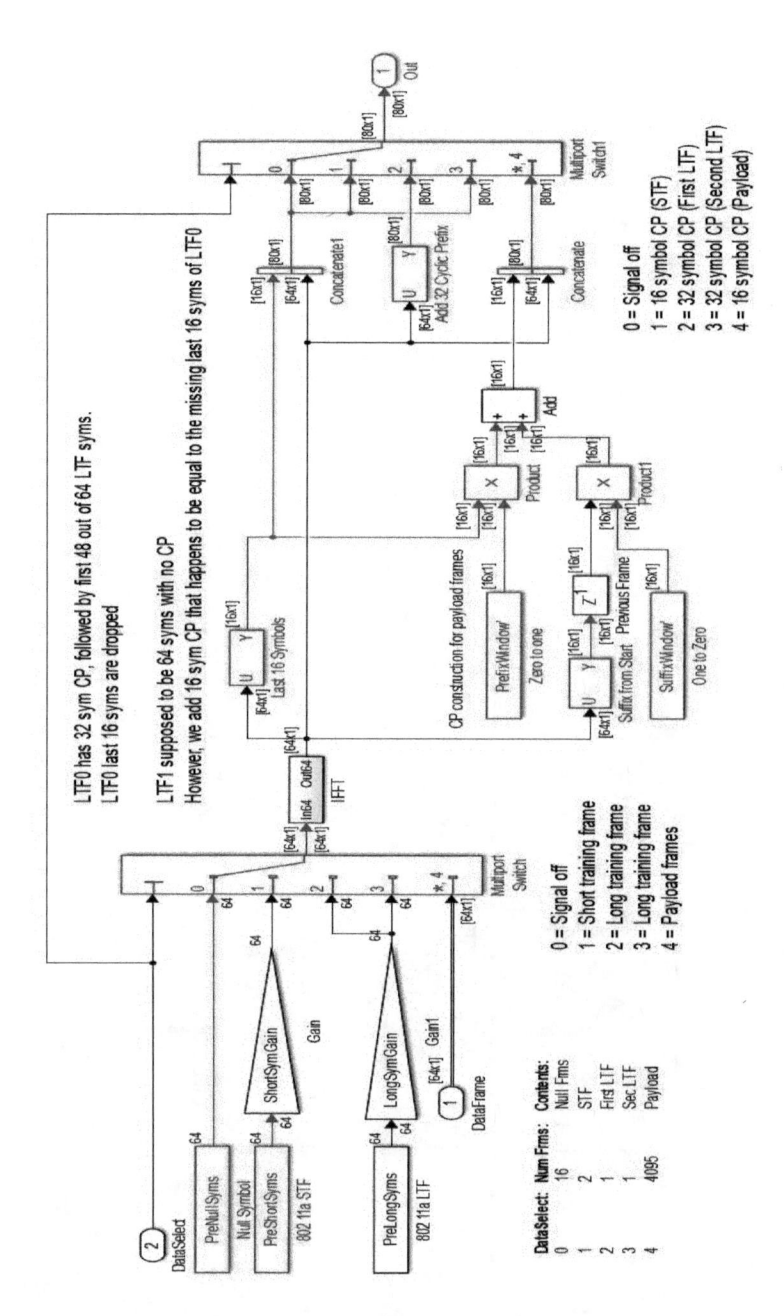

CHAPTER 13. IEEE802.11A DIGITAL SIGNAL PROCESSING 535

Figure 13.12 IEEE802.11a transmitter preamble construct block.

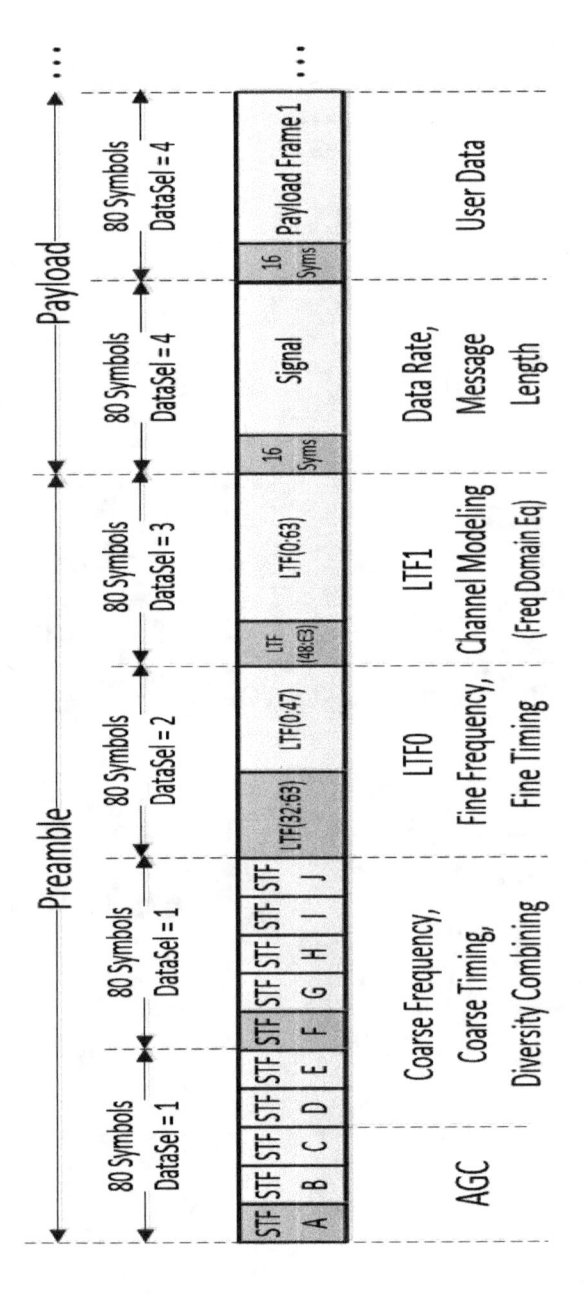

Figure 13.13 IEEE802.11a transmitter packet construction.

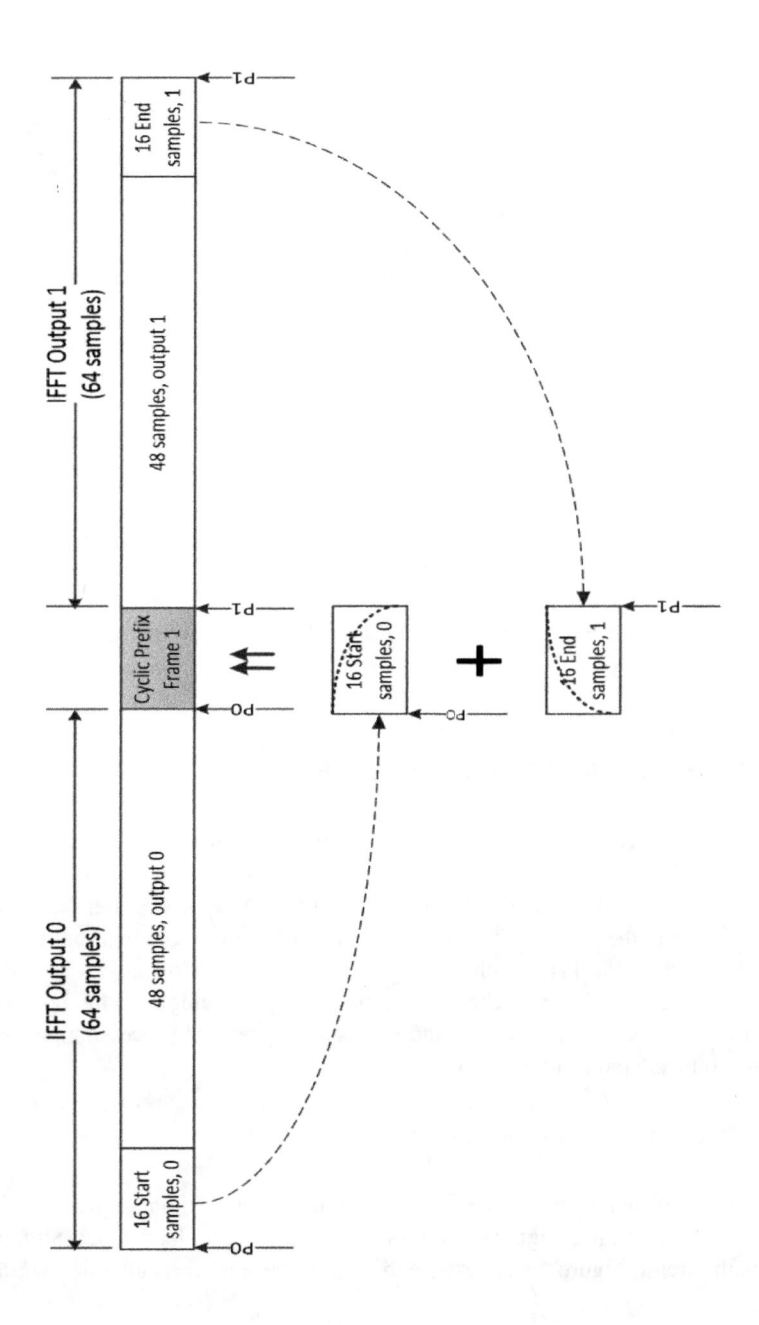

CHAPTER 13. IEEE802.11A DIGITAL SIGNAL PROCESSING 537

Figure 13.14 IEEE802.11a cyclic prefix construction for payload frames.

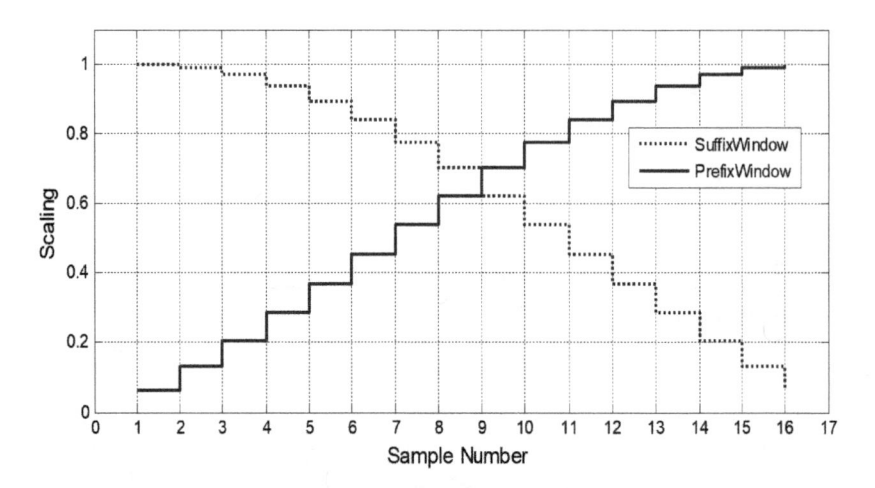

Figure 13.15 IEEE802.11a cyclic prefix scaling for payload frames.

the first 90° of a sine and cosine. We can do this because the data inside the CP is not used for anything.

13.6 IEEE802.11A RECEIVER DESIGN DETAILS

Figure 13.16 shows the complete IEEE802.11a Simulink Simulation, from transmitter in the upper left corner to transmit, receive error counter in the upper right corner. Let's take a close look at the block labeled Equalizer, Viterbi Decoder. The input signal enclosed in the first dotted input box is the 20 MHz frame samples aligned to the OFDMLatch pulse. Prior to this input, any residual phase offset is removed by the phase corrector multiplier. The signal labeled OFDM Samples is the boundary between the receiver front-end DSP and the back-end symbol processing. We will back up and tackle the front-end DSP first.

13.6.1 Receiver Front-End Signal Path

The front-end signal path includes coarse frequency correction, signal filtering, subsampling, and timing alignment blocks in Figure 13.5. In the detailed Simulink signal path circuit, Figure 13.17, extracted from the top-level simulation diagram,

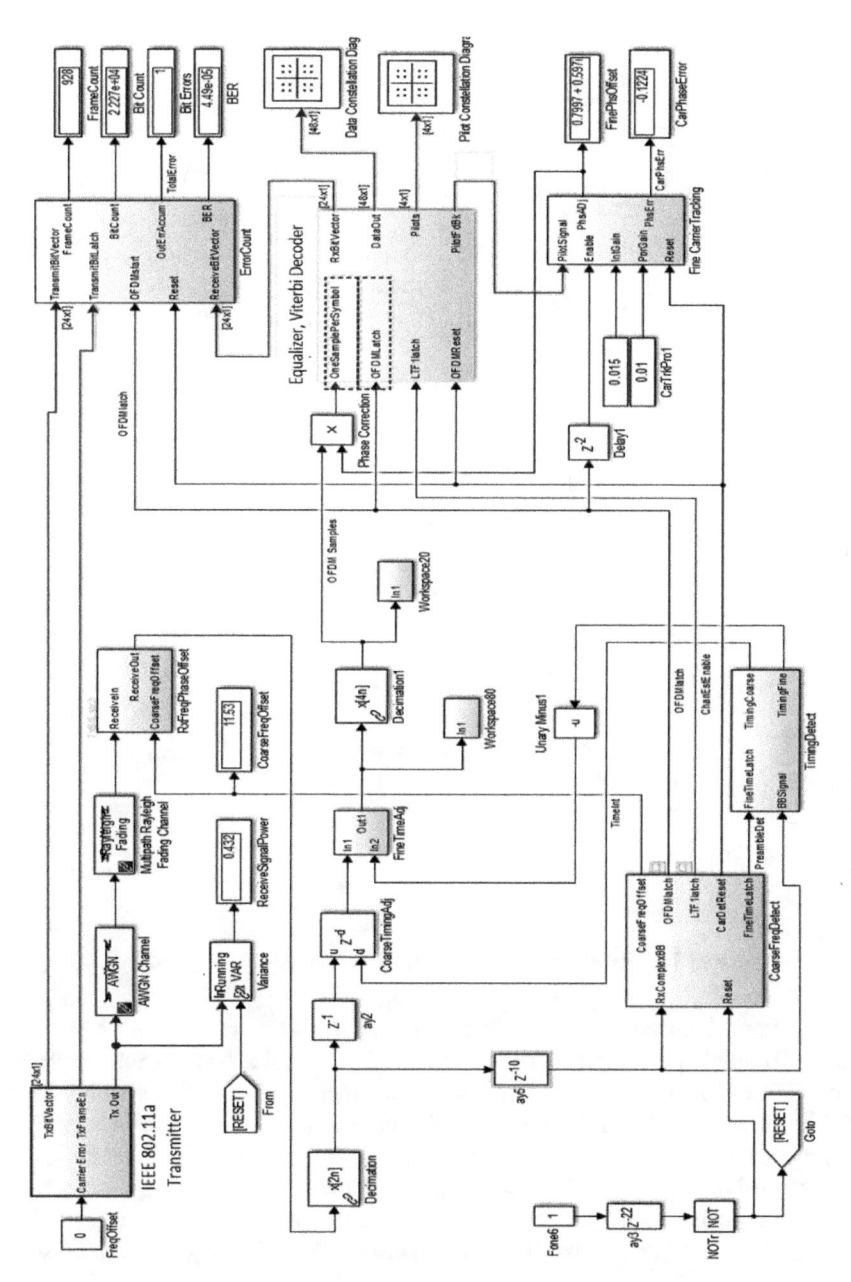

CHAPTER 13. IEEE802.11A DIGITAL SIGNAL PROCESSING 539

Figure 13.16 IEEE802.11a simulation, top level.

Figure 13.16, the received OFDM is frequency offset corrected and time aligned by estimates listed in Table 13.4.

Table 13.4 IEEE802.11a Receiver Front End Signal Estimations

Estimation	Accuracy	Comment
Crse freq offset	±200 Hz	Carrier tracking reduces this to 0
Crse frm timing	±3 samples	Preamble detect pulse posistioning
Int frm timing	±1 sample	Based on LTF0 correlation
Frac frm timing	±0.01 samples	Interpolation of LTF0 correlation

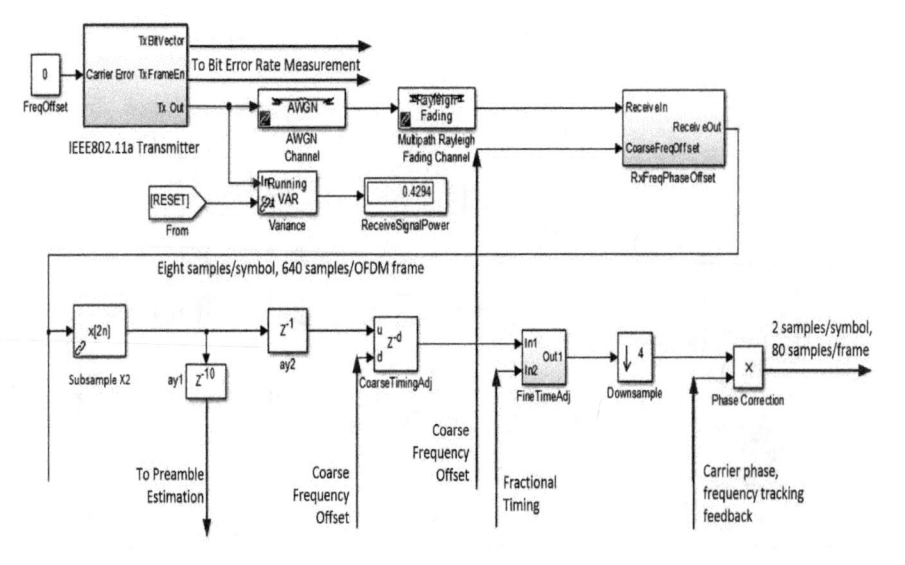

Figure 13.17 Receiver signal alignment and front end, Simulink diagram.

After the coarse frequency correction block (RxFreqPhaseOffset block) the received complex baseband signal is filtered and subsampled by two (Subsample X2 block). The subsampling filter response is shown in Figure 13.18. This filter has a Kaiser response; a special feature of the Kaiser response is the beta parameter. Beta trades off the frequency response rolloff rate with passband ripple. The MATLAB code to design and display this filter is shown below.

The input and output sampling bandwidths (one half sample rate) of the subsample by 2 filter are also shown in Figure 13.18. Removing as much noise as

CHAPTER 13. IEEE802.11A DIGITAL SIGNAL PROCESSING

possible from ±40 MHz to ±80 MHz is an important responsibility of this filter. The BER will be degraded if subsampling aliases this noise down into the range inside ±40, i.e., the range occupied by the OFDM signal.

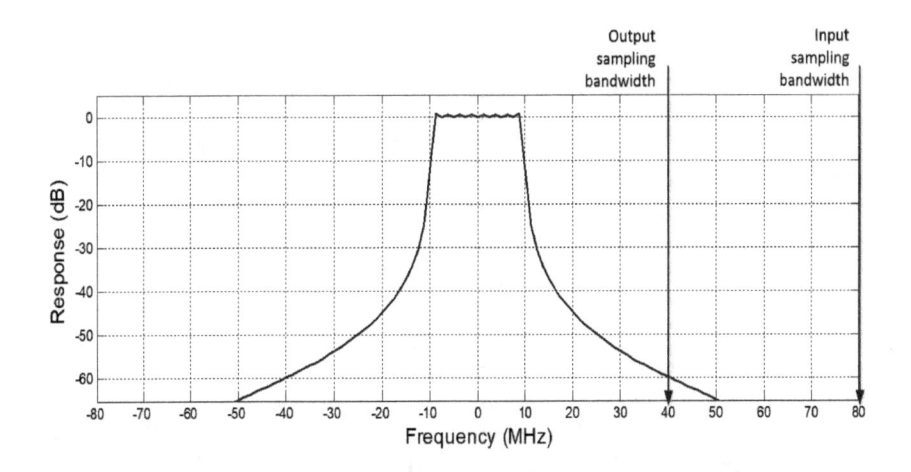

Figure 13.18 IEEE802.11a receiver input X2 subsampling filter.

```
% Subsample by 2 filter N = 128; % Filter Order
Fs = 160; % Sampling Frequency (MHz)
Fc = 9.6; % Cutoff Frequency (MHz)
Beta = 0.2; % Kaiser Window Parameter
% Create the window vector for the design algorithm
win = kaiser(N+1, Beta);
% Calculate the coefficients using the FIR1 function.
b = fir1(N, Fc/(Fs/2), 'low', win, 'scale');
RxFilt = dfilt.dffir(b);
RxCoef = RxFilt.Numerator;
Response = 20*log10(abs(fftshift(fft(RxCoef))));
ph = plot(Response);
set(ph,'LineWidth',1.5);
set(gca,'YLim',[-65 5]);
set(gca,'XLim',[1 length(Response)]);
set(gca,'XTick',[1:8:length(Response)]);
```

```
scale = (-Fs/2): (Fs/16): (Fs/2);
set(gca,'XTickLabel',scale);
xh = xlabel('Frequency (MHz)');
yh = ylabel('Response (dB)');
```

Figure 13.19 shows the power spectrum of the input signal (top trace), subsample X2 filter output (center trace) and subsample X4 output (bottom trace). The received signal $E_b N_0$ is set to 10 dB. This explains why the unfiltered input signal at the top is almost swamped by noise. Note that the four pilot signals exceed the channel signal amplitudes (set higher than normal so we can see where they are). Also note the zero channels in the center, to avoid receiver DC offsets. The two end channels are also zero to help to fit the signal in the 20 MHz sampling bandwidth of the bottom trace.

The X2 filter output provides considerable noise reduction in preparation for timing correction. The coarse and fine timing blocks implement integer and fractional sample timing correction, respectively. Packet detect must be very accurate so that, after the coarse and fine timing corrections are made, the OFDM Latch pulse lines up as close as possible with the actual received frame boundaries. When this line up is accurate, the 20 MHz transmitter IFFT output frames and receiver FFT input frames are the same.

Finally, in Figure 13.19, the X4 subsample output is at the FFT 20 MHz input sample rate in preparation for input to the demodulator FFT.

13.6.2 Coarse Frequency Estimation

Coarse frequency estimation is completely contained in Figure 13.23, below the heavy dashed line. Figure 13.20 is a simplified coarse frequency estimation timing diagram.

Referring to Figure 13.20, with no noise, the complex correlation of STFs by conjugate STFs cancels the modulated pattern, leaving only the frequency offset phase ramp. Nonzero frequency offsets will cause a phase change between STFs proportional to the frequency offset. In Figure 13.20, this phase change is represented by the complex number CorrAvgBB.

Variable *atan* is the angle (arc tangent) of CorrAvgBB. This represents the amount of phase, in radians, traversed by the carrier frequency offset in 256 samples or 3.2 usec at 80 MHz sample rate. Radians per 3.2 usec is converted to cycles/second by the two multipliers just prior to the exponential averaging filter; see also Equation (13.1). Notice at the top of Figure 13.20 that the CorrAvgBB stabilizes at the start of

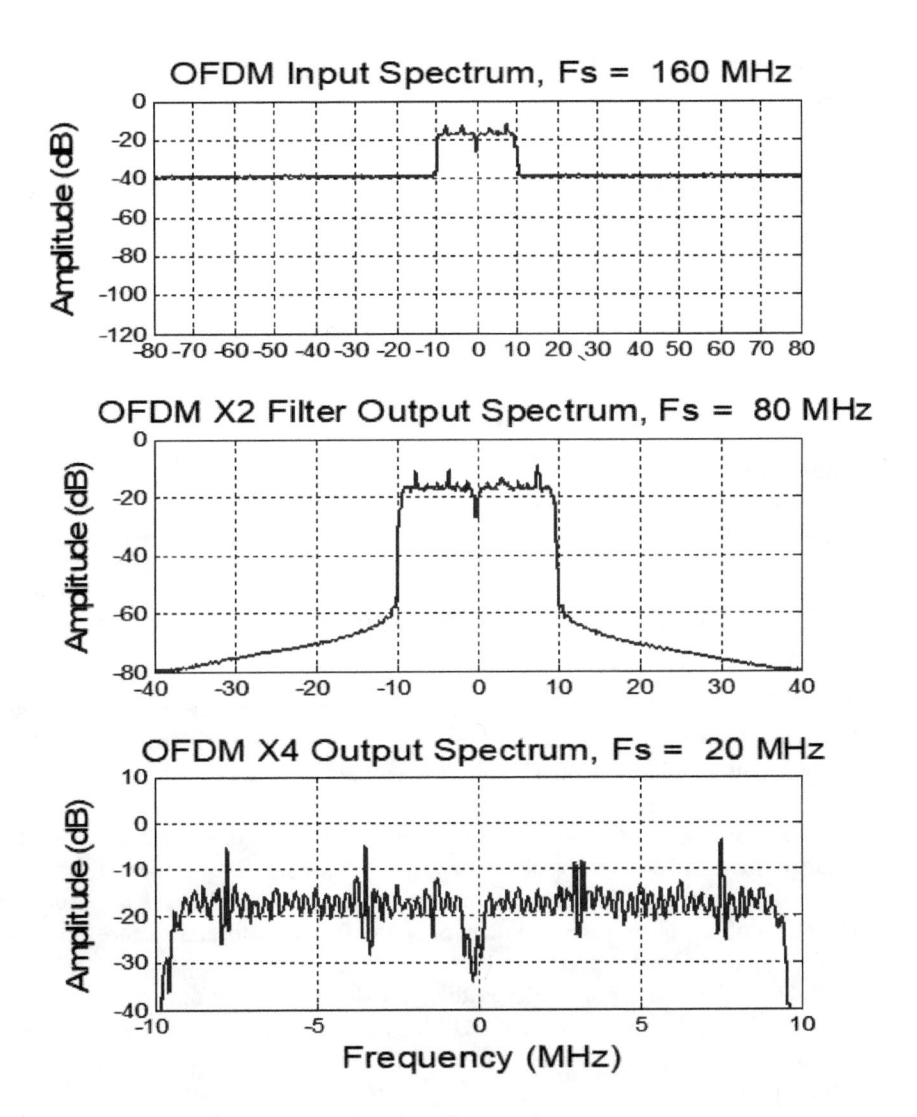

Figure 13.19 IEEE802.11a receiver sequence of spectrums.

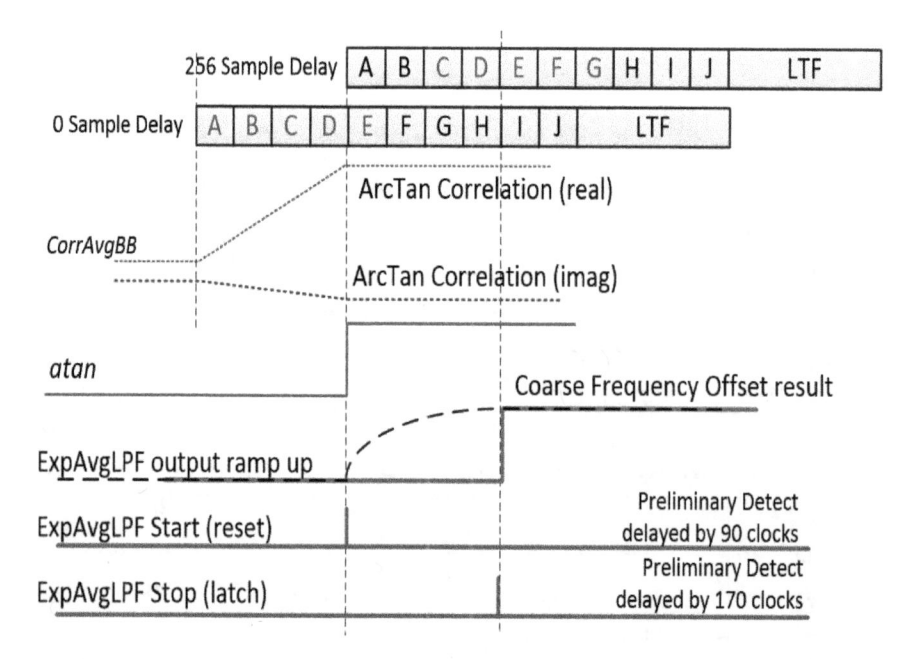

Figure 13.20 IEEE802.11a coarse frequency offset timing, 80 MHz sample rate.

the STF A times SFT E correlation. This starts a correlation plateau that lasts for four STFs. ExpAvgLPF, Figure 13.21, is an exponential averaging filter (Section 9.2.3) that attempts to remove as much noise as possible from this plateau. As shown in Figure 13.23, a delayed by 90 samples version of the Preliminary Detect pulse resets and runs this filter. A further delayed by 170 Preliminary Detect pulse latches the result. ExpAvgLPF runs for 80 clocks or 1 usec. Table 13.5 shows some test results of the carrier frequency offset circuit discussed here. Equation (13.1) calculates coarse frequency offset. Notice how the numerator and denominator units cancel to leave units of cycles/second or Hertz.

Depending on the noise level, this open-loop coarse frequency offset measurement is only accurate to about $\pm 200 Hz$[3]. Back-end pilot tone carrier tracking reduces this to zero. The top level front-end block RxFreqPhaseOffset applies the estimated frequency offset to the received complex baseband signal. This estimate must be applied before the end of the 10 undelayed preamble STFs.

[3] If 200 Hz seems like a lot, consider that for a 4 microsecond long OFDM frame, 200 Hz perturbs the received phase by (200 cycles / second)*(4e-6 seconds) = 0.0008 cycles, a tiny amount of phase.

CHAPTER 13. IEEE802.11A DIGITAL SIGNAL PROCESSING   545

$$\left(\frac{\arctan(\text{radians})}{1}\right)\left(\frac{1\ cycle}{2\pi\ radians}\right)\left(\frac{F_s samples}{\sec}\right)\left(\frac{1}{256\ samples}\right) \quad (13.1)$$

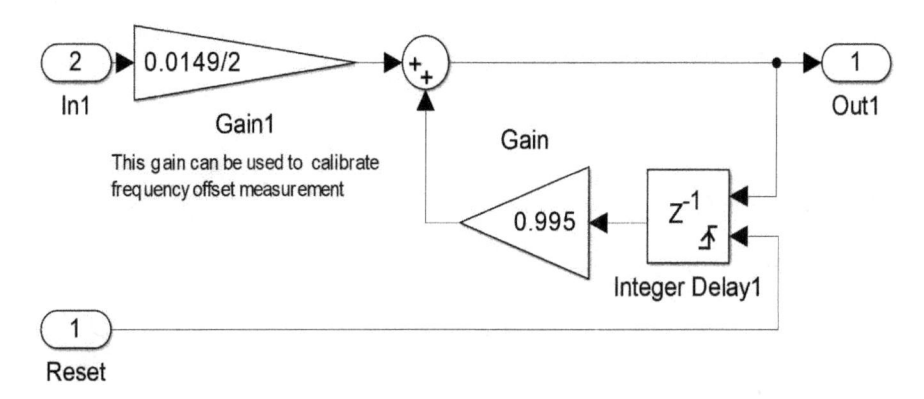

Figure 13.21 Exponential averaging lowpass filter.

Table 13.5 Coarse Frequency Offset Estimates at Six Test Frequencies and Four Different Receive Noise Levels

	Actual Frequency Offet (Hz)				
EbN0	-6000	-3000	0	3000	6000
inf dB	-5,945	-2,996	2.5	3,001	5,950
13.14 dB	-5,885	-3,099	-103	3,092	6,047
10.14 dB	-5,860	-3,146	-154	3,126	6,073
7.14 dB	-5,830	-3,232	-248	3,157	6,104

13.6.3 Preamble Detection and Alignment

Figure 13.22, top, shows the preamble component boundaries as defined by IEEE 802.11a. Figure 13.22, bottom, shows the preamble boundaries that our preamble processing uses. The LTF0 and LTF1 patterns are carefully defined and known a priori at the receiver.

If we consider 4 usec for every received OFDM frame, then Figure 13.22 top LTF0 (long training field) pattern runs into LTF1 (see arrow showing where this happens). The bottom part of Figure 13.22 shows how our preamble detect circuit reinterprets LTF0 and LTF1 to have separate fixed 80 symbol patterns for each one. There is no change in the LTF0 and LTF1 data, just a redefinition of boundaries to make detection easier. Figure 13.23 shows the Simulink circuit that implements the coarse frame timing we will describe. Primary input is the complex baseband signal labeled BBsignal (baseband signal) in Figure 13.23. One more input, LastFrame, is a pulse that indicates the end of the multiframe packet transmission. This resets the preamble detection process. The three outputs of Figure 13.23 are:

PreliminaryDet: One sample pulse indicating preamble detect.
PreambleDetOffset: Integer samples between PreliminaryDet and actual preamble start.
CarrierFreqOffset: Coarse estimate of frequency offset from zero.

13.6.4 Coarse Frame Timing

Figure 13.24, top, shows 10 STFs lettered A through J. This represents the start of a received preamble. Right below the top, the preamble is delayed by one STF (64 samples at 4× oversampling; that is, 80 MHz sampling rate). The incoming and delayed STFs are correlated to form the L1filt signal. L2filt is formed the same way, except a five STF delay is used. L2filt is then subtracted from L1filt to form the peak L1L2 in the center of the figure. The block labeled TimingDet in Figure 13.23 delays L1L2 by four STFs to generate PreliminaryDet when the crossover point, shown in Figure 13.24, is reached (this is just a technique for finding the halfway point down the L1L2 slope).

Note in the top right corner of Figure 13.23 how the last 320 samples of L1L2 are stored in a buffer. The buffer overlap is 319 so each new sample pushes the existing samples up by one (i.e., samples at index 2 to 320 shift to positions 1 to 319 and the new sample is placed at index 320, assuming we used the convention of counting from 1 instead of 0). As L1L2 ramps up, the MaxIdx (the index of the maximum value, not the maximum value itself) of the buffer will be 320, the most recent buffer entry. When L1L2 starts to ramp down, the MaxIdx will go down. The first decreasing L1L2 will make MaxIdx = 319, the second decreasing L1L2 will make MaxIdx = 318, and so on.

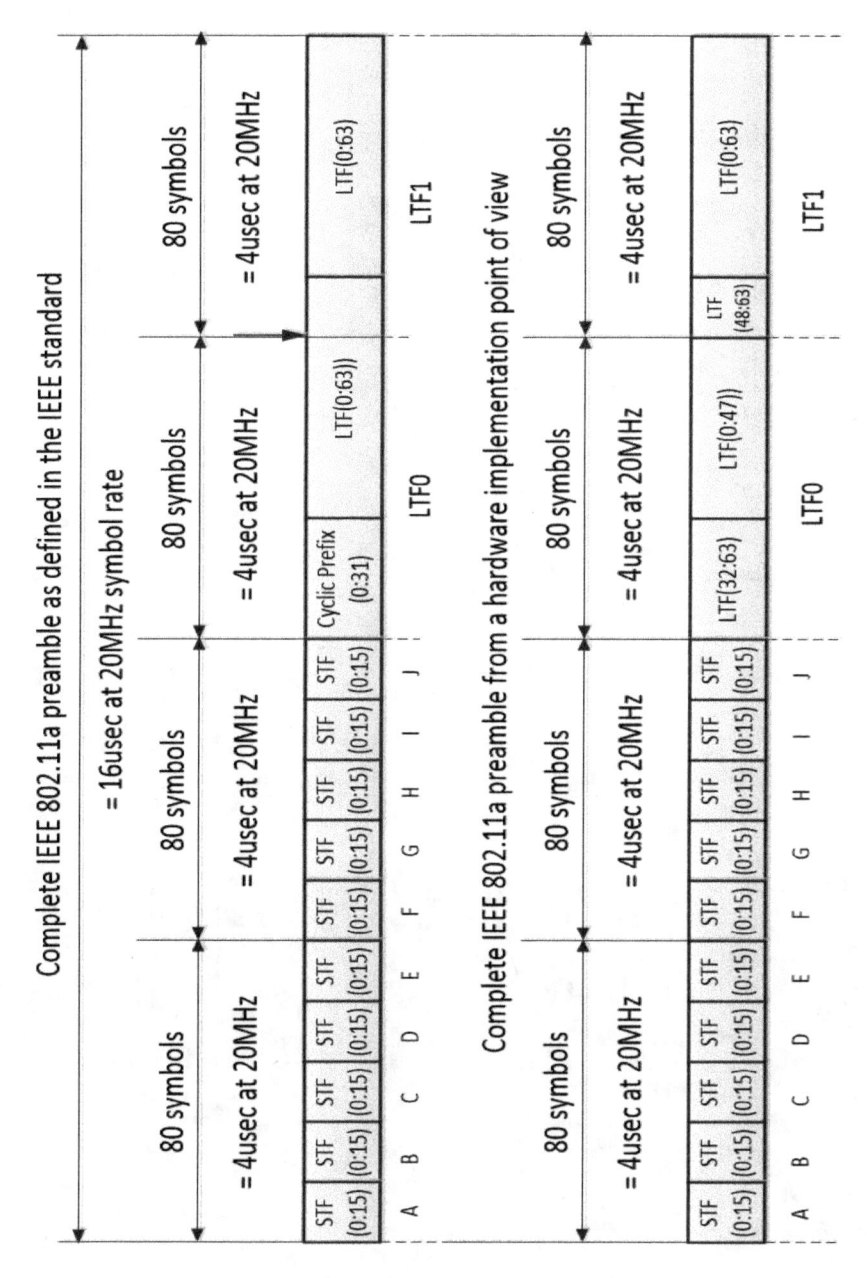

CHAPTER 13. IEEE802.11A DIGITAL SIGNAL PROCESSING 547

Figure 13.22 IEEE802.11a preamble structure.

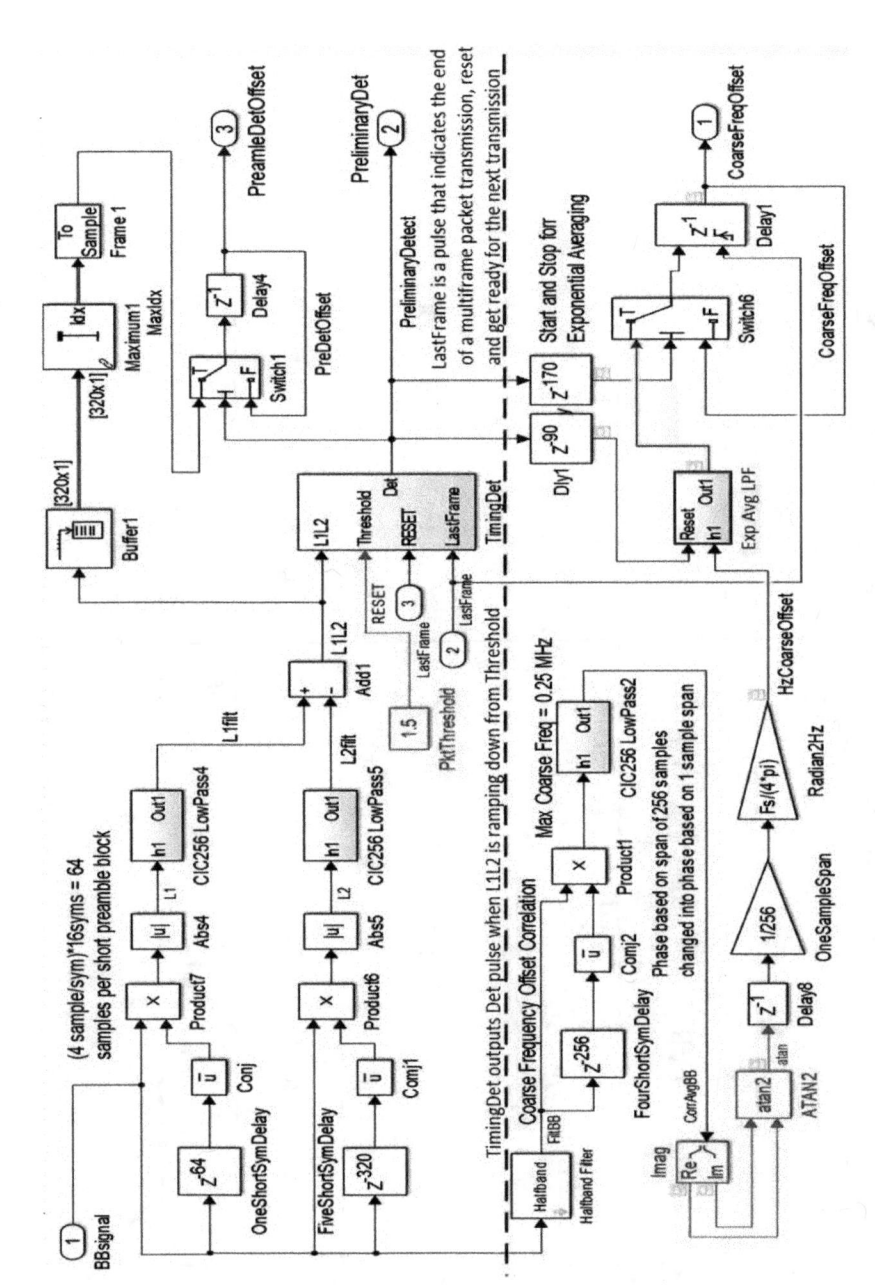

Figure 13.23 IEEE802.11a preliminary processing, coarse timing and circuit.

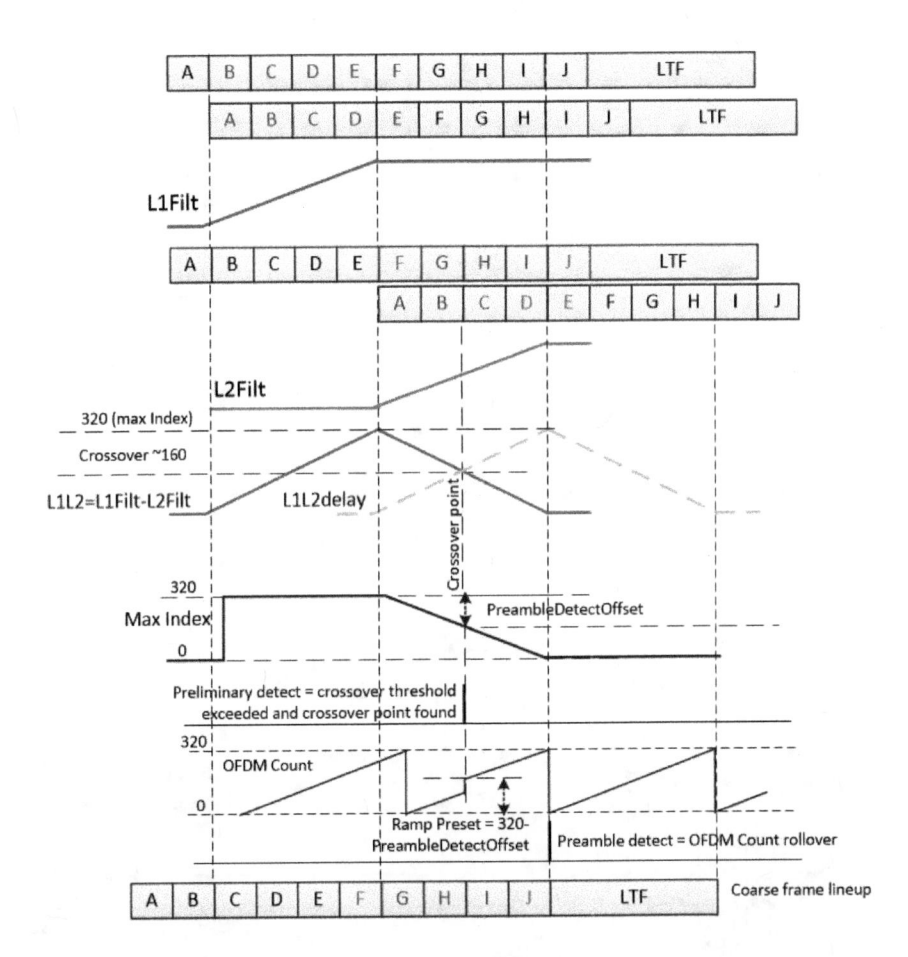

CHAPTER 13. IEEE802.11A DIGITAL SIGNAL PROCESSING

Figure 13.24 IEEE802.11a preliminary preamble detect basic operation.

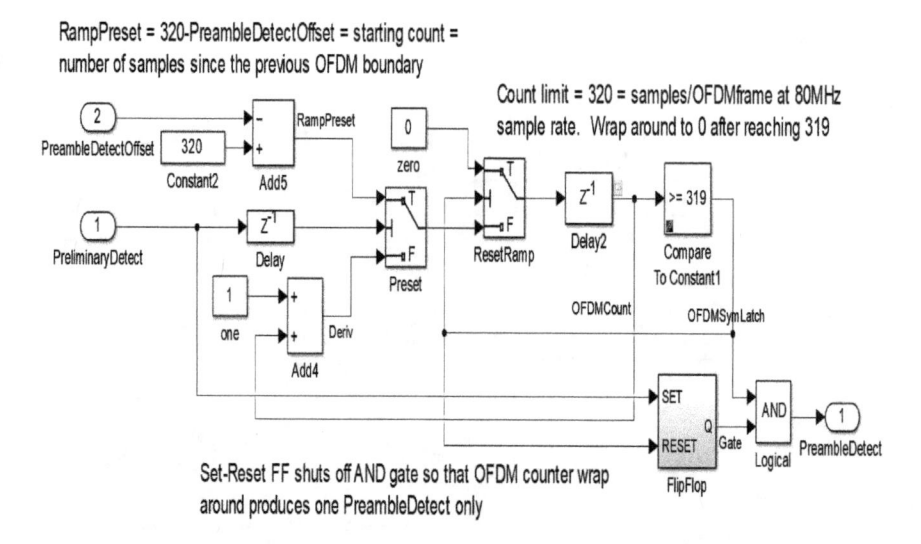

550   *Software Defined Radio: Theory and Practice*

Figure 13.25 IEEE802.11a OFDM frame counter, preamble synchronizer.

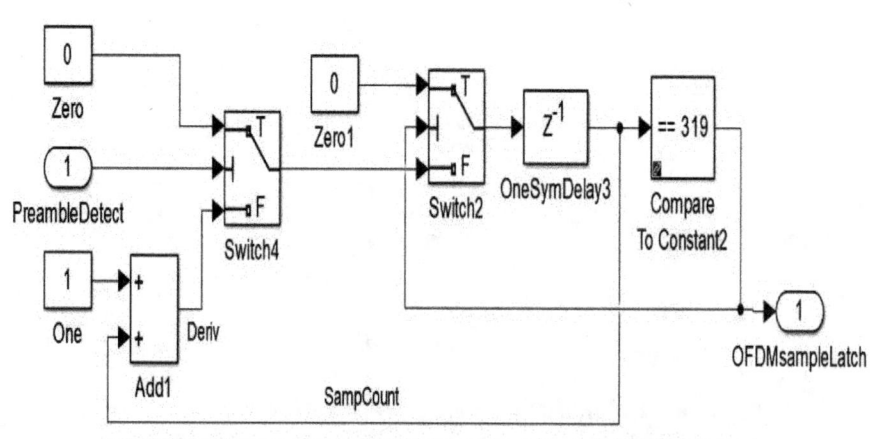

Figure 13.26 IEEE802.11a OFDM symbol counter.

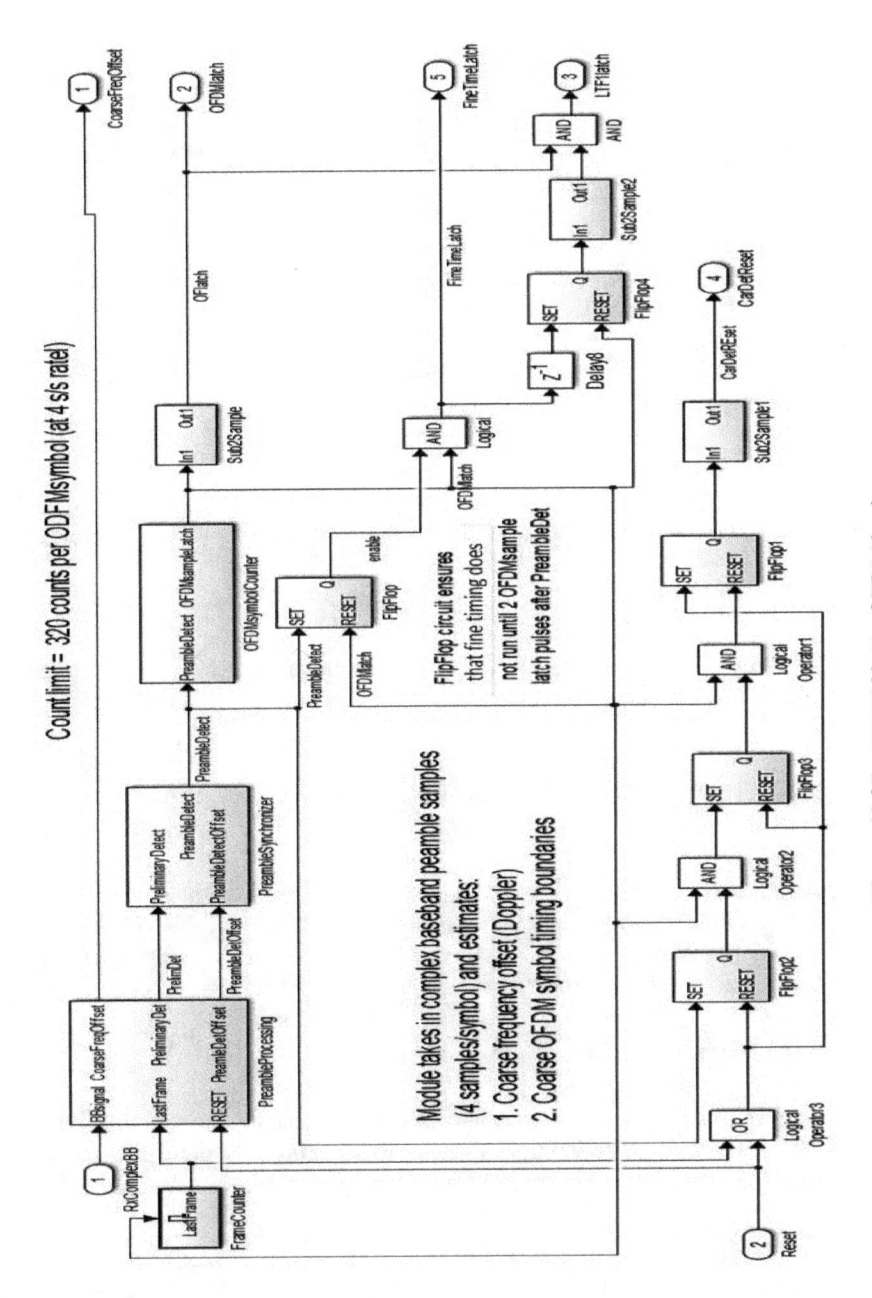

CHAPTER 13. IEEE802.11A DIGITAL SIGNAL PROCESSING 551

Figure 13.27 IEEE802.11a OFDM latch generator.

Thus, when PreliminaryDet goes high,[4] MaxIdx is latched to indicate the number of samples between the actual peak (index 320 before L1L2 started going down) and the peak value index at the time of the PrelininaryDet pulse. This coarse timing measurement is in integer samples and is called PreambleDetectOffset.

As shown in Figure 13.24 and described above, PreambleDetectOffset is the number of samples, counting down, between the 320-sample peak and the time the PreliminaryDetect goes high. To find a coarse frame boundary, the OFDM counter in Figure 13.25 simply counts back to the 320-sample peak by preloading Delay2, below, to (320 - PreambleDetectOffset). Counting up from that number to 320 recreates the time between the peak and the PreliminaryDetect and gets us back to an STF boundary. After that one official PreambleDetect pulse is generated.

Notice the block labeled PktThreshold in the center of Figure 13.23. The TimingDet block is activated only when L1L2 > PktThreshold. Then TimingDet outputs a preliminary preamble detect pulse when L1L2 ramps down halfway. The PktThreshold attempts to reject L1L2 signals that look like peaks but are really just noise. The setting of PktThreshold is a critical step in a cellular base station; see [2].

13.6.4.1 OFDM Counter

The OFDM counter in Figure 13.26 is reset by the Preamble Detect pulse and simply outputs a counter wraparound pulse, with coarse frame accuracy, at the start of every received frame.

As described, the PreambleDetect Pulse has been lined up with the end of an STF field. In Figure 13.24 we can see the Preamble Detect (bottom of figure) lining up with STFs J, I, and F. At the very bottom of the figure, the Preamble Detect pulse is correctly lined up with the first LTF. Thus, before the OFDM frames get to the fine timing circuits, there is a fixed delay to ensure the correct coarse lineup. This delay is part of the design and does not need adjustment.

Figure 13.27 shows how the blocks described above work together to produce OFDMlatch and other critical timing signals.

13.6.5 Fine Frame Timing

Figure 13.27 generates a fine time latch frame boundary coarse timing pulse; however, it is not accurate enough. Instead of adjusting this pulse in time, we remeasure the integer error between the generated pulse and the actual start of the

---

4   Note that PreliminaryDet pulses high when the crossover point is reached. This is the output of the TimingDet block.

received frame. This is applied to the signal path as Integer OFDM Timing in Figure 13.17.

For fractional fine timing, the receive signal is shifted in time by a fine timing interpolator for a more accurate lineup between the received OFDM frames and the frame boundary pulse. This is applied to the signal path as Fractional OFDM Timing in Figure 13.17.

Figure 13.29 shows the fine timing estimator Simulink circuit. Starting at the top left, the last 320 samples (one OFDM frame) of BBsignal input are serial to parallel converted and correlated with the LTF0 known reference pattern. The reference pattern is arranged as shown in Figure 13.22, lower. LTF0 is only used for fine timing here. The correlation between the LTF0 reference pattern and the received, coarse adjusted, frame is a complex number; its magnitude is calculated and the last 14 results stored as a timing profile. We only want the timing profile at the end of the frame (to see the alignment error) so it is latched by FineTimeLatch (FineTimeLatch is derived directly from the OFDM Counter in Figure 13.25).

The index of the highest magnitude in the 14-element timing profile is used to indicate the integer error between the frame boundary pulse (this pulse is coincident with FineTimeLatch) and the timing profile peak index.

In Figure 13.28, consider the integer frame boundary pulse (second from bottom) lined up incorrectly with actual frame boundaries. This frame boundary pulse comes from the preamble detect circuit OFDM counter circuit Figure 13.26. Notice how the incorrect frame boundary pulse latches the timing profile buffer. The timing profile buffer clearly indicates that the frame boundary pulse is five samples late. This will be the coarse (integer) timing error. The fractional timing error comes from a consideration of the peak timing correlation and the samples on either side of it. The fractional timing error is calculated in Figure 13.30. Because the two outside correlations are less than the center, we assume an inverse parabolic fine timing peak occurs. We find its offset from the center sample to get the fractional timing error. This is calculated by the block labeled FineTimeCalc in Figure 13.29. Note that, in Figure 13.29, FineTimeCalc input In3vec is a 3-element vector containing $C_E n$, $C_O(n)$, and $C_L(n)$ from Figure 13.30.

13.6.6 OFDM Receiver Back End

Figure 13.31 shows the IEEE802.11a receiver back-end components. Recall that the OFDM frame alignment process, shown in Figure 13.17, produces a carefully refined integer and fractional shift to line up the incoming frames with the OFDMLatch. This line-up is critical for the follow-on symbol equalization and detection circuits

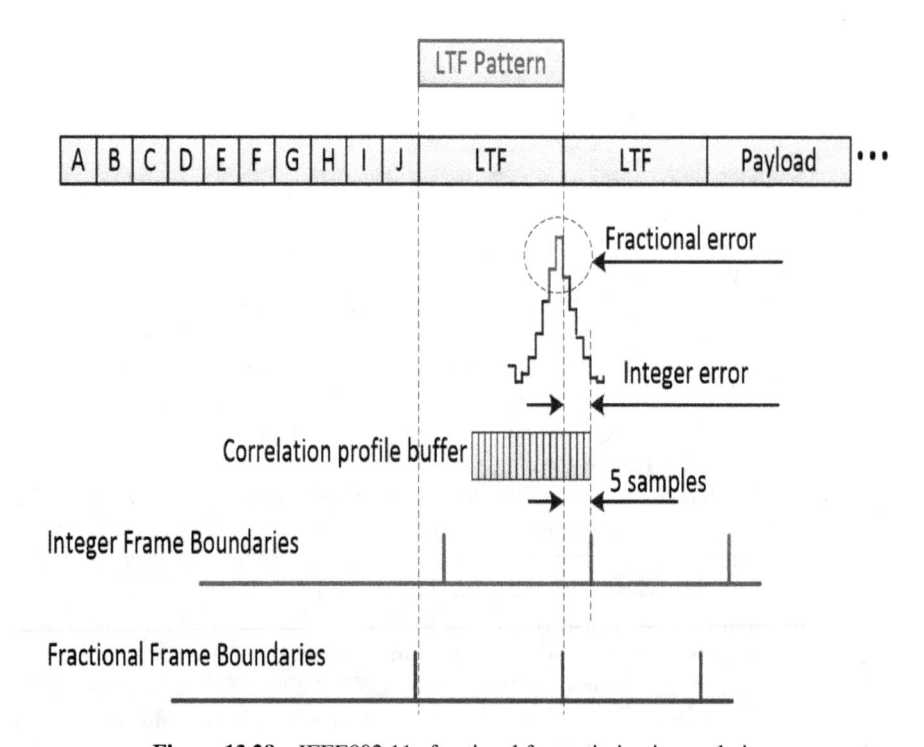

Figure 13.28 IEEE802.11a fractional frame timing interpolation.

to work properly. There is no continuous symbol tracking so there is only one chance to get the frame alignment correct.

The block input signal, from 4× subsampler, in Figure 13.31, comes from the output of the downsample by 4 in Figure 13.17. The phase correction multiply is repeated in both figures. Figure 13.31 is connected to right side output of Figure 13.17. Symbol detection, inside the block labeled Equalizer, Viterbi Decoder in Figure 13.16, generates the final estimates of the received symbol values.

13.6.6.1 Phase and Frequency Correction

Ideally, the RxFreqPhaseOffset block in Figure 13.17 generates a counteracting complex sine wave to remove most (but not all) carrier frequency offset from the received signal. The symbols input to the transmitter IFFT and output from the receiver FFT are generally QAM modulated (BPSK for this example). Frequency

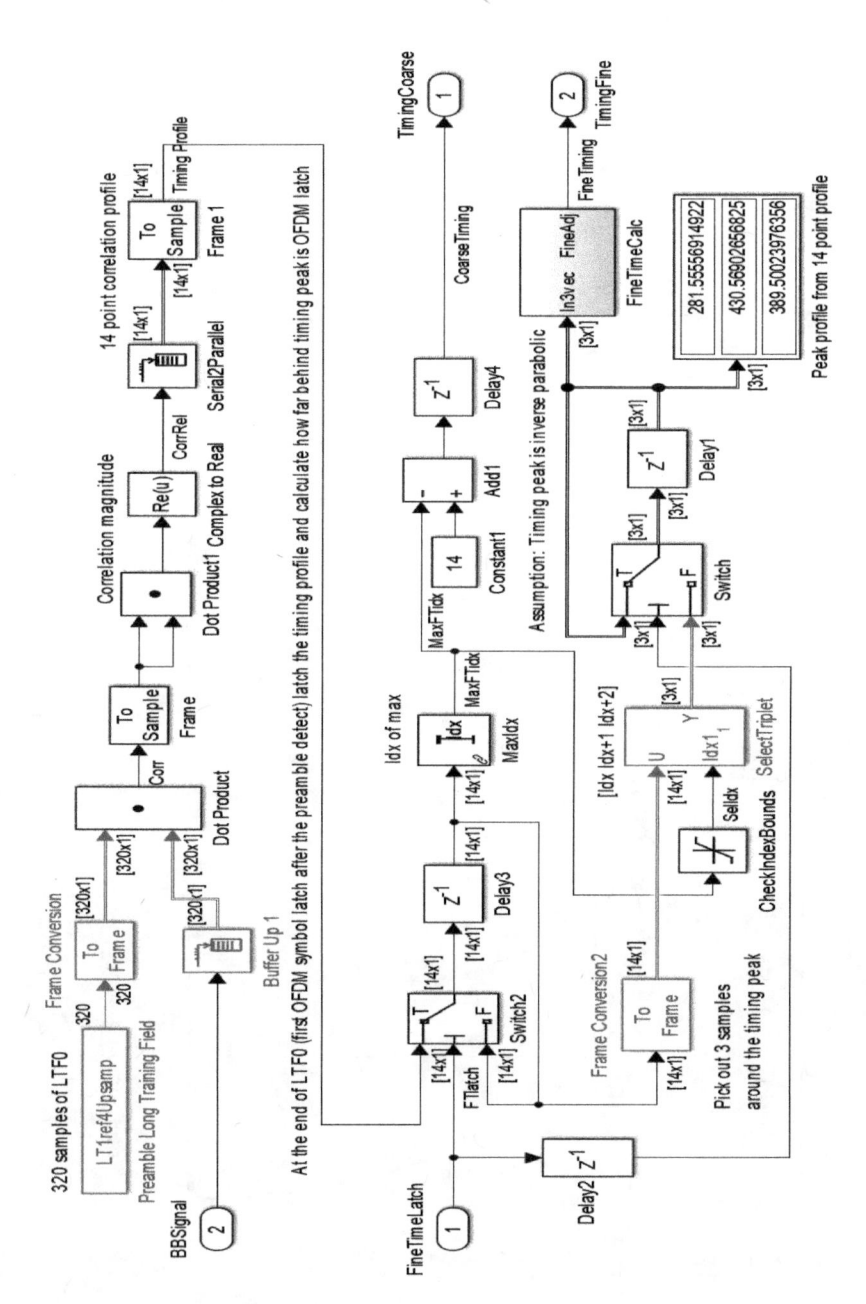

CHAPTER 13. IEEE802.11A DIGITAL SIGNAL PROCESSING 555

Figure 13.29 IEEE802.11a integer and fractional frame timing circuit.

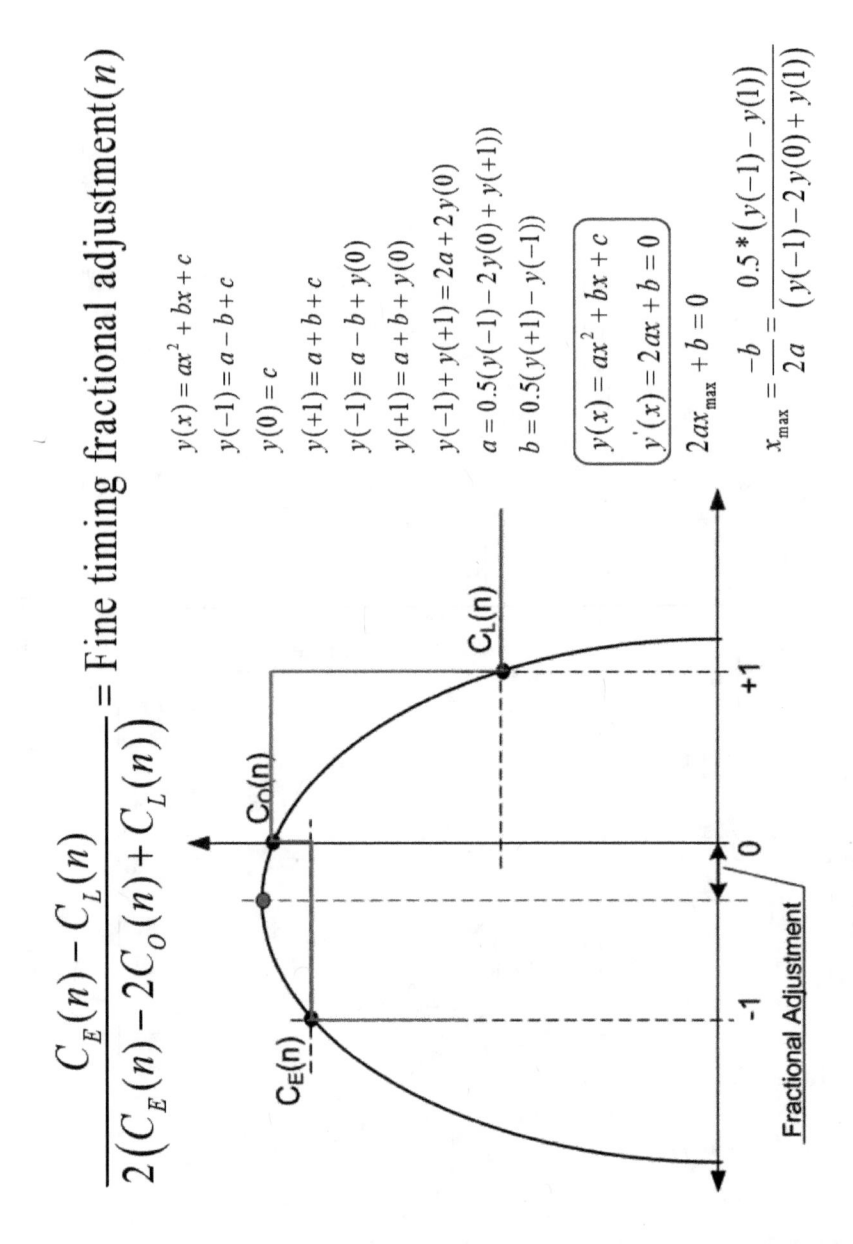

556   Software Defined Radio: Theory and Practice

$$\frac{C_E(n) - C_L(n)}{2(C_E(n) - 2C_O(n) + C_L(n))} = \text{Fine timing fractional adjustment}(n)$$

$y(x) = ax^2 + bx + c$

$y(-1) = a - b + c$

$y(0) = c$

$y(+1) = a + b + c$

$y(-1) = a - b + y(0)$

$y(+1) = a + b + y(0)$

$y(-1) + y(+1) = 2a + 2y(0)$

$a = 0.5(y(-1) - 2y(0) + y(+1))$

$b = 0.5(y(+1) - y(-1))$

$$\boxed{\begin{array}{l} y(x) = ax^2 + bx + c \\ y'(x) = 2ax + b \end{array}}$$

$2ax_{max} + b = 0$

$$x_{max} = \frac{-b}{2a} = \frac{0.5*(y(-1) - y(1))}{(y(-1) - 2y(0) + y(1))}$$

Figure 13.30 IEEE802.11a fractional frame timing calculations.

CHAPTER 13. IEEE802.11A DIGITAL SIGNAL PROCESSING

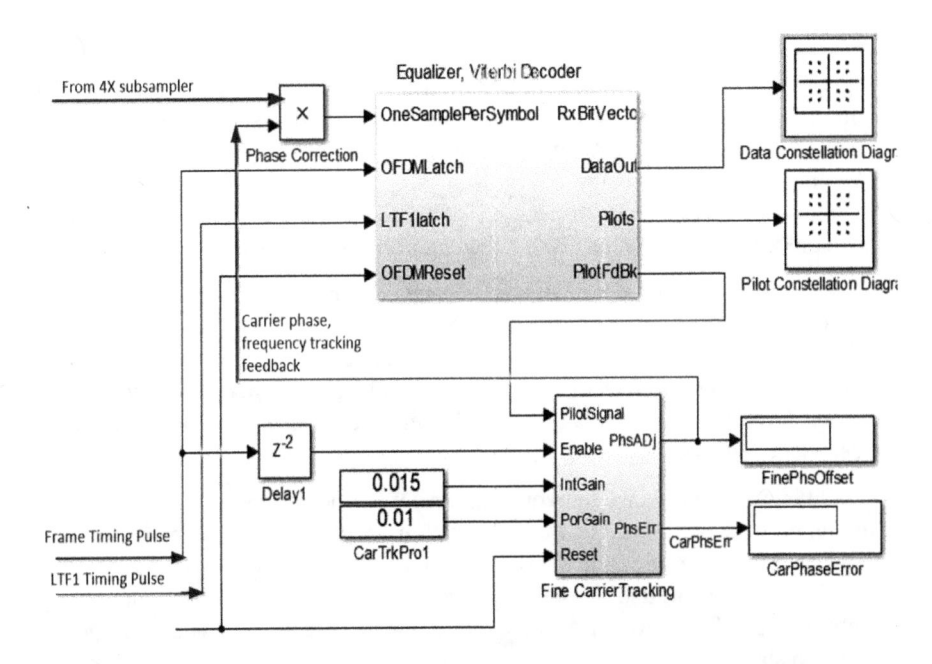

Figure 13.31 OFDM receiver symbol rate processing back end, 20 MHz sample rate.

offset not removed by the RxFreqPhaseOffset block will cause their constellation to rotate, as described in Chapter 11.

The Fine Carrier Tracking block in Figure 13.31 inputs the pilot signal average and generates a separate counteracting complex sine wave to remove all remaining carrier frequency offset so that the Equalizer, Viterbi Decoder block inputs a complex baseband signal with frequency and phase offset reduced to zero.

13.6.6.2 Equalizer, Viterbi Decoder

Figure 13.32 shows the Simulink components inside the block labeled Equalizer, Viterbi Decoder in Figure 13.31. The primary signal processing functions are listed below. Note that the QAM soft decision and Viterbi decoding are both Simulink Communications System Toolbox built-in components. Testing these built-in blocks in simulation and then autogenerating VHDL code from the simulation, using the Simulink code generator, can be a huge engineering time saver.

FDE Initial: One time calculation of received phase and amplitude correction based on LTF1.
FDE Continuous: Equalize frame by frame based on received pilot tones.
QAM Soft Decision: Calculate received symbol log likelihood ratio soft decisions.
Viterbi Decoding: Calculate received symbol hard decisions based on Viterbi decoder traceback.

Figure 13.33 shows the equalization DSP. These important circuits are the reason why the IEEE802.11a works so well in a multipath environment. The input, center left in Figure 13.33, is a serial stream of complex time-domain samples from the signal alignment and subsampling front end, Figure 13.17. The output, lower right, is a parallel frame of 48 FFT output symbols, equalized and ready for soft decision slicing.[5]

The receiver equalization circuits, in Figure 13.33, sends the input sample stream to an upper path, for computing the FDE initial equalizer coefficients, and a lower path for equalizing the received data. The upper path stores (latches) a frame only at the end of LTF1. The lower path stores every frame after LTF1. The FDE0 equalizer coefficients are only computed one time during LTF1.

---

5 Recall Figure 13.1, the transmitter example, where the IFFT input symbols are arranged in the frequency-domain on the left side. The transmitter IFFT output consists of time-domain samples. The receiver does the reverse: takes in time-domain samples and outputs a frame of frequency-domain symbols, matching (hopefully) the transmitter's IFFT input.

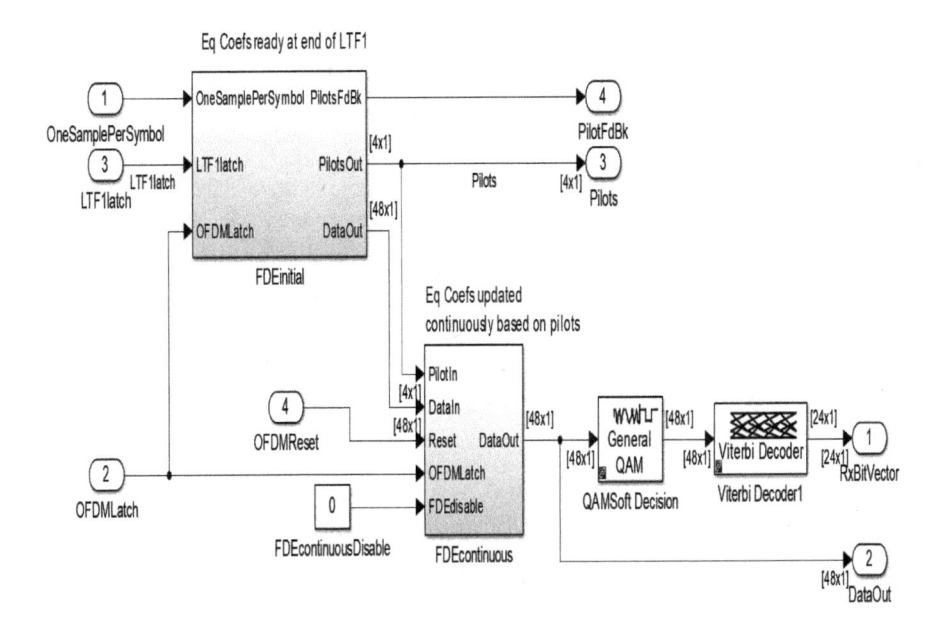

CHAPTER 13. IEEE802.11A DIGITAL SIGNAL PROCESSING

Figure 13.32 OFDM equalizer, Viterbi decoder.

The latch pulse inputs for LTF1 and for every succeeding frame come from the fine timing circuits and are shown in the lower left of Figure 13.31. Thanks to the fine timing function, these frames very closely match the transmitter's IFFT output frames. In each path, the cyclic prefix is removed and 64 samples at the FFT input are converted to 64 frequency-domain samples arranged as in Table 13.3.

13.6.6.3 FDE Initial

The FDE initial function uses the fixed and known a priori symbol pattern in LTF1. The top center of Figure 13.33 shows the multiplication between the received LTF1 and the stored conjugate reference symbols of LTF1. Due to the conjugate, the LTF1 pattern is removed and the remaining FDE Initial equalizer coefficients are simply the phase and amplitude error imparted by the channel. Phase shifts that are due to the known pattern are rotated back to zero by the conjugate of the pattern so any leftover phase has to be phase error, likewise for the amplitude error. Because the received LTF1 vector was latched and stays constant for the entire packet, the

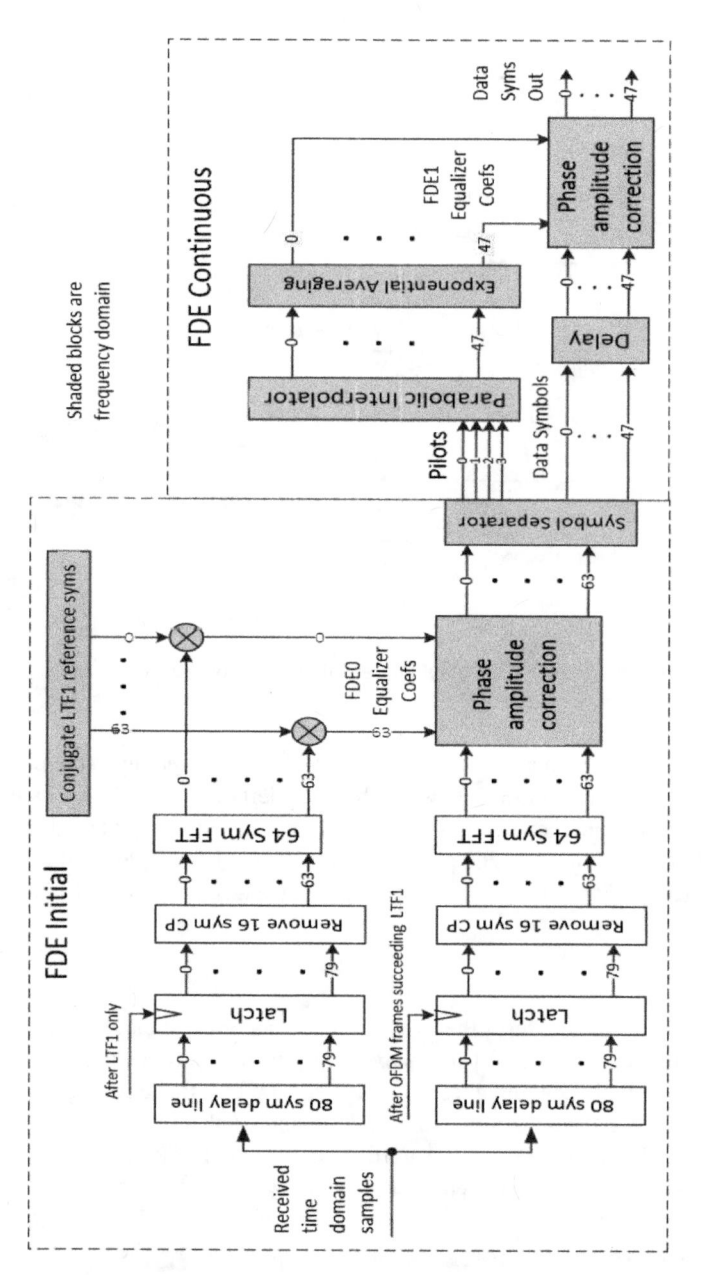

Figure 13.33 IEEE802.11a equalization circuits.

CHAPTER 13. IEEE802.11A DIGITAL SIGNAL PROCESSING

FDE initial equalizer coefficients stay constant and are applied (in the lower path, phase amplitude correction) to every received vector after LTF1. Figure 13.34 is the Simulink diagram for FDE Initial (circuits contained in block labeled FDEinitial in Figure 13.32).

The circuits inside the PhsAmpCorrection block, shown in Figure 13.35, use FDE Initial equalizer coefficients to correct each received symbol. As shown in Figure 13.35, each complex symbol is first rotated by the conjugate phase error (conjugate rotates in the opposite direction of the error). Then the received symbol amplitude is multiplied by the inverse amplitude error.

FDE Initial block operation is shown in Figure 13.36. As the receiver FFT output, these four diagrams are in the frequency-domain. Recall that the transmitter IFFT input consisted of symbols arranged in the frequency-domain. There is no additional meaning to the term frequency-domain used here, just a vector of symbols to be passed up to additional communications layers. Below is a description of each of the four traces:

Trace 1 LTF1 vector (frame) of 64 distorted received symbols (latched by LTF1 timing pulse).
Trace 2 Product of received LTF1 and conjugate reference LTF1.
Trace 3 OFDM vector subsequent to LTF1.
Trace 4 Equalized OFDM vector subsequent to LTF1. BPSK symbols have imaginary part zero. Also note 4 pilot symbols.

13.6.6.4 FDE Continuous

IEEE 802.11a is intended to perform well in RF channels with multipath and Doppler. These channels can change propagation characteristics rapidly. Thus, FDE Initial equalization based on a one-time comparison between received and known LTF1 may only be effective at the beginning of the received packet. The FDE Continuous function provides continuous channel estimation and correction by using the four pilot symbols present in every OFDM frame after the preamble.

Circuits in the left dashed block of Figure 13.33 are for FDE Initial, discussed previously. Circuits in the right dashed box are for FDE Continuous, discussed here. Between the blocks, the Symbol Separator in the lower center of Figure 13.33 strips off the four pilots and 48 coded data symbols and discards the zero symbols (see Table 13.3). The 48 coded data symbols are to be phase and amplitude-corrected adaptively every frame by FDE Continuous. Both FDE Initial and FDE Continuous use the same phase and amplitude corrector circuit; see Figure 13.35.

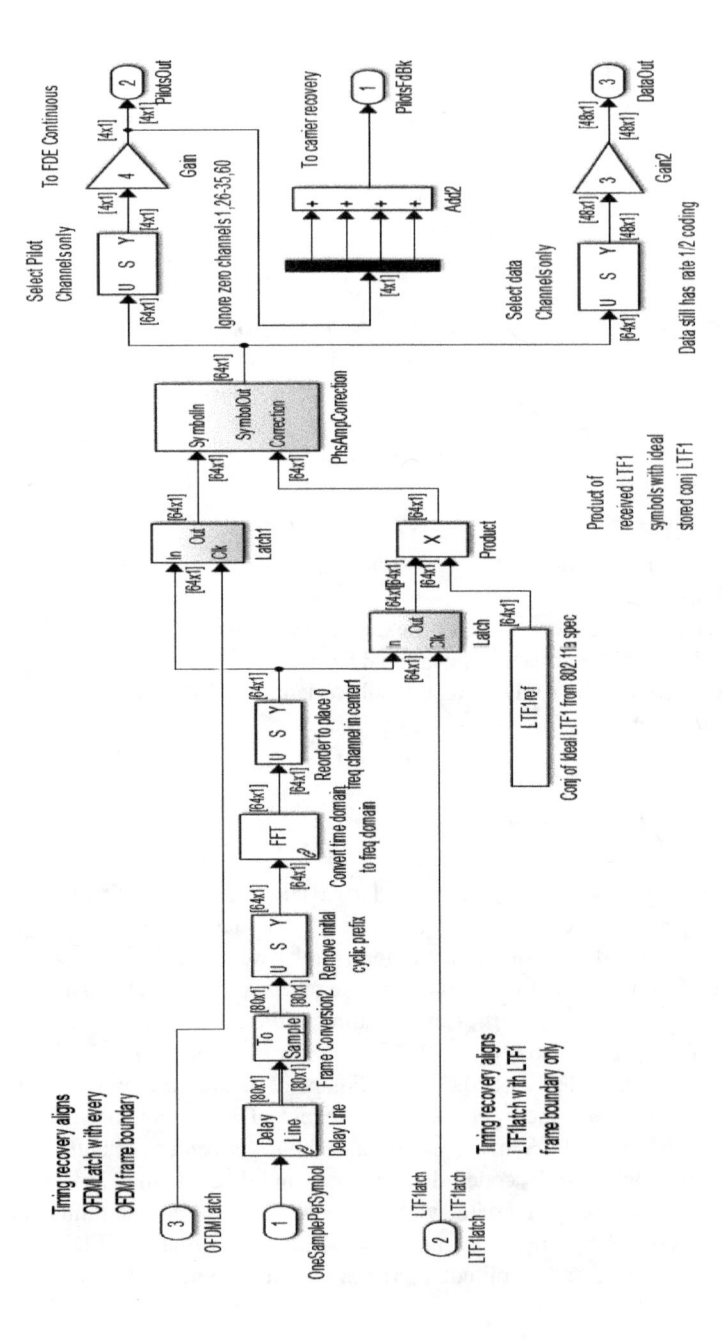

Figure 13.34 Receiver block, FDE initial Simulink diagram.

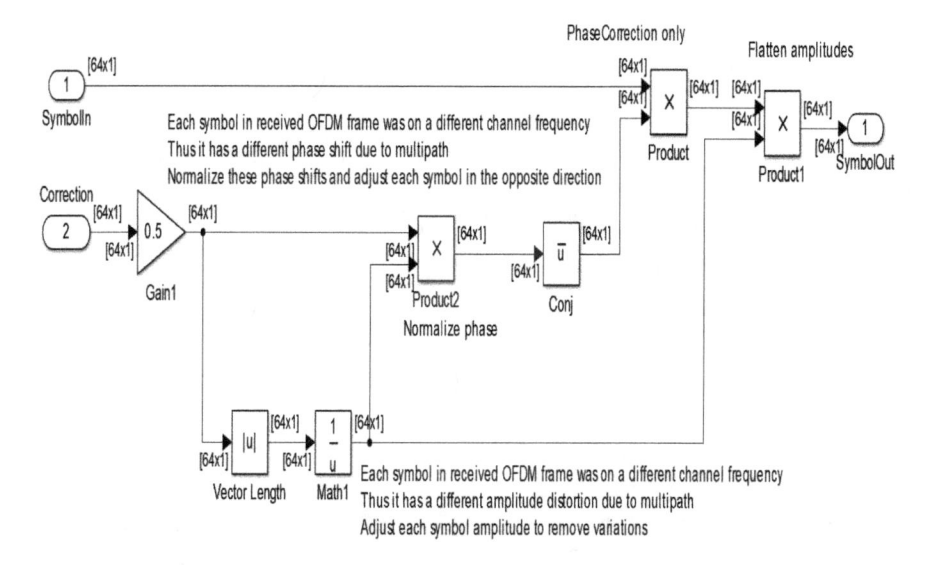

CHAPTER 13. IEEE802.11A DIGITAL SIGNAL PROCESSING 563

Figure 13.35 Receiver block, FDE initial continuous corrector.

Unlike FDE Initial, the calculation of the 48 continuous equalizer coefficients is updated for every received frame. This update is based solely on the four received pilots. Each pilot signal is subject only to channel distortion existing at its center frequency. Distortion measurement at all 48 data channels is interpolated parabolically from the four pilots (parabolic was chosen here because it is simple; however, other interpolation techniques could be used and may be more accurate).

As shown in Figure 13.37, there are two independent identical interpolators. The top interpolator uses pilots 1, 2, and 3 to estimate channels 1 to 24 and the lower interpolator uses pilots 2, 3, and 4 to estimate channels 25 to 48. These interpolators are the same; only the channel relative distances are different.

Figure 13.38 is an idealized model showing the location of pilots 1, 2, and 3 in relation to the 24 channels that they are being used to estimate. The vertical axis value of the three pilots on the hypothesized parabolic curve are shown as dots and are known. The horizontal axis pilot values are assigned as [-1, 0, +1] because we only need the relative distance between pilots and unknown channels. The unknown channel responses, to be estimated, are shown as small boxes. Estimates and relative distances based on pilots 2, 3, and 4 will have a similar curve.

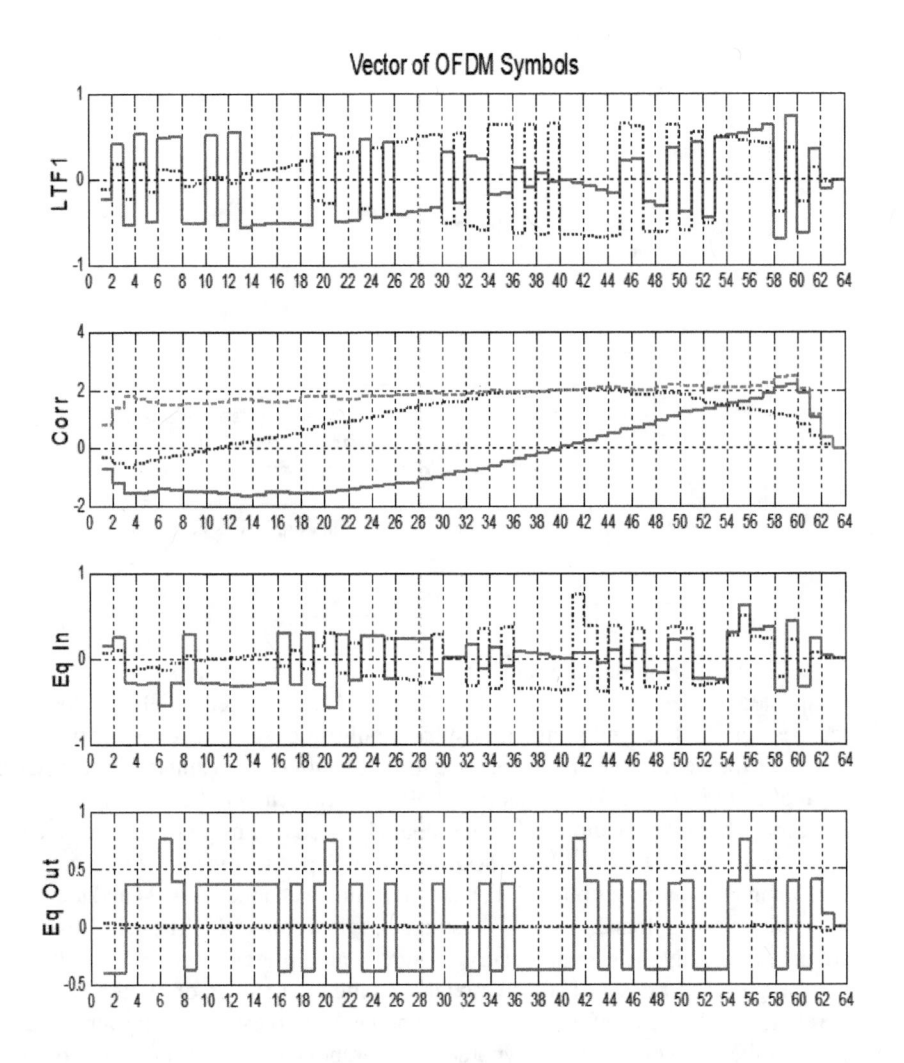

Figure 13.36 FDE initial equalization result; dotted lines are the imaginary part.

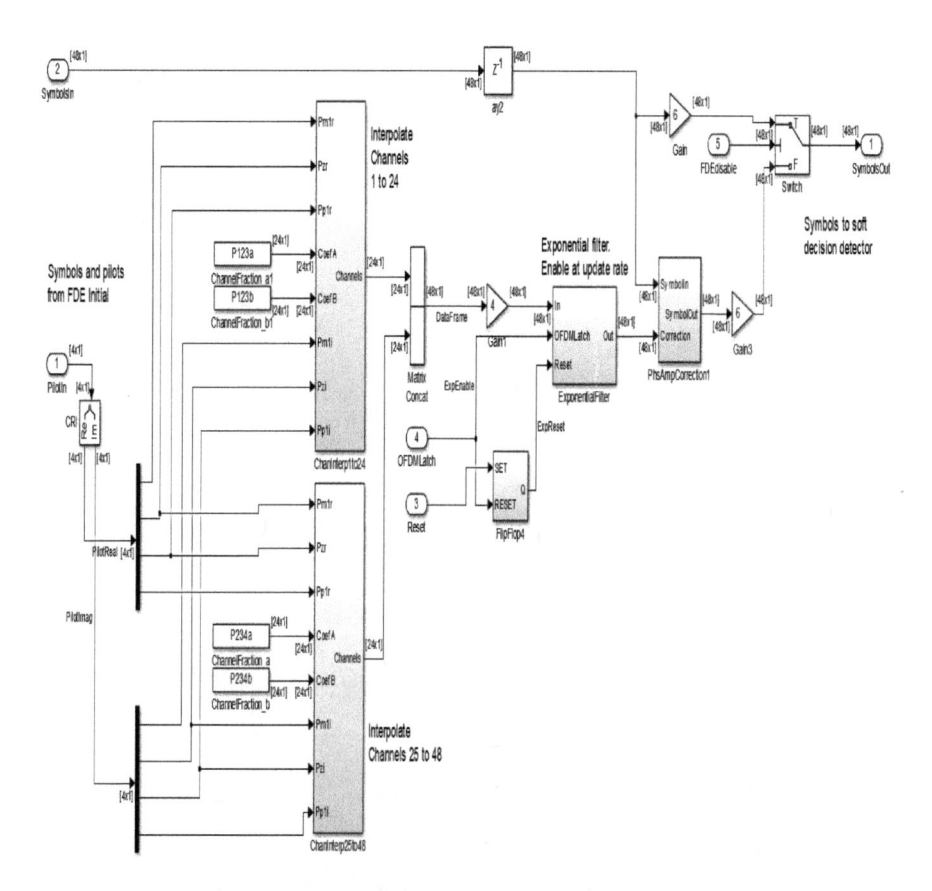

CHAPTER 13. IEEE802.11A DIGITAL SIGNAL PROCESSING 565

Figure 13.37 Receiver block, FDE continuous Simulink diagram.

Actual relative distances along the horizontal axis are shown in Tables 13.6 and 13.7. These are the numbers used for P123b and P234b in Figure 13.37. P123a = P123b squared and P234a = P234b squared. Note that the distances between the relative fractional positions are not always the same because the number of channels between pilots are not the same. Note that zero channels are ignored. Compare the following tables with Table 13.3.

If $y(k)$ is the estimated response for channel k and $x(k)$ is the corresponding fractional distance along the horizontal axis then we hypothesis that every channel, including the pilots, lies on the second-order curve defined in Equation (13.2). The

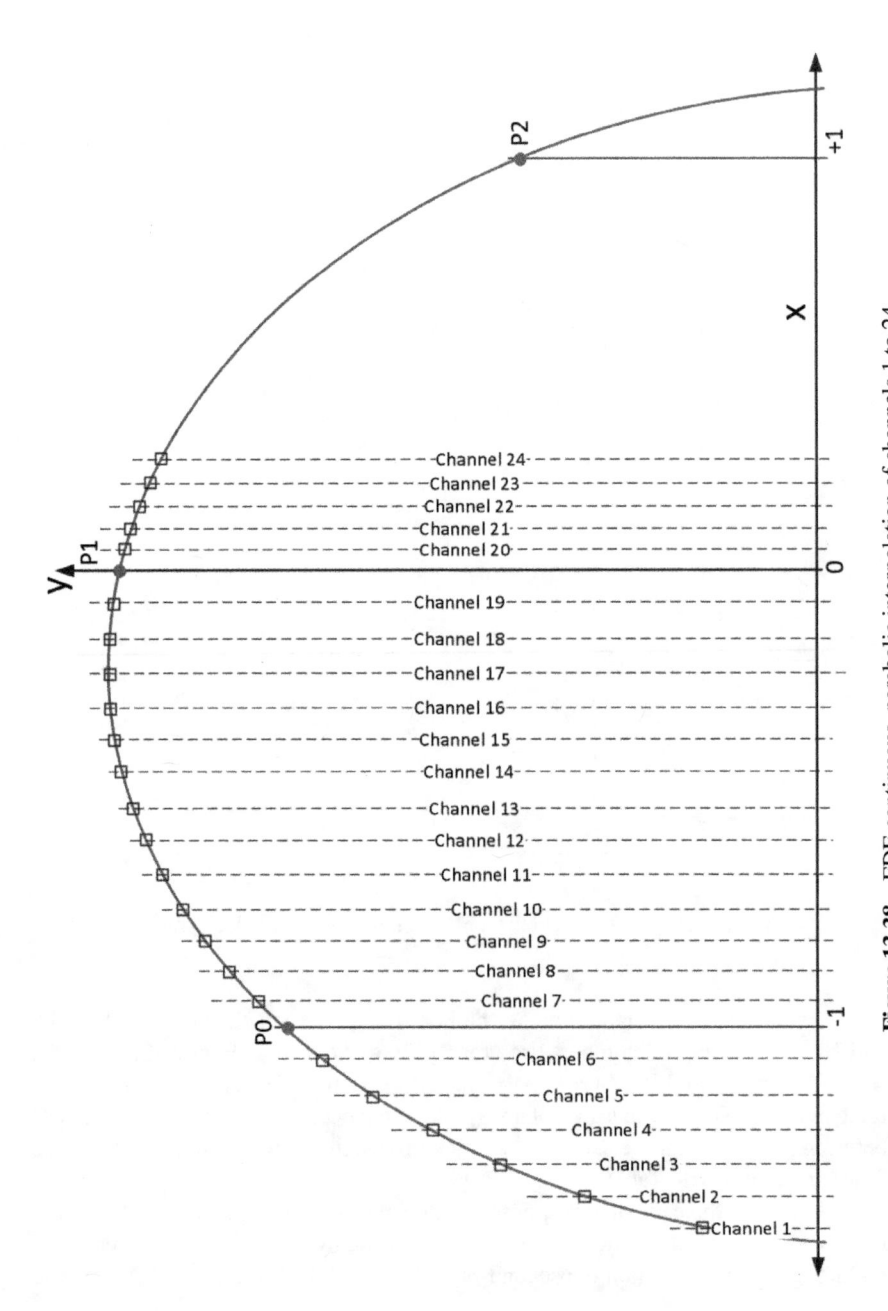

Figure 13.38 FDE continuous, parabolic interpolation of channels 1 to 24.

Table 13.6 Continuous FDE Relative Distances, Data Channels 1 to 24

Channel	Use	P123b
2	Data 1	-20/14
3	Data 2	-19/14
4	Data 3	-18/14
5	Data 4	-17/14
6	Data 5	-16/14
7	Data 6	-15/14
8	Pilot 1	n.a.
9	Data 7	-13/14
10	Data 8	-12/14
11	Data 9	-11/14
12	Data 10	-10/14
13	Data 11	-9/14
14	Data 12	-8/14
15	Data 13	-7/14
16	Data 14	-6/14
17	Data 15	-5/14
18	Data 16	-4/14
19	Data 17	-3/14
20	Data 18	-2/14
21	Data 19	-1/14
22	Pilot 2	n.a.
23	Data 20	1/21
24	Data 21	2/21
25	Data 22	3/21
36	Data 23	4/21
37	Data 24	5/21

Table 13.7 Continuous FDE Relative Distances, Data Channels 25 to 48

Channel	Use	P234b
38	Data 25	-5/21
39	Data 26	-4/21
40	Data 27	-3/21
41	Data 28	-2//21
42	Data 29	-1/21
43	Pilot 3	n.a.
44	Data 30	1/14
45	Data 31	2/14
46	Data 32	3/14
47	Data 33	4/14
48	Data 34	5/14
49	Data 35	6/14
50	Data 36	7/14
51	Data 37	8/14
52	Data 38	9/14
53	Data 39	10/14
54	Data 40	11/14
55	Data 41	12/14
56	Data 42	13/14
57	Pilot 4	n.a.
58	Data 43	15/14
59	Data 44	16/14
60	Data 45	17/14
61	Data 46	18/14
62	Data 47	19/14
63	Data 48	20/14

CHAPTER 13. IEEE802.11A DIGITAL SIGNAL PROCESSING   569

coefficient values are calculated from plugging in the measured complex response at points $[-1, P_0]$, $[0, P_1]$ and $[+1, P_2]$. These are input on the right side, center, of Figure 13.37.

$$\begin{aligned} y(k) &= ax(k)^2 + bx(k) + c \\ a &= 0.5P_0 - P_1 + 0.5P_2 \\ b &= 0.5P_2 - 0.5P_0 \\ c &= P_1 \end{aligned} \quad (13.2)$$

In Equation (13.2) constants are stored in the MATLAB code for each pilot response, position, and position squared, $y(k)$, $x(k)$, and $x^2(k)$. These are called P123a, P123b, P234a, P234b in Figure 13.37. The coefficients a, b, c in Equation (13.2) are recalculated for each new OFDM frame because the pilot responses can change frame to frame due to the channel. Finally, the two sets of 24 channel estimations are combined into one 48-element coefficient vector. In the center of Figure 13.37, each interpolated channel estimate is individually exponentially filtered to form the set of FDE Continuous equalizer coefficients; see Section 9.2.3. The exponential filter is updated every new OFDM frame. If $x(n)$ is one of the 48 equalizer channel estimates, then the exponential filter $y(n)$ output is shown in Equation (13.3). For $\alpha = 0.5$, each new filtered coefficient is half the current coefficient plus the sum of all the previous outputs in various states of exponential decay. Equation (13.3) shows a few steps of this recursion.

$$\begin{aligned} y(0) &= \alpha x(0) \\ y(1) &= \alpha x(1) + \alpha^2 x(0) \\ y(2) &= \alpha x(2) + \alpha^2 x(1) + \alpha^3 x(0) \\ y(3) &= \alpha x(3) + \alpha^2 x(2) + \alpha^3 x(1) + \alpha^4 x(0) \\ &\dots \end{aligned} \quad (13.3)$$

Figure 13.39 is a demonstration of one FDE continuous output symbol, not a frame but one symbol in a series of 126 frames. Symbol 12 was chosen for no particular reason. The first trace shows symbol 12 prior to FDE continuous equalization. Note that frame to frame varying amplitude and phase distortion. The second trace is the frame-to-frame equalizer correction calculated for symbol 12 (calculated from pilots 0, 1 and 2 as described previously). Finally, the third trace is the correction in the second trace applied to the distorted signal in the first trace. Note that symbol 12 in the first trace was already equalized by the FDE Initial process. However, due to the presence of Doppler and multipath, the channel is changing over time' thus the need for FDE Continuous.

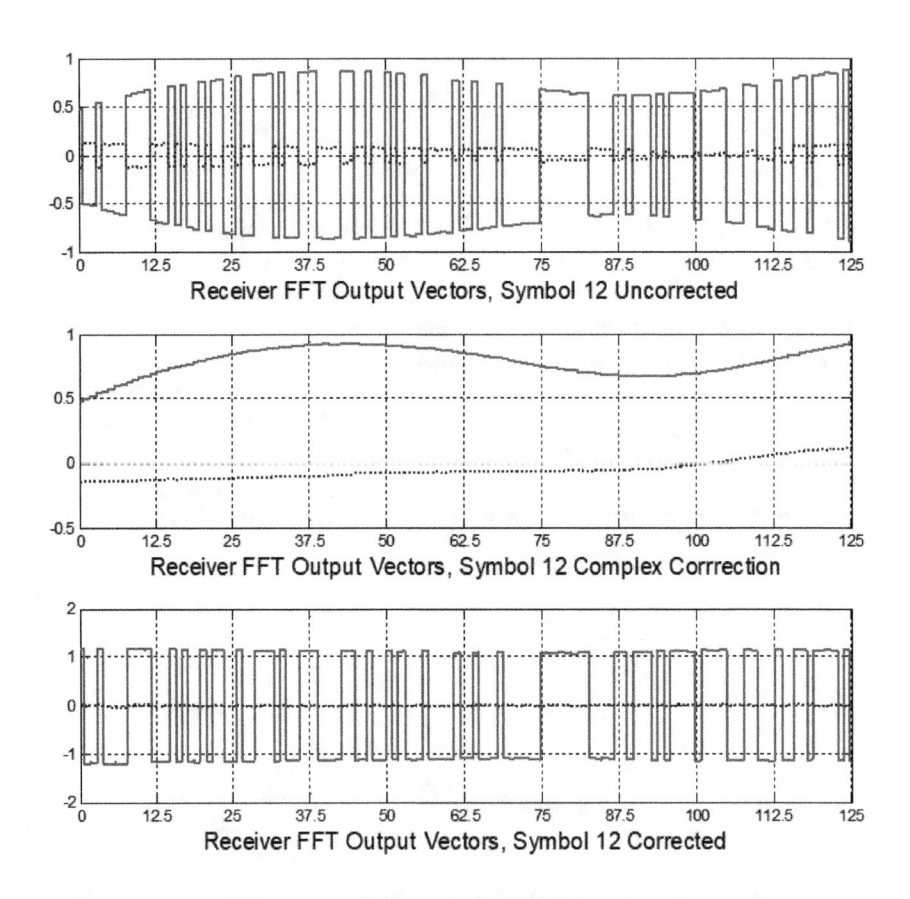

Figure 13.39 FDE continuous equalization, sym 12; dotted lines are imaginary.

13.6.6.5 Carrier Tracking

IEEE802.11a OFDM has the advantage that only the pilot symbols are needed to track out the last vestige of carrier frequency offset. Thus, carrier tracking is simpler than, for example, APSK carrier tracking described in Section 12.5.1. The four pilot symbols are simply constants with a value of $1 + j0$ each. Thus, they can be average together, as in the right center of Figure 13.34, and input to the Fine Carrier Tracking block as PilotFeedback; see Figure 13.31.

Figure 13.40 shows an actual pilot constellation diagram on the left. Because we are trying to force the pilot average onto the real axis, the imaginary axis tends to measure the pilot error. The real axis is discarded because it primarily measures the pilot power. The pilot imaginary part is filtered by the same proportion-integral (PI) loop filter discussed in Section 12.5.1. The table in Figure 13.40 shows the signed integer output of the PI loop filter converted to a stored integer format.[6] The look-up table angle is an address so it is based on the stored integer.

The complex sine/cosine output of the carrier tracking block is input to the Phase and Frequency Correction block to counteract any frequency and/or phase offset remaining from the downconversion to complex baseband process.

13.7 IEEE802.11A SYSTEM TESTING

13.7.1 Simulation Testing

Figure 13.16 is the Simulink top-level diagram. Note that in the top right corner, there is an Error Count test block. This calculates a delay match and compares 24 symbol transmit data vectors with 24 symbol receive data vectors. BER results can easily be verified for an AWGN channel.[7]

13.7.2 SDR Hardware Testing

The ADALM-Pluto SDR will be used for this test description. We will make use of MathWorks object-oriented ADALM-Pluto transmit and receive interfaces. Figure 13.41 shows the setup. The ADALM-Pluto has two included antennas, tuned for

---

6 Simulink stored integer is the binary equivalent of the signed two's complement input. For example, three bit fixed point two's complement -1 has a binary format of 111. The stored integer output will be 7 instead of -1.
7 Verified means compared with theoretical BER results; see Section 3.2.4. Other impairments, such as multipath fading, are much more difficult to theoretically predict.

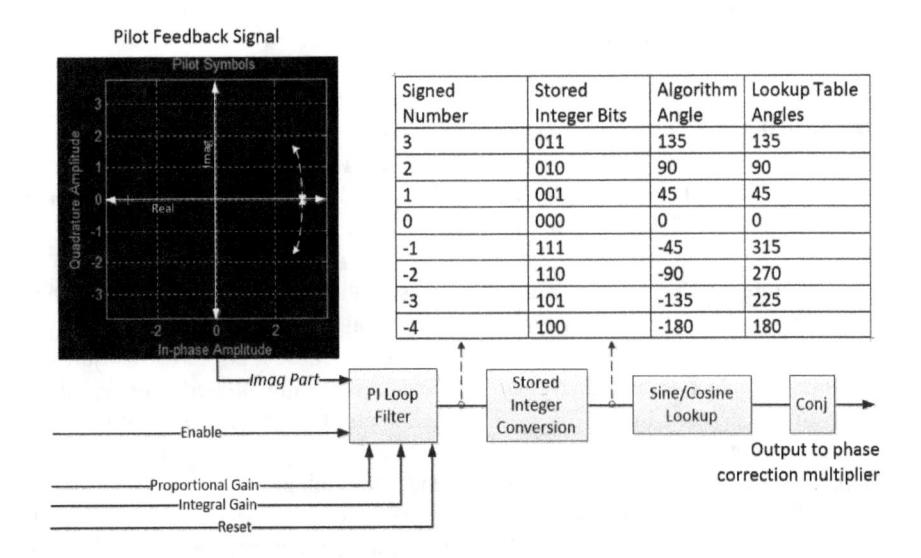

572   *Software Defined Radio: Theory and Practice*

Signed Number	Stored Integer Bits	Algorithm Angle	Lookup Table Angles
3	011	135	135
2	010	90	90
1	001	45	45
0	000	0	0
-1	111	-45	315
-2	110	-90	270
-3	101	-135	225
-4	100	-180	180

Figure 13.40   IEEE802.11a carrier tracking theory.

the 2.4 GHz default frequency. These should be installed prior to turning on the transmitter. This default frequency is in the unlicensed ISM band so radiating small amounts of power from the transmit antenna is no problem.

13.7.2.1   Transmit Setup

Figure 13.42 is an excerpt of the upper left corner of Figure 13.16, the IEEE802.11a simulation discussed in this chapter. Note the samples to workspace[8] collection block, Workspace160. This results in a sample file, OFDMbaseband.mat. This file stores two sample vectors, BBreal and BBimag.

The MATLAB code for transmitting the IEEE802.11a test signal is similar to the code discussed in Section 12.6.1.1 except the sampling rate is different and APSKbaseband.mat replaces OFDMbaseband.mat. As mentioned, because MATLAB is not a multithreaded program, we have to start the transmitter in repeat mode and then we can focus on the receiver.

---

8   MATLAB workspace is the top-level screen that comes up when starting MATLAB.

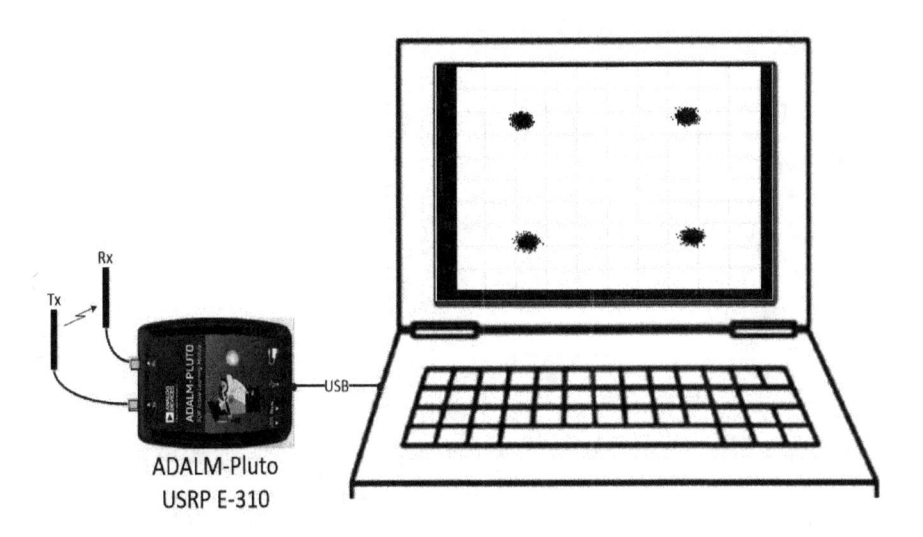

CHAPTER 13. *IEEE*802.11A *DIGITAL SIGNAL PROCESSING* 573

Figure 13.41 IEEE802.11a SDR system test configuration.

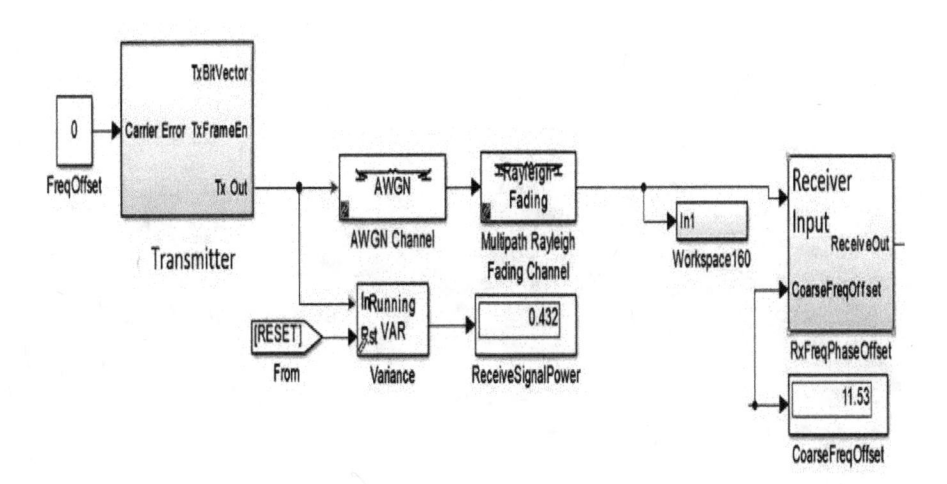

Figure 13.42 IEEE802.11a transmit signal storage.

```
close all
format compact
Fs160 = 160;
Fs40 = 40;
TxLength =84000; %Tx buffer space needed
TxStart = 1;
TxEnd = TxStart + TxLength-1;
%IEEE 802.11a transmit test signal, 160 MHz sampling rate
load OFDMbaseband % contains BBreal BBimag

%Reduce sample rate from 160 to 40 MHz
%This is to accommodate AD9363 sample rate limitations
HB1 = firhalfband(12,0.25); %First downsample filter
HB2 = firhalfband(18,0.25); %Second downsample filter
BBreal80 = downsample(filter(HB1,1,BBreal),2);
BBimag80 = downsample(filter(HB1,1,BBimag),2);
BBreal40 = downsample(filter(HB2,1,BBreal80),2);
BBimag40 = downsample(filter(HB2,1,BBimag80),2);

TxRl = 0.25*BBreal40(TxStart:TxEnd);
TxIg = 0.25*BBimag40(TxStart:TxEnd);
TxComplex = complex(TxRl, TxIg);
tx = sdrtx('Pluto',...
'Gain',-20,...
'BasebandSampleRate',Fs40*10ˆ6,...
'ShowAdvancedProperties',true);
%Download transmit signal samples (40 MHz) into AD9363
%TxComplex contains one IEEE802.11a, 512 frame packet
%This is repeated continuously
tx.transmitRepeat(TxComplex);
%To stop transmiiter type "tx.release" in workspace
```

13.7.2.2 Receive Setup

The MATLAB code for receiving the IEEE802.11a test signal is similar to the code discussed in Section 12.6.1.2, except, as for the transmitter, the sampling rate is

CHAPTER 13. IEEE802.11A DIGITAL SIGNAL PROCESSING

different. The MATLAB code below setups up an rx object to control the ADALM-Pluto receiver and then uploads samples from the receiver. Sample uploads are stored in the MATLAB record RxSamples.

The samples received this way could be further processed by the rest of the IEEE802.11a preamble detection and packet processing code. The Simulink version of this was discussed at length in this chapter. Although the SDR transmitter is in repeat mode and cannot be changed on the fly, other system aspects such as antenna placement can be tested. The effect of receiver code changes can be easily tested.

```
Fs40 = 40; %Over all sampling rate, tx and rx
RxLength = 2^17;

rx = sdrrx('Pluto',...
'SamplesPerFrame',RxLength, ...
'BasebandSampleRate',Fs40*10^6);

%Two ways to learn about the rx object:
%Type "rx" to see radio setups.
%Type methods(rx) to see object functions, examples:
%rx.release stops the receiver
%rx.info returns version information
%rx() samples transfered from SDR as int16
%The ADLAM-Pluto SDR is a fixed-point device
RxSamples = double(rx());
```

13.7.2.3 Results

Figure 13.43 shows the transmit and receive sample records along with their power spectrums. The transmit and receive spectrum should be the same since there is very little RF propagation loss. There are a couple of parameters in the SDR receive object that can be tweaked, for example, receive gain and AGC. A well-organized MATLAB receive program provides flexibility to try different receive parameters, for example, carrier tracking loop gains.

Note the bandwidth restriction in the last graph of Figure 13.43. This was described in detail in Section 12.6.1.4.

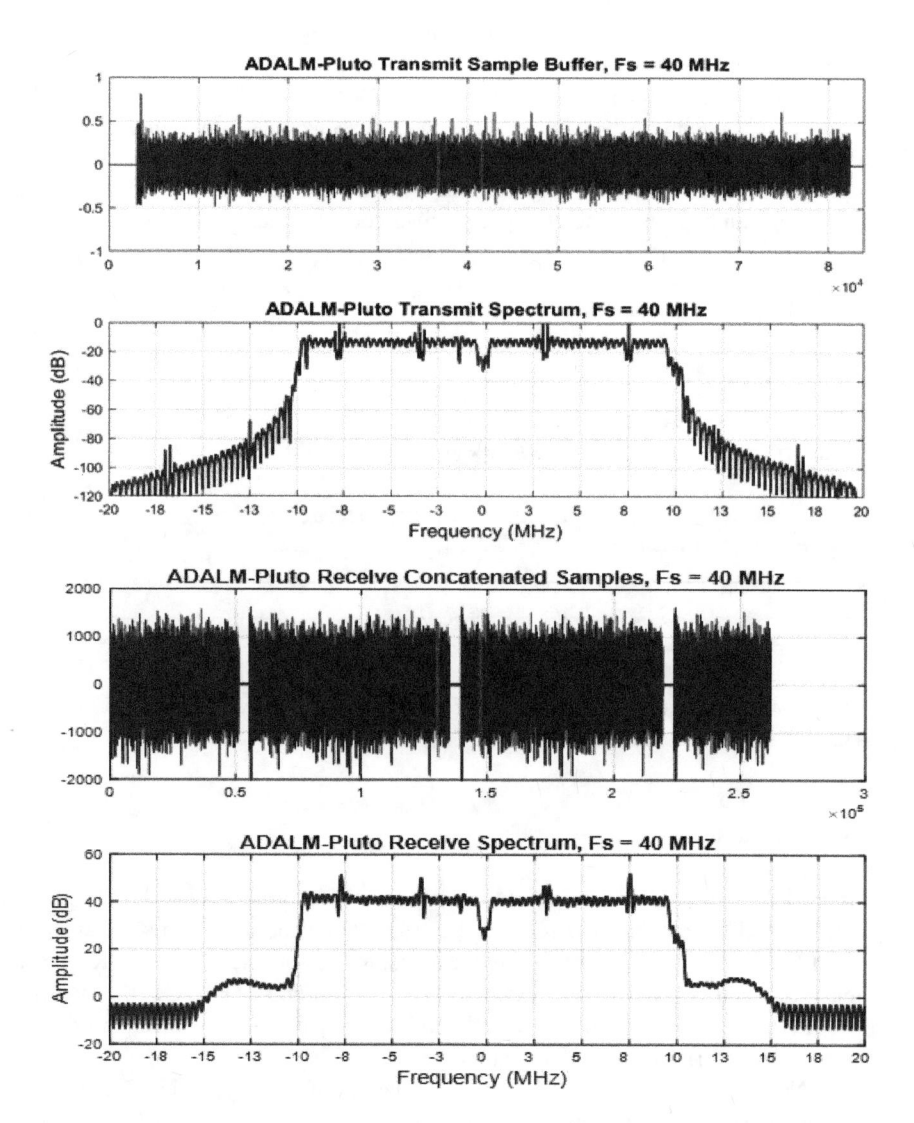

Figure 13.43 IEEE802.11a transmission and reception, ADALM-Pluto SDR.

REFERENCES

[1] ANSI/IEEE. ANSI/IEEE Std 802.11a-1999. *Std 802.11a*, 1999.

[2] Cisco. *Configuring Receiver Start of Packet Detection Threshold*. Release 8.1.

Chapter 14

More Fundamentals

14.1 INTRODUCTION

This last chapter is a collection of topics that deserve more detailed explanation than they received in the previous chapters. Hopefully, these will help round out the reader's software defined radio education.

14.2 FIXED POINT NUMBER FORMATS

Here we discuss the Simulink fixed point attribute (FPA) indicators. Commonly used forms are shown in Table 14.1. In this table, N indicates the total number of bits and M indicates the fractional part. For our DSP purposes, $N > M$. There are other ways to specify FPAs; however, since this book relies on Simulink, we will focus on their nomenclature.

For DSP, two's Complement is a commonly used binary representation. Given $N = 8$ bits, Table 14.2 shows the numeric ranges of two's complement, natural binary, and offset binary. Sometimes DAC manufacturers will use an offset binary input format. However, modem design and simulation described in this book rely on two's complement.

Two's complement is a good fit to the amplitude range of many DSP signals. Another advantage is that only adders are needed in hardware: $X - Y$ is calculated as $X + (-Y)$. To flip the sign of a two's complement number, simply invert the bits and add 1.

A caution is that many signals, such as voice, can have large amplitude peaks that may exceed the range of the chosen M and N. Two's complement overflow

Table 14.1 Examples of Simulink Fixed Point Attributes

FPA	Sign	Integer Bits	Fractional Bits	Comment
$ufixN\_EnM$	n.a.	$(N-1)\ldots(N-M)$	$(M-1)\ldots 0$	Unsigned integer, fraction
$sfixN\_EnM$	$N-1$	$(N-2)\ldots(N-M)$	$(M-1)\ldots 0$	Signed integer, fraction
$ufixN$	n.a.	$(N-1)\ldots 0$	n.a.	Unsigned integer
$sfixN$	$N-1$	$(N-2)\ldots 0$	n.a.	Signed integer
$Boolean$	n.a.	n.a.	n.a.	Single binary bit

CHAPTER 14. MORE FUNDAMENTALS   581

causes wraparound; for example, 127 wraps around to -128. This change does not match the actual signal peak excursion and thus causes distortion. Simulink offers a hard limiting option on operations, such as multipliers, that can easily overflow. Hard limiting would, for example, force signal excursion above 127 to equal 127 and likewise for negative signals. However, when hard limiting circuits are implemented in FPGA hardware, extra resources will be required.

Table 14.2 Two's Complement, Natural Binary, and Offset Binary Numeric Ranges, $N = 8$

Decimal	Two's Complement	Natural Binary	Offset Binary
255	n.a.	1111 1111	n.a.
254	n.a.	1111 1110	n.a.
253	n.a.	1111 1101	n.a.
⋮	⋮	⋮	n.a.
129	n.a.	1000 0001	n.a.
128	n.a.	1000 0000	n.a.
127	0111 1111	0111 1111	1111 1111
126	0111 1110	0111 1110	1111 1110
125	0111 1101	0111 1101	1111 1101
⋮	⋮	⋮	⋮
2	0000 0010	0000 0010	1000 0010
1	0000 0001	0000 0001	1000 0001
0	0000 0000	0000 0000	1000 0000
-1	1111 1111	n.a.	0111 1111
-2	1111 1110	n.a.	0111 1110
-3	1111 1101	n.a.	0111 1101
⋮	⋮	⋮	⋮
-126	1000 0010	n.a.	0000 0010
-127	1000 0001	n.a.	0000 0001
-128	1000 0000	n.a.	0000 0000

Table 14.3 shows a couple examples of a $ufix18\_En10$ FPA. In words, this is unsigned, fixed point, 18 total bits with 10 fractional bits. Note that there is an extra integer bit, weight = 2^7, available due to the lack of a sign bit.

Table 14.4 shows positive and negative examples of a $sfix22\_En18$ FPA. In words, this is signed, fixed point, 22 total bits with 18 fractional bits. Note the negative weighting of the sign bit.

Table 14.3 $ufix18\_En10$ Two's Complement Unsigned Fixed-Point Examples

Function	Bit No.	Weight	Example 1		Example 2	
Integer	15	2^7	1	128	0	0
	14	2^6	0	0	1	64
	13	2^5	1	32	1	32
	12	2^4	1	16	0	0
	11	2^3	0	0	0	0
	10	2^2	0	0	1	4
	9	2^1	1	2	1	2
	8	2^0	0	0	0	0
Fractional	7	2^{-1}	0	0	0	0
	6	2^{-2}	0	0	1	0.5
	5	2^{-3}	0	0	0	0
	4	2^{-4}	0	0	0	0
	3	2^{-5}	0	0	1	0.03125
	2	2^{-6}	1	0.015625	0	0
	1	2^{-7}	0	0	0	0
	0	2^{-8}	1	0.00390625	0	0
			Sum	178.01953125	Sum	102.53125

CHAPTER 14. MORE FUNDAMENTALS

Table 14.4 $sfix22\_En18$ Two's Complement Signed Fixed-Point Examples

Function	Bit No.	Weight	Example 1		Example 2	
Sign	21	-2^3	0	0	1	-8
Integer	20	2^2	0	0	1	4
	19	2^1	1	2	0	0
	18	2^0	1	1	1	4
Fractional	17	2^{-1}	0	0	0	0
	16	2^{-2}	0	0	1	0.25
	15	2^{-3}	1	0.125	1	0.125
	14	2^{-4}	0	0	0	0
	13	2^{-5}	0	0	0	0
	12	2^{-6}	0	0	0	0
	11	2^{-7}	0	0	0	0
	10	2^{-8}	0	0	0	0
	9	2^{-9}	0	0	1	0.001953125
	8	2^{-10}	1	0.00097656	0	0
	7	2^{-11}	0	0	0	0
	6	2^{-12}	1	0.00024414	0	0
	5	2^{-13}	0	0	0	0
	4	2^{-14}	0	0	0	0
	3	2^{-15}	1	0.000030518	0	0
	2	2^{-16}	1	0.000015259	0	0
	1	2^{-17}	0	0	0	0
	0	2^{-18}	0	0	0	0
			Sum	3.126266479	Sum	-2.623046875

14.3 COMPLEX NUMBER REVIEW

In the time domain, complex numbers consist of real and imaginary components that indicate amplitude and phase. Real numbers consist of one component only and do not indicate phase. In the frequency domain, a complex signal is centered around zero with different positive and negative frequency parts. This is called a complex baseband signal. A real signal has identical positive and negative frequency parts. This is called a passband signal. These are general observations for bandlimited signals.

14.3.1 Euler's Formula

Mathematicians use complex numbers to have a consistent system that includes numbers whose square is negative. These are called imaginary numbers and are indicated by the letter j, as in $\sqrt{-4} = 2j$; sometimes you will see $\sqrt{-4} = 2i$[1]. Mathematicians define the complex plane to include those special numbers. A line is extended up from the origin for positive imaginary numbers and down for negative imaginary numbers. A complex number is a combination of real and imaginary; that is, $a + jb$ plotted at $+a$ on the horizontal real axis and $+b$ on the vertical imaginary axis.

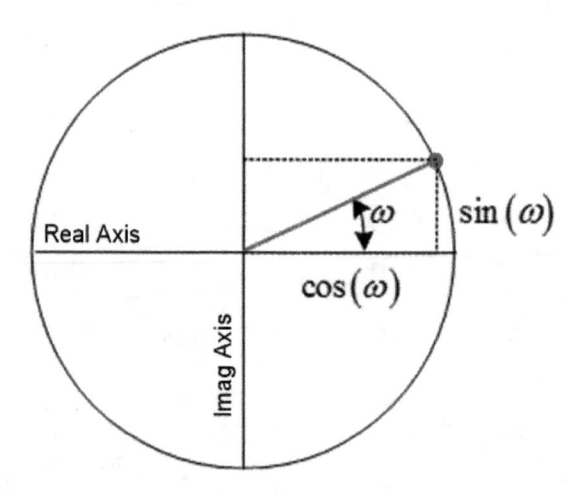

Figure 14.1 A unit circle illustration of Euler's formula.

In Euler's famous formula, $e^{j\omega} = cos(\omega) + jsin(\omega)$, the sine and cosine terms are not simply added, like on a number line, but are plotted on two orthogonal number lines, i.e., points in a complex plane. Figure 14.1 shows the complex plane with a circle of constant radius of $cos^2(\omega) + jsin^2(\omega) = 1$ for ω in the range $[0, 2\pi)$.

Let's change the exponent in Euler's formula to ω_0, a linear function of time (i.e., a ramp). Now we have $e^{j\omega_0 t} = cos(\omega_0 t) + jsin(\omega_0 t)$. This is a complex exponential that rotates CCW. We can also have a negative frequency complex

---

[1] Engineers tend to use j and mathematicians tend to use i. The letter j or i can also be considered a 90° rotation operator.

CHAPTER 14. MORE FUNDAMENTALS 585

exponential that rotates CW: $e^{-j\omega_0 t} = cos(\omega_0 t) - j sin(\omega_0 t)$. Figure 14.3 shows the real (x-axis) and imaginary (y-axis) components of the positive frequency complex exponential as well as the phase rotating with time (z-axis) on a unit circle. Note that the phase starts on the left at $0 + j1$. The phase goes around the unit circle 2.5 times, ending at $0 - j1$.

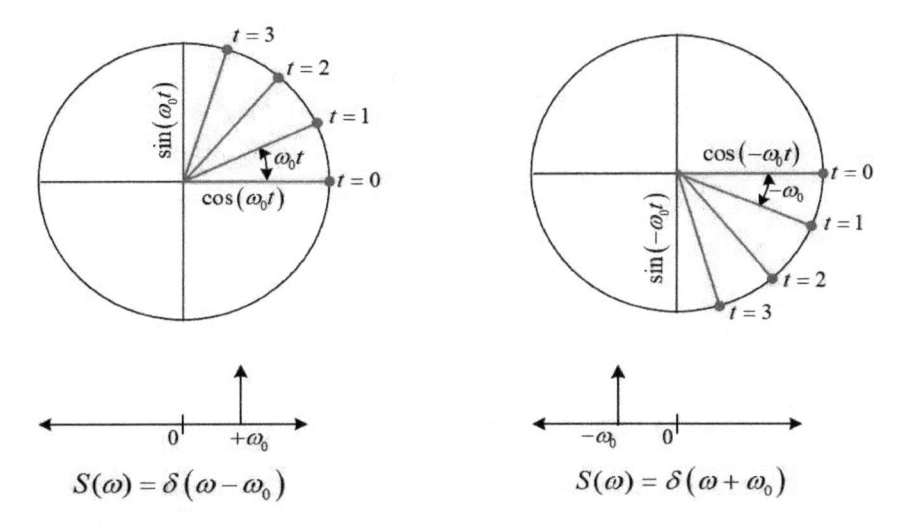

Figure 14.2 Positive and negative constant frequency exponentials on unit circles.

Considering the frequency domain, in Figure 14.2, the left unit circle exponential has a spectrum consisting of a single positive frequency impulse, $S(\omega) = \delta(\omega - \omega_0)$, and the right has a spectrum consisting of a single negative frequency impulse, $S(\omega) = \delta(\omega + \omega_0)$. The δ operator is an infinite impulse only at $\delta(0)$. For $S(\omega) = \delta(\omega - \omega_0)$, the infinite impulse is shifted from zero to $\omega = \omega_0$ because that is the frequency where there δ operator argument is 0. Likewise, for $S(\omega) = \delta(\omega + \omega_0)$, the infinite impulse is shifted from zero to $\omega = -\omega_0$. The impulse location is simply the frequency of the complex exponential.

14.3.2 Phase Modulation

Let's take the negative frequency complex exponential and phase modulate it with an arbitrarily changing $e^{j\theta(t)}$. The composite exponential is now:

$$e^{-j\omega_0 t} e^{j\theta(t)} = e^{-j(\omega_0 t - \theta(t))} \tag{14.1}$$

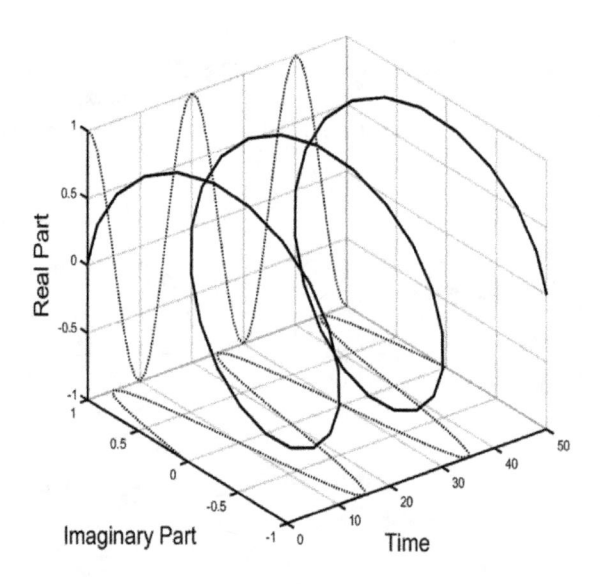

Figure 14.3 Phase progression of $e^{-j\omega_0 t} = cos(\omega_0 t) - j sin(\omega_0 t)$.

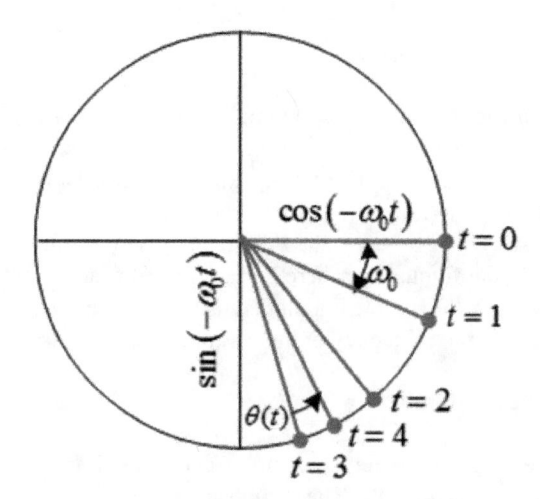

Figure 14.4 Unit circle illustration of complex exponential phase modulation.

CHAPTER 14. MORE FUNDAMENTALS

Figure 14.4 shows five time steps of a complex exponential undergoing phase modulation. Note that changes in $\theta(t)$ can change the direction of rotation. Recall from Figure 14.2 that changes in direction correspond to movements from positive to negative (or vice-versa) on the frequency-domain axis for a complex phase modulated signal..

The top trace of Figure 14.5 shows time and frequency-domain snippets of a real phase modulated signal centered around 0 Hz. Notice that the negative frequencies mirror the positive frequencies and provide no new information. The lower view shows a complex signal. Now the positive and negative signal frequencies do not have to be mirror images. Sometimes we refer to the complex signal as doubling the bandwidth of the real signal. This is because the real and imaginary components can carry independent modulation. The top trace of Figure 14.5 does not have this feature.

Transmitters often generate the complex baseband waveform in DSP and then shift it in frequency to an intermediate frequency such as ω_0 in Figure 14.6. Now the imaginary part is redundant and can be discarded. With no imaginary part, we have a real only signal that must have a mirror frequency image. This provides a real signal to send to the DAC that drives the RF power amplifier. Note from Figure 14.6 that the third trace (real) has the same information as the first trace (complex).

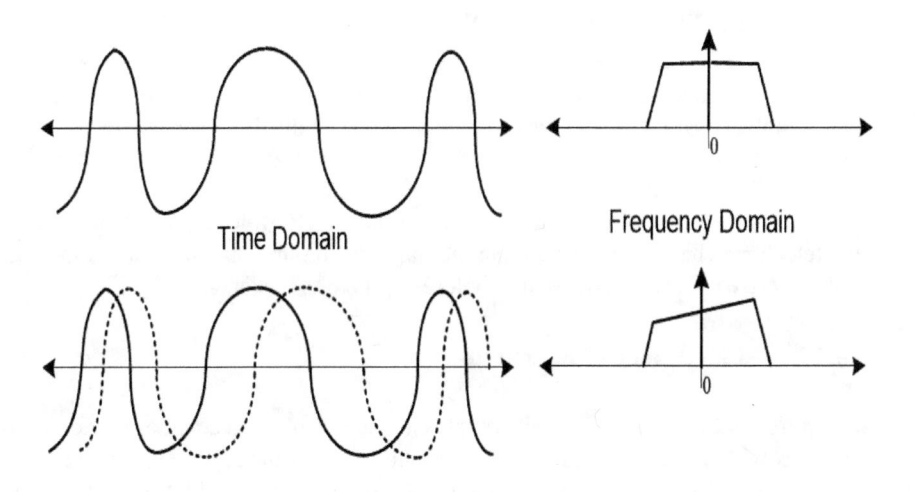

Figure 14.5 Time and frequency-domain difference between real and complex signals.

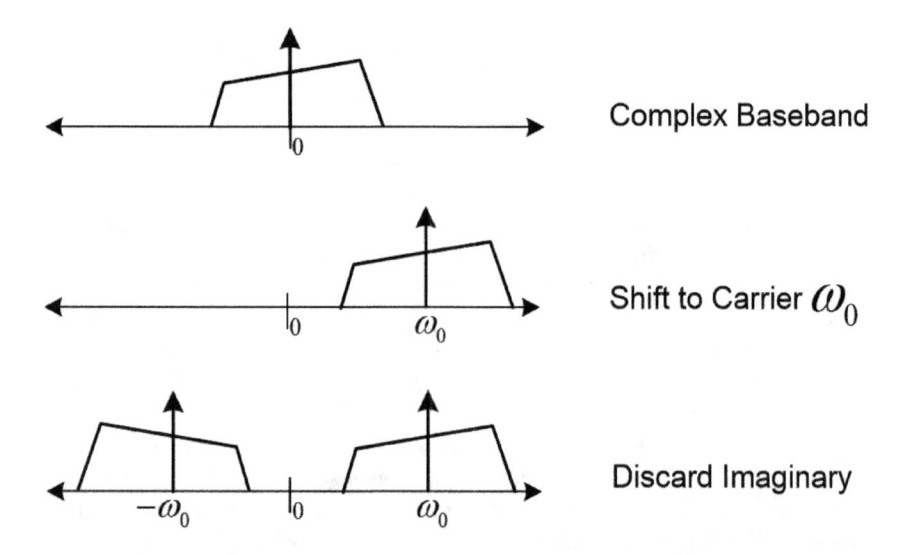

Figure 14.6 Complex baseband to real passband, frequency-domain.

14.4 AMPLITUDE COMPANDING

The term companding is a combination of the words compression and expansion. For an example, consider the ordinary voice telephone channel. The channel's dynamic range is the difference between the highest amplitude signal that can be transmitted without clipping and the channel noise floor. We can improve the channel dynamic range by compressing the amplitude range of the talker's voice signal at the input to the telephone channel and expanding the range (using the same nonlinear function, only inverted) at the channel output before it gets to the listener.

14.4.1 μ-Law Analog Companding

Equation (14.2) shows the overall compression and expansion equations, referred to as μ-law with $\mu = 255$. This is used in the United States and Japan.[2] There are many applications. For example, a microphone signal has a very wide dynamic range. Compression will make it easier to transmit wirelessly. Expansion at the receiver

---

2 Many other countries use A-Law companding, a similar technique.

CHAPTER 14. MORE FUNDAMENTALS

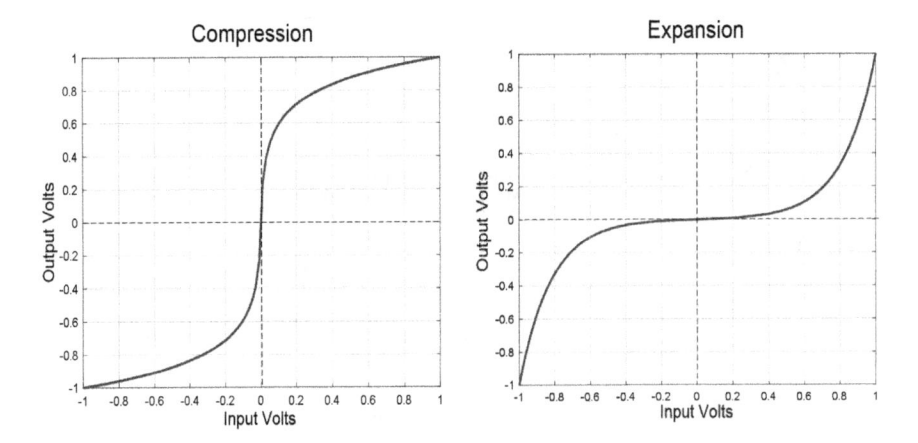

Figure 14.7 μ-law analog companding gain characteristic.

can restore the dynamic range. Companding thus allows transmission of a signal with a large dynamic range (e.g., voice) over a smaller dynamic range channel.

$$y(t) = \text{sgn}(x(t)) \left(\frac{Log_e \left(1 + \mu |x(t)|\right)}{Log_e \left(1 + \mu\right)} \right)$$

$$x(t) = \text{sgn}(y(t)) \left(\frac{(1+\mu)^{|y(t)|} - 1}{\mu} \right) \quad (14.2)$$

$$\text{sgn}(x) = \begin{cases} +1 & x > 0 \\ 0 & x = 0 \\ -1 & x < 0 \end{cases}$$

14.4.2 μ-Law Digital Companding

Nonlinear analog companding circuits can be very tricky to design. Getting consistent results over temperature could be especially difficult. In some cases, it may be easier to sample an entire input dynamic range using 12 bits and then compress digitally down to 8 bits. This will make the sampled signal compatible with AT&T T-carrier multiplexed transmission systems, which are designed around 8 bits per sample.

Figure 14.9 shows the positive half of the digital compression characteristic. The input segments the eight intervals on the horizontal axis. Inside each segment

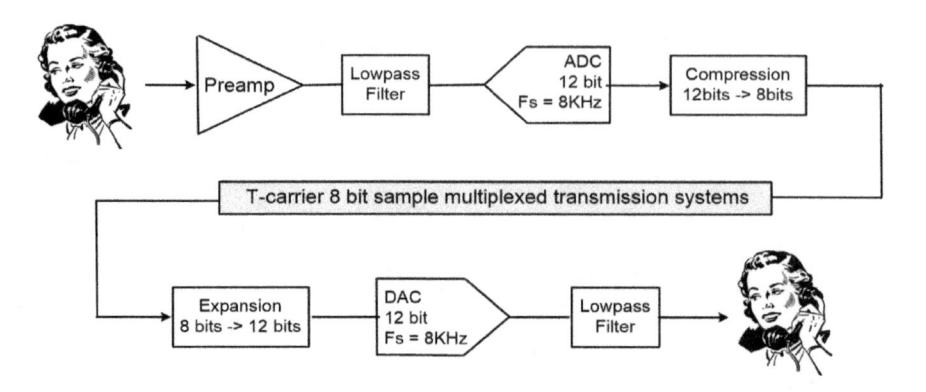

Figure 14.8 μ-law digital telephony application.

identifier are one of 16 quantization identifiers to complete the output amplitude specification. The eight-bit sample is constructed as shown in Equation (14.3).

$$[\text{Sign bit} \quad \text{3 bit Segment ID} \quad \text{4 bit Quantization ID}] \qquad (14.3)$$

14.5 POWER AMPLIFIERS AND PAPR

As we have seen, the back end of a transmitter and the front end of a receiver require analog signal processing. An analog signal has dynamic range, the ratio between the peak signal power and the lowest usable signal power. The analog processing components, primarily amplifiers, also have dynamic range. This section is concerned with both and how they interact.

14.5.1 Peak to Average Power Ratio

Imagine attending a lecture in a large auditorium that has a public address system. The electrical signal that is generated by the microphone goes to a power amplifier (PA) and then to loudspeakers. The human voice signal has an overall average power along with large peak power excursions; see Figure 14.10. To avoid distorting (i.e., clipping) the peaks, the power amplifier must have a wide dynamic range. For example, say, average PA input power is 0 dBm (1 milliwatt; see Chapter 2) and the largest peak power input is +10 dBm. Say the PA gain is +30 dB. The average PA

CHAPTER 14. MORE FUNDAMENTALS   591

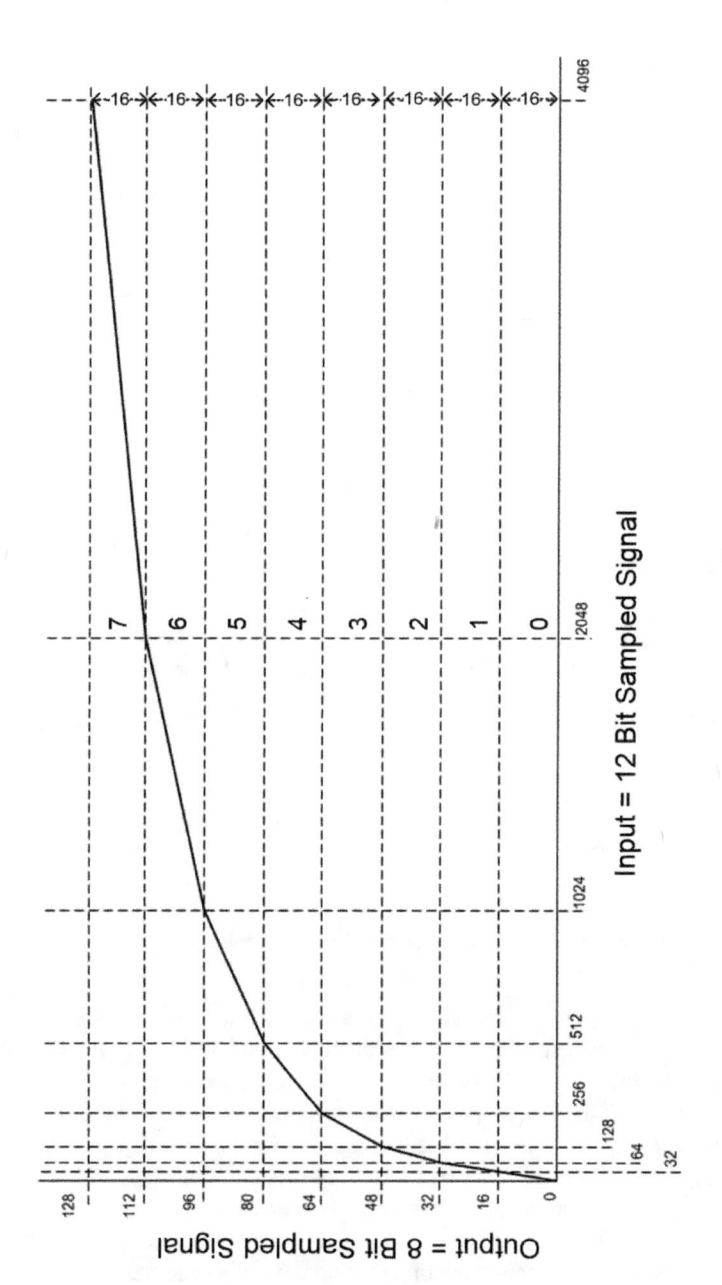

Figure 14.9 μ-law digital transfer characteristic

output power will be +30 dBm (1 watt) and peaks will occur up to +40 dBm (10 watts). This means that a 10 watt PA is needed to accurately reproduce the peak power events. If these are infrequent, most of the time the +10 dB headroom is not used and the PA efficiency is lowered.

Efficiency might not matter much for an audio frequency PA, but for an RF PA in a satellite, for example, it is an important factor. Some data signals, such as MSK, are constant envelope, meaning that they only have average power, no peaks. PAs for these signals can be very efficient since they require no headroom. A high PA efficiency in a satellite means less weight, less battery power and longer satellite lifespan.

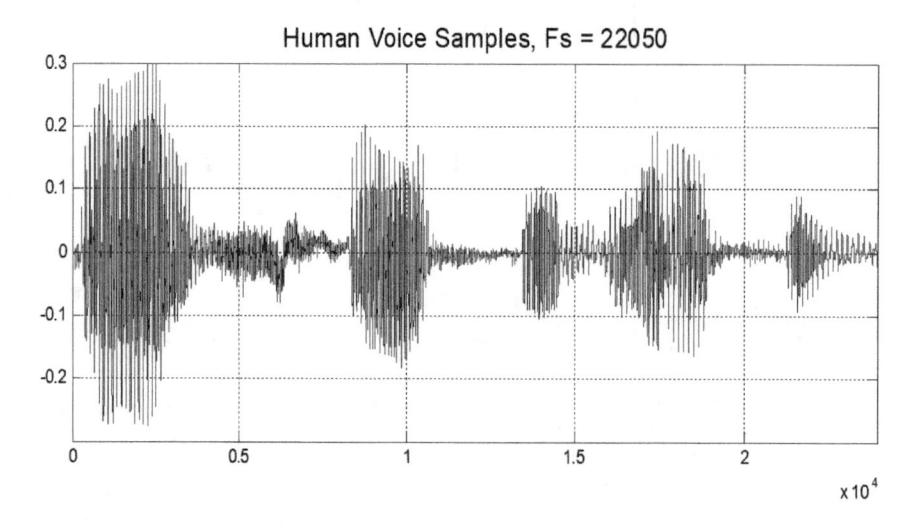

Figure 14.10 Human voice samples, note high random peaks.

Peak to average power ratio (PAPR) is simply the ratio, in dB, of maximum peak power to average power. A more detailed look at peak to average power is provided by the CCDF (complementary cumulative distribution function). In Figure 14.11, a CCDF example, average signal power is normalized to 0 dBm. The horizontal scale represents various peak magnitudes above average magnitude of 0 dB. The probability of each peak magnitude is shown on the vertical axis. For the Gaussian CCDF, for example, the probability of a peak power event of +6 dB above average is about 0.045. A Gaussian noise CCDF is often included on these curves as a point of reference. As described in Section 3.2.6, baseband QPSK is generally

CHAPTER 14. MORE FUNDAMENTALS

raised cosine (RC) filtered prior to frequency upconversion and transmission. The RC filter has an excess bandwidth parameter between 0 and 1. Here we choose as examples 0.25 and 0.5. Excess bandwidth = 0.25 results in a higher PAPR QPSK signal than excess bandwidth = 0.5. Although 0.5 has reduced the PAPR, there can be other considerations, such as more difficult symbol timing recovery in the receiver.

The Gaussian noise amplitude distribution, as shown in the top plot of Figure 14.12, is very peaky. In fact, the tails of the Gaussian distribution go to infinity. Thus, it is impossible to design a PA to have zero probability of clipping Gaussian noise. Note that this has to do with the Gaussian noise amplitude, not the frequency distribution of Gaussian noise. Gaussian noise frequency distribution is often uniform (called white).

For comparing PAPR as a single number, we must have a consistent probability. For example, in Figure 14.11, we decide on 10^{-4} and report that the QPSK 0.25 PAPR is 5.4 dB and the Gaussian PAPR is 11.8 dB. 10^{-4} is commonly used for this purpose.

Figure 14.12 shows the time samples used to calculate Figure 14.11. Notice that the Gaussian samples are indeed very peaky. Below is some of the MATLAB code used to generate both figures.

```
papr = 0:.25:12; % dB vector of horizontal axis PAPR
sampsPerSym = 8;
L = 100000;
hQAMMod = comm.GeneralQAMModulator;
% Setup a QPSK constellation
hQAMMod.Constellation = [1 1i -1 -1i];
alph = 0.3; % raised cosine excess bandwidth
h = fdesign.pulseshaping(sampsPerSym,...
'Raised Cosine','N,Beta',64,alph);
Hd = design(h);
rcosFlt = Hd.Numerator;
data = randi((0 3),L,1);
% Rotate QPSK from axis to corners of a square
yy = step(hQAMMod, data)*(exp(j*pi/4));
% Filter with Raised Cosine for better spectrum control
y0 = filter(rcosFlt, 1, upsample(yy, sampsPerSym));

%Calculate the CCDF
% P4=power of each sample of vector x
```

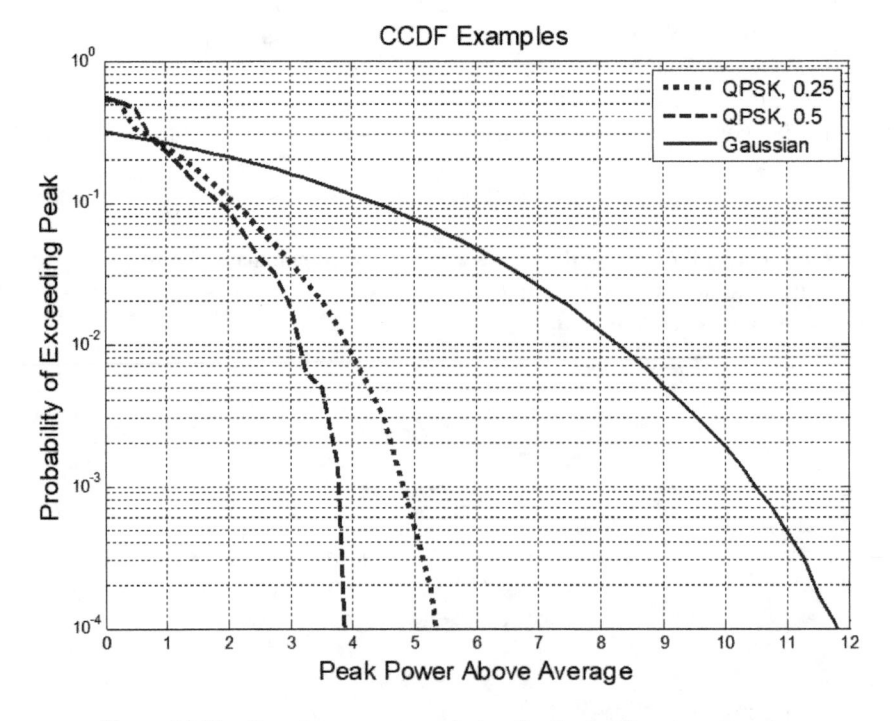

Figure 14.11 Complementary cumulative distribution function examples.

CHAPTER 14. MORE FUNDAMENTALS 595

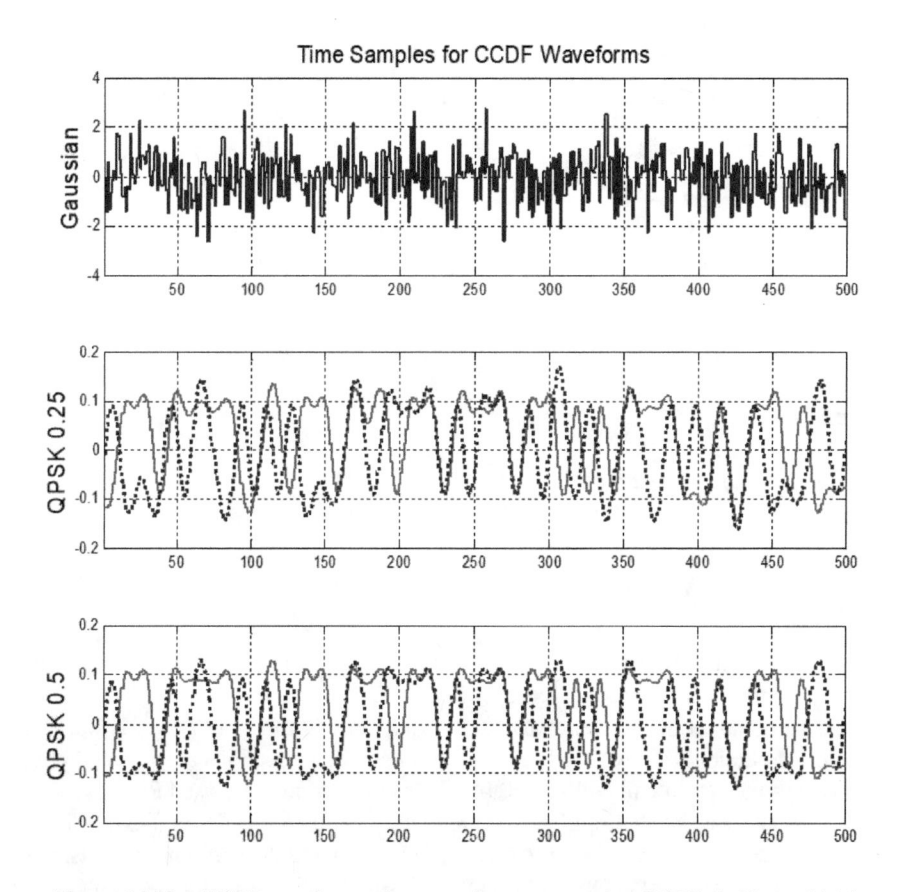

Figure 14.12 CCDF waveforms, time samples correspond to CCDFs in Figure 14.11.

```
P4 = real(y0.*conj(y0));
P4ratio = P4 /mean(P4); % power/average power
P4dB = 10*log10(P4ratio);
y4 = zeros(1,length(papr));
for k = 1:length(papr)
% For each papr, measure number
% of samples exceeding papr(i)
y4(k) = length(find(P4dB   papr(k)))/length(y0);
end

% PAPR of Gaussian Noise
randn('state',0); % reset random number generator
M = 100000; % number of samples
x_gauss = randn(M,1); % signal = Gaussian noise
y_gauss = zeros(1,length(papr));
% power/average power (average power = 1)
Pratio = x_gauss.^2;
PdB = 10*log10(Pratio);
for k = 1:length(papr)
y_gauss(k) = length(find(PdB > papr(k)))/M;
end
```

14.5.2 Power Amplifiers

In simple terms, power amplifiers (PA) efficiency is the power in watts delivered to the output load divided by the power in Watts required at the PA DC power input terminals. Low efficiency usually means more of the input power is wasted as heat. PA efficiency is extremely important in some applications such as cell phone handsets and satellite downlinks. Table 14.5 shows commonly used PA classes.

A one transistor, simplified PA circuit used for class A and C is shown in Figure 14.13. Class A is often used in low-cost products that require minimum parts, good fidelity and have plenty of available power. Class C is often used for FM transmitters. An LC output filter removes frequency spurs caused by switching. A push-pull PA circuit used for class AB is shown in Figure 14.14. Both transistors are biased very close to conduction (we can say 0 Vdc bias because they are close to conduction at 0.6 Vdc but not actually conducting). Class AB is often used for high-quality audio amplifiers and linear RF amplifiers. Finally, Figure 14.15 shows the input voltage (solid) and collector current (dashed) for each of the three classes.

CHAPTER 14. MORE FUNDAMENTALS

Table 14.5 Three Different PA Classes

PA Class	DC Bias	Typical Efficiency	Comment
A	Large	30%	Best linearity, worst efficiency
AB	Zero	40-50%	Still linear with good efficiency Used for AM
C	Negative	70%	Not linear, only used for FM Requires bandpass output filter Best efficiency

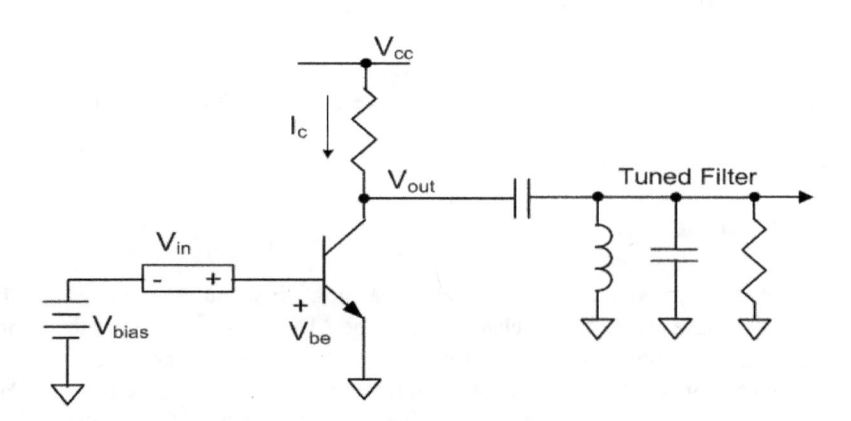

Figure 14.13 Simplified class A or C power amplifier circuit.

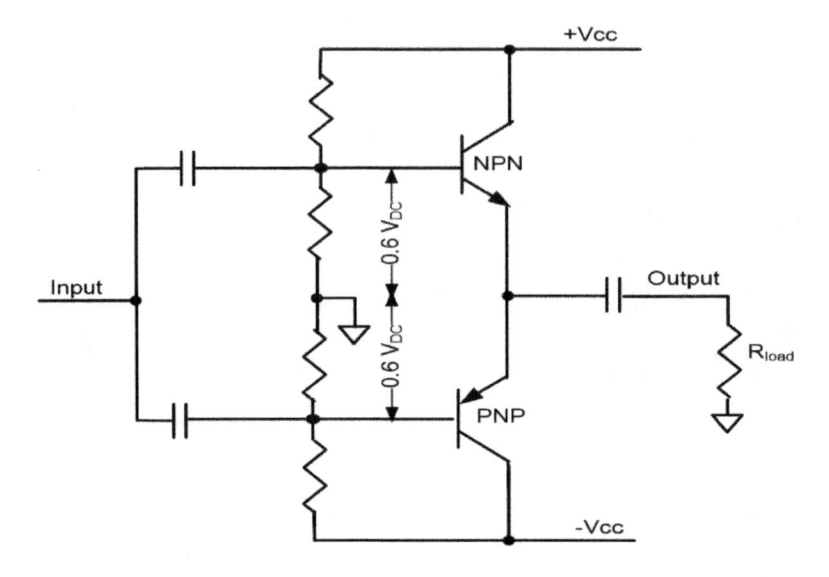

Figure 14.14 Simplified class AB power amplifier circuit.

14.6 SAMPLES PER SYMBOL QUESTION

The SDR modem back end; see Figure 12.13, operates on symbols at the symbol tracking output. These have already been time-aligned with correct symbol boundaries and are commonly represented by 2 or 4 samples per symbol. In this section we explore that choice.

14.6.1 $M = 2$ or 4?

Over the transmission channel, we have an analog RF signal. At the receiver ADC output, we have a sampled baseband signal. The ADC sampling rate and the symbol are usually not integer multiples of each other. So we assume noninteger samples per symbol at the ADC output. After downconversion to complex baseband and resampling, we need integer M samples per symbol. Finally, at the equalizer output and decision detect we generally have one sample per symbol.[3]

Here we examine the question of M = digital receiver samples per symbol. Components of the digital receiver back end (see Section 12.2) commonly operate

---

3 Interestingly, the digital receiver can be thought of as a smart downsampler.

CHAPTER 14. MORE FUNDAMENTALS

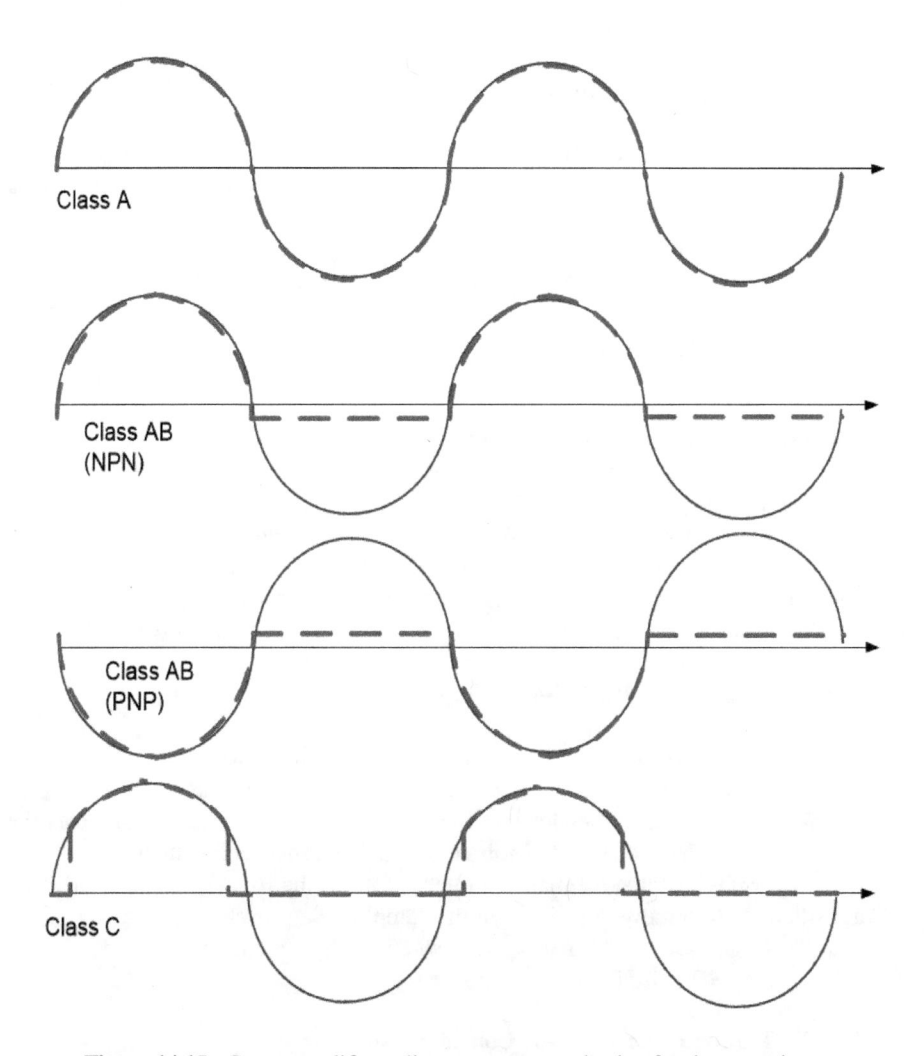

Figure 14.15 Power amplifier collector current conduction for sine wave input.

at the symbol rate. Each symbol is represented by $M = 2$ or 4 samples per symbol. In most cases, $M = 2$ requires less circuitry; however, that is not the only important consideration. Other factors affecting the choice of M are listed below.

14.6.1.1 Interpolator Accuracy

For our examples, the sample interpolator (also called resampler) in Figure 12.1 is the source of the M digital receiver samples per symbol. The sample stacker formats these samples into vectors for parallel processing in the matched filter, timing error detector, and equalizer input.

The heart of the resampler is the Farrow structure interpolator circuit; see Figure 12.5. This circuit is only suitable for a baseband signal (i.e., a signal centered at zero frequency). The interpolator adjusts symbol timing by affecting a variable sample delay. Ideally, the frequency response should be flat for any value of delay in the range $[-0.5\ 0.5]$ ($[0\ 1.0]$ is also used). As shown in Figure 14.16, the frequency response varies quite a bit with delay.

In Figure 14.16, the Nyquist frequency limit is on the right side of the horizontal scale. For $M = 2$, we have $F_{symbol} = F_s/4$, for $M = 4$, we have $F_{symbol} = F_s/8$. That is why $M = 4$ requires less processing bandwidth and achieves more interpolator accuracy. For $F_{symbol} = F_s/4$, the sampling rate is higher and thus the fixed bandwidth desired signal takes up less of the sampling bandwidth. Maximum variation is less than 1 dB for $M = 4$ but increases to 3 dB for $M = 2$.

14.6.2 Resampler Output Noise Floor

Figure 14.17 shows the desired signal (dotted curve) and noise floor (shaded boxes) at the resampler output for $M = 2$ and 4. Noise is shown uniformly distributed and is the same total power for each choice of M. Note that $M = 4$ doubles the bandwidth, which halves the total noise power in the signal bandwidth between 0 and $F_{Nyquist}$. Subsequently, the matched filter preserves the $[0, F_{symbol}/2]$ bandwidth and attempts to remove any signal in the sampling bandwidth between $F_{symbol}/2$ and the Nyquist bandwidth $F_{Nyquist}$. For $M = 4$, the remaining inband noise floor is half the $M = 2$ noise floor.

14.6.3 Timing Error Detector Considerations

Section 10.1.2 has a discussion of $M = 2$ or 4 for various choices of timing error detection and symbol time tracking. In particular, timing recovery based on timing tones at frequency $\pm F_{symbol}/2$ must operate at $M = 4$.

CHAPTER 14. MORE FUNDAMENTALS 601

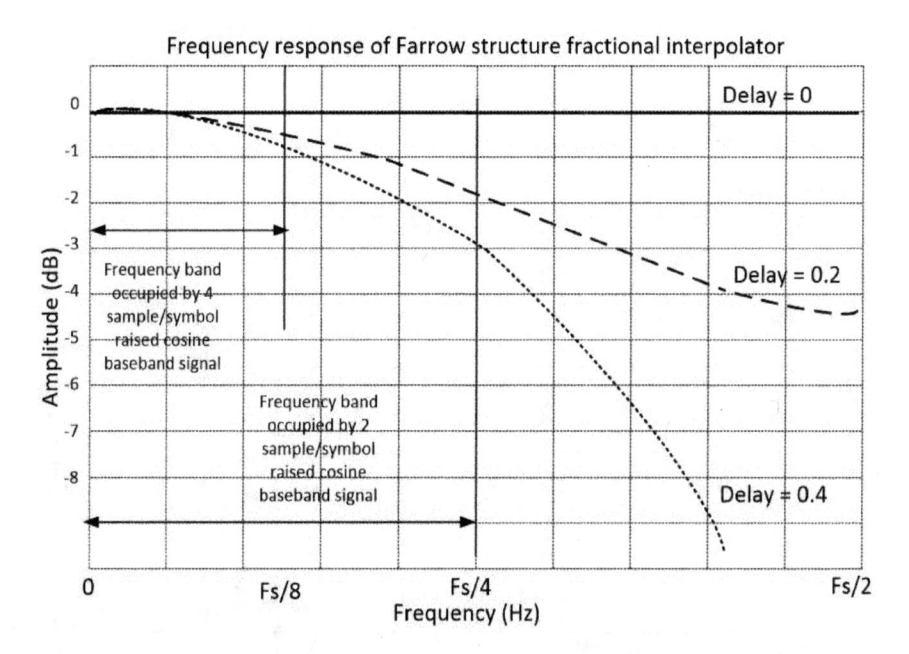

Figure 14.16 Fractional interpolator error versus frequency for various delays.

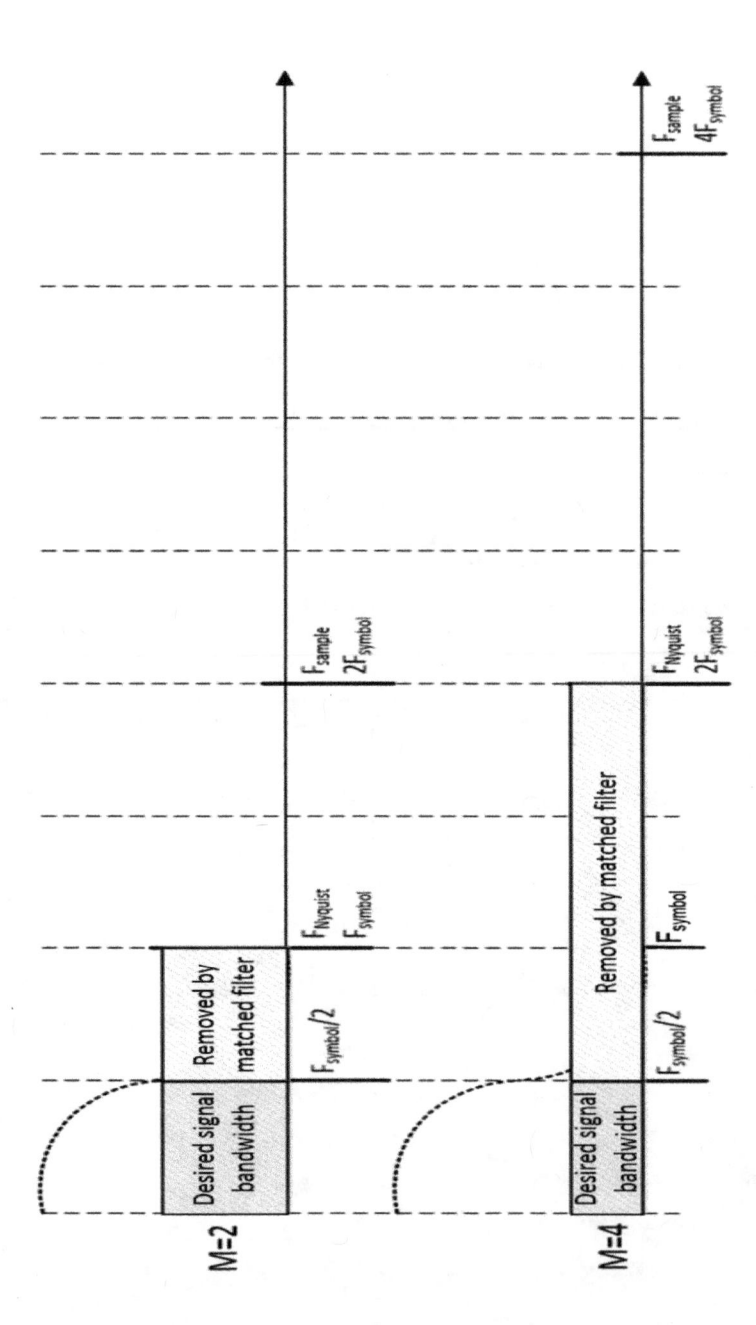

Figure 14.17 Resampler output spectrum showing noise distribution for $M = 2$ and 4.

CHAPTER 14. MORE FUNDAMENTALS

14.6.4 Equalizer Considerations

Matched filter output samples drive the equalizer input, 2 or 4 per symbol. Equalizers generally output one sample/symbol, i.e., the constellation point. Therefore, the equalizer output is subsampled, from either 2 or 4, to one sample per symbol. The final equalizer output sample rate is thus F_{symbol} and the final bandwidth is limited to $\pm F_{symbol}/2$.

Looking back to Figure 14.17, for either $M = 2$ or 4, all noise above frequency $\pm F_{symbol}/2$ is aliased back into the band $\pm F_{symbol}/2$ in the equalizer. However, the equalizer itself is an adaptable lowpass filter. Imagine that the equalizer removes noise above $F_{symbol}/2$ in Figure 14.17. Then the amount of noise, after downsampling to $F_{sample} = F_{symbol}$, will tend to be less for $M = 4$. See the DSP downsampler description in Section 9.2.8

14.7 DECISION DETECTORS

This section presents a theoretical explanation of two important decision detector approaches.

14.7.1 Maximum Likelihood Detector

Consider a transmit symbol set $S_0 = -1$ and $S_1 = 1$. When either of these are transmitted through a channel with additive white Gaussian noise (AWGN), their received probability distributions are shown in Figure 14.18. The average value (center) of each Gaussian distribution is set to the associated transmit symbol. Note that we are making the assumption that the channel adds noise without changing the signal amplitude; that is, S_1 and S_0 have the same amplitude at the transmitter and receiver.

Say for a particular symbol the value $r_{rec} = x$. We can compare the probability that S_0 was transmitted given x was received to the probability that S_1 was transmitted given x was received. The largest of these reveals the likely transmitted symbol. If the transmit symbols are equally likely, then this is the same as slicing the received signal around zero. In a practical system, the transmit symbols can be scrambled to make their probabilities equal. After detection, the receiver unscrambles them. This is known as a maximum likelihood detector (ML).

Equation (14.4) shows the calculation of a $log_{ML}()$ also called a log-likelihood function, $L_{ML}(r_{rec})$. By using the natural logarithm (ln), we eliminate the exponentials and end up with a simple function that compares the distances between r_{rec} and both S_0 and S_1. The resulting sign indicates the most probable hard decision.

$$L_{ML}(r_{rec}) = \ln\left(\frac{p(r_{rec}|S_1)}{p(r_{rec}|S_0)}\right) = \ln\left(\frac{\frac{1}{\sigma\sqrt{2\pi}}e^{\frac{-(r_{rec}-S_1)^2}{2\sigma^2}}}{\frac{1}{\sigma\sqrt{2\pi}}e^{\frac{-(r_{rec}-S_0)^2}{2\sigma^2}}}\right)$$

$$= \ln\left(e^{\frac{-(r_{rec}-S_1)^2}{2\sigma^2}}\right) - \ln\left(e^{\frac{-(r_{rec}-S_0)^2}{2\sigma^2}}\right) = \frac{-(r_{rec}-S_1)^2}{2\sigma^2} + \frac{(r_{rec}-S_0)^2}{2\sigma^2}$$

(14.4)

Practically speaking, if $L_{ML}(r_{rec})$ is greater than zero, then S_1 was most likely transmitted; if $L_{ML}(r_{rec})$ is less than zero, then S_0 was most likely transmitted.

14.7.2 Maximum A Posteriori Detector

Maximum a posteriori (MAP) detection is an optimum detection technique if the probability of transmission of each symbol is the same. If this is not true, we must take into account the probability difference. Start with Bayes rule:

$$p(S|r) = \frac{p(r|S)p(S)}{p(r)}$$

$$\frac{p(S_1|r)}{p(S_0|r)} = \frac{\frac{p(r|S_1)p(S_1)}{p(r)}}{\frac{p(r|S_0)p(S_0)}{p(r)}} = \frac{p(r|S_1)p(S_1)}{p(r|S_0)p(S_0)} \quad (14.5)$$

Taking the natural log of Equation (14.5) gives us the maximum a posteriori log-likelihood ratio, L_{MAP}. This is the likelihood ratio based on a posteriori receive signal observation plus the a priori symbol probability ratio. A posteriori information is observed; that is, after the event has occurred. A priori information is known before the event; for example, the probability distribution of symbol transmission. If a priori information $p(S_1) = p(S_0)$, then $L_{MAP} = L_{ML}$.

CHAPTER 14. MORE FUNDAMENTALS

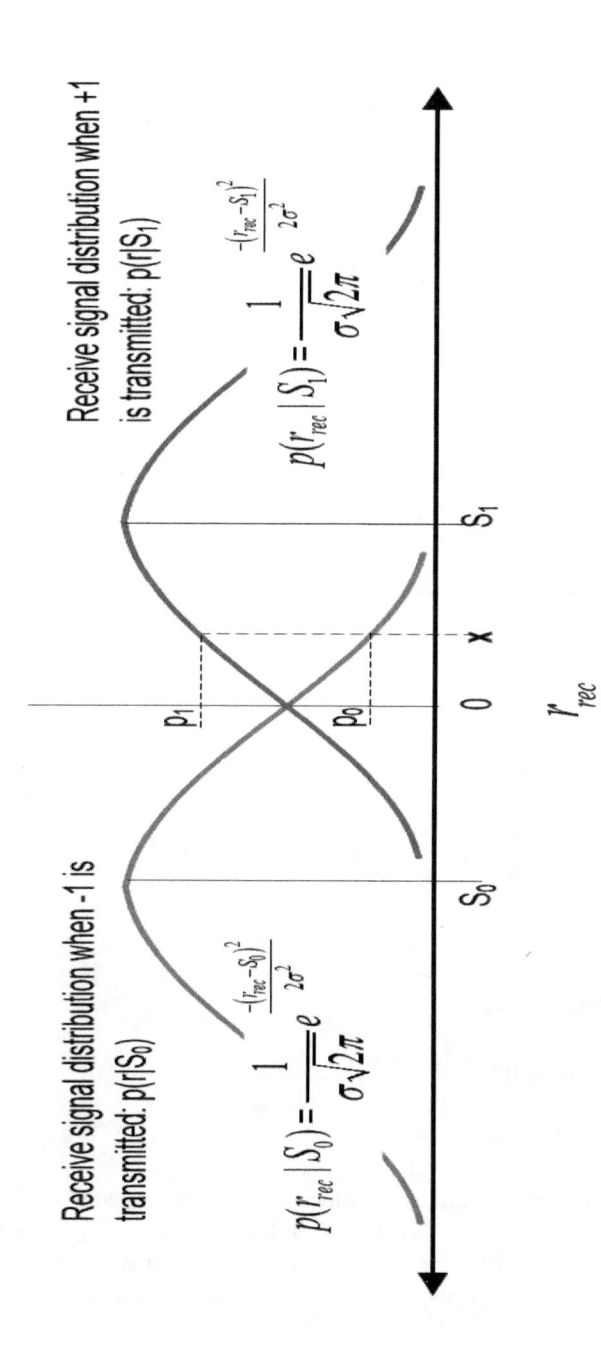

Figure 14.18 Likelihood functions for $S_0 = -1$ and $S_1 = 1$ transmit symbols.

$$L_{MAP} = \ln\left(\frac{p(S_1|r)}{p(S_0|r)}\right) = \ln\left(\frac{p(r|S_1)}{p(r|S_0)}\right) + \ln\left(\frac{p(S_1)}{p(S_0)}\right) \qquad (14.6)$$

Figure 14.19 shows how the logarithm of a ratio is plus for a numerator greater than the denominator and minus for vice versa. The idea is to show how the sign of the log-likelihood ratio serves as an antipodal hard decision.

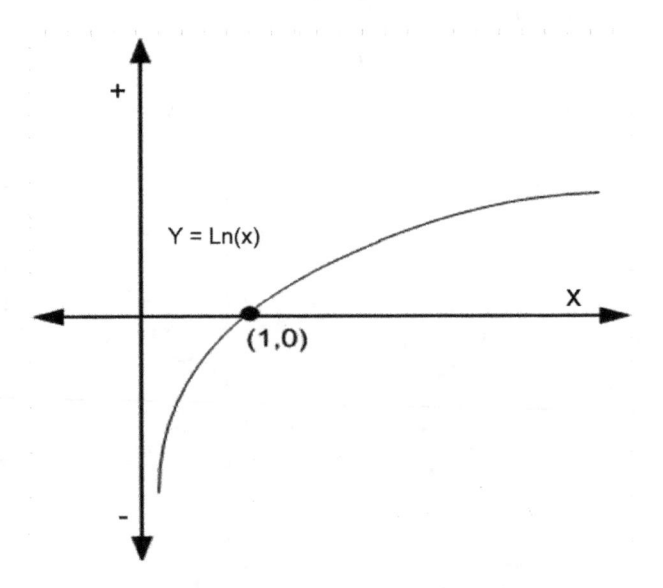

Figure 14.19 The log likelihood function.

14.8 PREAMBLE NOTES

The purpose of this section is to briefly study various preamble designs.

14.8.1 Existing Preambles

We start with three designs already implemented and described in detail earlier in this book. Two of these, ADSB Packet Preamble and IEEE802.11a Preamble, are based on specifications from the ICAO (International Civil Aviation Organization) and from the IEEE, respectively. The third, APSK Packet Preamble, is a homebrew

CHAPTER 14. MORE FUNDAMENTALS   607

preamble designed to facilitate BER testing. Descriptions have already been provided in the referenced chapters.

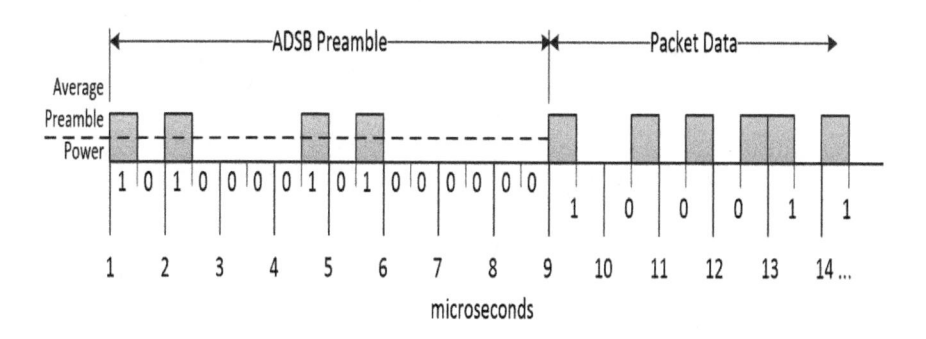

Figure 14.20 ADSB packet preamble, Chapter 11.

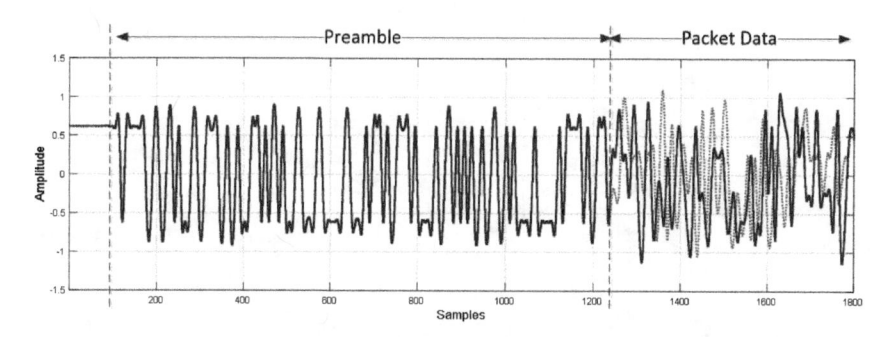

Figure 14.21 APSK packet preamble, Chapter 12.

Table 14.6 presents a list of generic preamble operational steps in time order and how they are carried out in the three preambles above. Many other details are contained in the appropriate chapters. A note about the APSK preamble: the random data pattern was not tested for its correlation properties. A more robust design would use a Barker code, or some sequence similar, with known good correlation properties.[4]

---

4   Ideal correlation properties mean the correlator outputs 1 for an exact lineup and 0 for any other alignment between received data and reference pattern. Known good correlation properties come close to that.

Table 14.6 Generalized Receiver Preamble Processing Steps and Examples

Step	Process	Description	ADSB	APSK	IEEE820.11a
0	AGC	Adjust receiver gain	AD9363 Fast AGC	Not implemented in simulation	Short training fields A,B,C
1	Initial signal detection	Determine preamble presense	Noncoherent OOK pattern detect	Timing predetermined	Differentially coherent STF
2	Coarse frequency, timing est.	Estimate	Pattern detect OOK resists Doppler	Carrier frequency, symbol timing run during preamble	STF D,...J
3	Fine frequency, timing est.	Estimate symbol boundary	n.a.	Same as 2	Long training field 0 (LTF0) finds frame boundaries
4	Start of message (SOM)	Line up customer data	Pattern detect has found this	Not needed for BER test	Payload after 3 OFDM frames

CHAPTER 14. MORE FUNDAMENTALS 609

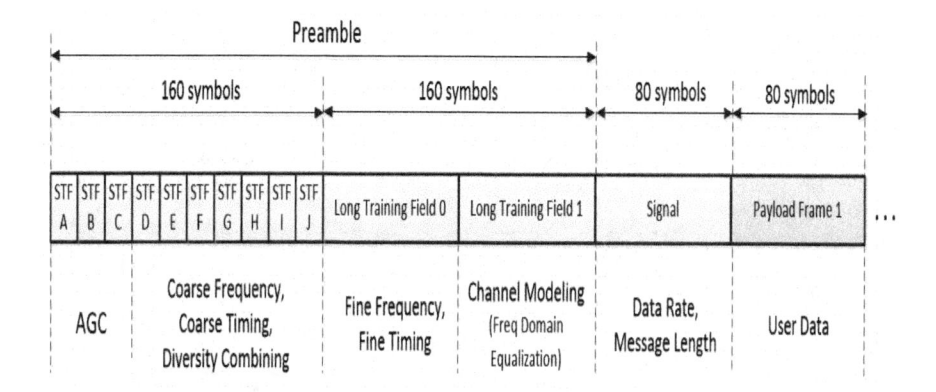

Figure 14.22 IEEE802.11a preamble, Chapter 13.

14.8.2 Frequency Domain Detection

A fourth design, not implemented in this book but perhaps useful to some readers, is frequency-domain detection.

Preamble initial AGC and packet present data pattern can be generated to form a certain FFT pattern at the receiver. For example, MSK modulated data pattern: [1 1 0 0 1 1 0 0 1 1 0 0 ...] will generate tones at carrier frequency offset + $[-F_{sym}/4, 0, F_{sym}/4]$, in the complex baseband FFT. This is shown in Figure 14.23.

After a series of FFT exponential accumulations,[5] the binary template is shifted bin by bin over a range of bins between the limits shown in Equation (14.7) below; N is the number of FFT bins. For each bin shift, the SNR is calculated as the average magnitude of the bins selected by ones in the binary template divided by the average magnitude of bins not selected by the binary template. This SNR will peak when the binary template matches bins with the desired three tones corresponding to $[-F_{sym}/4, 0, F_{sym}/4]$. The FFT frequency corresponding to the required shift will be a coarse estimate of the carrier frequency offset.

$$\left[\left\lfloor\frac{N}{2} - \left(\frac{NF_{DopplerMax}}{F_{samp}}\right)\right\rfloor \quad \left\lfloor\frac{N}{2} + \left(\frac{NF_{DopplerMax}}{F_{samp}}\right)\right\rfloor\right] \quad (14.7)$$

A better Doppler frequency estimate can be had using the sliding window structure of Figure 14.24. While the series of FFT magnitudes indicate preamble

---
5 The exponential accumulator output is: m(new data) + (m-1)(current output); see Section 9.2.3.

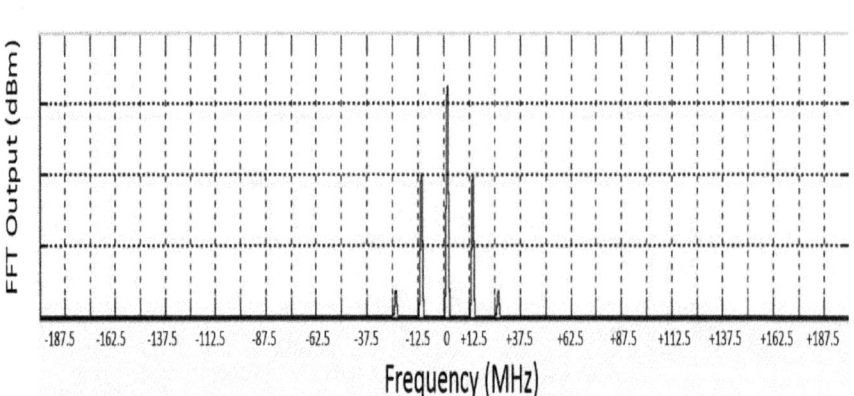

610   *Software Defined Radio: Theory and Practice*

Figure 14.23   FFT of received frequency domain packet preamble.

presence the phases of subsequent max bins, $b_{center}(k)$, (k is the FFT number) from the complex (nonaccumulated) FFT output can be stored. Phase changes between subsequent FFT complex max bin values can indicate Doppler frequency. The result is shown in Equation (14.8) below. A few of these estimates can be averaged for a more stable estimate.

$$\omega_{offset} = \left(\frac{b_{center}(k) - b_{center}(k-1)}{T_{samp} N_{new}} \right) \frac{rad}{\sec} \qquad (14.8)$$

CHAPTER 14. MORE FUNDAMENTALS

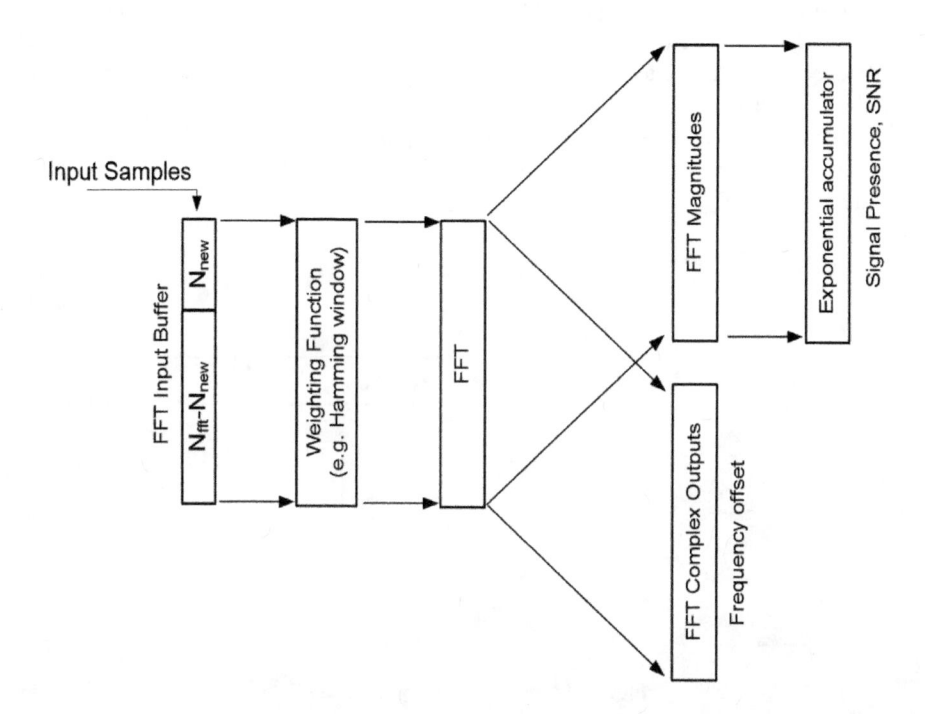

Figure 14.24 Frequency domain packet preamble.

14.9 DOPPLER DETAILS

This section collects a few aspects of the Doppler effect that were not covered earlier.

14.9.1 Doppler Time Dilation

Receive time dilation is an effect caused by nonzero transmitter-receiver velocity. Consider a single square pulse from a fixed transmitter to a receiver moving in a straight line away from the transmitter at v meters per second. Let's call d = distance between the transmitter and receiver at the leading edge of the transmit pulse and $d + vT_t$ = distance between the transmitter and receiver at the trailing edge of the transmit pulse. T_t is the time duration of the pulse.

The RF propagation times at the leading and trailing edges of transmit pulse are, respectively:

$$\frac{d}{c}, \quad \left(\frac{d + vT_t}{c}\right) \tag{14.9}$$

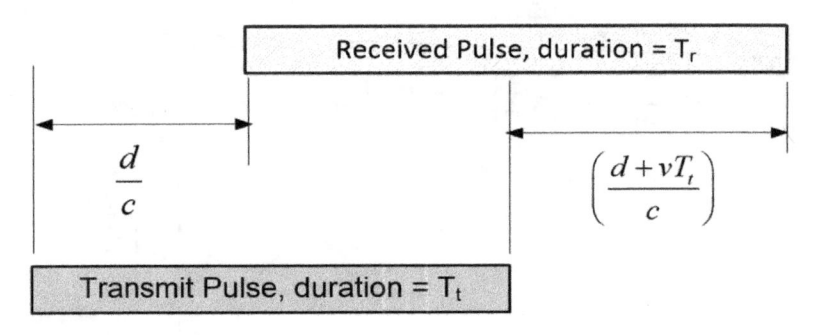

Figure 14.25 Basic Doppler time dilation.

Note that c is the speed of light in meters per second. As shown in Figure 14.25, the received pulse is stretched in time relative to the transmit pulse. The dilated time duration of pulse at the receiver is:

$$T_t + \frac{vT_t}{c} = T_t\left(1 + \frac{v}{c}\right) \tag{14.10}$$

This effect is cumulative in that a series of N pulses will take time $NT_t\left(1 + \frac{v}{c}\right)$ to receive. To cope with this effect, either the receive symbol time algorithm will

CHAPTER 14. MORE FUNDAMENTALS

need to be very agile or the transmit symbol timing will need to precompensate by shortening the transmit symbols. Prior knowledge of the received speed and trajectory may help also.

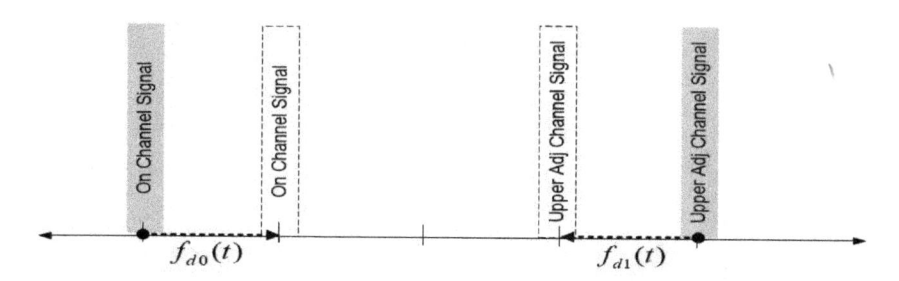

Figure 14.26 Doppler double adjacent channel frequency shift.

14.9.2 Adjacent Channel Interference

Under certain conditions, Doppler carrier frequency shift can cause a double adjacent channel interference. The example in Figure 14.27 shows an aircraft moving toward the on-channel transmitter and away from the upper adjacent channel transmitter. Because Doppler shift caused by moving towards the source is added and Doppler shift caused by moving away is subtracted, the on-channel carrier moves closer to the adjacent channel carrier by $f_{d1}(t) + f_{d0}(t)$. Figure 14.26 shows how the Doppler shift shown here can double the adjacent channel interference. The shifted carriers are in the dotted boxes. The Doppler shift on both carriers is shown exaggerated for illustration.

14.10 LOW COST SDR

The RTL-SDR, shown in Figure 14.29, is a very low-cost way to get started in SDR. Recall our ADSB project, shown in Figure 11.2. The first step was to install Simulink "Communications Toolbox Support Package for Analog Devices ADLAM-Pluto Radio." Likewise, to use the RTL-SDR, we install "Communications Toolbox Support Package for RTL-SDR Radio." Note that the RTL-SDR has no transmitter[6]

---

6 An RTL-SDR publication that deserves special mention is "The Hobbyist's Guide to the RTL-SDR: Really Cheap Software Defined Radio" by Carl Laufer. All kinds of useful information and application examples are presented.

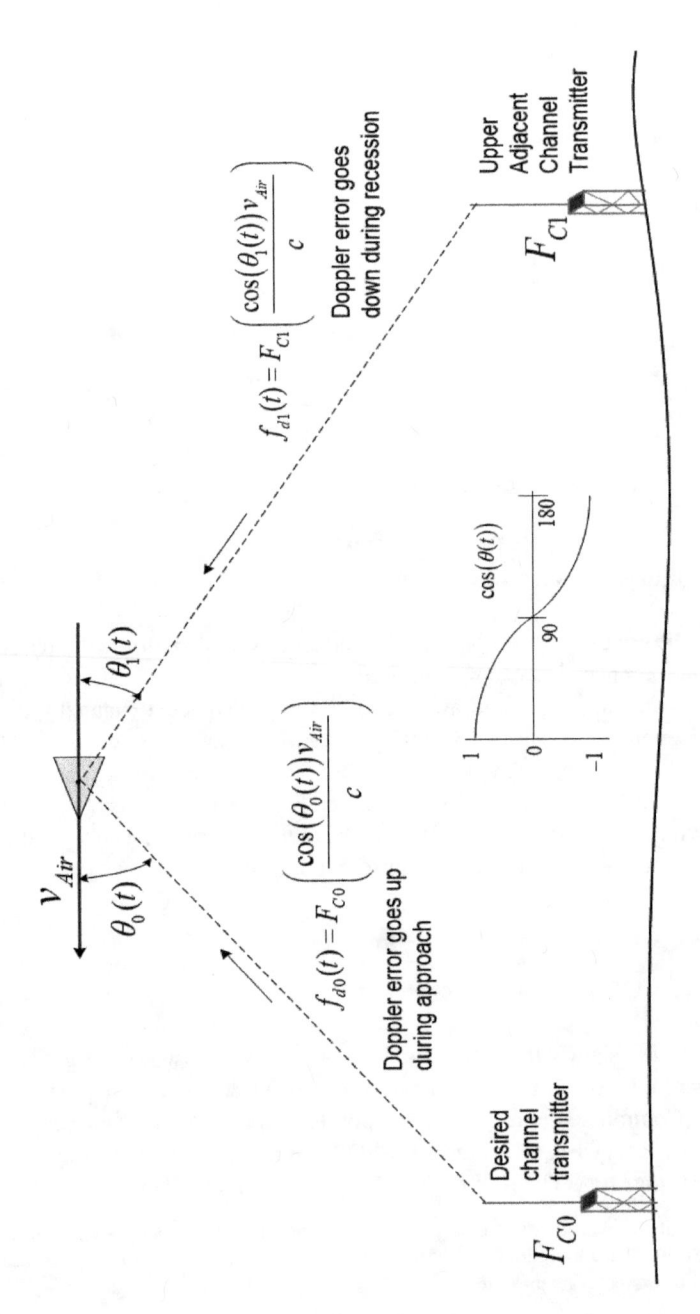

Figure 14.27 Doppler double adjacent channel interference example.

CHAPTER 14. MORE FUNDAMENTALS

The RTL-SDR is an HDTV receiver that was repurposed for SDR. The RTL-SDR can monitor VHF and UHF bands and can process up to 2 MHz receive bandwidth. There are many packaging choices and prices for the RTL-SDR.

As a receiver only, the RTL-SDR is commonly used to monitor distant stations. SDR Sharp is a software tool that works well with the RTL-SDR for analyzing and cataloging receive signals; see Figure 14.28. Note the signal demodulation choices in the upper left corner of Figure 14.28. Also note the OFDM signal blocks on either side of the main spectrum. These are for HD radio.

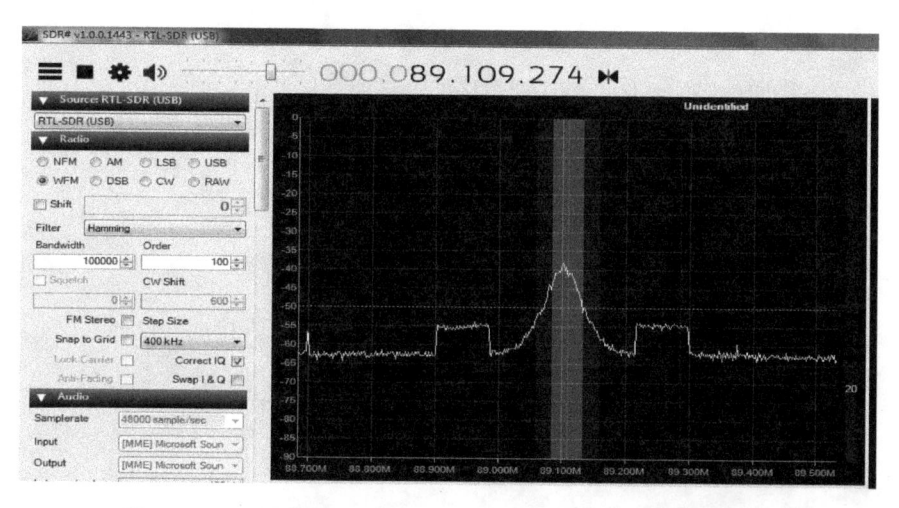

Figure 14.28 SDR Sharp receiving FM radio through an RTL-SDR.

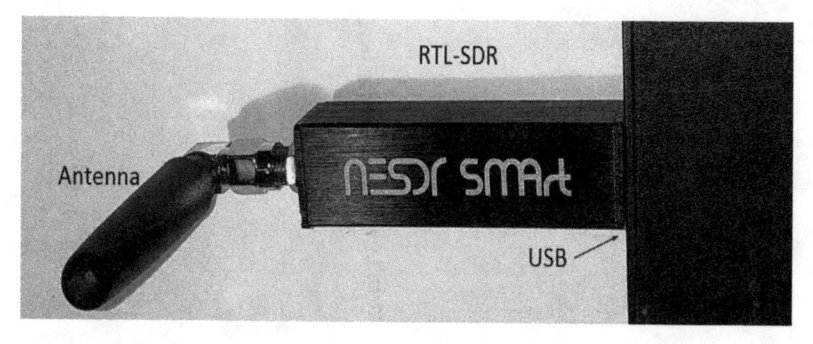

Figure 14.29 A low-cost RTL-SDR receiver plugged into a PC USB port.

The Figure 14.30 block diagram shows the R820T tuner chip that has a frequency range of 24 to 1,766 MHz. There are other tuner chips with different tuning ranges, for example, the E4000 tuner goes all the way to 2.3 GHz. Figure 14.30, matches many, but not all, of the RTL-SDRs sold.

CHAPTER 14. MORE FUNDAMENTALS 617

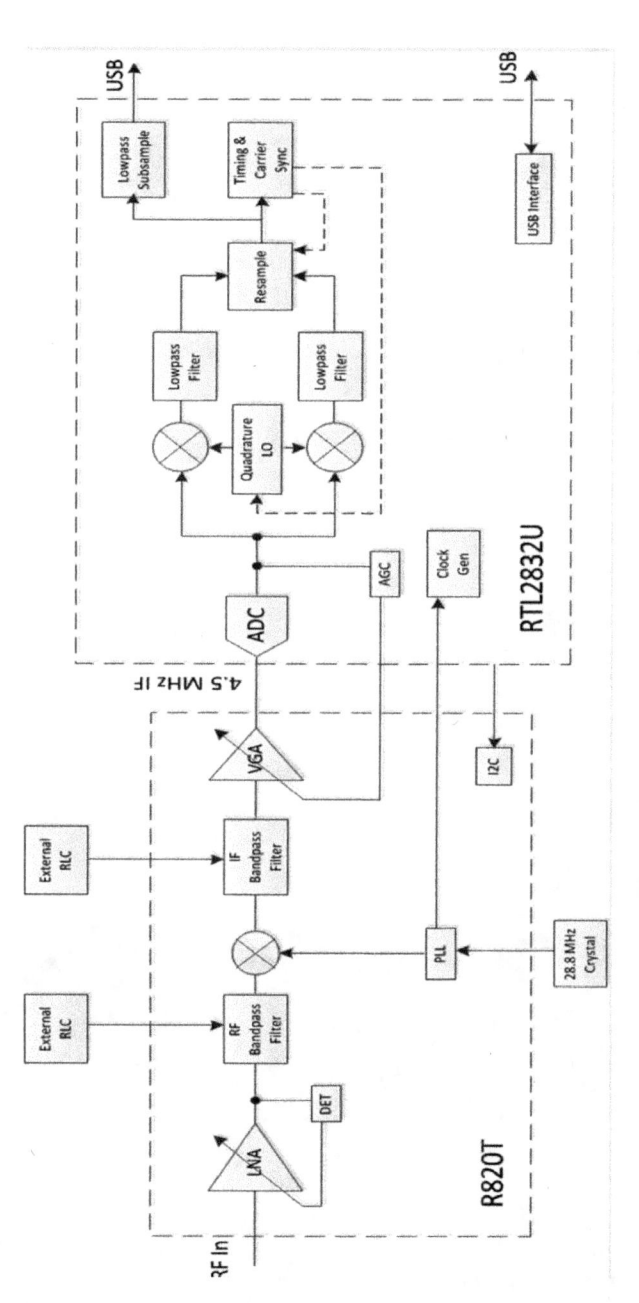

Figure 14.30 RTL-SDR typical block diagram.

About the Author

John M. Reyland has 25 years of experience in digital communications design for both commercial and military applications. Dr. Reyland holds a PhD in electrical engineering from the University of Iowa, an MSEE from George Mason University, and a BSEE from Texas A&M University. He has presented numerous seminars on digital communications in both academic and industrial settings. In addition to writing this book, he is heavily involved in FM broadcast station planning and design.

Index

$E_b N_0$, 542
$E_s N_0$, 209
1-bit ADC, 305
16APSK, 407, 470
16QAM, 15
4/3 sampling, 265
64QAM, 82
8PSK, 209

Absorption, 144
ACM, 38
Active filters, 259
ADALM-Pluto, 511, 575
Adaptive equalizer, 181
ADC dither, 301
ADC noise, 275
ADC noise floor, 274
ADC output spur, 302
ADC sampling, 295
Adjacent channel, 58, 524
ADLAM-Pluto, 513
Admiral Byrd, 141
ADSB acronym, 439
ADSB antenna, 445
ADSB packet, 447
AGC, 268
AGC control, 271
AGC threshold, 273
Aircraft position, 460
Aliased, 184
Aliasing, 266, 297

Altitude, 460
AM envelope, 282
AM modulation index, 47
Amplifier nonlinearity, 243
Amplifiers, 241
Amplitude, 53
Amplitude error, 282
Amplitude imbalance, 283
Amplitude modulation, 244
Analog filters, 396
Angular offset, 430
Antenna, 24
Antenna gain, 14
Antenna placement, 575
Antialias, 278
Antialias BPF, 266
Antipodal, 90
Antipodal mapping, 45
Aperture jitter, 304, 375
Arbitrary frequency response, 363
Arctangent, 385, 418
Average polarity, 408
Averaging length, 345
AWGN, 37

Back end, 553
Balanced signal, 131
Band-limited, 288, 375
Bandpass filter, 288
Bandpass frequency, 349
Bandwidth efficiency, 38, 44, 66, 108, 519
Bandwidth measures, 108
Bandwidth-limited, 204
Baseband remodulator, 492
Basis vectors, 53
Bathtub-shaped, 149

Bell, 91
Bessel, 87
BIBO, 339
BIBO stability, 341
BIBO stable, 345
Bit energy, 56
Block code, 205, 211
Block coder, 158
Block mode, 513
Boltzmann, 23
Boltzmann constant, 246
BPSK, 45
Branch metrics, 217
Broadcast FM, 87
BT, 108
Burst mode, 442
Butterworth filters, 325

Carrier frequency, 44, 554
Carrier power, 257
Carrier shift, 119
Carrier tracking, 342, 489
Carson's rule, 86
Cascade, 248
Cascade integrator comb, 343
CDMA, 155
Center frequency, 297
Center sample, 407
CFA inputs, 475
Channel coding, 204
Channel decoder, 217
Channel estimation, 561
Channel matrix, 166
Chebyshev 1, 363
Chebyshev 2, 363
Chipping sequence, 155
CIC filter, 346
CIC frequency response, 346
Circular polarization, 130
Circular polarized, 133
Clarke's model, 148
Class C, 106
Clipping, 399
Clipping level, 400
Closed-loop, 406
Code vector, 207

Coding gain, 224, 228
Coefficients, 191
Coherence bandwidth, 145, 175
Coherence time, 150
Complementary pair, 342
Complex analytic, 367
Complex baseband, 34, 112, 299, 305, 375, 383, 468, 520
Complex constellation, 419
Complex correlation, 98, 157
Complex exponential, 585
Complex LO, 63, 377
Complex numbers, 584
Complex phasor, 331
Complex plane, 584
Complex signal, 587
Complex sine, 332
Composite IP3, 243
Composite noise, 247
Concatenated coding, 226
Condition number, 167
Confidence levels, 217
Conjugate phase, 561
Conjugate poles, 349
Conjugate reference, 559
Constant envelope, 53, 61, 291
Constellation, 558
Constellation agnostic, 406
Constellation point, 54, 498
Constellation spin, 486
Continuous equalization, 569
Continuous mode, 442
Continuous phase, 94
Continuous time, 295
Convergence, 192, 391
Convolution, 73, 296, 353
Convolutional coding, 215, 216, 228
Convolutional decoder, 219, 226
Cophased, 157
Cost function, 190
Costas loop, 431
Covariance, 155
CPM, 93
Cross-correlation, 191
Cross-modulation, 243
Cross-product, 130

Index 623

Crystal oscillators, 257
CurrentFraction, 475
Cutoff frequency, 325, 354
Cyclic prefix, 75, 520, 530, 559

DAC reconstruction, 325
Damping factor, 423
dBd, 14
dBi, 13
DBPSK, 63
dBu, 13
DC offset, 281, 282
Decision boundary, 54
Decision directed, 184
Decision feedback, 185
Decorrelated, 169
Deinterleaver, 226
Delay spread, 145, 146
Denominator parameters, 349
Differential distortions, 300
Differential nonlinearity, 300
Differentially continuous, 110
Digital to analog, 295
Digital video, 467
Dirac delta, 295
Dispersion directed, 184
Doppler, 1
Doppler shift, 147
Doppler spectrum, 149
Downsample, 184, 358
DSP multipliers, 398
DVB-S2, 39, 467
Dynamic range, 275
Dynamic response, 489

EbN0, 209
EIRP, 14
Electric field, 129
Encoder state, 215
Encoding trellis, 215
Energy, 15, 53
Energy efficiency, 291
Entropy, 198
Equalizer, 190, 192
Equalizer correction, 569
Equalizer taps, 179

Error correcting, 208
Error detecting, 208
Error performance, 275
Error variance, 397
Euclidean distance, 218
Euclidian, 218
Euler's formula, 584
Evading Shannon, 165
EVM, 16
Excess bandwidth, 58
Excess bandwidth parameter, 184
Exponential Averaging, 342
Exponential factors, 70
Exponential filter, 569

F bit, 462
Farrow structure, 600
Fast Fourier transform, 520
Fast hopping, 163
FDE continuous, 561
FDE initial, 558, 559
Feedback error, 184
FFT averaging, 302
FFT precoding, 84
Field strength, 13
Fifth order, 243
FilterDesigner, 363
Filtered spectrum, 266
Fine timing, 472, 542, 553, 559
Fixed environment, 144
Fixed point, 350
Flash ADC, 317, 318
Fourier transform, 332
Fraction precision, 398
Fractional bits, 398
Fractional distance, 565
Fractional sample, 475
Fractional timing error, 553
Frame boundary, 552
Frame timing, 546
Frequency discriminator, 93
Frequency diversity, 154, 163
Frequency domain, 296, 558, 585
Frequency drift, 257
Frequency error, 257, 489
Frequency folding, 297

Frequency limits, 257
Frequency modulation, 86
Frequency offset, 415, 427, 486, 528, 529, 542
Frequency plan, 278
Frequency planning, 261
Frequency response, 340, 349
Frequency response cycle, 350
Frequency selective, 145, 146
Fresnel zone, 142
Friis equation, 136
Front end, 472
FSK, 89
Full response, 108

G over T, 25
Gain reduction, 268
Gardner, 406
Gardner timing, 407
Gate function, 354
Gaussian amplitude, 400
Gaussian distribution, 54
Gaussian filter, 107, 108
Generator matrix, 205
Geometrically, 318
Gibbs ringing, 354
GMSK, 107
GNU Radio Companion, 7
Godard timing, 415
Gradient, 191
Gray coding, 209, 506
Ground plane, 131
Group delay, 278, 325
GSM cellular, 108

Half symbol, 408
Half-wavelength, 130, 446
Halfband characteristic, 354
Halfband filter, 350
Halfband lowpass, 350
Hamming, 218
Hamming block, 159
Hang threshold, 273
Hard decision, 218, 431, 500
Hedy Lamarr, 163
Helix, 131
High Q, 415

Highpass, 281
Hilbert transform, 363, 367
Horizontally polarized, 130

Ideal ADC, 300
IDFT, 69
IEEE802.11a receive, 528
IF passband, 255
IF sampling, 375
Image frequency, 250, 278
Image rejection, 251
Imaginary numbers, 584
Imaginary part, 418, 489
Imbalance correction, 286
Impedance matching, 276
Impulse response, 145, 179, 339, 354
Impulse train, 296
Infinite sum, 339
Inherently stable, 354
Input spectrum, 418
Insertion loss, 258
Integral feedback, 483
Integral nonlinearity, 300
Integrator ramp, 309
Intercept point, 243
Interleaving matrix, 158
Intermediate frequencies, 249
Intermediate frequency, 305, 468
Intermodulation, 324
Intermodulation distortion, 242, 268
Interpolated parabolically, 563
Interpolating DAC, 325
Interpolation, 325, 414
Intersample, 472, 476
Intersymbol interference, 107, 146
Inverse amplitude, 561
Ionosphere, 140
Ionospheric skip, 155
IQ imbalance, 282
ISI, 175
ISM, 519
Isotropic, 13

Jammer, 120
Joules, 15, 53

Index

Kaiser window, 354
Kelvin, 22
Knife-edge, 142

Latent images, 350
Latitude, 460
Least significant, 500
Lempel-Ziv, 200
Line-of-sight, 136
Linear amplifier, 242
Linear regression, 179
LMS solution, 187
LNA, 24
LNA gain, 282
LO frequency, 306
Local oscillator, 51, 256, 276, 305
Local oscillators, 419
Logarithm, 11
Long training, 532
Longitude, 460
Longitude zones, 463
Longitudinal, 130
Lookup tables, 377
Loop filter, 483
Loop parameters, 490
Lower sideband, 418
Lowpass filter, 259, 340
LTE, 78

Magnetic field, 129
Matched filter, 217, 407, 408, 468, 477
Matrix inverse, 286
Maximum ratio, 157
Message bits, 158
MIMO, 166
MIMO training, 169
Minimum distance, 208
Missing output codes, 317
Mixer linearity, 278
Mixers, 249
Modified Gardner, 406, 408
Modulation bandwidth, 86
Modulation frequency, 86
Modulation index, 86, 87
Multilevel symbols, 101
Multipath, 75, 558

Multipath mitigation, 154
Multiply-accumulate, 398

Narrowband FM, 94
NBFM, 86
NCO transfer, 489
Negative frequency, 585
NewFrame, 530
NewSym, 529
NextSample, 476
Noise, 15
Noise bandwidth, 424
Noise enhancement, 184, 185
Noise factor, 24, 245, 247
Noise figure, 245
Noise floor, 279, 312, 600
Noise power, 247, 274
Noise reduction, 542
Noise shaping, 311
Noise spectrum, 256
Noncausal, 353
Nonlinear characteristic, 243
Normally distributed, 275
Numerically controlled, 423
Nyquist bandwidth, 33
Nyquist criteria, 32
Nyquist zone, 266, 297, 334, 350

Objective function, 286
OFDM, 68
OFDM equalizer, 76
OFDM frames, 528
OFDM generator, 530
OFDM latch, 528, 553
On channel, 524
Op-amps, 259
Optimal estimator, 187
Optimum solution, 187
Orthogonal, 584
Orthogonal carrier, 73
Orthogonal multiplex, 520
Orthogonality principle, 187
Output intercept, 243
Overdetermined system, 179, 189
Overflow flag, 442

PA nonlinearity, 291
Packet preamble, 468
Parabolic, 553
Parallel processing, 600
Parameter beta, 354
Parity check, 207
Partial response, 110
Passive circuit, 246
Path loss, 138
Path metrics, 221
Pattern detector, 447
Pattern noise, 415
Peak to average, 400
Peak to average power, 44
Percent BW, 44
Phase advance, 46, 334
Phase cancelation, 139
Phase coherence, 283
Phase correction, 554
Phase deviation, 86
Phase discontinuity, 533
Phase error, 282, 435
Phase locked loop, 256
Phase memory, 94, 110
Phase modulation, 86
Phase noise, 257
Phase offset, 558
Phase transitions, 93
Phone lines, 181
Pilot average, 571
Pilot constellation, 571
Pilot responses, 569
Pilot signal, 563
Pipeline delays, 318
Plastic pipe, 446
Polarity indicator, 306
Pole location, 341
Pole radius, 349
Poles, 336
Polynomial notation, 209
Postcursor, 175
Power amplifiers, 243
Power efficiency, 44, 291
Power spectrum, 191, 542
Power-conserving, 304
Power-limited, 204

Poynting vector, 130
PPM, 431
Preamble, 470, 524
Preamble chips, 447
Preamble detection, 116, 447
Preamble frames, 530
Preamble processing, 545
Preamble samples, 451
Preamble structure, 523
PreambleDetectOffset, 552
Precoding, 110
Precoding type 1, 111
Precoding type 2, 112
Precursor, 175
Predetection filter, 116, 479
PreliminaryDetect, 552
Preselector, 250, 265, 278
Preselector filter, 257, 263
PresetInt load, 475
Priority encoder, 317
Processing gain, 299
Propagation, 138
Propagation characteristics, 561
Propagation path, 147
Proportional feedback, 483
Pseudonoise, 119
Pulse forming, 48

Q function, 54
QAM, 66
QPSK, 48
QPSK Costas, 435
Quality factor, 257, 349
Quantizing error, 397, 398
Quantizing noise, 299, 302, 397
Quantizing step, 397
Quarter-wave, 131

Radiation environments, 101
Radio horizon, 141
Radio ID, 441
Raised cosine, 33, 58, 177, 414
Rake receiver, 157
Random event, 198
Random variable, 198
Rank-deficient, 166

Index

Rayleigh, 151
Rayleigh fading, 148
Receive bandwidth, 513
Receive modem, 528
Receive object, 513, 575
Receive parameters, 513, 575
Received syndrome, 207
Receiver configuration, 275
Receiver noise, 274
Receiver sensitivity, 274
Reciprocal, 14
Reciprocal mixing, 257
Reconfiguration, 4
Redundancy, 200
Reed-Solomon, 226
Reflected power, 17
Reflection coefficient, 17
Reflectors, 147
Refraction, 141
Regression coefficients, 179
Regression matrix, 179
Remodulated, 497
Repeat mode, 510, 572
Replicate, 356
Resampler, 407
Residue, 318
Resonant frequency, 17
Resonator peak, 349
Resonator poles, 349
Resonator transfer function, 349
Response time, 424
RF input attenuator, 271
RF link budgets, 27
Rician, 151
RMS, 17
RMS delay, 145
Robust preamble, 431
RTL-SDR, 613

S-curve, 407
Sample and hold, 309, 375
Sample interpolator, 472, 600
Sample repeating, 358
Sample space, 198
Sample stacker, 472, 479, 600
Sample vector, 443, 476

Sample zero, 418
Sampling intervals, 304
Sampling jitter, 304
Sampling rate, 334
Sampling spectrum, 297
Satellite, 25, 248
Satellite communications, 467
SC-FDMA, 81, 82
SDR receiver, 4
SDR transmitter, 4
Second order, 483
Self-information, 198
Sensitivity pattern, 131
Serial to parallel, 520
SFDR, 302
Shannon channel, 37
Shannon-Hartley, 34
Shape factor, 258
Sidebands, 33, 86
Signal alignment, 558
Signal envelope, 242
Signal-to-noise, 275, 299
Simulink, 8, 356
Sinc function, 354
Skip distance, 140
Slicer, 185
Slope distortion, 300
Slow hopping, 163
SNR degradation, 304
Soft decision, 217, 419, 435, 453, 558
Software development, 5
Solar cells, 304
Source coding, 200
Source data, 201
Space-time, 164
Spatial diversity, 155, 164
Speakeasy, 2
Spectral broadening, 148
Spectral mask, 524
Spectral null, 184, 185
Spectral regrowth, 243
Spectral shape, 523
Specular reflection, 138
Spread spectrum, 120
Spur, 13
Spur energy, 302

Spur frequencies, 255
Spurious spectral, 302
Spurs, 255
Squitter, 439
Standard deviation, 54, 400
Standard filter, 358
State diagram, 215, 226
Step size, 497
Stopband, 356
Stored integer, 571
Strictly monotonic, 317
Subsample by 2, 383
Subsample output, 542
Superposition, 215
Surviving path, 220
Symbol boundaries, 403, 468
Symbol decision, 501
Symbol energy, 56
Symbol equalization, 553
Symbol phase, 403
Symbol separator, 561
Symbol timing, 476
Symbol tracking, 403
Symbol values, 554
SymbolLatch, 476
SymbolReady, 486
Symmetric constellations, 470
Sync pattern, 119
Sync pulses, 33
Syndrome, 207
Systematic, 215
Systematic code, 205

Tap coefficients, 190
Temperature, 24
Thermal noise, 246
Third order, 243
Three-dimensional, 187
Time delay, 363
Time delay variation, 363
Time derivative, 86
Time diversity, 157
Time domain, 82, 522, 558
Time epochs, 304
Time-domain, 406
Time-invariant, 145

Timing error, 404
Timing tone, 415, 418
Tone frequency, 92
Tone modulation, 71
Toroidal, 131
Trace back, 221
Tracking loop, 489
Transition band, 363
Transition tracking, 406, 408
Transmission line, 246
Transversal equalizers, 185
Transversal filter, 181
Transverse wave, 129
Trellis coding, 38, 228
Trellis diagram, 215, 219
Tuning frequency, 265
Tuning range, 250
Two's complement, 399
Two-port, 247
Two-tone test, 324

Unbalanced, 21
Uncorrelated, 63
Undersampling, 304
Undesired image, 263
Ungerboeck, 228
Uniformly distributed, 398
Unit circle, 297, 332, 334, 342
Unit delay, 331, 332
Unit energy, 53
Upsample block, 356
Upsampling, 356

Variable compression, 244
Variance, 275
Velocity factor, 445
Vertically polarized, 129, 130
Vietnam War, 141
Viterbi, 99, 215
Viterbi algorithm, 221
Viterbi Decoder, 554
Viterbi decoding, 558
Voltage-driven, 300
Voyager, 226
VSWR, 17
VU, 14

Index

Waveguide, 246
Wavelength, 138
Weighting function, 533
Wideband FM, 86
Wireless LAN, 276

Z plane, 332
Z-transform, 332
Zero crossings, 415
Zero-order hold, 295
Zeros, 336
ZIF tuning, 278

Artech House Mobile Communications Library

William Webb, Series Editor

3G CDMA2000 Wireless System Engineering, Samuel C. Yang

3G Multimedia Network Services, Accounting, and User Profiles, Freddy Ghys, Marcel Mampaey, Michel Smouts, and Arto Vaaraniemi

5G and Satellite RF and Optical Integration, Geoff Varrall

5G-Enabled Industrial IoT Networks, Amitava Ghosh, Rapeepat Ratasuk, Simone Redana, and Peter Rost

5G New Radio: Beyond Mobile Broadband, Amitav Mukherjee

5G Spectrum and Standards, Geoff Varrall

802.11 WLANs and IP Networking: Security, QoS, and Mobility, Anand R. Prasad and Neeli R. Prasad

Achieving Interoperability in Critical IT and Communications Systems, Robert I. Desourdis, Peter J. Rosamilia, Christopher P. Jacobson, James E. Sinclair, and James R. McClure

Advances in 3G Enhanced Technologies for Wireless Communications, Jiangzhou Wang and Tung-Sang Ng, editors

Advances in Mobile Information Systems, John Walker, editor

Advances in Mobile Radio Access Networks, Y. Jay Guo

Artificial Intelligence in Wireless Communications, Thomas W. Rondeau and Charles W. Bostian

Broadband Wireless Access and Local Network: Mobile WiMax and WiFi, Byeong Gi Lee and Sunghyun Choi

CDMA for Wireless Personal Communications, Ramjee Prasad

CDMA RF System Engineering, Samuel C. Yang

CDMA Systems Capacity Engineering, Kiseon Kim and Insoo Koo

Cell Planning for Wireless Communications, Manuel F. Cátedra and Jesús Pérez-Arriaga

Cellular Communications: Worldwide Market Development, Garry A. Garrard

Cellular Mobile Systems Engineering, Saleh Faruque

Cognitive Radio Interoperability through Waveform Reconfiguration, Leszek Lechowicz and Mieczyslaw M. Kokar

Cognitive Radio Techniques: Spectrum Sensing, Interference Mitigation, and Localization, Kandeepan Sithamparanathan and Andrea Giorgetti

The Complete Wireless Communications Professional: A Guide for Engineers and Managers, William Webb

EDGE for Mobile Internet, Emmanuel Seurre, Patrick Savelli, and Pierre-Jean Pietri

Emerging Public Safety Wireless Communication Systems, Robert I. Desourdis, Jr., et al.

From LTE to LTE-Advanced Pro and 5G, Moe Rahnema and Marcin Dryjanski

The Future of Wireless Communications, William Webb

Geospatial Computing in Mobile Devices, Ruizhi Chen and Robert Guinness

GPRS for Mobile Internet, Emmanuel Seurre, Patrick Savelli, and Pierre-Jean Pietri

GSM and Personal Communications Handbook, Siegmund M. Redl, Matthias K. Weber, and Malcolm W. Oliphant

GSM Networks: Protocols, Terminology, and Implementation, Gunnar Heine

GSM System Engineering, Asha Mehrotra

Handbook of Land-Mobile Radio System Coverage, Garry C. Hess

Handbook of Mobile Radio Networks, Sami Tabbane

Handbook of Next-Generation Emergency Services, Barbara Kemp and Bart Lovett

High-Speed Wireless ATM and LANs, Benny Bing

Implementing Full Duplexing for 5G, David B. Cruickshank

In-Band Full-Duplex Wireless Systems Handbook, Kenneth E. Kolodziej, editor

Inside Bluetooth Low Energy, Second Edition, Naresh Gupta

Interference Analysis and Reduction for Wireless Systems, Peter Stavroulakis

Interference and Resource Management in Heterogeneous Wireless Networks, Jiandong Li, Min Sheng, Xijun Wang, and Hongguang Sun

Internet Technologies for Fixed and Mobile Networks, Toni Janevski

Introduction to 3G Mobile Communications, Second Edition, Juha Korhonen

Introduction to 4G Mobile Communications, Juha Korhonen

Introduction to Communication Systems Simulation, Maurice Schiff

Introduction to Digital Professional Mobile Radio, Hans-Peter A. Ketterling

An Introduction to GSM, Siegmund M. Redl, Matthias K. Weber, and Malcolm W. Oliphant

Introduction to Mobile Communications Engineering, José M. Hernando and F. Pérez-Fontán

Introduction to OFDM Receiver Design and Simulation, Y. J. Liu

An Introduction to Optical Wireless Mobile Communications, Harald Haas, Mohamed Sufyan Islim, Cheng Chen, and Hanaa Abumarshoud

Introduction to Radio Propagation for Fixed and Mobile Communications, John Doble

Introduction to Wireless Local Loop, Broadband and Narrowband, Systems, Second Edition, William Webb

IS-136 TDMA Technology, Economics, and Services, Lawrence Harte, Adrian Smith, and Charles A. Jacobs

Location Management and Routing in Mobile Wireless Networks, Amitava Mukherjee, Somprakash Bandyopadhyay, and Debashis Saha

LTE Air Interface Protocols, Mohammad T. Kawser

Metro Ethernet Services for LTE Backhaul, Roman Krzanowski

Mobile Data Communications Systems, Peter Wong and David Britland

Mobile IP Technology for M-Business, Mark Norris

Mobile Satellite Communications, Shingo Ohmori,
   Hiromitsu Wakana, and Seiichiro Kawase

*Mobile Telecommunications Standards: GSM, UMTS, TETRA, and
   ERMES,* Rudi Bekkers

Mobile-to-Mobile Wireless Channels, Alenka Zajić

*Mobile Telecommunications: Standards, Regulation, and
   Applications,* Rudi Bekkers and Jan Smits

Multiantenna Digital Radio Transmission, Massimiliano Martone

Multiantenna Wireless Communications Systems, Sergio Barbarossa

*Multi-Gigabit Microwave and Millimeter-Wave Wireless
   Communications,* Jonathan Wells

Multiuser Detection in CDMA Mobile Terminals, Piero Castoldi

OFDMA for Broadband Wireless Access, Slawomir Pietrzyk

Practical Wireless Data Modem Design, Jonathon Y. C. Cheah

The Practitioner's Guide to Cellular IoT, Cameron Kelly Coursey

*Prime Codes with Applications to CDMA Optical and Wireless
   Networks,* Guu-Chang Yang and Wing C. Kwong

Quantitative Analysis of Cognitive Radio and Network Performance,
   Preston Marshall

QoS in Integrated 3G Networks, Robert Lloyd-Evans

Radio Resource Management for Wireless Networks, Jens Zander and
   Seong-Lyun Kim

*Radiowave Propagation and Antennas for Personal Communications,
   Third Edition,* Kazimierz Siwiak and Yasaman Bahreini

RDS: The Radio Data System, Dietmar Kopitz and Bev Marks

Resource Allocation in Hierarchical Cellular Systems, Lauro
   Ortigoza-Guerrero and A. Hamid Aghvami

RF and Baseband Techniques for Software-Defined Radio,
   Peter B. Kenington

RF and Microwave Circuit Design for Wireless Communications,
   Lawrence E. Larson, editor

Sample Rate Conversion in Software Configurable Radios,
   Tim Hentschel

Signal Processing Applications in CDMA Communications, Hui Liu

Signal Processing for RF Circuit Impairment Mitigation, Xinping Huang, Zhiwen Zhu, and Henry Leung

Smart Antenna Engineering, Ahmed El Zooghby

Software-Defined Radio for Engineers, Travis F. Collins, Robin Getz, Di Pu, and Alexander M. Wyglinski

Software Defined Radio for 3G, Paul Burns

Software Defined Radio: Theory and Practice, John M. Reyland

Spectrum Wars: The Rise of 5G and Beyond, Jennifer A. Manner

Spread Spectrum CDMA Systems for Wireless Communications, Savo G. Glisic and Branka Vucetic

Technical Foundations of the IoT, Boris Adryan, Dominik Obermaier, and Paul Fremantle

Technologies and Systems for Access and Transport Networks, Jan A. Audestad

Third-Generation and Wideband HF Radio Communications, Eric E. Johnson, Eric Koski, William N. Furman, Mark Jorgenson, and John Nieto

Third Generation Wireless Systems, Volume 1: Post-Shannon Signal Architectures, George M. Calhoun

Traffic Analysis and Design of Wireless IP Networks, Toni Janevski

Transmission Systems Design Handbook for Wireless Networks, Harvey Lehpamer

UMTS and Mobile Computing, Alexander Joseph Huber and Josef Franz Huber

Understanding Cellular Radio, William Webb

Understanding Digital PCS: The TDMA Standard, Cameron Kelly Coursey

Understanding WAP: Wireless Applications, Devices, and Services, Marcel van der Heijden and Marcus Taylor, editors

Universal Wireless Personal Communications, Ramjee Prasad

Virtualizing 5G and Beyond 5G Mobile Networks, Larry J. Horner, Kurt Tutschku, Andrea Fumagalli, and ShunmugaPriya Ramanathan

WCDMA: Towards IP Mobility and Mobile Internet, Tero Ojanperä and Ramjee Prasad, editors

Wi-Fi 6: Protocol and Network, Susinder R. Gulasekaran and Sundar G. Sankaran

Wireless Communications in Developing Countries: Cellular and Satellite Systems, Rachael E. Schwartz

Wireless Communications Evolution to 3G and Beyond, Saad Z. Asif

Wireless Intelligent Networking, Gerry Christensen, Paul G. Florack, and Robert Duncan

Wireless LAN Standards and Applications, Asunción Santamaría and Francisco J. López-Hernández, editors

Wireless Sensor and Ad Hoc Networks Under Diversified Network Scenarios, Subir Kumar Sarkar

Wireless Technician's Handbook, Second Edition, Andrew Miceli

For further information on these and other Artech House titles, including previously considered out-of-print books now available through our In-Print-Forever® (IPF®) program, contact:

Artech House
685 Canton Street
Norwood, MA 02062
Phone: 781-769-9750
Fax: 781-769-6334
e-mail: artech@artechhouse.com

Artech House
16 Sussex Street
London SW1V 4RW UK
Phone: +44 (0)20 7596-8750
Fax: +44 (0)20 7630-0166
e-mail: artech-uk@artechhouse.com

Find us on the World Wide Web at: www.artechhouse.com